# Analytical Chemistry for Technicians

## Third Edition

# Analytical Chemistry for Technicians

## Third Edition

## John Kenkel

## LEWIS PUBLISHERS

A CRC Press Company
Boca Raton   London   New York   Washington, D.C.

## Library of Congress Cataloging-in-Publication Data

Kenkel, John.
    Analytical chemistry for technicians / by John V. Kenkel. — 3rd ed.
        p. cm.
    Includes index.
    ISBN 1-56670-519-3 (alk. paper)
    1. Chemistry, Analytic. 1. Title.

QD75.22 .K445 2002
543—dc21                                                                      2002029654

This material is based upon work supported by the National Science Foundation under Grant Nos. DUE9751998 and DUE9950042.

**Visit the CRC Press Web site at www.crcpress.com**

# Dedication

To my wife, Lois, and daughters Sister Emily, Jeanie, and Laura.
For your love, joy, faith, and eternal goodness.
May God's graces and blessings be forever yours.

# Preface

This third edition of *Analytical Chemistry for Technicians* is the culmination and final product of a series of four projects funded by the National Science Foundation's Advanced Technological Education Program and two supporting grants from the DuPont Company. The grant funds have enabled me to utilize an almost limitless reservoir of human and other resources in the development and completion of this manuscript and to vastly improve and update the previous edition. A visible example is the CD that accompanies this book. This CD, which was not part of the previous editions, provides, with a touch of humor, a series of real-world scenarios for students to peruse while studying the related topics in the text.

One very important resource has been the Voluntary Industry Skill Standards for entry-level chemistry laboratory technicians published by the American Chemical Society in 1997. These standards consist of a large number of competencies that such technicians should acquire in their educational program prior to employment as technicians. While many of these competencies were fortuitously addressed in previous editions, many others were not. It was a resource that I consulted time and time again as the writing proceeded.

The grant funds enabled me to enroll in ten American Chemical Society and Pittcon short courses since 1995. Often taught by industrial chemists, these courses were key resources in the manuscript's development.

Another important resource was simply the communications I have had with my colleagues in both industry and academe. Early on, for example, I was able to spend several days at two different DuPont industrial plants to see firsthand what chemical laboratory technicians in these plants do in their jobs. I came away with written notes and mental pictures that were very insightful and useful. I also communicated more regularly with chemists and technicians in my local area, especially when I had specific questions concerning the use of various equipment and techniques in their laboratories. Finally, I have had a network of field testers and reviewers (enabled through the grant funding) for this work. This was a resource that was not available to such an in-depth degree for the previous editions.

Some major changes resulted from all of this. New chapters on physical testing methods and bioanalysis, both written by individuals more suited than I am for this task, are perhaps the most noticeable changes. In addition, we provide in this new edition a series of over 50 workplace scenes, sideboxes with photographs of technicians and chemists working with the equipment or performing the techniques discussed in the text at that point. In addition, a laboratory information management system (LIMS) has been created for students to use when they perform the experiments in the text. Besides these, there have been numerous consolidations, additions, expansions, and deletions of many other topics. I am confident that the product you now hold in your hands and the accompanying support material is the most up-to-date and appropriate tool that I am personally capable of providing for your analytical chemistry educational needs.

**John Kenkel**
Southeast Community College
Lincoln, Nebraska

# Acknowledgments

Partial support for this work was provided by the National Science Foundation's Advanced Technological Education (ATE) Program through grant DUE9950042. Partial support was also provided by the DuPont Company through their Aid to Education Program. Any opinions, findings, and conclusions or recommendations expressed in this material are those of the authors and do not necessarily reflect the views of the National Science Foundation (NSF) or the DuPont Company.

This book is the major product of the ATE project funded by NSF. The following individuals were fully dedicated to assisting with this project, often in two or more categories, and contributed significantly and untiringly to the book and associated products:

Paul Kelter, University of North Carolina–Greensboro (UNCG)
John Amend, Montana State University
Kirk Hunter, Texas State Technical College–Waco
Onofrio Gaglione, New York City Technical College (CUNY), retired
Don Mumm, Southeast Community College–Lincoln
Ken Chapman, Cardinal Workforce Developers, LLC
Paul Grutsch, Athens Area Technical College
Susan Marine, Miami University Middletown
Karen Wosczyna-Birch, Tunxis Community College
Janet Johannessen, County College of Morris
Bill McLaughlin, University of Nebraska–Lincoln
Connie Murphy, The Dow Chemical Company
Sue Rutledge, Southeast Community College

The following gave some assistance to one or more of the aspects of the project, including field testing, reviewing, workshop participation, experiment development, serving on the National Visiting Committee, etc.:

Ildy Boer, County College of Morris
David Baker, Delta College
Gunay Ozkan, Community College of Southern Nevada
Ray Turner, Roxbury Community College
Pat Cunnif, Prince George's Community College
Fran Waller, Air Products and Chemicals
Dan Martin, LABSAF Consulting
Joe Rosen, New York City Technical College (CUNY)
Robert Hofstader, formerly of the American Chemical Society
Marc Connelly, formerly of the American Chemical Society
Naresh Handagama, Pellissippi State Technical College

Linda Sellers-Hann, Del Mar College
Jon Schwedler, ITT Technical Institute

A special acknowledgment goes to my artist David Jané, whose expertise was very important to the project.

Students at the University of North Carolina–Greensboro and at the University of Nebraska–Lincoln also assisted with the project, and students at Southeast Community College endured drafts of the book as a course textbook and offered corrections and inspired content revisions and additions.

Many people, too numerous to name, assisted with the acquisition of the workplace scenes, including those pictured in the scenes and others.

The personnel at the National Science Foundation deserve particular recognition. These include Frank Settle, who influenced the direction of the project early on; Vicki Bragin, program officer for most of the grant period; Iraj Nejad, who served during the final year of the project; and Liz Teles, who has directed the ATE Program from the beginning.

Special acknowledgment also goes to the personnel at CRC Press/Lewis Publishers for their support and hard work on behalf of this and past projects.

Finally, the author wishes to acknowledge his family, to whom the book is dedicated, for the love and understanding so graciously given during the entire writing period and the Divine Master for the gifts and talents so freely bestowed.

# The Author

**John Kenkel** is a chemistry instructor at Southeast Community College (SCC) in Lincoln, Nebraska. Throughout his 25-year career at SCC, he has been directly involved in the education of chemistry-based laboratory technicians in a vocational program presently named Laboratory Science Technology. He has also been heavily involved in chemistry-based laboratory technician education on a national scale, having served on a number of American Chemical Society (ACS) committees, including the Committee on Technician Activities and the Coordinating Committee for the Voluntary Industry Standards project. In addition to these, he has served a 5-year term on the ACS Committee on Chemistry in the Two-Year College, the committee that organizes the two-year college chemistry consortium conferences. He was the chair of this committee in 1996.

Mr. Kenkel has authored several popular textbooks for chemistry-based technician education. Two editions of *Analytical Chemistry for Technicians* preceded this current edition, the first published in 1988 and the second in 1994. In addition, he has authored four other books: *Chemistry: An Industry-Based Introduction* and *Chemistry: An Industry-Based Laboratory Manual*, both published in 2000–2001; *Analytical Chemistry Refresher Manual*, published in 1992; and *A Primer on Quality in the Analytical Laboratory*, published in 2000. All were published through CRC Press/Lewis Publishers.

Mr. Kenkel has been the principal investigator for a series of curriculum development project grants funded by the National Science Foundation's Advanced Technological Education Program, from which four of his seven books evolved. He has also authored or coauthored four articles on the curriculum work in recent issues of the *Journal of Chemical Education* and has presented this work at more than twenty conferences since 1994.

In 1996, Mr. Kenkel won the prestigious National Responsible Care Catalyst Award for excellence in chemistry teaching, sponsored by the Chemical Manufacturer's Association. He has a master's degree in chemistry from the University of Texas in Austin (1972) and a bachelor's degree in chemistry from Iowa State University (1970). His research at the University of Texas was directed by Professor Allen Bard. He was employed as a chemist from 1973 to 1977 at Rockwell International's Science Center in Thousand Oaks, California.

# Safety in the Analytical Laboratory

The analytical chemistry laboratory is a very safe place to work. However, that is not to say that the laboratory is free of hazards. The dangers associated with contact with hazardous chemicals, flames, etc., are very well documented, and as a result, laboratories are constructed and procedures are carried out with these dangers in mind. Hazardous chemical fumes are, for example, vented into the outdoor atmosphere with the use of fume hoods. Safety showers for diluting spills of concentrated acids on clothing are now commonplace. Eyewash stations are strategically located for the immediate washing of one's eyes in the event of accidental contact of a hazardous chemical with the eyes. Fire blankets, extinguishers, and sprinkler systems are also located in and around analytical laboratories for immediately extinguishing flames and fires. Also, a variety of safety gear, such as safety glasses, aprons, and shields, is available. There is never a good excuse for personal injury in a well-equipped laboratory where well-informed analysts are working.

While the pieces of equipment mentioned above are now commonplace, it remains for the analysts to be well informed of potential dangers and of appropriate safety measures. To this end, we list below some safety tips of which any laboratory worker must be aware. This list should be studied carefully by all students who have chosen to enroll in an analytical chemistry course. *This is not intended to be a complete list, however.* Students should consult with their instructor in order to establish safety ground rules for the particular laboratory in which they will be working. Total awareness of hazards and dangers and what to do in case of an accident is the responsibility of the student and the instructor.

1. Safety glasses must be worn at all times by students and instructors. Visitors to the lab must be appropriately warned and safety glasses made available to them.
2. Fume hoods must be used when working with chemicals that may produce hazardous fumes.
3. The location of fire extinguishers, safety showers, and eyewash stations must be known.
4. All laboratory workers must know how and when to use the items listed in number 3.
5. There must be no unsupervised or unauthorized work going on in the laboratory.
6. A laboratory is never a place for practical jokes or pranks.
7. The toxicity of all the chemicals you will be working with must be known. Consult the instructor, material safety data sheets (MSDSs), safety charts, and container labels for safety information about specific chemicals. Recently, many common organic chemicals, such as benzene, carbon tetrachloride, and chloroform, have been deemed unsafe.
8. Eating, drinking, or smoking in the laboratory is never allowed. Never use laboratory containers (beakers or flasks) to drink beverages.
9. Shoes (not open-toed) must always be worn; hazardous chemicals may be spilled on the floor or feet.
10. Long hair should always be tied back.

11. Mouth pipetting is *never* allowed.
12. Cuts and burns must be immediately treated. Use ice on new burns and consult a doctor for serious cuts.
13. In the event of acid spilling on one's person, flush thoroughly with water immediately. Be aware that acid–water mixtures will produce heat. Removing clothing from the affected area while water flushing may be important, so as to not trap hot acid–water mixtures against the skin. Acids or acid–water mixtures can cause very serious burns if left in contact with skin, even if only for a very short period of time.
14. Weak acids (such as citric acid) should be used to neutralize base spills, and weak bases (such as sodium carbonate) should be used to neutralize acid spills. Solutions of these should be readily available in the lab in case of emergency.
15. Dispose of all waste chemicals from the experiments according to your instructor's directions.
16. In the event of an accident, report immediately to your instructor, regardless of how minor you perceive it to be.
17. Always be watchful and considerate of others working in the laboratory. It is important not to jeopardize their safety or yours.
18. Always use equipment that is in good condition. Any piece of glassware that is cracked or chipped should be discarded and replaced.

It is impossible to foresee all possible hazards that may manifest themselves in an analytical laboratory. Therefore, it is very important for all students to listen closely to their instructor and obey the rules of their particular laboratory in order to avoid injury. Neither the author of this text nor its publisher assumes any responsibility whatsoever in the event of injury.

# Contents

## 5 Applications of Titrimetric Analysis

6  Introduction to Instrumental Analysis

7  Introduction to Spectrochemical Methods

## 9   Atomic Spectroscopy

## 10   Other Spectroscopic Methods

## 13 High-Performance Liquid Chromatography

## 14  Electroanalytical Methods

## 15  Physical Testing Methods

# 1

# Introduction to Analytical Science

## 1.1 Analytical Science Defined

Imagine yourself strolling down the aisle in your local grocery store to select your favorite foods for lunch. You pick up a jar of peanut butter, look at the label, and read that there are 190 mg of sodium in one serving. You think to yourself: "I wish I knew how they knew that for sure." After picking up the lunch items you want, you proceed to the personal hygiene aisle to look for toothpaste. Again you look at the label and notice that the fluoride content is 0.15% weight per volume (w/v). "How do they know that?" you again ask. Finally, you stop by the pharmaceutical shelves and pick up a bottle of your favorite vitamin. Looking at the label, you see that there are 1.7 mg of riboflavin in every tablet and marvel at how the manufacturer can know that that is really the case.

There is a seemingly endless list of example scenarios like the one above that one can think of without even leaving the grocery store. We could also visit a hardware store and look at the labels of cleaning fluids, adhesives, paint or varnish formulations, paint removers, garden fertilizers, and insecticides and make similar statements. Although you may question how the manufacturers of these products know precisely the content of their products in such a quantitative way, you yourself may have undertaken exactly that kind of work at some point in your life right in your own home. If you have an aquarium, you may have come to know that it is important to not let the ammonia level in the tank get too high, and you may have purchased a kit to allow you to monitor the ammonia level. Or you may have purchased a water test kit to determine the pH, hardness, or even nitrate concentration in the water that comes from your tap. You may have a soil test kit to determine the nitrate, phosphate, and potassium levels of the soil in your garden. Then you think: "Gee, it's actually pretty easy." But when you sit down and read the paper or watch the evening news, you are baffled again by how a forensic scientist determines that a criminal's DNA was present on a murder weapon, or how someone determined the ammonia content in the atmosphere of the planet Jupiter without even being there, or how it can be possible to determine the ozone level high above the North Pole.

The science that deals with the identification and quantification of the components of material systems such as these is called **analytical science** . It is called that because the process of determining the level of any or all components in a material system is called **analysis**. It can involve both physical and chemical processes. If it involves chemical processes, it is called **chemical analysis** or, more broadly, **analytical chemistry**. The sodium in the peanut butter, the nitrate in the water, and the ozone in the air in the above scenarios are the substances that are the objects of analysis. The word for such a substance is **analyte** , and the word for the material in which the analyte is found is called the **matrix** of the analyte.

Another word often used in a similar context is the word "assay." If a material is known by a particular name and an analysis is carried out to determine the level of that named substance in the material, the analysis is called an **assay** for that named substance. For example, if an analysis is being carried out to determine what percent of the material in a bottle labeled "aspirin" is aspirin, the analysis is called an

# CHARACTERIZING A MATERIAL

A combination of the qualitative and quantitative analysis of a material or matrix is sometimes called **characterizing** the material. A total analysis such as this might involve a
complete reporting of the properties of a material as well as the identity and quantity of
component substances. For example, a company that manufactures a perfume might characterize
its product as having a particular fragrance, a particular staying power, and a particular feel on
the skin, but it may also report the identity of the ingredients and the quantity of each. Characterization of the raw materials used to make a product as well as the final product itself, and even
the package in which the product is contained, is often considered a very important part of a
manufacturing effort because of the need to assure the product's quality.

assay for aspirin. In contrast, an *analysis* of the aspirin would imply the determination of other minor
ingredients in addition to the aspirin itself.

The purpose of this book is to discuss in a systematic way the techniques, methods, equipment, and
processes of this important, all-encompassing science.

## 1.2   Classifications of Analysis

Analytical procedures can be classified in two ways: first, in terms of the goal of the analysis, and second,
in terms of the nature of the method used. In terms of the goal of the analysis, classification can be based
on whether the analysis is qualitative or quantitative. **Qualitative analysis** is identification. In other
words, it is an analysis carried out to determine only the identity of a pure analyte, the identity of an
analyte in a matrix, or the identity of several or all components of a mixture. Stated another way, it is
an analysis to determine *what* a material is or what the components of a mixture are. Such an analysis
does not report the amount of the substance. If a chemical analysis is carried out and it is reported that
there is mercury present in the water in a lake and the quantity of the mercury is not reported, then the
analysis was a qualitative analysis. **Quantitative analysis**, on the other hand, is the analysis of a material
for *how much* of one or more components is present. Such an analysis is undertaken when the identity
of the components is already known and when it is important to also know the quantities of these
components. It is the determination of the quantities of one or more components present per some
quantity of the matrix. For example, the analysis of the soil in your garden that reports the potassium
level as 342 parts per million (ppm) would be classified as a quantitative analysis. The major emphasis
of this text is on quantitative analysis, although some qualitative applications will be discussed for some
techniques. See Workplace Scene 1.1.

Analysis procedures can be additionally classified into procedures that involve physical properties, wet
chemical analysis procedures, and instrumental chemical analysis procedures. **Analysis using physical
properties** involves no chemical reactions and at times relatively simple devices (although possibly
computerized) to facilitate the measurement. Physical properties are especially useful for identification,
but may also be useful for quantitative analysis in cases where the value of a property, such as specific
gravity or refractive index (Chapter 15), varies with the quantity of an analyte in a mixture.

**Wet chemical analysis** usually involves chemical reactions or classical reaction stoichiometry, but no
electronic instrumentation beyond a weighing device. Wet chemical analysis techniques are classical
techniques, meaning they have been in use in the analytical laboratory for many years, before electronic
devices came on the scene. If executed properly, they have a high degree of inherent accuracy and
precision, but they take more time to execute.

# WORKPLACE SCENE 1.1

In most chemical process industries, both qualitative and quantitative analyses are performed on many varieties of company products and the raw materials that go into these products. Some of the qualitative tests require a simple mixing of the test sample with a reagent to produce a color change. One such test can be run, for example, to confirm the contents of drums of tribasic calcium phosphate, which is a raw material for some pharmaceutical products. The test sample is dissolved in water, acidified, and then tested with a molybdate solution. A yellow precipitate indicates that the material is indeed tribasic calcium phosphate.

Eric Niedergeses examines a test tube for the presence of a yellow precipitate to confirm that the contents of a drum is tribasic calcium phosphate. Test tubes representing other drums are in the test tube rack on the bench in front of Eric.

**Instrumental analysis** can also involve chemical reactions, but it always involves modern sophisticated electronic instrumentation. Instrumental analysis techniques are high-tech techniques, often utilizing the ultimate in complex hardware and software. While sometimes not as precise as a carefully executed wet chemical method, instrumental analysis methods are fast and can offer a much greater scope and practicality to the analysis. In addition, instrumental methods are generally used to determine the minor constituents or constituents that are present in low levels, rather than the major constituents of a sample. We discuss wet chemical methods in Chapters 3 and 5. Chapter 15 is concerned with physical properties; Chapters 7 to 14 involve specific instrumental methods.

## 1.3 The Sample

A term for the material under investigation is **bulk system**. The bulk system in the case of analyzing toothpaste for fluoride is the toothpaste in the tube. The bulk system in the case of determining the ammonia level in the water in an aquarium is all the water in the aquarium.

When we want to analyze a bulk system such as these in an analytical laboratory, it is usually not practical to literally place the entire system under scrutiny. We cannot, for example, bring all the soil found in a garden into the laboratory to determine the phosphate content. We therefore collect a representative *portion* of the bulk system and bring this portion into the laboratory for analysis. Hence, a portion of the water in a lake is analyzed for mercury and a portion of the peanut butter from the jar is analyzed for sodium. This portion is called a **sample**. The analytical laboratory technician analyzes these samples by subjecting them to certain rigorous laboratory operations that ultimately result in the identity or quantity of the analyte in question. The key is that the sample must possess all the characteristics of the entire bulk system with respect to the analyte and the analyte concentration in the system. In other words, it must be a **representative sample** —it must truly represent the bulk system. There is much to discuss with respect to the collection and preparation of samples, and we will do that in Chapter 2.

## 1.4   The Analytical Strategy

The process by which an analyte's identity or the concentration level in a sample is determined in the laboratory may involve many individual steps. In order for us to have a coherent approach to the subject, we will group the steps into major parts and study each part individually. In general, these parts vary in specifics according to what the analyte and analyte matrix are and what methods have been chosen for the analysis. In this section, we present a general organizational framework for these parts; in later chapters we will proceed to build upon this framework for each major method of analysis to be encountered. Let us call this framework the **analytical strategy** .

There are five parts to the analytical strategy: 1) obtain the sample, 2) prepare the sample, 3) carry out the analysis method, 4) work up the data, and 5) calculate and report the results. These are expressed in the flow chart in Figure 1.1. The terminology used in Figure 1.1 and the various steps in the carrying out of the method may be foreign to you now, but they will be discussed as we progress through the coming chapters.

## 1.5   Analytical Technique and Skills

If the label on a box of Cheerios™ states that there are 22 g of carbohydrates in each serving, how does the manufacturer know with certainty that it is 22 g and not 20 or 25 g? If the label on a bottle of rubbing alcohol says that it is 70% isopropyl alcohol, how does the manufacturer know that it is 70% and not 65 or 75%? The answers have to do with the quality of the manufacturing process and also with how accurately the companies' quality assurance laboratories can measure these ingredients. But much of it also has to do with the skills of the technicians performing the analyses.

An analytical laboratory technician is a person with a special mind-set and special skills. He or she must be a thinking person—a person who pays close attention to detail and never waivers in his or her pursuit of high-quality data and results, even in simple things performed in the laboratory. We say that he or she possesses **good analytical technique** or **good analytical skills** .

Quality is emphasized because of the value and importance that are usually riding on the results of an analysis. Great care must be exercised in the lab when handling the sample and all associated materials. Contamination or loss of a sample through avoidable accidental means cannot be tolerated. The results of a chemical analysis could affect such ominous decisions as the freedom or incarceration of a prisoner on trial, whether to proceed with an action that could mean the loss of a million dollars for an industrial company, or the life or death of a hospital patient.

Students should develop a kind of psychology for functioning in an analytical laboratory—a psychology that facilitates good techniques and skills. One must always stop and think before proceeding with a new step in the procedure. What might happen in this step that would cause contamination or loss of the sample? A simple example would be when stirring a solution in a beaker with a stirring rod. You may wish to remove the stirring rod from the beaker when going on to the next step. However, if you stop

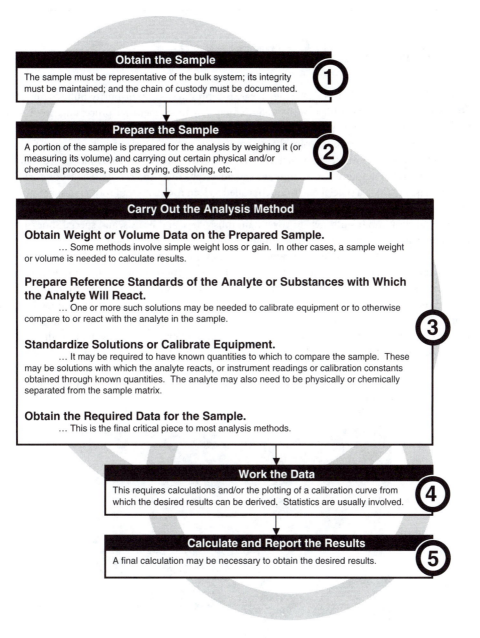

**Obtain the Sample**

The sample must be representative of the bulk system; its integrity must be maintained; and the chain of custody must be documented.

①

**Prepare the Sample**

A portion of the sample is prepared for the analysis by weighing it (or measuring its volume) and carrying out certain physical and/or chemical processes, such as drying, dissolving, etc.

②

**Carry Out the Analysis Method**

**Obtain Weight or Volume Data on the Prepared Sample.**
… Some methods involve simple weight loss or gain. In other cases, a sample weight or volume is needed to calculate results.

**Prepare Reference Standards of the Analyte or Substances with Which the Analyte Will React.**
… One or more such solutions may be needed to calibrate equipment or to otherwise compare to or react with the analyte in the sample.

**Standardize Solutions or Calibrate Equipment.**
… It may be required to have known quantities to which to compare the sample. These may be solutions with which the analyte reacts, or instrument readings or calibration constants obtained through known quantities. The analyte may also need to be physically or chemically separated from the sample matrix.

**Obtain the Required Data for the Sample.**
… This is the final critical piece to most analysis methods.

③

**Work the Data**

This requires calculations and/or the plotting of a calibration curve from which the desired results can be derived. Statistics are usually involved.

④

**Calculate and Report the Results**

A final calculation may be necessary to obtain the desired results.

⑤

**FIGURE 1.1**  Flow chart of the analytical strategy.

and think in advance, you would recognize that you need to rinse wetness adhering to the rod back into the beaker as you remove it. This would prevent the loss of that part of the solution adhering to the rod. Such a loss might result in a significant error in the determination. See Workplace Scene 1.2.

Quality of technique in an analytical laboratory is so important that there are laws governing the actions of the scientists in such a laboratory. These laws, passed by Congress and placed into the Federal Registry in the late 1980s, are known as **good laboratory practices**  (GLPs). The GLP laws address such things as labeling, record keeping and storage, documentation and updating of laboratory procedures (known as **standard operating procedures**  (SOPs)), and the laboratory **protocols**, which are formal

# WORKPLACE SCENE 1.2

S ACHEM Inc., located in Cleburne, Texas, is a producer of high-purity bulk chemicals for companies that have high-purity requirements in their chemical processing. Because the products are of high purity, laboratory operations to assure the quality of the products (quality assurance operations) involve the determination of trace levels of contaminants. Contamination of laboratory samples and materials is of special concern in cases like this because an uncommonly small amount of contaminant can adversely affect the results. The laboratory work therefore takes place in a special environment called a clean room. A clean room is a space in which extraordinary precautions are taken to avoid the slightest contamination. Laboratory personnel wear special clean room suits, nets to cover hair, mustaches, and beards, and special shoes, gloves, and safety glasses to minimize possible contamination.

One of SACHEM's products is tetramethylammonium hydroxide (TMAH), which is sold to semiconductor industries. Suspended particles in TMAH solutions could cause severe mechanical damage to the electronic devices manufactured by their customers. The determination of the particle content in such solutions is therefore critical. It is performed with a laser-equipped particle counter, which provides 70% detection efficiency. The counting must take place in a clean room because tiny airborne particles can land in the solutions and give them a false high reading. A class 1000 environment is required in this case, which means that the count of particles in the air that are greater than or equal to 0.5 $\mu$m in diameter must be less than 1000 per cubic foot. Typically, a customer's specification for TMAH solutions is less than 100 particles per milliliter for particles greater than or equal to 0.5 $\mu$m in diameter.

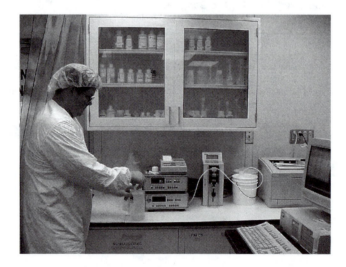

Paul Plumb of SACHEM Inc. counts particles in the ultrapure solutions of TMAH by a laser-equipped particle counter in the clean room. Notice the hair net and special lab coat.

written documents defining and governing the work a laboratory is performing. They also address such things as who has authority over various aspects of the work, who has authority to change SOPs, and the processes by which they are changed. The GLPs also allow for **audits**, or regular inspections of a laboratory by outside personnel to ensure compliance with the regulations. GLP regulations receive much attention in analytical laboratories. See Appendix 1.

# No Place for Mush Heads

The *Paper Chase* is the name of a movie that debuted in 1973. It is the story a student's pursuit of a law degree at a high-profile Ivy League university. The movie stars John Hausman as a distinguished and intimidating law professor with a reputation for ruthlessness in the classroom and for striking horrific fear in the minds of students because of his demands for hard work and for accuracy and fine detail in oral responses to classroom questions. The character's name is Professor Kingsfield. The viewer gets a taste of his demeanor early on as the students meet the professor at the first class session. With all the students seated quietly in their seats, a somber Professor Kingsfield enters from the side door of the classroom, strolls confidently to the front table, stares out at the class with penetrating intimidation, and with his booming voice strikes immediate fear in the minds of the students with these words: "You come into my classroom with a skull full of mush, but you will leave thinking like a lawyer."

It is clear from the beginning that this law school has one of the keenest reputations in the country for turning out high-quality lawyers. And this reputation is not lost on Professor Kingsfield as he makes it very clear from the beginning that his class is a no-nonsense class and then goes on to demand accuracy and fine detail.

There is a very good analogy here to learning analytical chemistry skills and technique. The analytical laboratory is no place for mush heads. The development of good laboratory technique and skills is absolutely essential to success on the job and for the success of a company's endeavor. A professor of analytical chemistry might say: "You come into my laboratory with a skull full of mush, but you will leave thinking like an analytical chemist."

## 1.6   The Laboratory Notebook

As indicated in the last section, one item addressed in the GLP regulations is record keeping. As such, it is considered a legal document. Indeed, good record keeping is central to good analytical science. Not only must data obtained for samples and analytes be recorded, but it must be recorded with diligence, and with considerable thought being given to integrity and purpose.

Accordingly, an analytical laboratory will usually have strict guidelines with respect to laboratory notebooks. The following typifies what these guidelines might be:

I. General guidelines
   A. All notebooks must begin with a table of contents. All pages must be numbered, and these numbers must be referenced in the table of contents. The table of contents must be updated as projects are completed and new projects are begun.
   B. All notebook entries will be made in ink. Use of graphite pencils or another erasable writing instrument is strictly prohibited.
   C. No data entries will be erased or made illegible. If an error was made, a single line is drawn through the entry. Do not use correction fluid. Initial and date corrections and indicate why the correction was necessary.
   D. Under no circumstances will the notebook be taken or otherwise leave the laboratory unless there is data to be recorded at a remote site, such as at a remote sampling site, or unless special permission is granted by the supervisor.

    E.  The following notebook format should be maintained for each project undertaken: 1) title and date, 2) purpose or objective statement, 3) data entries, 4) results, and 5) conclusions. Each of these are explained below. In each case, write out and underline the words "title and date," "objective," "data," etc., to clearly identify the beginning of each section.

    F.  Make notebook entries for a given project on consecutive pages where practical. Begin a new project on the front side of a new page. You may skip pages only in order to comply with this guideline.

    G.  Draw a single diagonal line through blank spaces that consist of four or more lines (including any pages skipped according to guideline F above). These spaces should be initialed and dated.

    H.  Never use a highlighter in a notebook.

    I.  Each notebook page must be signed, dated, and possibly witnessed.

II.  Title and date

    A.  All new experiments will begin with the title of the work and the date it is performed. If the work was continued on another date, that date must be indicated at the point the work was restarted.

    B.  The title will reflect the nature of the work or shall be the title given to the project by the study director.

III.  Purpose or objectives statement

    A.  Following the title and date, a statement of the purpose or objective of the work will be written. This statement should be brief and to the point.

    B.  If appropriate, the SOPs will be referenced in this statement

IV.  Data entries

    A.  Enter data into the notebook as the work is being performed. This means that loose pieces of paper used for intermediate recordings are prohibited. Entries should be made in ink only.

    B.  If there is any deviation from the SOPs, permission must be obtained from the study director and this must be thoroughly documented by indicating exactly what the deviation was and why it occurred.

    C.  The samples analyzed must be described in detail. Such descriptions may include the source of the sample, what steps were taken to ensure that it represents the whole (reference SOPs if appropriate), what special coding may be assigned, and what the codes mean. If the codes were recorded in a separate notebook (such as a field notebook), this notebook must be cross-referenced.

    D.  Show the mathematical formulas utilized for all calculations and also a sample calculation.

    E.  Construct data tables whenever useful and appropriate.

    F.  Both numerical data and important observations should be recorded.

    G.  Limit attachments (chart recordings, computer printouts, etc.) to one per page. Clear tape or glue may be used. Do not use staples. Only one fold in attachments is allowed. Do not cover any notebook entries with attachments.

V.  Results

    A.  The results of the project, such as numerical values representing analysis results, should be reported in the notebook in table form if appropriate. Otherwise, a statement of the outcome is written, or if a single numerical value is the outcome, then it is reported here. In order to identify what is to be reported as results, consider what it is the client wants and needs to know.

VI.  Conclusion

    A.  After results are reported, the experiment is drawn to a close with a brief concluding statement indicating whether the objective was achieved.

Two sample pages from the notebook of a student performing Experiment 6 are shown in Figure 1.2. Notice that each page is signed with the student's name and dated, as suggested by letter I of the general guidelines above. This is often required in industry for patent protection.

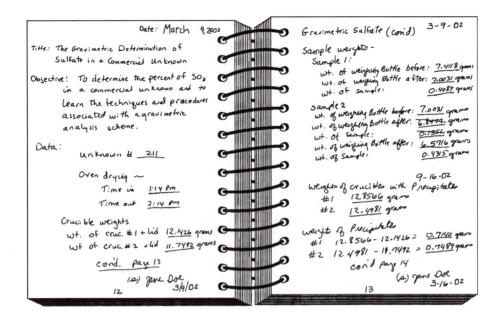

**FIGURE 1.2**   Sample pages from a laboratory notebook that a student is using for Experiment 6 in this text.

## 1.7   Errors, Statistics, and Statistical Control

From the discussion thus far, it is clear that the need for accuracy in the laboratory is an important issue. If the analytical results reported by a laboratory are not accurate, everything a company or government agency strives for may be in jeopardy. If the customer discovers an error in the results of a laboratory analysis, especially through painful means, the trust the public has placed in the entire enterprise is lost. For example, if a baby dies due to nitrate contamination in drinking water that a city's health department had determined through laboratory work to be safe, that department, indeed the entire city government, is liable. In this worst-case scenario, some employees would likely lose their jobs and perhaps even be brought to justice in a court of law.

The most important aspect of the job of the chemical analyst is to assure that the data and results that are reported are of the maximum possible quality. This means that the analyst must be able to recognize when the test instrument is breaking down and when a human error is suspected. The analyst must be as confident as he or she can be that the readout from an instrument does in fact indicate a true readout as much as is humanly possible. The analyst must be familiar with error analysis schemes that have been developed and be able to use them to the point where confidence and quality is assured.

### 1.7.1   Errors

Errors in the analytical laboratory are basically of two types: determinate errors and indeterminate errors. **Determinate errors**, also called **systematic errors**, are errors that were known to have occurred, or at least were determined later to have occurred, in the course of the lab work. They may arise from avoidable sources, such as contamination, wrongly calibrated instruments, reagent impurities, instrumental malfunctions, poor sampling techniques, errors in calculations, etc. Results from laboratory work in which avoidable determinate errors are known to have occurred must be rejected or, if the error was a calculation error, recalculated.

Determinate errors may also arise from unavoidable sources. An error that is known to have occurred but was unavoidable is called a **bias**. Such an error occurs each time a procedure is executed, and thus its effect is usually known and a correction factor can be applied.

**Indeterminate errors**, also called **random errors**, on the other hand, are errors that are not specifically identified and are therefore impossible to avoid. Since the errors cannot be specifically identified, results arising from such errors cannot be immediately rejected or compensated for as in the case of determinate errors. Rather, a statistical analysis must be performed to determine whether the results are far enough "off-track" to merit rejection.

Statistics establish quality limits for the answers derived from a given method. A given laboratory result, or a sample giving rise to a given result, is considered "good" if it is within these limits. In order to understand how these limits are established, and therefore how it is known if a given result is unacceptable, some basic knowledge of statistics is needed. We now present a limited treatment of elementary statistics.

## 1.7.2   Elementary Statistics

The procedure used to determine whether a given result is unacceptable involves running a series of identical tests on the same sample, using the same instrument or other piece of equipment, over and over. In such a scenario, the indeterminate errors manifest themselves in values that deviate, positively and negatively, from the mean (average) of all the values obtained. Given this brief background, let us proceed to define some terms related to elementary statistics.

1. **Mean.** In the case in which a given measurement on a sample is repeated a number of times, the average of all measurements is called the mean. It is calculated by adding together the numerical values of all measurements and dividing this sum by the number of measurements. In this text, we give the mean the symbol m. The **true mean**, or the mean of an infinite number of measurements (the entire population of measurements), is given the symbol $\mu$, the Greek letter mu.

2. **Deviation.** How much each measurement differs from the mean is an important number and is called the deviation. A deviation is associated with each measurement, and if a given deviation is large compared to others in a series of identical measurements, the proverbial red flag is raised. Such a measurement is called an **outlier.** Mathematically, the deviation is calculated as follows:

$$d = |m - e| \qquad (1.1)$$

   in which d is the deviation, m is the mean, and e represents the individual experimental measurement. (The bars refer to absolute value, which means the value of d is calculated without regard to sign.)

3. **Standard deviation.** The most common measure of the dispersion of data around the mean is the standard deviation:

$$s = \sqrt{\frac{\left(d_1^2 + d_2^2 + d_3^2 + \ldots\right)}{(n-1)}} \qquad (1.2)$$

   The term n is the number of measurements, and n − 1 is referred to as the number of **degrees of freedom**. The term s represents the standard deviation. The significance of s is that the smaller it is numerically, the more precise the data (the more the measurements are "bunched" around the mean). For an infinite number of measurements (where the mean is $\mu$), the standard deviation is symbolized as $\sigma$ (Greek letter sigma) and is known as the **population standard deviation**. An infinite number of measurements is approximated by 30 or more measurements.

4. **Relative standard deviation.** One final deviation parameter is the relative standard deviation (RSD). It is calculated by dividing the standard deviation by the mean and then multiplying by 100 or 1000:

$$RSD = \frac{s}{m} \qquad (1.3)$$

and

$$\text{relative \% standard deviation} = \text{RSD} \times 100 \tag{1.4}$$

and

$$\text{relative parts per thousand standard deviation} = \text{RSD} \times 1000 \tag{1.5}$$

Relative standard deviation relates the standard deviation to the value of the mean and represents a practical and popular expression of data quality.

## Example 1.1

The following numerical results were obtained in a given laboratory experiment: 0.09376, 0.09358, 0.09385, and 0.09369. Calculate the relative parts per thousand standard deviation.

### Solution 1.1

We must calculate both the mean and the standard deviation in order to use Equations (1.3) and (1.5). First, the mean, m:

$$m = \frac{0.09376 + 0.09358 + 0.09385 + 0.09369}{4} = 0.09372$$

Next, the deviations:

$$d_1 = |0.09372 - 0.9376| = 0.00004$$
$$d_2 = |0.09372 - 0.09358| = 0.00014$$
$$d_3 = |0.09372 - 0.09385| = 0.00013$$
$$d_4 = |0.09372 - 0.09369| = 0.00003$$

Then, the standard deviation:

$$s = \sqrt{\frac{(0.00004)^2 + (0.00014)^2 + (0.00013)^2 + (0.00003)^2}{(4-1)}}$$

$$= 1.14 \times 10^{-4} = 1.1 \times 10^{-4}$$

Finally, to get the relative parts per thousand standard deviation:

$$\text{RSD} = \frac{s}{m} \times 1000 = \frac{1.14 \times 10^{-4}}{0.09372} \times 1000 = 1.2$$

## 1.7.3 Normal Distribution

For an infinite data set (in which the symbols $\mu$ and $\sigma$ as defined in Section 1.7.2 apply), a plot of frequency of occurrence vs. the measurement value yields a smooth bell-shaped curve. It is referred to as bell-shaped because there is equal drop-off on both sides of a peak value, resulting in a shape that resembles a bell. The peak value corresponds to $\mu$, the population mean. This curve is called the **normal distribution curve** because it represents a normal distribution of values for any infinitely repeated measurement. This curve is shown in Figure 1.3.

The normal distribution curve is a picture of the precision of a given data set. The more points there are bunched around the mean and the sharper the drop-off away from the mean, the smaller the standard

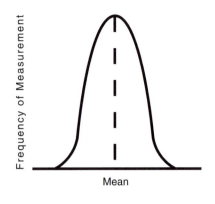

**FIGURE 1.3**    The normal distribution curve.

deviation and the more precise the data. It can be shown that approximately 68.3% of the area under the curve falls within 1 standard deviation from the mean, approximately 95.5% of the area falls within 2 standard deviations from the mean, and approximately 99.7% of the area falls within 3 standard deviations from the mean.

### 1.7.4    Precision, Accuracy, and Calibration

We have made references in the foregoing discussion to the precision of data, or how precise the data are. We have also made reference to the accuracy of data. **Precision** refers to the repeatability of a measurement. If you repeat a given measurement over and over and these measurements deviate only slightly from one another, within the limits of the number of significant figures obtainable, then we say that the data are precise, or that the results exhibit a high degree of precision. The mean of such data may or may not represent the real value of that parameter. In other words, it may not be accurate. **Accuracy** refers to the correctness of a measurement, or how close it comes to the correct value of a parameter.

For example, if an analyst has an object that he or she knows weighs exactly 1.0000 g, the accuracy of a laboratory balance (weight measuring device) can be determined.* The object can be weighed on the balance to see if the balance will read 1.0000 g. If a series of repeated weight measurements using this balance are all between 0.9998 and 1.0002 g, we say the balance is both precise and accurate. If, on the other hand, a series of repeated weight measurements using this balance are all between 0.9983 and 0.9987 g, we say that the balance is precise, but not accurate. If repeated weight measurements using this balance are all between 0.9956 and 0.9991 g, the data are neither precise nor accurate. Finally, if repeated weight measurements using this balance are all between 0.9956 and 1.0042 g, such that the mean is 1.0000 g, then the balance is not precise but it appears to be accurate. These facts on accuracy and precision are illustrated further in Figure 1.4.

If it is established that a measuring device provides a value for a known sample that is in agreement with the known value to within established limits of precision, that device is said to be calibrated. Thus, **calibration** refers to a procedure that checks the device to confirm that it provides the known value. An example is an analytical balance, as discussed above. Sometimes the device can be electronically adjusted to give the known value, such as in the case of a pH meter that is calibrated with solutions of known pH. However, calibration can also refer to the procedure by which the measurement value obtained on a device for a known sample becomes known. An example of this is a spectrophotometer, in which the absorbance values for known concentrations of solutions become known. We will encounter all of these calibration types in our studies.

---

*Standard weights certified by the National Institute of Standards and Technology, NIST, are available.

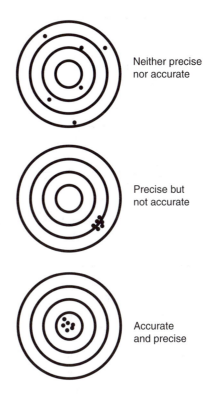

Neither precise
nor accurate

Precise but
not accurate

Accurate
and precise

**FIGURE 1.4** Illustration of precision and accuracy.

## 1.7.5 Statistical Control

A given device, procedure, process, or method is usually said to be in **statistical control** if numerical values derived from it on a regular basis (such as daily) are consistently within 2 standard deviations from the established mean, or the most desirable value. As we learned in Section 1.7.3, such numerical values occur statistically 95.5% of the time. Thus if, say, two or more consecutive values differ from the established value by more than 2 standard deviations, a problem is indicated because this should happen only 4.5% of the time, or once in roughly every 20 events, and is not expected two or more times consecutively. The device, procedure, process, or method would be considered out of statistical control, indicating that an evaluation is in order.

Similarly, if just one individual numerical value differs from the established mean by more than 3 standard deviations, a problem is also indicated because, as we also saw in Section 1.7.3, this should only occur 0.3% of the time, or once in every 333 events. Again, an evaluation is in order.

Analytical laboratories, especially quality assurance laboratories, will often maintain graphical records of statistical control so that scientists and technicians can note the history of the device, procedure, process, or method at a glance. The graphical record is called a **control chart** and is maintained on a regular basis, such as daily. It is a graph of the numerical value on the y-axis vs. the date on the x-axis. The chart is characterized by five horizontal lines designating the five numerical values that are important for statistical control. One is the value that is 3 standard deviations from the most desirable value on the positive side. Another is the value that is 3 standard deviations from the most desirable value on the negative side. These represent those values that are expected to occur only less than 0.3% of the time. These two numerical values are called the action limits because one point outside these limits is cause for action to be taken.

Additionally, two other horizontal lines are drawn at the values that are 2 standard deviations from the most desirable value, one on the positive side and one on the negative side. These represent those values that are expected to occur only 4.5% of the time. The fifth horizontal line is the desirable line itself. See Figure 1.5.

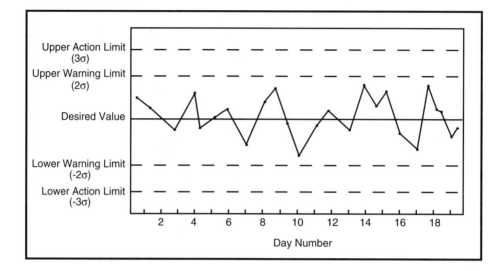

**FIGURE 1.5**   An example of a control chart showing a device, procedure, process, or method that is in statistical control because the numerical values are consistently between the warning limits.

Any device routinely checked for calibration can be monitored in this way. For example, an analytical balance can be tested with a known weight, the value of the known weight being the desirable value and the expected range of precision dictating the warning and action limits (Experiment 1).

A procedure or method may be checked by the use of a quality control solution (often called a **control**), a solution that is known to have a concentration value that should match what the procedure or method would measure. The known numerical value is the desirable value in the control chart. The numerical value determined for the control by the procedure or method is charted. The warning and action limits are determined by preliminary work done a sufficient number of times so as to ascertain the population standard deviation.

A process, such as a manufacturing process, may also be monitored with a control chart. In this case, the desirable value, warning limits, and action limits for the product of the manufacturing process is determined over time using materials and equipment that the scientists and engineers are confident provide an accurate picture of the product.

# Experiments

## Experiment 1: Assuring the Quality of Weight Measurements

Note: This experiment assumes that a permanent log and a quality control chart are constantly maintained for each analytical balance in use in the laboratory. Each day you use a given analytical balance and log in with your name and date. The following calibration check should be performed weekly on all balances. If, according to the log, the calibration of the balance you want to use has not been checked in over a week, perform this procedure. Review Section 3.3 for basic information concerning the analytical balance.

1. Obtain a certified 500-mg standard weight or other weight suggested by your instructor. Do not touch the weight, but handle it with tweezers, and never allow any water or other foreign material to touch it.
2. Check the calibration of the analytical balance you have chosen to use by weighing this standard weight on this balance. When finished, store the standard weight in the specified protected location.

3. Along with your name and date in the logbook, also record the measured weight of the standard weight.

4. Plot your measured weight on the control chart. If any irregularity is observed, report to your instructor.

## Experiment 2: Weight Uniformity of Dosing Units

Reference: General Test <905>, U.S. Pharmacopeia and National Formulary, USP 24–NF 19, 2000, p. 2000.

1. Randomly collect ten ibuprofen tablets from a bottle of ibuprofen from a pharmacy.
2. Handling the tablets with tweezers, carefully weigh each on an analytical balance (see Section 3.3 and Appendix 1).
3. Calculate the mean, standard deviation, and relative percent standard deviation of this data set.
4. Evaluate the results from step 3. Comment on the uniformity of the tablet weight. Also note the milligrams of ibuprofen per tablet found on the label and compare this with your results. If the label value is less than the mean you calculated, give some possible reasons for this.

# Questions and Problems

1. Define analytical science, analysis, chemical analysis, analyte, matrix, and assay.
2. When is an analysis an assay and when is it not? Give examples of both.
3. Distinguish between qualitative analysis and quantitative analysis. Give examples.
4. Imagine that you are an analytical chemist and someone brings you an oily rag to analyze in order to identify the material on the rag. Is this a qualitative or quantitative analysis?
5. Distinguish between wet chemical analysis, instrumental analysis, and analysis using physical properties.
6. Imagine taking a tour of an industrial facility and having a particular laboratory being described to you as the "wet lab." What do you suppose is the kind of activity going on in such a laboratory?
7. When would you choose a wet chemical analysis procedure over an instrumental analysis procedure? When would you choose an instrumental analysis procedure over a wet chemical analysis procedure?
8. What are the five steps in the analytical strategy?
9. What is a sample? What does it mean to obtain a sample?
10. What does it mean to prepare a sample?
11. What is an analytical method, and how does it fit into the total analysis process?
12. What does it mean to carry out the analytical method?
13. What happens after a chemist acquires data from an analytical method?
14. What does it mean to say that a laboratory worker has good analytical technique?
15. Why should a stirring rod that is removed from a beaker containing a solution of the sample being analyzed be rinsed back into that solution with distilled water?
16. Explain GLP and SOP.
17. What sort of things do the GLP regulations address?
18. Why must good laboratory technique apply to laboratory notebooks as well as to the handling of laboratory equipment and chemicals?
19. What constitutes data and results as recorded in a laboratory notebook?
20. Why should notebook data entries always be made in ink and never erased or otherwise made unintelligible?
21. Given the care with which laboratory equipment (balances, burets, instruments, etc.) is calibrated at the factory, why should the chemical analyst worry about errors?

22. Distinguish between determinate and indeterminate errors.
23. Define bias.
24. An analyst determines that the analytical balance he used in a given analytical test is wrongly calibrated. Is this a determinate or an indeterminate error? Explain.
25. A student determines the weight of an object on an analytical balance to be 12.2843 g. The actual weight, unknown to the student or to anyone else, is 12.2845 g. Is the error in the student's measurement determinate or indeterminate? Explain.
26. Can a person really trust the results of a laboratory analysis to be accurate? Explain.
27. Distinguish clearly between accuracy and precision.
28. A given analytical test was performed five times. The results of the analysis are represented by the following values: 6.738, 6.738, 6.737, 6.739, and 6.738%. Would you say that these results are precise? Can you say that they are accurate? Explain both answers.
29. A given analytical test was performed five times. The results of the analysis are represented by the following values: 37.23, 32.91, 45.38, 35.22, and 41.81%. Would you say that these results are precise? Can you say that they are accurate? Explain both answers.
30. Suppose the correct answer to the analysis represented in number 28 above is 6.923%. What can you say now about the precision and accuracy?
31. Calculate the standard deviation and the relative standard deviation for the following data:

| Measurement No. | Value (g) |
| --- | --- |
| 1 | 16.7724 |
| 2 | 16.7735 |
| 3 | 16.7722 |
| 4 | 16.7756 |
| 5 | 16.7729 |
| 6 | 16.7716 |
| 7 | 16.7720 |
| 8 | 16.7733 |

32. A series of eight absorbance measurements using an atomic absorption spectrophotometer are as follows: 0.855, 0.836, 0.848, 0.870, 0.859, 0.841, 0.861, and 0.852. According to the instrument manufacturer, the precision of the absorbance measurements using this instrument should not exceed 1% relative standard deviation. Does it in this case?
33. Why is the relative standard deviation considered a popular and practical expression of data quality?
34. Explain this statement: A relative standard deviation of 1% can be achieved in this experiment.
35. What laboratory analysis results might cause a batch of raw material at a manufacturing plant to be rejected from potential use in the plant process? Explain.
36. How can a quality control chart signal a problem with a routine laboratory procedure?

# 2

# Sampling and Sample Preparation

## 2.1 Introduction

The strategy for analyzing a bulk system for the identity or quantity of an analyte involves the five steps that were summarized in Figure 1.1. These were: 1) obtain a sample of the bulk system, 2) prepare this sample for the analytical method to be used, 3) execute the method chosen for the analysis, 4) work the data, and 5) calculate and report the results. A great deal of detail is presented in this book, and in other books, concerning step 3. The reason for this is the large variety of methods that are available and routinely used. Each of these methods must be studied separately and in detail if we are to understand specifically how each one produces the result that is sought.

However, no analytical method, no matter how simple or sophisticated, no matter how specialized or routine, no matter how easy or difficult, and no matter how costly, will produce the correct result if the sample is not correctly obtained and prepared. The first two steps of the analytical strategy are therefore on at least equal footing with the analytical method in terms of importance to the end result. So although the topics of sampling and sample preparation are given the space of only one chapter in this book, their critical importance should not go unnoticed. Quality sampling and sample preparation is crucial to the success of an analysis.

## 2.2 Obtaining the Sample

As stated in Chapter 1, a laboratory analysis is almost always meant to give a result that is indicative of a concentration in a large system. For example, a farmer wants an analysis result to represent fertilizer needs in an entire 40-acre field. A pharmaceutical manufacturer wants an analysis result to represent the concentration of an active ingredient in each tablet in 80 cases of its product, each case containing three dozen bottles of 100 tablets each. A governmental environmental control agency wants a single laboratory analysis to represent the concentration of a toxic chemical in every cubic inch of soil within 5 miles of a hazardous waste dump site.

The critical part of any sampling task is to obtain a sample that represents the bulk system as well as possible. As stated in Chapter 1, the sample must possess all the characteristics of the entire bulk system with respect to the analyte and the analyte concentration in the system. In other words, it must be a **representative sample** —it must truly represent the bulk system. Whatever concentration level is found for a given component of a sample is also then taken to be the concentration level in the entire system. For example, in order to analyze the water in a lake for mercury, a bottle is filled with the water and is then taken into the laboratory for analysis. If the mercury level in this sample is determined to be 12 parts per million (ppm), then, assuming the sample is representative of the entire lake, the entire lake is assumed to be 12 ppm in mercury also.

Obviously, there are different degrees of difficulty and different sampling modes involved with obtaining samples for analysis, depending on the type of sample to be gathered, whether the source of the sample is homogeneous, the location of and access to the system, etc. For example, obtaining a sample of blood from a hospital patient is completely different from obtaining a sample of coal from a train car full of coal.

A blood sample will depend on the part of the body from which it is taken, the time of the day, the patient's dietary habits, and whether the patient is on medication that could affect the analysis.

With the coal sample it is important to recognize that the coal held in a train car may not be homogeneous, and a sampling scheme that takes this into account must be implemented, so that a proper sample preparation scheme can be planned.

The key word in any case is "representative." A laboratory analysis sample must be representative of the whole so that the final result of the chemical analysis represents the entire system that it is intended to represent. If there are variations in composition, such as with the coal example above, or at least suspected variations, small samples must be taken from all suspect locations. If results for the entire system are to be reported, these small samples are then mixed and made homogeneous to give the final sample to be tested. Such a sample is called a **composite sample** . In some cases, analysis on the individual samples may be more appropriate. Such samples are called **selective samples** .

Consider the analysis of soil from a farmer's field. The farmer wants to know whether he needs to apply a nitrogen-containing fertilizer to his field. It is conceivable that different parts of the field could provide different types of samples in terms of nitrogen content. Suppose there is a cattle feed lot nearby, perhaps uphill from part of the field and downhill from another part of the field such that runoff from the feed lot affects part of the field but not the other part. If the farmer wishes to have an analysis report for the field as a whole, then the sample taken should include combined portions from all parts of the field that may be different (a composite sample) so that it will truly represent the field as a whole. Alternatively, two selective samples could be taken, one from above the feed lot and one from below the feed lot, so that two analyses are performed and reported to the farmer. These would be referred to as selective samples. At any rate, one wants the results of the chemical analysis to be correct for the entire area for which the analysis is intended.[*]

Consider the analysis of the leaves on a tree for pesticide residue. The tree grower wants to know whether the level of pesticide residue on the leaves indicates that the tree needs another pesticide application. Once again, the analyst must consider all parts of the tree that might be different. Leaves at the top, in the middle, and at the bottom should be sampled (one can imagine differences in application rates at the different heights); leaves on the outside and leaves close to the trunk should be sampled; and perhaps there would also be a difference between the shady side and sunny side of the tree. All leaf samples are then combined for a composite sample.

Consider the analysis of a blood sample for alcohol content (imagine that a police officer suspects a motorist to be intoxicated). The problem here is not sampling different locations within a system, but rather a time factor. The blood must be sampled within a particular time frame in order to demonstrate intoxication at the time the motorist was stopped.

If it is thought that a bulk system is homogeneous for a particular component, then a **random sample** is taken. This would be just one sample taken from one location at random in the bulk system. An example would be when determining the level of active ingredient in a pharmaceutical product stored in boxes of individual bottles in a warehouse. Having no reason to assume a greater concentration level in one bottle or box than in another, a sample is chosen at random. See also Workplace Scene 2.1.

Other designations for samples are **bulk sample, primary sample, secondary sample, subsample, laboratory sample** , and **test sample** . These terms are used when a sample of a bulk system is divided, possibly a number of times, before actually used in an analysis. For example, a water sample from a well

---

[*]This example is offered for illustration only. Modern precision agriculture utilizes geographical positioning systems (GPS) for sampling and for determining fertilizer needs. See, for example, the following Web site: http://edis.ifas.ufl.edu/BODY SS402 (as of April 22, 2002).

# WORKPLACE SCENE 2.1

Treatment of municipal water with chlorine and ammonia results in the formation of chloramines, a long-lasting disinfectant. Too much ammonia, however, enhances nitrification by bacteria in the water, which, in turn, increases the nitrate and nitrite levels. High nitrate and nitrite levels in drinking water is a health hazard, particularly for infants.

To monitor the nitrate and nitrite levels in water treated with chlorine and ammonia, water samples must be obtained regularly from water distribution systems and plant process sites and taken to a laboratory for analysis.

Allison Trentman and B.J. Kronschnabel of the City of Lincoln, Nebraska, Water Treatment Plant Laboratory take samples of drinking water from a distribution system sampling site.

may be collected in a large bottle (bulk sample or primary sample) from which a smaller sample is acquired by pouring into a vial to be taken to the laboratory (secondary sample, subsample, or laboratory sample) and then poured into a beaker (another secondary sample or subsample) before a portion is finally carefully measured into a flask (test sample) and diluted to make the sample solution.

The problems associated with sampling are unique to every individual situation. The analyst simply needs to take all possible variations into account when obtaining the sample, so that the sample taken to the laboratory truly does represent what it is intended to represent.

## 2.3   Statistics of Sampling

A consideration of statistics is required in a discussion of sampling because of the randomness with which samples are acquired. Just as random, indeterminate errors associated with laboratory work (Chapter 1) are dealt with by statistics because there is no other way to deal with them, so also with sampling errors. We may take pains to see that a sample randomly acquired is representative, but the compositions of a series of such samples taken from the same system always vary to some unknown extent.

A novice might think that a lab analyst obtains a single sample from a bulk system, analyzes it one time in the laboratory, and reports the answer to this one analysis as the analysis results. If variances in the sampling and lab work are both insignificant, these results may be valid. However, due to possible large variances in both the sampling and the lab work, such a result cannot be considered reliable. In Chapter 1, we indicated that the correct procedure is to perform the analysis many times and deal with the variances with statistics.

The fact that sampling introduces a second statistics problem means that we must consider taking a large number of samples and deal with the results with statistics just as we perform a laboratory analysis several times and deal with those results with statistics. For example, it can be shown that if a measurement system generates data with a standard deviation of 10 ppm, and we need to know an average concentration to ±5 ppm with 95% confidence, then we must perform the analysis 16 times. If the sampling variance is high but the lab analysis variance is low, then we must measure 16 samples each one time. If the lab analysis variance is high but the sampling variance is low, then we must measure one sample 16 times. If both the sampling variance and the lab analysis variance are high, then we must measure 16 samples each 16 times.

Chemists want to have as low a variance (or standard deviation) as possible for the greatest accuracy. If it is not possible to have a low enough standard deviation to suit the need, then the number of measurements (either the number of samples, the number of lab analyses, or both) must be increased. If increasing the number of measurements is not desirable (due to an increased workload or expense, etc.), then the analyst must live with a larger error.

## 2.4   Sample Handling

The importance of a high-quality representative sample has already been noted. How to obtain the sample and what to do with it once it reaches the laboratory are obviously important factors. But the handling of the sample between the sampling site and the laboratory is often something that is given less than adequate consideration. The key concept is that the sample's integrity must be strictly maintained and preserved. If a preservative needs to be added, then someone needs to be sure to add it. If maintaining integrity means refrigerating the sample, then it should be adequately refrigerated. If there is a specified holding time, then this should be accurately documented. If preventing contamination means a particular material for the storage container (e.g., glass vs. plastic), then that material should be used.

### 2.4.1   Chain of Custody

It is very important to document who has handled the sample, what responsibility each handler has at various junctures between the sampling site and the laboratory, and what actions the sample handler has taken relating to sample integrity while the sample was in his or her custody. In other words, the **chain of custody** must be maintained and documented. A sample can have a number of custodians along the way to the laboratory. A sample of lake water may be taken by a sampling technician at the site. The sampling technician may give it to a driver who transports it to the analysis site. A shipping and receiving clerk may log in the sample and give it to a subordinate who takes it to the laboratory. Along the way, this sample is in the hands of five different handlers: the sampling technician, the driver, the shipping and receiving clerk, the subordinate, and the laboratory technician. Each should maintain documentation of his or her activity and duties, and copies of the chain of custody should be filed (see Figure 2.1).

### 2.4.2   Maintaining Sample Integrity

It is important for all sample custodians to maintain the sample in its original physical and chemical condition so that it remains representative of the bulk system in terms of the analyte identity and concentration. Possible changes to avoid are: 1) loss of sample matrix or solvent through evaporation or other means, 2) loss of analyte through evaporation, chemical reaction, temperature effects, bacterial

**FIGURE 2.1**   A representation of the chain of custody for a water sample from a remote site, as described in the text. The point is that this chain of custody needs to be documented. (From Kenkel, J., *A Primer on Quality in the Analytical Laboratory*, CRC Press/Lewis Publishers, Boca Raton, FL, 2000.)

# AN EXAMPLE OF SAMPLE PRESERVATION: ENVIRONMENTAL WATER[*]

Environmental water samples to be analyzed for metals are best stored in quartz or Teflon™ containers. However, because these containers are expensive, polypropylene containers are often used. Borosilicate glass may also be used, but soft glass should be avoided because it can leach traces of metals into the water. If silver is to be determined, the containers should be light absorbing (dark colored). Samples should be preserved by adding concentrated nitric acid so that the pH of the water is less than two. The iron in well water samples, for example, will precipitate as iron oxide upon exposure to air and would be lost to the analysis if not for this acidification.

Environmental water samples to be analyzed for phosphate are not stored in plastic bottles unless kept frozen, because phosphates can be absorbed onto the walls of plastic bottles. Mercuric chloride, used as a preservative and acid (such as the nitric acid suggested for metals above), should *not* be used unless total phosphorus is determined. All containers used for water samples to be used for phosphate analysis should be acid rinsed, and commercial detergents containing phosphates should not be used to clean sample containers or laboratory glassware.

Environmental water samples analyzed for nitrate should be analyzed immediately after sampling. If storage is necessary, samples may be stored for up to 24 h at $4^{\circ}$C, or indefinitely, at $4^{\circ}$C, if sulfuric acid is added (2 mL of concentrated acid per liter). In such samples, however, nitrate and nitrite cannot be differentiated.

---

[*]Source: *Standard Methods for the Examination of Water and Wastewater*, 19th edition, APHA, AWWA, 2000.

---

effects, etc., 3) contamination with an additional analyte (or another chemical that would interfere with the analysis) through contact with other matrices or chemical systems, and 4) moisture absorption or adsorption by exposure to humid air.

To avoid loss by evaporation, the sample should be sealed in its container. For liquid samples, the container should be filled to the top with no headspace. If any matrix component is volatile, it may need to be stored at a temperature lower than room temperature and care should be taken when opening the container in the laboratory. Care should be taken not to lose any sample when transferring to another container, filter, etc. It may be important to not leave any portion of the sample behind (such as on the container walls) when transferring. It is also important to consider the cleanliness of the container or the laboratory equipment that comes into contact with the sample.

To avoid loss of analyte by chemical reaction or by temperature or bacterial effects, it may be necessary to take extraordinary precautions specific to the analyte itself. Such precautions may include adding a preservative to the sample, maintaining specific conditions of temperature, humidity, or other environmental conditions, avoiding sunlight or oxygen, etc. In addition, laboratory equipment that comes into contact with the sample in the preparation process should be clean and free of material that would remove analyte or add contaminant.

# WORKPLACE SCENE 2.2

Materials purchased for the purpose of mixing with other materials in the preparation of pharmaceutical products are called **raw materials**. Pharmaceutical companies often purchase solid raw materials as powders in large drums. Quality assurance laboratories require samples of the material in the drums for the purpose of performing quality tests to see if the raw materials meet the specifications required for the company's products. Sampling procedures must assure that the drums are not contaminated in the process of taking the sample, and if the material is hazardous, the sampling technicians must protect themselves appropriately. For example, magnesium hydroxide is an ingredient found in some heartburn medications. This strong base is purchased as a powder in large drums. Because it is a strong base, it is hazardous, particularly as a powder, because powders can become airborne and become a hazard for breathing.

Kelly White, pharmaceutical sampling technician, takes a sample of magnesium hydroxide from a drum of this strong base. Notice the extraordinary personal protective equipment and contamination protection.

Given all these facts, sampling technicians take extraordinary precautions in the process of taking samples. A special room with a positive air pressure is used; special personal protective equipment, including particulate respirators and sleeve covers, is used; and anticontamination precautions, such as the use of hair, beard, and shoe covers, are taken.

Other matrices or chemical systems that come into contact with a sample include the sample container, spatulas, scoops, grinders, mixers, filters, etc. Precautions should be taken to assure that these do not unintentionally affect or alter the composition of the sample in any way.

A good sample custodian is one who has all the analytical laboratory skills outlined in Chapter 1 (see Section 1.5) and utilizes them in executing any analytical laboratory operation. A conscientious and careful laboratory technician will succeed in maintaining sample integrity. See Workplace Scene 2.2.

## 2.5   Sample Preparation: Solid Materials

Except for the moon and the planet Mars, no extraterrestrial body has had pieces of its mass directly examined by scientists in an earthly laboratory. This means that there has been no laboratory sample preparation scheme performed on samples of solid matter from any of the other planets, their moons, comets, or asteroids. And yet we read repeatedly about how scientists have been able to surmise the chemical composition of the solid materials on the surfaces of these celestial bodies following, for example, the close approach of or the landing of a spacecraft loaded with special scientific equipment.

Obviously, the technology exists for obtaining analytical results without special preparation and analysis in a laboratory. However, at the present time there is no acceptable substitute for direct laboratory examination of samples if we want the kind of accuracy and confidence we have come to expect. All conventional methods for analysis of solid materials require one or more of the following preparation activities before an analytical method can be properly executed: 1) particle size reduction, 2) homogenization and division, 3) partial dissolution, and 4) total dissolution. Let us briefly discuss each of these individually.

## 2.5.1   Particle Size Reduction

If a solid sample is to be efficiently dissolved, either partially or totally, the solvent must have intimate contact with small particles, since without contact there can be no dissolution. The reason is simply that a solvent will only have contact with the outer surface of a particle and not the entire particle. Thus it would be ideal if a solid's particle size can be reduced to that of a fine powder, if possible, so that the solvent contact can be as complete as possible. Thus, sample preparation activities such as crushing, milling, grinding, or pulverizing are common. If the solid sample is of a state that reducing to a powder is not possible (such as with a polymer film or piece of organic tissue), then particle size reduction may take the form of cutting, chopping, blending, etc. Thus, crushers, ball mills, mortars and pestles, scissors, and blenders are common items used in sample preparation in analytical laboratories.

## 2.5.2   Sample Homogenization and Division

The amount of solid sample called for in an analysis procedure may be less, sometimes considerably less, than the amount of sample at hand. In that case, a portion of the larger sample must be taken. Obtaining a portion that is representative of the larger sample (and therefore representative of the entire bulk system) requires that the larger sample be homogeneous. Particle size reduction procedures may not leave a sample homogeneous. For example, large dirt clods from part of a dried soil sample may have been reduced to powder through crushing procedures, but this powder, from which a test sample must be taken, may not be sufficiently homogeneous without first undergoing a mixing procedure. Thus, a mixing procedure and a procedure by which a sample is divided to create the required portions may follow particle size reduction. See Workplace Scene 2.3. In addition, to ensure that the solid particles are of uniform size, they may be passed through a sieve (see Chapters 3 and 15 for descriptions of sieves).

## 2.5.3   Solid–Liquid Extraction

An analyte may be present in one material phase (either a solid or liquid sample) and, as part of the sample preparation scheme, be required to be separated from the sample matrix and placed in another phase (a liquid). Such a separation is known as an **extraction** —the analyte is extracted from the initial phase by the liquid and is deposited (dissolved) in the liquid, while other sample components are insoluble and remain in the initial phase. If the sample is a solid, the extraction is referred to as a **solid–liquid extraction**. In other words, a solid sample is placed in the same container as the liquid and the analyte is separated from the solid because it dissolves in the liquid while other sample components do not.

The process can take one of two forms. In one, the sample and liquid are shaken (or otherwise agitated) together in the same container, the resultant mixture filtered, and the filtrate, which then contains the analyte, collected. In the other, fresh liquid is continuously cycled over a period of hours through the solid sample via a continuous evaporation–condensation process (that usually does not require an extra filtration step), and the liquid is collected. This latter method is known as a **Soxhlet extraction**. Soxhlet extraction will be discussed in more detail in Chapter 11.

# WORKPLACE SCENE 2.3

Many fertilizers sold to gardeners and farmers consist of a variety of granular chemicals that are blended together at the manufacturing facility. Often, the various granules have different colors reflecting the fact that they have different chemical compositions. Analytical laboratory technicians that test samples of these fertilizers for the various constituents must transform the obviously nonhomogeneous samples into homogeneous ones. At the Nebraska State Agriculture Laboratory, this is done by grinding the samples in a grinding mill. A grinding mill has a rapidly rotating blade, much like a conventional kitchen blender. The mill pulverizes the samples and makes them homogeneous.

Tai Van Ha of the Nebraska State Agriculture Laboratory inspects a sample of fertilizer that is obviously not homogeneous. There are various light- and dark-colored granules of fertilizer in the bag.

## 2.5.4 Other Extractions from Solids

Extractions from solid samples may also be performed with supercritical fluids. A **supercritical fluid** is the physical state of a substance that is neither a liquid or a gas but something in between, meaning its properties change to a point different from either the liquid phase or the gas phase of that substance. For example, the solvent properties of supercritical fluids are better than those of their corresponding gas phase and their viscosities (resistance to flow) are considerably less than those of their corresponding liquid phase. The supercritical fluid state is achieved at a high temperature and extremely high pressure, exceeding the so-called critical temperature and pressure for that substance.

Because their solvent properties are very good and their viscosities are very low, supercritical fluids can be used for very efficient extraction of analytes from solid phase samples. The solid phase sample is held in a tube or cartridge and the supercritical fluid made to flow through (minimal pressure required). The fluid with the analyte is then made to flow through a trap solvent. The analyte dissolves in this solvent and the fluid reverts back to the gas phase.

Finally, extractions from solids can be performed by heating followed by solvent trapping. Such a procedure is known as **thermal extraction**.

**TABLE 2.1**    Solubility Rules for Ionic Compounds in Water

| Compound Class | Soluble? | Exceptions |
|---|---|---|
| Nitrates | Yes | None |
| Acetates | Yes | Silver acetate is sparingly soluble |
| Chlorides and bromides | Yes | Chlorides and bromides of Ag, Pb, and Hg are insoluble; chlorides and bromides of Pb are soluble in hot water |
| Sulfates | Yes | Sulfates of Ba and Pb are insoluble; sulfates of Ag, Hg, and Ca are slightly soluble |
| Carbonates | No | Carbonates of Na, K, and $NH_4$ are soluble |
| Phosphates | No | Phosphates of Na, K, and $NH_4$ are soluble |
| Chromates | No | Chromates of Na, K, $NH_4$, and Mg are soluble |
| Hydroxides | No | Hydroxides of Na, K, and $NH_4$ are soluble; hydroxides of Ba, Ca, and Sr are slightly soluble |
| Sulfides | No | Sulfides of Na, K, $NH_4$, Ca, Mg, and Ba are soluble |
| Sodium salts | Yes | Some rare exceptions |
| Potassium salts | Yes | Some rare exceptions |
| Ammonium salts | Yes | Some rare exceptions |
| Silver salts | No | Silver nitrate and perchlorate are soluble; silver acetate and sulfate are sparingly soluble |

## 2.5.5   Total Dissolution

In the event that extraction is not useful or feasible, total dissolution of a sample may be required. Total dissolution involves the proper choice of solvent—one that will indeed dissolve the total sample. For solid samples, this will usually involve the use of water or acid–water mixtures, sometimes fairly concentrated acid solutions. For this reason, it is useful to summarize the application of water and the most common laboratory acids:

**Water**. It should come as no surprise that ordinary water can be an excellent solvent for many samples. Due to its extremely polar nature, water will dissolve most substances of likewise polar or ionic nature. Obviously, then, when samples are composed solely of ionic salts or polar substances, water would be an excellent choice. An example might be the analysis of a commercial iodized table salt for sodium iodide content. A list of solubility rules for ionic compounds in water can be found in Table 2.1.

**Hydrochloric acid**. Strong acids are used frequently for the purpose of sample dissolution when water will not do the job. One of these is hydrochloric acid, HCl. Concentrated HCl is actually a saturated solution of hydrogen chloride gas, fumes of which are very pungent. Such a solution is 38.0% HCl (about 12 *M*). Hydrochloric acid solutions are used especially for dissolving metals, metal oxides, and carbonates not ordinarily dissolved by water. Examples are iron and zinc metals, iron oxide ore, and the metal carbonates of which the scales in boilers and humidifiers are composed. Being a strong acid, it is very toxic and must be handled with care. It is stored in a blue color-coded container.

**Sulfuric acid**. An acid that is considered a stronger acid than HCl in many respects is sulfuric acid, $H_2SO_4$. When sulfuric acid contacts clothing, paper, etc., one can see an almost instantaneous reaction—paper towels turn black and disintegrate; clothing fibers become weak and holes readily form. Concentrated sulfuric acid is about 96% $H_2SO_4$ (about 18 *M*), the remainder being water, and is a clear, colorless, syrupy, dense liquid. It reacts violently with water, evolving much heat, and so water solutions of sulfuric acid must be prepared cautiously, often to include a means of cooling the container. Its sample dissolution application is limited mostly to organic material, such as vegetable plants. It is not as useful for metals because many metals form insoluble sulfates. It is the solvent of choice for the Kjeldahl analysis (Chapter 5) of such materials as grains and products of grain processing. It is also used to dissolve aluminum and titanium oxides on airplane parts. It is stored in a yellow color-coded container.

# WORKPLACE SCENE 2.4

Radioactive plutonium isotopes emit alpha particles. The amount of radioactive plutonium in a sample can be measured by alpha spectroscopy, a technique for counting the alpha radiation. The technique is used at the Los Alamos National Laboratory (LANL) in New Mexico in order to monitor employees for exposure.

Sample preparation is rather involved. A sample of urine or fecal matter is obtained and treated with calcium phosphate to precipitate the plutonium from solution. This mixture is then centrifuged, and the solids that separate are dissolved in 8 M nitric acid and heated to convert the plutonium to the +4 oxidation state. This nitric acid solution is passed through an anion exchange column, and the plutonium is eluted from the column with a hydrochloric–hydroiodic acid solution. The solution is evaporated to dryness, and the sample is redissolved in a sodium sulfate solution and electroplated onto a stainless steel planchette. The alpha particles emitted from this electroplated material are measured by the alpha spectroscopy system, and the quantity of radioactive plutonium ingested is calculated. Approximately 2000 samples per year are prepared for alpha spectroscopy analysis. The work is performed in a clean room environment like that described in Workplace Scene 1.2.

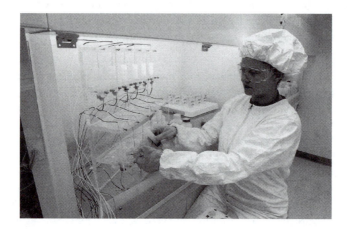

Stephanie Boone, a technician at Los Alamos National Laboratory, Chemistry Division Bioassay Program, prepares samples to be electroplated for alpha spectroscopy analysis. The hair net and special lab coat are required for a clean room environment.

**Nitric acid**. Another acid that has significant application is nitric acid, $HNO_3$. This acid is also very dangerous and corrosive and is aptly referred to as an oxidizing acid. This means that a reaction other than hydrogen gas displacement (as with HCl) occurs when it contacts metals. Frequently, oxides of nitrogen form in such a reaction, and noxious brown, white, and colorless gases are evolved. Concentrated $HNO_3$ is 70% $HNO_3$ (16 *M*) and is used for applications where a strong acid with additional oxidizing power is needed. These include metals such as silver and copper as well as organic materials such as in a wastewater sample. Nitric acid will turn skin yellow after only a few seconds of contact. It is stored in a red color-coded container. See Workplace Scene 2.4.

**Hydrofluoric acid**. An acid that has some very useful and specific applications, but is also very dangerous, is hydrofluoric acid, HF. This acid reacts with skin in a way that is not noticeable at first, but becomes quite serious if left in contact for a period of time. It has been known to be especially serious if trapped against the skin or after diffusing under fingernails. Treatment of this is difficult and painful. Concentrated HF is about 50% HF (26 *M*). It is an excellent solvent for silica ($SiO_2$)-based materials such as sand, rocks, and glass. It can also be used for stainless steel alloys. Since it dissolves glass, it must be stored in plastic containers. This is also true for low pH solutions of fluoride salts.

**Perchloric acid**. Another important acid for sample preparation and dissolution is perchloric acid, $HClO_4$. It is an oxidizing acid like $HNO_3$, but is considered to be even more powerful in that regard when hot and concentrated. It can be used for metals, since metal perchlorates are highly soluble, but it is most useful for more difficult organic samples, such as leathers and rubbers, often in combination with nitric acid. It can also be used for stainless steels and other more stable alloys. Commercial $HClO_4$ is 72% $HClO_4$ (12 *M*). It is a very dangerous acid, especially when hot and in contact with organic matter, and should be used only in a fume hood designed for the collection of its vapors. Contact with alcohols, including polymeric alcohols such as cellulose, and other oxidizable materials should be avoided due to the potential for explosions.

**Aqua regia**. An acid mixture that is prepared by mixing one part concentrated $HNO_3$ with three parts concentrated HCl is called aqua regia. This mixture is among the most powerful dissolving agents known. It will dissolve the very noble metals (gold and platinum) as well as the most stable of alloys.

Table 2.2 summarizes these acids and their properties. Occasionally, organic liquids are also used for total sample dissolution. Common laboratory organic solvents are described in Section 2.6.

**TABLE 2.2**    Various Laboratory Acids and Their Properties

| Name | Formula | Description | Example Uses |
|---|---|---|---|
| Water | $H_2O$ | Clear, colorless liquid with low vapor pressure, highly polar | Dissolving polar and ionic compounds |
| Hydrochloric acid | HCl | Commercially available concentrated solution is 38% (12 *M*) HCl; evolves pungent fumes and must be handled in fume hood | Dissolving some metals and metal ores |
| Sulfuric acid | $H_2SO_4$ | Commercially available concentrated solution is 96% (18 *M*) $H_2SO_4$; a dense, syrupy liquid; reacts on contact with skin and clothing; evolves much heat when mixed with water | Organic samples, such as for Kjeldahl analysis (see Chapter 5); also oxides of Al and Ti |
| Nitric acid | $HNO_3$ | Commercially available concentrated solution is 70% (16 *M*) $HNO_3$; reacts with clothing and skin (turns skin yellow); evolves thick brown and white fumes when in contact with most metals | Dissolving noble metals (e.g., copper and silver) and also some organic samples |
| Hydrofluoric acid | HF | Commercially available concentrated acid is 50% (26 *M*) HF; must be stored in plastic containers, since it attacks glass; very damaging to skin | Dissolving silica-based materials and stainless steel |
| Perchloric acid | $HClO_4$ | Commercially available concentrated solution is 72% (12 *M*) $HClO_4$ | Dissolving difficult organic samples and stable metal alloys |
| Aqua regia | — | A mixture of concentrated $HNO_3$ and HCl in the ratio of 1:3 $HNO_3$:HCl | Dissolving highly unreactive metals, such as gold |

## 2.5.6  Fusion

For extremely difficult samples, a method called fusion may be employed. Fusion is the dissolving of a sample using a molten inorganic salt, generally called a flux, as the solvent. This flux dissolves the sample and, upon cooling, results in a solid mass that is then soluble in a liquid reagent. The dissolving power of the flux is mostly due to the extremely high temperatures (usually 300 to 1000°C) required to render most inorganic salts molten.

Additional problems arise within fusion methods, however. One is the fact that the flux must be present in a fairly large quantity in order to be successful. The measurement of the analyte must not be affected by this large quantity. Also, while a flux may be an excellent solvent for difficult samples, it will also dissolve the container to some extent, creating contamination problems. Platinum crucibles are commonly used, but nickel, gold, and porcelain have been successfully used for some applications.

Probably the most common fluxes are sodium carbonate ($Na_2CO_3$), lithium tetraborate ($Li_2B_4O_7$), and lithium metaborate ($LiBO_2$). Fluxes may be used by themselves or in combination with other compounds, such as oxidizing agents (nitrates, chlorates, and peroxides). Applications include silicates and silica-based samples and metal oxides.

For dissolving particularly difficult metal oxides, the acidic flux potassium pyrosulfate ($K_2S_2O_7$) may be used.

## 2.6  Sample Preparation: Liquid Samples, Extracts, and Solutions of Solids

Even though a sample may be dissolved in a liquid, it may not be ready for the method chosen. There may be other sample components present that may interfere, or the solvent may not be suitable for the chosen method. Thus original liquid solutions and the extracts and solutions resulting from the procedures described in Section 2.5 may require separation or dilution–concentration procedures.

### 2.6.1  Extraction from Liquid Solutions

In Section 2.5, we described separation procedures in which analytes are extracted from solid samples via contact with liquid solvents that selectively dissolve the analyte and leave other components undissolved or unextracted. There are several methods by which analytes can be extracted from liquid matrices as well.

In **liquid–liquid extraction**, the liquid solution containing the analyte (usually a water solution) is brought into contact with a liquid solvent (usually a nonpolar organic solvent) that is immiscible with the first solvent. The container is usually a separatory funnel (Figure 2.2). Since the two solvents are immiscible, there are two liquid layers in the separatory funnel. Shaking the separatory funnel brings the two solvents into intimate contact such that the analyte then moves from the original (first) solvent to the new (second) solvent. Being immiscible, the two layers can then be separated from each other by allowing the two layers to drain one at a time through the stopcock at the bottom of the funnel. The desired solution is then carried forward to the next step. A more comprehensive discussion of liquid–liquid extraction can be found in Chapter 11.

Table 2.3 presents a list of some nonpolar organic solvents that are most often used as extracting solvents for aqueous solutions.

In **solid phase extraction**, the liquid solution containing the analyte is passed through a cartridge containing a solid sorbent. The sorbent retains either the analyte or its matrix. If it is the analyte that is retained by the sorbent, the cartridge is first washed with a solvent that will sweep away interferences. The analyte is then eluted with a stronger solvent and collected. If it is the matrix that is retained by the sorbent, the analyte is eluted immediately and collected.

In a **purge-and-trap** procedure, volatile analytes can be purged from a liquid sample by helium sparging, vigorous bubbling of helium through the sample. The helium–analyte mixture is guided through an adsorption cartridge in which the analytes are trapped on the adsorbent in a manner similar

**TABLE 2.3**    Some Nonpolar Organic Solvents That Are Often Used for Extracting Analytes from Aqueous Solutions

*Aliphatic hydrocarbons n-hexane, cyclohexane, and n-heptane*: Aliphatic hydrocarbons are nonpolar. Their solubility in water is virtually nil. They are less dense than water, and thus would be the top layer in a separatory funnel with a water solution. They are obviously poor solvents for polar compounds, but are very good for extracting traces of nonpolar solutes from water solutions. They are highly flammable and have a low toxicity level.

*Methylene chloride*: This solvent is a slightly polar solvent also known as dichloromethane, $CH_2Cl_2$. Its solubility in water is 1.32 g/100 mL. It is denser than water (density = 1.33 g/mL); thus it would be the bottom layer when used with a water solution in a separatory funnel. It may form an emulsion when shaken in a separatory funnel with water solutions. It is not flammable and is considered to have a low toxicity level.

*Toluene*: Toluene ($C_6H_5$–$CH_3$) is a nonpolar aromatic liquid. It is flammable and has a density of 0.87 g/mL, and thus would be the top layer in the separatory funnel with water. It is only slightly soluble in water (0.47 g/L). Its acute toxicity is low.

*Diethyl ether*: Diethyl ether ($CH_3CH_2$–O–$CH_2CH_3$), also frequently referred to as ethyl ether and as ether, is a mostly nonpolar organic liquid that is highly volatile and extremely flammable, a dangerous combination. Also, explosive peroxides form with time. Precautions regarding storage to discourage peroxide formation include storage in metal containers in explosion-proof refrigerators. It should be disposed of after about 9 months of storage. Ether is only slightly soluble (approximately 60 g/L at 25°C) in water. Its density is 0.71 g/mL; therefore it would be the top layer in a separatory funnel with water. Since it is volatile, it is easily evaporated from extraction fractions. Its acute toxicity is low.

*Chloroform*: Chloroform (trichloromethane, $CHCl_3$) is a nonflammable organic liquid with low miscibility with water (approximately 7.5 g/L at 25°C). It shows carcinogenic effects in animal studies and should be avoided when another solvent would do as well. Vapors should not be inhaled, and contact with the skin should be avoided.

**FIGURE 2.2**    Omer Elbasheer at Southeast Community College, Lincoln, Nebraska, holds a separatory funnel containing two immiscible liquid layers. The lower layer has green food coloring added for contrast.

to solid phase extraction. The analytes are then desorbed by heating the cartridge. The gases that come off the cartridge are typically analyzed by gas chromatography. (Refer to Workplace Scene 12.1.)

## 2.6.2  Dilution, Concentration, and Solvent Exchange

The analyte in a sample may be too concentrated or too dilute for the chosen method. If it is too concentrated, a dilution with a compatible solvent may be performed. The dilution should be performed with volumetric glassware and with good analytical technique so that the dilution factor is known and accuracy is not diminished.

If the analyte is too dilute for the chosen method, its concentration may be increased in any number of ways. One is a controlled evaporation of the solvent (such that the factor by which its concentration is increased is known). Another is to perform an extraction that results in a smaller solution volume for the same quantity of analyte. Another is to evaporate the analyte solution to dryness and then reconstitute (i.e., redissolve) with a smaller volume of solvent.

The reconstitution can occur with a different solvent, a process known as **solvent exchange** . A different solvent may be more compatible with the chosen method.

### 2.6.3   Sample Stability

In the event that the sample cannot be carried forward through the analytical method procedure because the analyte is not stable (e.g., it might be thermally unstable and would decompose at the higher temperature used in the method procedure), it may be important to protect the sample from decomposition in some way or to derivatize the analyte. Protection from decomposition might mean storing the sample (e.g., a biological sample) in a refrigerator, protecting it from light, protecting it from exposure to air or humidity (i.e., store in a desiccator), etc. Derivatizing the analyte means to chemically convert it to a form (a derivative) that is stable so that the quantity of the analyte can be determined indirectly by analyzing for the derivative.

## 2.7   Reagents Used in Sample Preparation

A **reagent** is a substance used in a chemical reaction in an analytical laboratory because of its specific applicability to a given system or procedure. Reagents used in sample preparation must be wholly applicable to the preparation undertaken in terms of identity and purity. For example, if a given procedure calls for certified American Chemical Society (ACS) grade methyl alcohol, then certified ACS grade methyl alcohol should be used and not methyl alcohol of a lesser grade (lower purity). In addition, this purity must not have diminished since the reagent was purchased. If a given item is designated with a particular **shelf life**, then the analyst must be aware of the date beyond which the composition of a reagent can no longer be trusted.

There are a number of different purity grades of chemicals that are available. Some of the more common are listed in Table 2.4.

TABLE 2.4   Examples of Purity Grades of Chemicals

*Primary standard*: A specially manufactured analytical reagent of exceptional purity for standardizing solutions and preparing reference standards.

*ACS certified*: A reagent that meets or exceeds the specifications of purity put forth by the American Chemical Society. The certificate of analysis is on the label.

*Certified reagent*: A reagent that meets the standards of purity established by the manufacturer. The certificate of analysis is on the label.

*USP/NF*: Reagents that meet the purity requirements of the U.S. Pharmacopeia (USP) and the National Formulary (NF). Generally of interest to the pharmaceutical profession, these specifications may not be adequate for reagent use.

*Spectro grade or spectranalyzed*: Solvents of suitable purity for use in spectrophotometric procedures. A certificate of analysis is on the label.

*High-performance liquid chromatography (HPLC) grade*: Solvents of suitable purity for use in liquid chromatography procedures.

*Practical*: Chemicals of sufficiently high quality to be suitable for use in some syntheses. Organic chemicals of practical grade may contain small amounts of intermediates, isomers, or homologs.

*Technical*: Chemicals of reasonable purity for applications that have no official standard for purity.

## 2.8   Labeling and Record Keeping

Samples gathered and solutions prepared by laboratory personnel must be properly labeled at the time of sampling or preparation. In addition, a complete record of the sampling or preparation should be maintained. Sound quality assurance practices include a notebook record where one can find the source and concentration of the material used, the identity and concentration of the standard being prepared, the name of the analyst who prepared it, the specific procedure used, the date it was sampled or prepared, and the expiration date for any stored solutions. The reagent label should have a clear connection to the notebook record. A good label includes an ID number that matches the notebook record, the name of the material and its concentration, the date, the name of the analyst, and the expiration date. See Workplace Scene 2.5.

# Experiments

### Experiment 3: A Study of the Dissolving Properties of Water, Some Common Organic Liquids, and Laboratory Acids

Introduction: In this experiment, the miscibility of water with some common organic solvents and the dissolving power of water, hydrochloric acid, sulfuric acid, and nitric acid will be studied. First, small amounts of water will be mixed with roughly equal amounts of the organic solvents, and the miscibilities will be observed. Then small volumes of water and the acids will be allowed to contact granules of some selected metals and other compounds, and their dissolving power will be noted. The objectives are to confirm some of the statements made in Tables 2.1, 2.2, and 2.3 and to discover the behavior of some solvents and acids not listed.

*Remember safety glasses and wear gloves.*

#### Part A:   A Study of the Miscibility of Water with Selected Organic Solvents

1. Obtain a test tube rack and 11 small test tubes. Add distilled water to each tube such that each is about a fourth to a third full. Place a piece of labeling tape on each.
2. Obtain samples of the following 11 organic liquids contained in individual small dropper bottles: n-hexane (or other alkane), acetonitrile, methylene chloride, acetone, toluene, methanol, diethyl ether, ethyl acetate, ethylbenzene, ethanol, and chloroform. Then label each of the test tubes from step 1 with the names, or an abbreviation of the names, of these liquids.
3. Add small amounts of each liquid indicated in step 2 to the water in the test tubes with the corresponding label. Stopper, shake, and observe the miscibility. Record the miscibility (miscible or immiscible) and polarity (polar or nonpolar) in a table such as the following in the data section of your notebook.

| Solvent | Miscibility | Polarity |
|---|---|---|
| n-Hexane | | |
| Acetonitrile | | |
| Methylene chloride | | |
| Acetone | | |
| Toluene | | |
| Methanol | | |
| Diethyl ether | | |
| Chloroform | | |
| Ethyl acetate | | |
| Ethylbenzene | | |
| Ethanol | | |

# WORKPLACE SCENE 2.5

T he receiving, unpacking, and aliquoting of various actinide materials is the primary function of sample management (SM) in the actinide analytical chemistry group at Los Alamos
National Laboratory. Upon receipt of a sample shipment, the contents of each container
must be closely validated with the labeling on the container and with the paperwork that accompanies the shipment. Any discrepancy must be immediately rectified.

A wide variety of radioactive materials are received into the group, including a number of
matrices, e.g., metal, oxides, liquids, etc. Once the material is introduced into an argon inert
atmosphere glove box, a series of aliquots or sample cuts is made for the type of analysis requested.
The cut weights are predetermined by the various analysts using different analytical techniques.
An inert atmosphere is often required because many of the actinide metals are air and moisture
sensitive and will react with air to degrade the original sample integrity. Accuracy in sample cutting
cannot be underestimated. Sample cuts vary from 10 mg to 1.0 g. Quick turnout to satisfy
programmatic needs and requirements is essential. A discard path for all analyzed aqueous and
solid sample residue, plus unused excess material, is identified in sample management. Another
important function in SM is to systematically and expeditiously return this material to its original
owner for cementation disposal, recovery, or reprocessing.

Manuel Gonzales, a technician with the Los Alamos National Laboratory in New Mexico, inspects a sample
container, validating proper labeling against its contents.

Compare the results in your table with what was stated in Table 2.3 for miscibility with water and
polarity for some of the liquids.

4. Optional: Check out the liquids in the above table to see if they are miscible with each other rather
than with water. You might expect those that are polar to be immiscible with those that are nonpolar.
Check out, for example, the miscibility of acetone and methanol with those you identified as

nonpolar. Place your observations in a table and draw some appropriate conclusions regarding polarities.

5. Dispose of the contents of each test tube according to the directions of your instructor.

**Part B: A Study of the Solubility of Some Selected Inorganic Materials in Water, Acetic Acid, Hydrochloric Acid, Sulfuric Acid, and Nitric Acid**

1. Obtain small samples of $NaCl$, $CaCO_3$, $Fe_2O_3$, $Al_2O_3$, and the metals aluminum, zinc, iron, copper, and silver. Obtain around 20 small test tubes and a test tube rack. Place a piece of labeling tape on each tube.

2. Make a table or grid in the data section of your notebook as follows:

| Water or Acid | NaCl | CaCO₃ | Fe₂O₃ | Al₂O₃ | Al | Zn | Fe | Cu | Ag |
|---|---|---|---|---|---|---|---|---|---|
| Water | | | | | | | | | |
| Acetic | | | | | | | | | |
| Hydrochloric | | | | | | | | | |
| Sulfuric | | | | | | | | | |
| Nitric | | | | | | | | | |

3. Place very small granules of each of the nine materials indicated in step 1 individually into test tubes. Label each of the nine tubes with the chemical symbol or formula of the contents.

4. Fill each of the test tubes in step 2 halfway with distilled water. Now also place a symbol for water (such as a W or $H_2O$) on each label. After 2 min, carefully observe the contents of each tube to determine whether there has been a reaction. If there has been no visible reaction, stopper and shake, then observe whether the material has dissolved. Place an S (for soluble) in the grid if the material has either reacted or dissolved (or both) and an I (for insoluble) if nothing happened.

5. Obtain dropper bottles containing 50% (by volume) solutions of acetic acid, hydrochloric acid, sulfuric acid, and nitric acid. Repeat steps 2 and 3 using acetic acid in place of water, testing all those for which water gave an I. Then repeat with hydrochloric acid and sulfuric acid. Repeat finally with nitric acid, but only for those for which acetic, hydrochloric, and sulfuric acids gave an I.

6. Compare the results with statements of uses of water and these acids in Tables 2.1 and 2.2.

# Questions and Problems

1. What general tasks relating to the chemical analysis of a material sample must be performed prior to executing the chosen method of analysis?
2. Why are sampling and sample preparation procedures as crucial to the success of an analysis as the analytical method chosen?
3. What is a representative sample? Do not use the words "represent" or "representative" in your answer.
4. Under what circumstances would a selective sample be more appropriate than a composite sample?
5. How would you obtain a representative sample of each of the following?
   (a) Water in a creek
   (b) Polymer film manufactured by a chemical company
   (c) Aspirin in a bottle of aspirin tablets
   (d) Soil in your yard
   (e) Old paint on a building
   (f) Animal tissue being analyzed for pesticide residue
6. Differentiate between bulk sample, primary sample, secondary sample, subsample, laboratory sample, and test sample.

7. Why is statistics important in sampling?
8. Why should one be concerned about who has custody of a sample between the sampling site and the laboratory?
9. If a water sample is to be analyzed for trace levels of metals, why is a glass container for sampling and storage inappropriate?
10. If a sample to be analyzed in a laboratory is subject to alteration due to the action of bacteria, what can be done to preserve its integrity?
11. Look up in a reference book and explain exactly what is involved in taking and preserving a sample of soil to be analyzed for nitrate.
12. Why is it important to reduce the size of solid particles present in a sample?
13. Why is it important for a sample to be homogeneous before it is divided to create the test sample?
14. Why is the separation of an analyte from a sample matrix called an extraction?
15. What two forms may a solid–liquid extraction take? Why does one require filtering while the other may not?
16. What is a supercritical fluid? Why are supercritical fluids useful in solid–liquid extraction?
17. Identify water or a specific acid as the least drastic dissolving agent for each of the following:
    (a) A mixture of sodium chloride and sodium sulfate
    (b) Material from a silver mine
    (c) A sample of stainless steel
    (d) A sample of iron carbonate ore
    (e) Material from a gold mine
    (f) A sample of sand from a beach
    (g) Oxide on the surface of an airplane part
18. What concentrated acid is correctly described by each of the following?
    (a) A very dense, syrupy liquid
    (b) A liquid that can be absorbed through the skin
    (c) A liquid that turns skin yellow
    (d) A liquid that gives off unpleasant fumes
    (e) A liquid that is mixed with HCl to give aqua regia
    (f) A solvent for silver metal
    (g) A solvent for corrosion products on an airplane part
    (h) A solvent for sand
    (i) A solvent for stainless steel
    (j) A solvent for iron ore
    (k) A solvent for leather
    (l) A liquid that gets especially hot when mixed with water
    (m) An acid that turns paper towels black on contact
19. What is aqua regia?
20. Fill in the blanks with a specific example of the kind of sample dissolved by the acid indicated.
    (a) Hydrofluoric acid _____
    (b) Nitric acid _____
    (c) Sulfuric acid_____
    (d) Hydrochloric acid _____
    (e) Aqua regia _____
    (f) Perchloric acid _____
21. Fill in the blanks with the name of the acid described by the statement.
    (a) The bottle containing _____ has a red color code.
    (b) _____, when mixed with nitric acid in certain proportions, gives aqua regia.
    (c) When an open bottle of _____ sits next to an open bottle of ammonium hydroxide, thick white fumes form.
    (d) _____ gets especially hot when mixed with water.

    (e) _____ diffuses through skin and is especially bad when it gets under the fingernails.

    (f) The bottle containing _____ has a yellow color code.

    (g) _____ is used to dissolve the sample for the Kjeldahl analysis.

22. From the following list, choose those samples that would be best dissolved by HCl and those that would be best dissolved by $HNO_3$. Not all samples will be used.

    (a) NaCl

    (b) Iron ore

    (c) Gold metal

    (d) Lettuce leaves

    (e) Copper metal

    (f) Sand

    (g) Aluminum oxide

23. What is meant by fusion as a method for sample dissolution?

24. What is meant by the flux in a fusion procedure?

25. What materials must be used as containers for the flux or sample mixture in a fusion procedure? Explain.

26. Differentiate between solid–liquid extraction, liquid–liquid extraction, and solid phase extraction.

27. Give two examples of when an extraction, rather than total sample dissolution, is useful.

28. Describe a separatory funnel and tell for what and how it is used.

29. Define extraction as it is used in a chemical analysis laboratory.

30. Why are the solvents listed in Table 2.3 useful for extracting analytes from water samples?

31. What are the hazards associated with the use of diethyl ether and what specific precautions are taken?

32. Of the solvents listed in Table 2.3, which would be the top layer in a separatory funnel containing an aqueous solution? Explain what the solvent's density has to do with your answer.

33. List the organic solvents listed in Table 2.3 that are (a) toxic and (b) flammable.

34. What is a purge-and-trap procedure?

35. What must be done to a sample if it is too concentrated for the chosen method? What must be done if it is too dilute?

36. What is meant by solvent exchange?

37. Why must the quality of all reagents used in the laboratory be assured before use?

38. Perhaps the most common grade of chemical used in the laboratory is ACS certified. Explain what ACS certified refers to.

39. Can chemicals labeled as technical grade be used as standards in analytical procedures? Explain.

40. What designation of purity is used for solvents that are appropriate for spectrophotometric analysis? Liquid chromatography analysis?

41. How important is the label on a sample or reagent? What information should appear on the label of a sample to be analyzed? What information should appear in the notebook record?

# 3

# Gravimetric Analysis

## 3.1 Introduction

We now begin to discuss the parts of the analytical strategy designated "carry out the analytical method" and "work the data" (refer again to the analytical strategy flow chart given in Figure 1.1). As stated in Chapter 1, there is a wide variety of analytical methods, but they can all be classified in one of three categories: wet chemical methods, physical methods, and instrumental methods. We will systematically discuss each of these throughout the book. We will present a modified analytical strategy flow chart at the beginning of each discussion, with the parts designated "carry out the analytical method" and "work the data" highlighted.

The first of the two wet chemical methods we will consider is the gravimetric method of analysis. **Gravimetric analysis** is characterized by the fact that the measurement of weight is the primary measurement, usually the only measurement made on the sample, its components, or its reaction products. Thus, weight measurements are used in the calculation of the results and are often the only measurements used in this calculation. The analytical strategy flow chart specific to gravimetric analysis is given in Figure 3.1.

## 3.2 Weight vs. Mass

The most fundamental, and possibly the most frequent, measurement made in an analysis laboratory is that of weight (or mass). While we often speak of mass and weight in the same breath, it is of some importance to recognize that they are not the same. Mass is the quantity or amount of a substance being measured. This quantity is the same no matter where the measurement is made—on the surface of the Earth, in a spaceship speeding toward the moon, or on the surface of Mars. Weight is a measure of the Earth's gravitational force exerted on a quantity of matter. Weight is one way to measure mass. In other words, we can measure the quantity of a substance by measuring the Earth's gravitational effect on it.

Since nearly 100% of all weight measurements made in any analysis laboratory are made on the surface of the Earth, where the gravitational effect is nearly constant, weighing has become the normal method of measuring mass. Weighing devices are calibrated in grams, which is defined as the basic unit of mass in the metric system. Technically, however, analytical quantities are mass quantities and not weight quantities. Despite this, because the term weight is used almost universally to describe the amount of matter, this is the term we will use throughout this text to describe the amount of a material.

## 3.3 The Balance

The laboratory instrument built for measuring weight is called the **balance**. The name is derived from mechanical devices that utilize known weights to balance the object to be weighed across a fulcrum, like a teeter-totter. Most balances in use today are electronic, rather than mechanical, balances. An **electronic balance** is one that uses an electromagnet to balance the object to be weighed on a single pan. The older

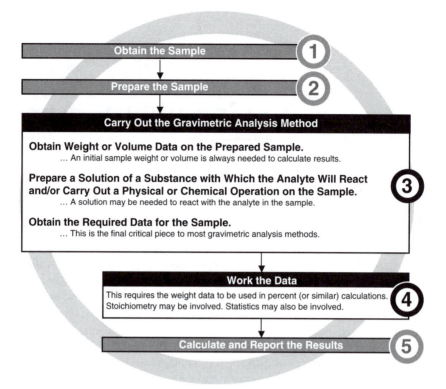

**FIGURE 3.1** The analytical strategy for gravimetric analysis methods.

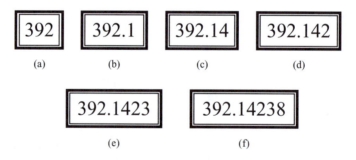

**FIGURE 3.2** Displays of balances read (a) to the nearest gram, (b) to the nearest tenth of a gram, (c) to the nearest hundredth of a gram, (d) to the nearest milligram, (e) to the nearest tenth of a milligram, and (f) to the nearest hundredth of a milligram.

mechanical balances also utilize a single-pan concept, but these also use the teeter-totter design, so they are mechanical and not electronic.

Different examples of laboratory work require different degrees of precision, i.e., different numbers of significant figures to the right of the decimal point. Thus, there are a variety of balance designs available, and this variety reflects the need for them. Some balances are read to the nearest gram; some are read to the nearest tenth or hundredth of a gram; some are read to the nearest milligram; and still others are read to the nearest tenth or hundredth of a milligram. See Figure 3.2.

Balances with few such significant figures to the right of the decimal point (zero to three) are often referred to as ordinary balances or top-loading balances (precision is ±100 to ±1 mg). A **top-loading balance** is an electronic ordinary balance with a pan on the top, as shown in Figure 3.3. The electronic

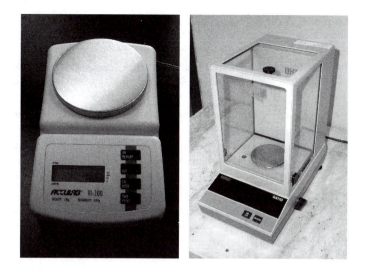

**FIGURE 3.3**    Left, a top-loading balance. Right, an analytical balance.

top loaders often have a tare feature. A chemical, for example, can be conveniently weighed on a piece of weighing paper without having to determine the weight of the paper. **Taring** means that the balance is simply zeroed with the weighing paper on the pan.

A balance that is used to obtain four or five digits to the right of the decimal point in the analytical laboratory is called the **analytical balance** (precision is ±0.1 or ±0.01 mg). The modern laboratory utilizes single-pan electronic analytical balances almost exclusively for this kind of precision. Figure 3.3 shows a typical modern electronic analytical balance. Notice that it is a single-pan balance with the pan enclosed. The chamber housing the pan has transparent walls for easy viewing. Sliding doors on the right and left sides and on the top make the pan accessible for loading and handling samples. The reason for enclosing the pan in this manner is to avoid the effect of air currents on the weight. Most modern analytical balances also have the tare feature mentioned above.

The operator of an analytical balance should keep in mind that this instrument is extremely sensitive and, therefore, must be handled carefully and correctly. Also, this discussion presumes that, if necessary, the sample has been dried (such as in a laboratory drying oven) and has been kept dry (such as by storing in a desiccator) prior to weighing.

## 3.4    Calibration and Care of Balances

It is important to keep the balances, especially analytical balances, and the immediate area clean. Spilled chemicals can react over time with the various finished surfaces on the balances and may inhibit their sensitive and accurate operation at some point. In addition, spills and debris can contaminate articles to be weighed if they are allowed to remain in the enclosed space or the immediate area. It is a good idea to hire a professional to clean a laboratory's balances on a regular basis.

Balances should also be calibrated on a regular basis. **Calibration** refers to the process by which a balance is checked to see if the weight obtained for an object is correct. To calibrate a balance, a known weight is used to see if that weight is displayed by the balance within an acceptable error range. If it is, the balance is said to be calibrated for that weight. If it is not, it is said to be out of calibration. Such known weights are available from the federal agency charged with providing standard reference materials, the National Institute of Standards and Technology (NIST), and must be stored in a dry, protected environment in the lab. NIST can provide an entire set of weights so that the calibration of a balance can be checked over a wide range of weights. If it is determined that a balance is out of calibration, it must be taken out of service and repaired or replaced.

# 3.5   When to Use Which Balance

The question of which balance to use under given circumstances, an ordinary balance or an analytical balance, remains. An analyst must be able to recognize when a weight measurement with four or five decimal places is needed (analytical balance) or when one with only one to three decimal places is needed (ordinary balance).

First, an analytical balance should *not* be used if it is not necessary to have the number of significant figures that this balance provides. It does not make sense to use a highly sensitive instrument to measure a weight that does not require such sensitivity. This includes a weight measurement when the overall objective is strictly qualitative or when quantitative results are to be reported to two significant figures or less. Whether the weight measurements are for preparing solutions to be used in such a procedure, for obtaining an appropriate weight of a sample for such an analysis, etc., they need not be made on an analytical balance, since the outcome either will be only qualitative or does not necessarily require more than one or two significant figures.

Second, if the results of a quantitative analysis are to be reported to three or more significant figures, then weight measurements that enter directly into the calculation of the results should be made on an analytical balance so that the results of the analysis can be correctly reported after the calculation is performed.

Third, if a weight is only incidental to the overall result of an accurate quantitative analysis, then it need not be made on an analytical balance. This means that if the weight measurement to be performed has no bearing whatsoever on the quantity of the analyte tested, but is only needed to support the chemistry or other factor of the experiment, then it need not be measured on an analytical balance.

Fourth, if a weight measurement does directly affect the numerical result of an accurate quantitative analysis (in a way other than entering directly into the calculation of the results, which is the second case mentioned above), then it must be performed on an analytical balance.

These last two points require that the analyst carefully consider the purpose of the weight measurement and whether it will directly affect the quantity of analyte tested or the numerical value to be reported. While it is often required that weight measurements be made on an analytical balance so that the result can have the appropriate degree of precision, it is also sometimes true that weight measurements taken during such a procedure need not be made with an analytical balance. It depends on whether this weight enters into the calculation of the result and how many significant figures are desired in the result.

# 3.6   Details of Gravimetric Methods

Gravimetric analysis methods proceed with the following steps: 1) the weight or volume of the prepared sample is obtained, 2) the analyte is either physically separated from the sample matrix or chemically altered and its derivative separated from the sample matrix, and 3) the weight of the separated analyte or its derivative is obtained. The data thus obtained are then used to calculate the desired results.

## 3.6.1   Physical Separation Methods and Calculations

The method by which an analyte is physically separated from the matrix varies depending on the nature of the sample and what form the analyte is in relative to its matrix. For example, the analyte or its matrix may be sufficiently volatile so that one or the other can be separated by evaporation at a temperature attainable by laboratory ovens or burners. In that case, the analyte weight is measured either by sample weight loss if the analyte has been evaporated or directly if the matrix has been evaporated. In either case, the weight of a container may also be involved.

As another example, consider that the analyte may be able to be filtered or sieved (separated according to particle size) from the matrix. If the sample is a liquid suspension, its volume is first measured. This is followed by a filtering or wet sieving step to capture the desired solid material. The filtering or sieving medium or unit is weighed first, and thus the weight of the analyte is determined from the weight gain

## SIDEBOX 3.6.1

A desiccator is a storage container used either to dry samples or, more commonly, to keep samples and crucibles dry and protected from the laboratory environment once they have been dried by other means. The typical laboratory desiccator is shown here. However, any container that can be sealed, such as an ordinary coffee can, can be used. A quantity of water-absorbing chemical, called the desiccant, is placed in the bottom of the container and will absorb all the moisture inside the sealed vessel, thus providing a dry environment. A good commercially available desiccant is called Drierite™ (anhydrous $CaSO_4$). This substance can be purchased in the indicating form, in which case a color change (from blue to pink) will be observed when the material is saturated with moisture. If the analyst suspects a desiccant to be saturated, it should be replaced. Drierite can be recharged by heating in a vacuum oven.

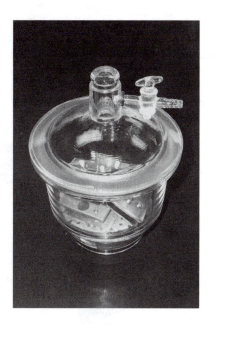

that occurs as a result of the filtering or sieving. If the sample is a solid, its weight is measured and the sample either dry-sieved (through a stack of nested metal sieves with different-sized holes) or dissolved such that insoluble matter is filtered. In the case of sieving, the weight gain of each sieve is the analyte weight for that particle size. In the case of insoluble matter, the weight gain of the filter is the analyte weight.

In each of these cases, the percent of the analyte is often calculated. The weight percent of an analyte in a sample is calculated using the definition of weight percent:

$$\text{weight \%} = \frac{\text{weight of part}}{\text{weight of whole}} \times 100 \qquad (3.1)$$

or

$$\text{\% analyte} = \frac{\text{weight of analyte}}{\text{weight of sample}} \times 100 \qquad (3.2)$$

Following are examples of gravimetric analysis when the analyte is physically separated from the matrix.

### 3.6.1.1 Loss on Drying

This is an example of a loss through volatilization (evaporation) under temperature conditions at which water would volatilize, hence the word "drying." The loss can occur by elevating the temperature of the sample to just above the boiling point of water (although a different temperature may be specified) or through desiccation. This is not necessarily limited to water, however, as any sample component that volatilizes in the case of elevated temperature would be included in the weight loss. Thus it is called "loss

# WORKPLACE SCENE 3.1

Loss on drying is a routine analytical procedure in the pharmaceutical industry. Special loss-on-drying equipment that includes a balance is available in which a shallow, round metal tray is filled with the solid granulated sample and placed on the balance in a short cylindrical enclosure, with a large electrical heating coil immediately above. As the coil heats, the sample weight is automatically monitored and the percentage loss on drying is calculated and displayed. When the calculated percentage is stable, the unit beeps and the percentage loss on drying for the sample is displayed.

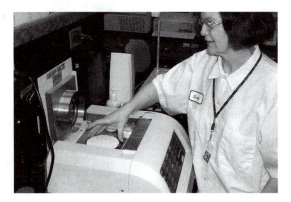

Evaline Robotham sets the sample tray in place in the loss-on-drying instrument. Notice the cylindrical enclosure with heating coil.

on drying" rather than the "determination of water or moisture by drying." If it is known that only water is lost in the drying step, then the latter would accurately describe it. For example, only water would be lost if the loss occurs through desiccation (placing the sample in a desiccator) and not by heating. In any case, the sample is usually contained in either a predried weighing bottle (see Section 3.7.1) or an evaporating dish and placed in an oven or desiccator for a specified period of time or to a constant weight. **Constant weight** refers to repeated drying steps until two consecutive weights agree to within a specified precision, such as having two consecutive weights that do not differ by more than 0.25%. The results are calculated as either the percent loss in weight or the percent moisture or water.

$$\% \text{ loss or water} = \frac{\text{weight loss or weight of water}}{\text{weight of sample}} \times 100 \tag{3.3}$$

$$= \frac{(\text{weight before drying} - \text{weight after drying})}{\text{weight before drying}} \times 100 \tag{3.4}$$

In the event that the tare feature of a balance is not used, the container weight must be subtracted before Equation (3.4) can be used. See Workplace Scene 3.1.

## Example 3.1

What is the percent loss on drying if a sample weighs 4.5027 g before drying and 3.0381 g after drying?

**Solution 3.1**

Utilizing Equation (3.4), we have

$$\% \text{ loss} = \frac{(4.5027 \text{ g} - 3.0381 \text{ g})}{4.5027 \text{ g}} \times 100$$

$$\% \text{ loss} = 32.527\%$$

Application of the loss-on-drying method is found in the pharmaceutical, food, and agricultural industries.[*]

### 3.6.1.2  Loss on Ignition

This is similar to loss on drying, except that an extremely high temperature, such as that of a Meker burner or muffle furnace, is used such that changes other than just moisture evaporation occur, a process known as **ignition**. The changes can still be just simple evaporation of water and other components that would volatilize at the temperature used, although some procedures require that the sample be dried first. The changes may also include chemical changes, for example, carbon dioxide evolution from carbonate materials, such as in cements or limestone, as in the following:

$$CaCO_3 + \text{heat} \rightarrow CO_2 (\uparrow) + CaO$$

In any case, the sample is usually contained in a crucible (prepared in advance by ignition without a sample) and placed in the oven or over the Meker burner for a specified period of time or to a constant weight, as in the loss on drying. The calculation is also similar to loss on drying, and the weight of the crucible will need to be subtracted if the tare feature of a balance is not used.

$$\% \text{ loss} = \frac{\text{weight loss}}{\text{weight of sample}} \times 100 \tag{3.5}$$

$$= \frac{(\text{weight before ignition} - \text{weight after ignition})}{\text{weight before ignition}} \times 100 \tag{3.6}$$

**Example 3.2**

What is the percent loss on ignition of a sample given the following data?

| | |
|---|---|
| Weight of empty crucible | 17.2884 g |
| Weight of crucible with sample | 17.9393 g |
| Weight of crucible with sample after ignition | 17.8190 g |

**Solution 3.2**

Utilizing Equation (3.6), we have

$$\% \text{ loss} = \frac{(17.9393 \text{ g} - 17.8190 \text{ g})}{(17.9393 \text{ g} - 17.2884 \text{ g})} \times 100$$

$$\% \text{ loss} = 18.48\%$$

Applications of loss on ignition are found in pharmaceutical, agricultural, cement, and rock and mineral industries.[**]

### 3.6.1.3  Residue on Ignition

The mineral particles and ash that remain after the ignition of a sample as described above may be important to calculate and report. The mineral and ash are known as the residue. In pharmaceutical

---

[*]See a method for loss on drying for pharmaceutical raw materials and formulations. Method #731 on p. 1954 of the U.S. Pharmacopeia-National Formulary, USP24-NF19 (2000).

[**]See Method #733 on p. 1954 of the U.S. Pharmacopeia-National Formulary, USP24-NF19 (2000).

applications, the sample is pretreated with sulfuric acid prior to ignition in order to consume the carbon in the sample; the mineral and ash are sometimes referred to as sulfated ash residues. The process usually must be repeated to obtained a constant weight.

In the calculation, the weight of the residue in the crucible is in the numerator and the sample weight in the denominator.

$$\% \text{ residue on ignition} = \frac{\text{weight of residue}}{\text{weight of sample}} \times 100 \tag{3.7}$$

### Example 3.3

Calculate the percent residue on ignition from the data in Example 3.2.

### Solution 3.3

$$\% \text{ residue} = \frac{(17.8190 \text{ g} - 17.2884 \text{ g})}{(17.9393 \text{ g} - 17.2884 \text{ g})} \times 100$$

$$\% \text{ residue} = 81.52\%$$

### 3.6.1.4 Insoluble Matter in Reagents

As part of a general analysis report, the labels on certified chemicals that are soluble in water indicate the percent of insoluble matter present (see example in Figure 3.4). This insoluble matter is determined by preparing a solution of the chemical, which sometimes involves heating to boiling in a covered beaker and maintaining a higher temperature for a specified time to effect dissolution. A certain weight of the chemical is used; this is the sample weight for the calculation. The solution is filtered through a tared or preweighed, fine-porosity sintered glass crucible. After being rinsed thoroughly with water, this crucible is then dried and weighed again. The results are reported as percent insoluble matter.

$$\% \text{ insoluble matter} = \frac{\text{weight of insoluble matter}}{\text{weight of sample}} \times 100 \tag{3.8}$$

### Example 3.4

Given the following data, what is the percent insoluble matter for the chemical being investigated?

| | |
|---|---|
| Weight of chemical used | 5.3008 g |
| Weight of empty filtering crucible | 18.0048 g |
| Weight of crucible with insoluble matter | 18.0192 g |

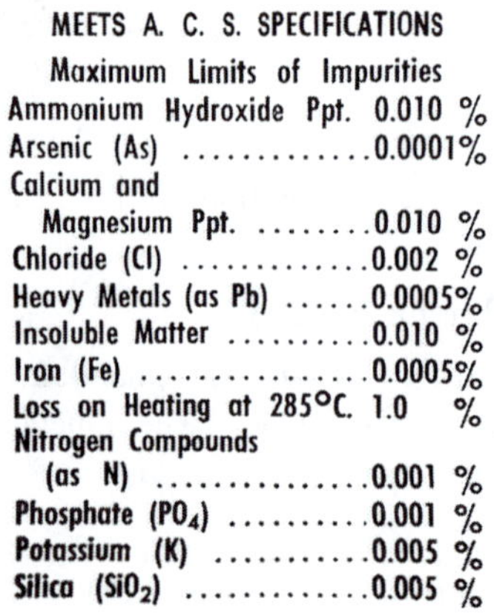

MEETS A. C. S. SPECIFICATIONS

Maximum Limits of Impurities
Ammonium Hydroxide Ppt. 0.010 %
Arsenic (As) . . . . . . . . . . . . .0.0001%
Calcium and
   Magnesium Ppt. . . . . . . . .0.010 %
Chloride (Cl) . . . . . . . . . . . . .0.002 %
Heavy Metals (as Pb) . . . . . .0.0005%
Insoluble Matter . . . . . . . . . .0.010 %
Iron (Fe) . . . . . . . . . . . . . . . .0.0005%
Loss on Heating at 285°C. 1.0   %
Nitrogen Compounds
   (as N) . . . . . . . . . . . . . . .0.001 %
Phosphate (PO₄) . . . . . . . . . .0.001 %
Potassium (K) . . . . . . . . . . .0.005 %
Silica (SiO₂) . . . . . . . . . . . . .0.005 %

FIGURE 3.4 A partial view of a general analysis report on a certified chemical bottle showing insoluble matter as 0.010%. Notice also the loss-on-heating report.

*Solution 3.4*

$$\% \text{ insoluble matter} = \frac{(18.0192 \text{ g} - 18.0048 \text{ g})}{5.3008 \text{ g}} \times 100$$

$$= 0.272\%$$

### 3.6.1.5  Solids in Water and Wastewater

When analyzing for solids in liquid samples, such as water and wastewater samples, the analysis is based on a volume of the sample rather than a weight. Thus, while this is considered an example of a gravimetric analysis because the weight of the solids is determined, a volume of the sample is measured rather than a weight. There are several categories of solids that may be determined.

For both water and wastewater samples, total solids and suspended solids may be determined. **Total solids** involve measuring a volume of the water or wastewater into a preweighed evaporating dish, evaporating the water using a drying oven, and obtaining the weight of the residue, which includes all solids, suspended and dissolved. To report results, the weight of the residue in milligrams is divided by the volume of the sample in liters (mg/L).

**Suspended solids** are solids from a measured volume of sample that can be trapped on a preweighed filter. The filter is prepared by washing, drying, and weighing. The sample is then filtered with the use of a vacuum and dried and weighed again. The weight gain of the filter is divided by the volume of sample used, and the results are again reported in milligrams per liter.

For wastewater samples, a number of additional solid determinations are typically performed, including total dissolved solids, volatile solids, and settleable solids. **Total dissolved solids** are the total solids minus the suspended solids and are determined in a manner similar to total solids, but after filtering out the suspended solids. Thus, a volume of the filtrate is measured into the evaporating dish and the water evaporated in a drying oven.

**Volatile solid** determination is similar to the loss on ignition discussed earlier. Either total solids or suspended solids can be analyzed for volatile solids. After following the procedure for total or suspended solids, as discussed above, the solids (in the preweighed evaporating dish or filter) are ignited in a muffle furnace. The volatile solids are the weight loss upon ignition divided by the volume of sample, with the results again reported in milligrams per liter.

For **settleable solids**, a sample of the wastewater is taken after the suspended solids are allowed to settle for a specified time. The solids in this sample are then measured in a manner similar to that for total solids. These are the nonsettleable solids. The settleable solids are then determined by subtracting from the total solids determined previously. See Workplace Scene 3.2.

### Example 3.5

A sample of lake water is tested for suspended solids and the following data are obtained. How many milligrams of suspended solids are there per liter of sample?

| | |
|---|---|
| Volume of water used | 100.00 mL |
| Weight of empty, dry filter | 0.1028 g |
| Weight of dried Gooch crucible after filtering the water | 0.3837 g |

*Solution 3.5*

$$\text{milligrams per liter} = \frac{\text{milligrams of solids on filter}}{\text{liters of water used}}$$

$$= \frac{(383.7 \text{ mg} - 102.8 \text{ mg})}{0.10000 \text{ L}} = 2809 \text{ mg}/\text{L}$$

# Workplace Scene 3.2

The laboratory analysts at the City of Lincoln, Nebraska, Wastewater Treatment Plant Laboratory routinely analyze wastewater plant influent and effluent samples, as well as industrial wastewater samples, for total solids, suspended solids, total volatile solids, and total suspended volatile solids. Other kinds of samples for this work can include samples from chemical toilets, grease traps, septic tanks, and belt presses. Belt press samples are samples of digested sludge that have had the water squeezed out with belt presses. All of these analyses are gravimetric, involving weight loss from heating or weight gain from filtering.

Cory Neth, an analyst for the City of Lincoln, Nebraska, Wastewater Treatment Plant Laboratory, removes a tray of crucibles from a muffle furnace. The crucibles are filtering crucibles used in the determination of suspended solids. The muffle furnace is used in the total suspended volatile solid test.

### 3.6.1.6   Particle Size by Analytical Sieving

Solid materials that have been crushed or that otherwise exist in variable particle sizes can be analyzed gravimetrically to determine the percentages of the various particle sizes — the particle size distribution. For this to occur, the various-sized particles must be separated from each other so that their individual weights can be measured. This is accomplished by sieving. A sieve is a brass or stainless steel cylinder approximately 2 in. tall and 5 to 12 in. in diameter, open on one end and having a plate with holes or wire mesh on the other end (see Figure 3.5). Often these sieves are nested, or stacked, as in Figure 3.5, in order to analyze for a number of different particle sizes at once, or obtain a particle size distribution profile. The individual sieves in the nest are weighed in advance and then are stacked in order, or descending aperture size, from top to bottom. The sample is introduced to the top sieve. The nested sieves are then shaken using a sieve shaker so that the particles fall through the stack until they reach a sieve through which they cannot pass due to the small size of the apertures in the sieve. Thus, particles of a particular range of sizes are found on a particular sieve. The final weight of each sieve is then determined and the percentage calculated. Often, the percentages are expressed as a cumulative percentage, as in the example below. See Chapter 15 for more details.

**FIGURE 3.5** Nested sieves as described in the text.

## Example 3.6

The following data are obtained from a particle size analysis. What is the percent of each size in the sample?

| | |
|---|---|
| Weight of sample | 5.82 g |
| Weight of individual sieves before sieving: | |
| #1 (4.0 mm) | 348.87 g |
| #2 (2.0 mm) | 346.09 g |
| #3 (1.0 mm) | 351.02 g |
| #4 (0.5 mm) | 350.88 g |
| Bottom collecting pan | 369.20 g |
| Weights of individual sieves after sieving: | |
| #1 | 349.55 g |
| #2 | 346.81 g |
| #3 | 354.37 g |
| #4 | 351.45 g |
| Bottom collecting pan | 369.70 g |

*Solution 3.6*

$$\% \text{ particle size} = \frac{(\text{weight of sieve after sieving} - \text{weight of empty sieve})}{\text{weight of sample}} \times 100$$

$$\% \text{ 4.0-mm size} = \frac{(349.55 \text{ g} - 348.87 \text{ g})}{5.82 \text{ g}} \times 100 = 12\% \text{ (cumulative 12\%)}$$

$$\% \text{ 2.0-mm size} = \frac{(346.81 \text{ g} - 346.09 \text{ g})}{5.82 \text{ g}} \times 100 = 12\% \text{ (cumulative 24\%)}$$

$$\% \text{ 1.0-mm size} = \frac{(354.37 \text{ g} - 351.02 \text{ g})}{5.82 \text{ g}} \times 100 = 57.6\% \text{ (cumulative 82\%)}$$

$$\% \ 0.5\text{-mm size} = \frac{(351.45 \text{ g} - 350.88 \text{ g})}{5.82 \text{ g}} \times 100 = 9.8\% \text{ (cumulative 91\%)}$$

$$\% \ < 0.5\text{-mm size} = \frac{(369.70 \text{ g} - 369.20 \text{ g})}{5.82 \text{ g}} \times 100 = 8.6\% \text{ (cumulative 100\%)}$$

Analytical sieving may also be done with the solids in a liquid suspension. In this case, it is called wet sieving, and the suspension is transferred to the nested sieves and rinsed through with solvent. The suspended particles are captured on the sieves in the same manner as described above.

Particle size analysis by analytical sieving is important in the agricultural industry for soil analysis and in the pharmaceutical industry for various raw materials (such as sugar spheres) and liquid suspensions. For a more detailed look at particle size analysis, see Chapter 15.

### 3.6.2   Chemical Alteration and Separation of the Analyte

The physical separation of an analyte from a sample so that its weight can be measured, which the above examples typify, can be difficult, if not impossible, to accomplish. It is important to recognize that in many cases there are no physical means by which such a separation can take place. For example, if you wished to determine the sodium sulfate present in a mixture with sodium chloride, it would not be possible to separate it from the sodium chloride by physical means to determine its weight.

However, a gravimetric procedure might still be used if a chemical reaction is employed to convert the analyte to another chemical form that is both able to be separated cleanly and able to be weighed accurately. In the example of determining sodium sulfate in the presence of sodium chloride, one can dissolve the mixture in water and precipitate the sulfate with barium chloride to form barium sulfate. Sodium chloride would not react.

$$Na_2SO_4 + BaCl_2 \rightarrow BaSO_4 (\downarrow) + NaCl$$

$$NaCl + BaCl_2 \rightarrow \text{no reaction}$$

The weight of the precipitate after filtering and drying can then be measured free of any influence from the NaCl and converted back to the weight of the analyte with the use of a gravimetric factor (see the next section) and its percent in the sample calculated. Examples are given in Section 3.6.4.

### 3.6.3   Gravimetric Factors[***]

A **gravimetric factor** is a number used to convert, by multiplication, the weight of one chemical to the weight of another. Such a conversion can be very useful in an analytical laboratory. For example, if a recipe for a solution of iron calls for 55 g of $FeCl_3$ but a technician finds only iron wire on the chemical shelf, he or she would want to know how much iron metal is equivalent to 55 g of $FeCl_3$ so that he or she could prepare the solution with the iron wire instead and have the same weight of iron in either case. In one formula unit of $FeCl_3$, there is one atom of Fe, so the fraction of iron(III) chloride that is iron metal is calculated as follows:

$$\text{fraction of iron(III) chloride that is iron} = \frac{\text{atomic weight of Fe}}{\text{formula weight of } FeCl_3} \tag{3.9}$$

$$= \frac{55.845 \text{ g / mol}}{162.203 \text{ g / mol}} = 0.34429 \text{ Fe / } FeCl_3$$

---

[***]See Appendix 3 for the stoichiometric basis for gravimetric factors.

The weight of iron metal that is equivalent to 55 g of iron(III) chloride can be calculated by multiplying the weight of iron(III) chloride by this fraction:

weight of iron equivalent to 55 g of iron(III) chloride = 55 g × 0.34429 = 19 g

This fraction is therefore a gravimetric factor because it is used to convert the weight of $FeCl_3$ to the weight of Fe, as shown.

In general, many gravimetric factors are calculated by simply dividing the atomic or formula weight of the substance whose weight we are seeking (substance sought) by the atomic or formula weight of the substance whose weight we already know (substance known), as we did in the above example. However, the calculation can be further complicated when the number of atoms in the symbol of the particular element or compound in the numerator does not match the number of atoms of the same element in the symbol or formula of the compound in the denominator. If we symbolize the atomic weight and formula weight used in the above by the corresponding chemical symbols and formulas, we can write the gravimetric factor for our example as follows:

$$\text{gravimetric factor} = \frac{\text{sought}}{\text{known}} = \frac{Fe}{FeCl_3} \tag{3.10}$$

The one element the numerator has in common with the denominator (the common element) is iron, and there is one atom (no subscripts) for iron in each, so there is no such complication in this case. When the number of atoms of a common element are *not* the same, we must multiply by coefficients to make them the same, as is done in balancing equations. To illustrate, consider an example in which we want to know the weight of NaCl that is equivalent to a given weight of $PbCl_2$. NaCl is the compound sought and $PbCl_2$ is the compound known. The common element is Cl and the numerator must be multiplied by 2 in order to equalize the number of Cl in the two formulas. In one formula unit of $PbCl_2$, there are two atoms of Cl.

$$\text{gravimetric factor} = \frac{\text{sought}}{\text{known}} = \frac{NaCl \times 2}{PbCl_2} \tag{3.11}$$

The general formula for a gravimetric factor is then as follows:

$$\text{gravimetric factor} = \frac{\text{atomic or formula weight of chemical sought} \times Q_S}{\text{atomic or formula weight of chemical known} \times Q_K} \tag{3.12}$$

In this formula, $Q_S$ is the balancing coefficient for the common element in the formula of the substance sought and $Q_K$ is the balancing coefficient for the common element in the formula of the substance known.

## Example 3.7

What is the gravimetric factor for converting a weight of AgCl to a weight of Cl?

### Solution 3.7

Substance sought    Cl

Substance known    AgCl

$$\text{gravimetric factor} = \frac{Cl \times 1}{AgCl \times 1} = \frac{35.4527 \text{ g/mol} \times 1}{143.321 \text{ g/mol} \times 1} = 0.247366 \text{ Cl/AgCl}$$

## Example 3.8

What is the gravimetric factor if a weight of Cl is to be determined from a weight of $PbCl_2$?

**Solution 3.8**

   Substance sought   Cl

   Substance known   $PbCl_2$

$$\text{gravimetric factor} = \frac{Cl \times 2}{PbCl_2 \times 1} = \frac{35.4527 \text{ g/mol} \times 2}{278.1 \text{ g/mol} \times 1} = 0.2550 \text{ Cl/}PbCl_2$$

## Example 3.9

If you have a weight of $AlCl_3$ and want to know the weight of $Cl_2$ that is equivalent to this, what gravimetric factor is required?

**Solution 3.9**

   Substance sought   $Cl_2$

   Substance known   $AlCl_3$

$$\text{gravimetric factor} = \frac{Cl_2 \times 3}{AlCl_3 \times 2} = \frac{70.9054 \text{ g/mol} \times 3}{133.340 \text{ g/mol} \times 2} = 0.797646 \text{ }Cl_2 \text{/}AlCl_3$$

In cases where there are two common elements, one of which is oxygen, the oxygen is ignored and the other element is balanced.

## Example 3.10

What gravimetric factor is required to calculate the weight of $SO_3$ from a weight of $BaSO_4$?

**Solution 3.10**

   Substance sought   $SO_3$

   Substance known   $BaSO_4$

Besides oxygen, sulfur is the common element.

$$\text{gravimetric factor} = \frac{SO_3 \times 1}{BaSO_4 \times 1} = \frac{80.064 \text{ g/mol} \times 1}{233.391 \text{ g/mol} \times 1} = 0.34305 \text{ }SO_3 \text{/}BaSO_4$$

## 3.6.4   Using Gravimetric Factors

As stated earlier, gravimetric factors are used to convert the weight of one chemical to the weight of another, as in the example cited at the beginning of Section 3.6.3. Below is another example of such a conversion.

## Example 3.11

How many grams of copper(II) sulfate pentahydrate are required to prepare a solution that has the equivalent of 0.339 g of copper dissolved?

**Solution 3.11**

The formula of copper(II) sulfate pentahydrate is $CuSO_4 \cdot 5H_2O$ and its formula weight is 249.686 g/mol.

   Substance sought   $CuSO_4 \cdot 5H_2O$

   Substance known   Cu

$$\text{gravimetric factor} = \frac{CuSO_4 \cdot 5H_2O \times 1}{Cu \times 1} = \frac{249.686}{63.546} = 3.9292$$

$$\text{grams of } CuSO_4 \cdot 5H_2O = \text{grams of copper} \times \text{gravimetric factor}$$

$$= 0.339 \times 3.9292 = 1.33 \text{ g of } CuSO_4 \cdot 5H_2O$$

In the case in which the analyte participates in a chemical reaction, the product of which is weighed, the weight of this product must be converted to the weight of the analyte before the percent can be calculated.

This can be done with the use of a gravimetric factor. Examples include the sulfate and iron determinations in the experiments at the end of this chapter (see Experiments 6 and 7).

**Example 3.12**

A sample that weighed 0.8112 g is analyzed for phosphorus (P) content by precipitating the phosphorus as $Mg_2P_2O_7$. If the precipitate weighs 0.5261 g, what is the percent of P in the sample?

***Solution 3.12***

$$\% \text{ analyte} = \frac{\text{weight of product} \times \text{gravimetric factor}}{\text{weight of sample}} \times 100$$

Substance sought     P

Substance known     $Mg_2P_2O_7$

$$\text{gravimetric factor} = \frac{30.9738 \text{ g} \times 2}{222.555 \text{ g}} = 0.278347$$

$$\% \text{ P} = \frac{0.5261 \text{ g} \times 0.278347}{0.8112 \text{ g}} \times 100 = 18.05\%$$

## 3.7 Experimental Considerations

### 3.7.1 Weighing Bottles

A weighing bottle is a small glass bottle with a ground glass top and stopper. It is used for containing, drying, weighing, and dispensing finely divided solids for all types of analysis, as suggested by their use in the laboratory experiments in this chapter and those that follow. For gravimetric analysis, weighing bottles can be used for containing samples in loss-on-drying experiments, and in other instances in which a sample must be dried before proceeding, and then conveniently weighed and dispensed later.

The drying takes place in a drying oven, an oven that can be set to temperatures up to 100°C and beyond. The weighing bottle, with sample contained, is usually placed in the oven, with lid ajar, in a small beaker, and the beaker covered with a watch glass, as in Figure 3.6.

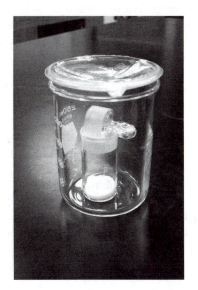

**FIGURE 3.6** A weighing bottle with sample, lid ajar, in small beaker, ready to be placed in a drying oven for the purpose of drying the contained sample.

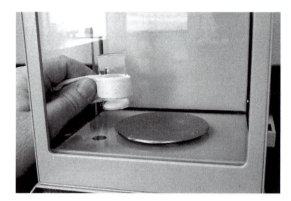

**FIGURE 3.7**    Handling a weighing bottle with a rolled-up piece of paper towel to avoid fingerprints.

## 3.7.2   Weighing by Difference

While dispensing and weighing a small amount of the dried sample contained in the weighing bottle, it is prudent to avoid contacting the sample with laboratory utensils or weighing paper in order to avoid loss of a portion of the sample, which might adhere to the utensil or paper. In such instances, weighing by difference is recommended.

**Weighing by difference** is the act of determining the weight of a sample by obtaining two weighing bottle weights, one before and one after dispensing the sample, and then subtracting the two weights. In the process, no utensil or weighing paper contacts the sample because the sample is shaken into the receiving vessel directly from the weighing bottle. If the two weights are determined on an analytical balance, it is desirable to avoid touching the weighing bottle with fingers from the time the first weight is obtained until the second weight is obtained. This avoids putting fingerprints on the weighing bottle that would add measurable weight to the weighing before the second weight is determined. One can conveniently avoid fingerprints by wearing clean gloves or by using a rolled-up paper towel, as shown in Figure 3.7.

## 3.7.3   Isolating and Weighing Precipitates

In a gravimetric analysis experiment in which a chemical reaction is used to obtain a precipitate, the manner in which the precipitate is isolated and weighed involves a filtration step, including a full rinse with distilled water and two weight measurements. The filtration can involve a filtering crucible or simply a piece of filter paper and a funnel. If a filtering crucible is used, its weight is measured before filtering and again after filtering, after the water solvent has been evaporated in a drying oven. The difference in the two weights is the weight of the precipitate. See Workplace Scene 3.3. If a piece of filter paper and funnel are used, the weight of the precipitate is obtained after disintegrating the filter paper by ignition with a burner flame. The filter paper must be ashless filter paper, meaning that it can leave no ash residue after the ignition.

The distilled water rinse mentioned in this discussion is necessary to rinse the filtering crucible or filter paper and precipitate free of any other soluble chemicals that might be present. The rinse is accomplished with the use of streams of distilled water from a squeeze bottle directed over the entire crucible or filter paper. The rinsing step is shown in Figure 3.8 for a filter paper and funnel example.

To disintegrate the filter paper, the rinsed filter paper containing the precipitate is folded and placed into a preweighed porcelain crucible and the crucible is heated, first cautiously with a cooler Bunsen burner flame, so as to dry and char the filter paper, and then with the full heat of a Meker burner. For this final ignition step, the crucible should be tilted to one side and partially covered with the crucible lid so that the burner flame can be directed to the bottom of the crucible, thus avoiding engulfing the

# Workplace Scene 3.3

Fertilizers sold to gardeners and farmers have a particular composition that is guaranteed by the manufacturer. One of the constituents for which there is a guaranteed percentage is sulfur. Technicians at the Nebraska State Agriculture Laboratory utilize a gravimetric analysis procedure to check the sulfur percentage. The soluble sulfates in the samples are extracted from weighed samples and precipitated as barium sulfate with a barium chloride solution. The precipitate settles to the bottom of the beaker while it is digested at a temperature just under the boiling point. After cooling, the precipitate is filtered out using a preweighed filtering crucible and a vacuum system. A gravimetric factor is used to convert the weight of barium sulfate to the weight of sulfur so that the percentage of sulfur can be determined.

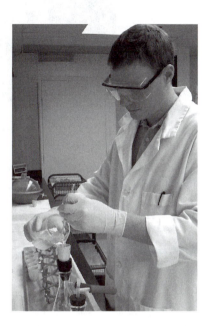

Wil Kastning of the Nebraska State Agriculture Laboratory filters barium sulfate precipitates using filtering crucibles and a vacuum system while performing a gravimetric analysis of fertilizers for sulfate content.

**FIGURE 3.8** Rinsing the precipitate and filter paper with distilled water from a squeeze bottle.

**FIGURE 3.9**    A crucible tilted to one side, with lid partially covering the opening. This allows the burner flame to be directed to the bottom of the crucible and air to circulate into the crucible.

entire crucible in flame and preventing air from getting inside the crucible. Air must circulate freely into the open crucible. See Figure 3.9 for the setup.

# Experiments

## Experiment 4: Practice of Gravimetric Analysis Using Physical Separation Methods

Note: Your instructor will assign one or more of the following parts to be completed in one or two lab sessions. Exercise good analytical technique according to the precision of weighing in each experiment. If an analytical balance is used, be diligent in avoiding small errors, such as fingerprints.

### Part A: Loss on Drying

***Introduction***

A number of experiments in this text require that a sample be dried before proceeding with the chosen method. Examples in this chapter include an unknown sulfate (see Experiment 6, step 1) and an unknown iron (see Experiment 7, step 1). An example in Chapter 4 is the KHP standard in Experiment 8 (see Experiment 8, step 1). In this experiment, the percent loss on drying of one of these materials will be determined.

Reference: General Test <731>, U.S. Pharmacopeia–National Formulary, USP 24–NF 19, 2000, p. 1954. *Remember safety glasses.*

1. Clean a shallow glass weighing bottle and lid, and dry it with the lid ajar (Figure 3.6) in a drying oven for 30 min at 105°C. After drying, avoid fingerprints. Allow to cool for 10 min and weigh on an analytical balance. Record the weight.
2. Add the sample to be dried to the weighing bottle. Record the exact weight, weighing bottle plus sample.
3. Place the weighing bottle with the sample, lid ajar, back in the drying oven and dry for the time specified in Experiment 6, 7, or 8.
4. Remove from the drying oven, place the lid on the bottle, allow to cool for 15 min in your desiccator, and then weigh.
5. Repeat steps 3 and 4 until you have a constant weight (successive weights agreeing to within 0.001 g). Calculate and report the loss on drying as a percentage. Store your sample in your desiccator if the experiment that utilizes this material (Experiment 6, 7, or 8) is to be performed.

### Part B: Loss on Ignition of Powdered Cement

Reference: ASTM 114–00, *Standard Test Method for Chemical Analysis of Hydraulic Cement*. As suggested in this document, the sample should not be a slag cement sample. Your instructor may choose to dispense a preanalyzed commercial unknown to you. In that case, your answer can be compared to the "right answer" unique to the unknown and grade.

*Consult your instructor on how to use a muffle furnace safely and remember safety glasses.*

1. Prepare a porcelain crucible and lid by cleaning with hot soapy water and a brush, followed by rinsing and drying with a towel. Then heat the crucible, with lid slightly ajar, in a muffle furnace set at 950°C for 30 min. This step is to ensure that any volatile material will be removed now rather than later, when the weight loss is critical. Allow to cool for 10 min on a heat resistant surface and then 10 min more in a desiccator. Then weigh the crucible with lid on an analytical balance.

2. If there is one, record the unknown number of your cement material. Using the analytical balance, weigh approximately 1.0 g of the cement (do not dry first) into the crucible, transferring with a spatula. If the balance has a tare feature, you may use it, but make sure you have the weight of the crucible with lid recorded.

3. Place the crucible with lid back into the muffle furnace, with lid ajar, for 30 min. Remove from the oven and allow to cool, as before, for a total of 20 min. Weigh on the analytical balance. Repeat this step, but for just 5 min in the muffle furnace.

4. If the two weights from step 3 differ by more than 2 mg, heat again for 5 min and cool and weigh again. Continue to heat, cool, and weigh until you have a constant weight (no weight loss greater than 2 to 3 mg).

5. Calculate the percent loss on ignition.

## Part C: Residue on Ignition of Carboxymethyl Cellulose, Sodium Salt

Reference: General Test <281>, U.S. Pharmacopeia–National Formulary, USP 24–NF 19, 2000.
*Remember safety glasses.*

1. Prepare a crucible and lid using hot soapy water and a brush, as in step 1 of Part 4B. Set up a ring stand with ring and clay triangle. Place the crucible on the clay triangle and adjust the height of the ring so that there is only a $^1/_2$ inch gap between the top of the Meker burner and the bottom of the crucible. Heat, with lid ajar, with the full heat of the Meker burner for 30 min. Cool in place for 5 min and for 15 min more in a desiccator. Weigh on an analytical balance.

2. Using an analytical balance, weigh 1 to 2 g of carboxymethyl cellulose, sodium salt, into the crucible.

3. Working in a fume hood, heat cautiously with a Bunsen burner at first (lid slightly ajar) until the sample is thoroughly charred. Use the same ring stand apparatus as before.

4. Cool and add 1 mL of concentrated sulfuric acid. Heat gently until white fumes are no longer evolved. Place the crucible with contents (with lid, but with lid removed) in a muffle furnace (also in the fume hood) set at 800°C and ignite for 1 h.

5. Remove the crucible and lid, set them on a heat-resistant surface, and cool for 5 min. Then place them in a desiccator for an additional 15 min. Weigh the crucible and lid.

6. Repeat steps 4 and 5 until you have successive weights that do not change by more than 2 mg. Calculate the percent residue on ignition.

## Part D: Insoluble Matter in Reagents

Reference: Official USP Monograph for Dibasic Sodium Phosphate, U.S. Pharmacopeia and National Formulary, USP 24, 2000, p. 1539.
*Remember safety glasses.*

1. Select anhydrous dibasic sodium phosphate, or other chemical, for which the label indicates the insoluble matter present to be as high as 0.4%. Note: The percent of insoluble matter must be high enough to provide enough matter to be collected and measured.

2. Clean a fine sintered glass filtering crucible by drawing a 50% nitric acid solution through it with the help of a vacuum. Follow this with a rinse with hot distilled water drawn through with a vacuum in the same way. Dry it in a drying oven at 105°C for 30 min. Allow it to cool for 15 min and weigh, avoiding fingerprints in the process. Store in a desiccator.

3. Weigh approximately 5 g of the phosphate on an analytical balance and place in a 250-mL beaker. Add 100 mL of hot distilled water and stir vigorously to dissolve.

4. Filter the solution through the filtering crucible with the help of a vacuum. Rinse the beaker with hot distilled water and pour the rinsings through the crucible, drawing the hot water through the crucible with the vacuum.

5. Dry the filtering crucible at 105°C for 2 h. Allow it to cool in your desiccator and then weigh it. Calculate the percent insoluble matter.

### Part E: Suspended Solids in Wastewater

Reference: *Standard Methods for the Examination of Water and Wastewater*, APHA-AWWA-WPCF. *Remember safety glasses.*

1. Obtain the filter to be used, and using forceps, place it on the vacuum filter support. Apply the vacuum and wash the filter liberally with distilled water from a squeeze bottle while drawing the water through it. Place it on a clean watch glass and dry it for 1 h at 103°C. Cool in a desiccator and weigh on an analytical balance.

2. Thoroughly clean the filter funnel. Using forceps to handle the filter, assemble the entire filtering apparatus and wet the filter. Proceed to filter 100.0 mL (graduated cylinder) of the sample, drawing the water through the filter with the vacuum. Rinse the graduated cylinder with distilled water, and add the rinsings to the filter.

3. Remove the filter from the apparatus, and place it back on the watch glass. Dry again at 103°C, cool, and weigh. Calculate the milligrams per liter of suspended solids.

### Part F: Particle Sizing

Note: Instructor and students may want to refer to Section 15.7 and the particle sizing experiment at the end of Chapter 15 for possible variations and additional data handling procedures.

Note also that a high-capacity balance (up to several kilograms) is required for this experiment. *Remember safety glasses.*

1. Obtain some sand, such as from a bag of playground sand, from a building materials supplier, or from a beach. Obtain test sieves of various aperture sizes appropriate for the sand and a bottom collection pan.

2. Clean the test sieves to make sure that there are no particles stuck in the apertures. Weigh each test sieve to the nearest 0.1 g and stack the sieves.

3. Tare a beaker or other container and weigh, to the nearest 0.1 g, a sample of the sand that would occupy about half the volume of the top sieve.

4. Pour the sample into the top sieve, place on a sieve shaker, and shake for 5 min. Obtain the new weight of each test sieve to the nearest 0.1 g.

5. Reassemble and shake for an additional 5 min and weigh again. If the weights have changed appreciably, reassemble and shake again. Repeat until you have a constant weight (within 0.5 g) for each test sieve.

6. Calculate the percentage of each particle size range retained in the sand sample. Also report the cumulative percentages.

## Experiment 5: The Percent of Water in Hydrated Barium Chloride

### Introduction

The water that is trapped within the crystal structure of some ionic compounds (the water of hydration) can be removed easily by heating. The amount of the water present in a given sample can be determined by weighing the sample before and after this heating. The weight loss that occurs is the weight of the water in the sample. The percent of water in the hydrate is calculated as it is in a loss-on-drying experiment.

In this experiment, the percent of water in $BaCl_2 \cdot 2H_2O$ will be determined. You will do the experiment in duplicate (run two determinations at the same time) and report two answers.

One problem encountered in this procedure is that a small amount of the $BaCl_2 \cdot 2H_2O$ may be evaporated along with the water when the sample is heated. This would obviously cause an error because the weight loss would include more than just the water. To ensure that this does not happen, do not heat the sample directly with the flame of a burner, but rather, use the radiant heat obtained from a flame-heated ceramic-centered wire gauze.

The theoretical percent of water in any hydrate can be calculated from atomic and formula weights. It is recommended that you perform this calculation and compare your results with this theoretically correct value. Try to account for any deviation. Since your sample of $BaCl_2 \cdot 2H_2O$ is "as received," it may have additional water adsorbed due to room humidity.

*Remember safety glasses.*

### Procedure

1. Set up two ring stands, each with one ring, one clay triangle, and a Bunsen burner under the ring. Place a porcelain crucible with lid (with both crucible and lid premarked as #1 or #2 with a heat-stable label) in each clay triangle and adjust the height of the rings so as to obtain the maximum heat with the burners at the bottom of the crucibles.
2. Heat the crucibles, with lids ajar, directly with the Bunsen burners for at least 20 min. During this time, the bottoms of the crucibles should get hot enough so as to glow with a dull red color. This step is to ensure that any volatile material will be removed now rather than later, when the weight loss is critical. At the end of the 20 min, allow the crucibles to cool in place for 5 min and then in your desiccator for an additional 15 min.
3. Weigh the samples of $BaCl_2 \cdot 2H_2O$ hydrate into the crucibles as follows. Using tongs to handle the crucibles, weigh the first crucible, with lid, on the analytical balance, and record the weight in your notebook. Then, with a spatula, add enough of the hydrate to the crucible so as to increase the weight by about 1 g. Also record this weight in your notebook. (Note: If your balance has a tare feature, you can use it, but be sure to have the weight of the empty crucible recorded.) Do the same with the second crucible. At this point, you can touch the crucibles with your fingers. Carry them back to your workspace.
4. Set up two ring stands with two rings each and one Bunsen burner each. In each setup, the lower ring should be 9 cm above the top of the burner and the upper ring should be placed 9 cm above the lower. Use a metric ruler for these measurements. Place a ceramic-centered wire gauze on each of the lower rings and clay triangles on the upper rings. Place the crucibles, with lids ajar, in the clay triangles. Proceed to heat the wire gauze with the full heat of the Bunsen burners for 30 min. The flame should be a tall, hot flame that covers the entire ceramic center of the gauze on the bottom side.
5. At the end of the 30 min, allow the crucibles to cool in place for 1 min and then in your desiccator for an additional 15 min. Handle them with tongs until after you have weighed them. Weigh each, with lids, on the analytical balance. Record the weights in your notebook.
6. Calculate the percent of water in the hydrate for each sample and report to your instructor.

## Experiment 6: The Gravimetric Determination of Sulfate in a Commercial Unknown

### Introduction

This experiment represents an example of a gravimetric analysis based on the stoichiometry of a precipitation reaction (as discussed in Sections 3.6.2 to 3.6.4), rather than a simple physical separation, such as heating. A sample containing a soluble sulfate compound, such as $Na_2SO_4$ or $K_2SO_4$, is dissolved, and the sulfate is precipitated with $BaCl_2$ as $BaSO_4$. The percent of $SO_3$ is determined using a gravimetric factor, as discussed in the text. Read the entire experiment completely before beginning the lab work. The procedure calls for you to run the experiment in duplicate. Your instructor may want you to run it in triplicate.

*Remember safety glasses.*

### Procedure
*Session 1*

1. Obtain a sample of unknown sulfate from your instructor. Record the unknown number in your notebook. Dry the sample for 1 h using a weighing bottle and drying oven.

2. While waiting for your sample to dry, prepare two porcelain crucibles by heating with Meker burners for a $^1/_2$ h. This can be done with the use of two ring stands and rings with clay triangles to support the crucibles. The lids should be ajar during the heating, and the tops of the burners should be about $^1/_2$ in. from the bottom of the crucibles. Both the crucibles and lids should be labeled with heat-stable labels prior to heating. If only one Meker burner is available per student, use a Bunsen burner for one of the crucibles and alternate burners after 15 min so that each crucible is heated for 15 min with each burner.

3. After the $^1/_2$ h has expired, turn off the burners and allow the crucibles and lids to cool in place for 5 min. Then, using crucible tongs, transfer them to a desiccator to continue cooling for 15 min more.

4. Continuing to handle the crucibles and lids with tongs; weigh each crucible with its lid on an analytical balance and record the weights in your notebook. All measurements of weight should be made as accurately as possible, i.e., they should be made on an analytical balance. Take special precautions to keep the original lids with the crucibles, since the lids and crucibles will be weighed together again later. Store in your desiccator until session 2.

5. After the 1-h drying time has expired, remove the unknown from the oven, place the lid on the weighing bottle, and place the weighing bottle in the desiccator to cool for 10 min. Thoroughly clean two 400-mL beakers. Weigh by difference two samples of your unknowns, each weighing around 0.4 g, into the two 400-mL beakers, labeled #1 and #2. The weights can be recorded in the data section of your notebook as follows:

| | |
|---|---|
| Sample 1 | Weight of weighing bottle before: _____ g |
| | Weight of weighing bottle after:_____ g |
| Sample 2 | Weight of weighing bottle before:_____ g |
| | Weight of weighing bottle after:_____ g |

Store the two beakers and the desiccator until session 2, or proceed to session 2 at the discretion of your instructor.

*Session 2*

6. Add to each of the two beakers 125 mL of water and 3 mL of concentrated HCl. Obtain two clean glass stirring rods and place one in each beaker. Stir the mixtures until the unknowns have dissolved. Leaving the stirring rods in the beakers, place watch glasses on the beakers and place both beakers on a hot plate. Place alongside these a third beaker containing about 130 mL of barium chloride stock solution (26 g/L of concentration) prepared by your instructor. Place a watch glass on this beaker also.

7. Carefully heat all solutions to boiling. Then remove from the hot plate, and while stirring, add 55 mL (use a 100-mL graduated cylinder) of the barium chloride to each of the unknowns. Use beaker tongs for handling hot beakers. The fine white precipitate, barium sulfate, forms almost immediately. The amount of barium chloride used is in excess of the amount required. This is to ensure complete precipitation of the sulfate in your unknown. Continue to stir both solutions for at least 20 sec. Again, leave the stirring rods in the beakers.

8. Adjust the setting on the hot plate so that the two remaining solutions remain very hot (but not quite boiling). If either sample boils, cool it for a moment and place it back on the hot plate. Allow the precipitate to digest for 1 h. **Digestion** refers to a procedure in which a sample is heated or stored for a period of time to allow a particular chemical or physical process to occur. In this case, a digestion step is required in order to change the small, finely divided particles of barium sulfate

precipitate into larger, filterable particles. Heating for a period of 1 h accomplishes this goal. During this period, you can proceed to gather and assemble the equipment needed for the remainder of the experiment and prepare your notebook for the data to be recorded.

9. Assemble a filtration system consisting of a ring stand, a funnel rack (or two small rings), two clean long-stem funnels, and two beakers or flasks to catch the filtrate. Obtain two pieces of Whatman #40 filter paper, fold, and place into the funnels, moistening with a little distilled water so they adhere to the funnel walls (Figure 3.8). Label the funnels to coincide with the labels on your beakers.

10. After the 1-h digestion period has expired, remove the beakers from the hot plate; be careful not to disturb the precipitate, which has settled to the bottom. The filtration step will be completed in less time if the supernatant liquid is filtered first, followed by the transfer of the precipitate. For each solution, using the stirring rod as a guide for the solution (to avoid splashing), transfer the supernatant liquid to the funnel, and allow it to filter. Be careful to transfer only solution #1 into funnel #1, etc. Then transfer the precipitate carefully, using the last 20 mL of supernatant to stir up the settled particles, while transferring with the supernatant. Add about 20 mL of hot distilled water and repeat. Repeat with one 10-ml portion and complete the transfer with the aid of a rubber policeman. Scrub the entire interior of the beaker with the policeman until no more white particles are visible. Rinse the policeman into the funnel with a stream of hot distilled water.

11. Rinse the precipitate with several portions of hot distilled water (Figure 3.8). To test for sufficient rinsing (elimination of all residual chloride from the excess barium chloride, HCl, and possibly the sample), collect some of the most recent washings in a test tube and add two drops of silver nitrate solution. If a white precipitate (cloudy appearance) forms, more rinsing is required before proceeding to step 13. Continue to rinse and test until the rinsings are clean.

12. After allowing the last rinsings to completely drain from the filter, carefully remove the filter paper from the funnel, fold, and place into the crucible with the same label. Place the lids on the crucibles. At this point, you will need a minimum of 60 min to complete the experiment. If time does not allow the completion of the experiment, store your crucibles outside of your desiccator until the next lab period.

*Session 3*

13. Set up your ring stands with burners as before, but in a fume hood. Place the crucibles, with lids ajar, on the clay triangles in the rings and heat slowly with a Bunsen burner to evaporate the water in the crucibles. Do not allow the filter paper to catch fire. If it does, you can conveniently extinguish the fire by momentarily covering the crucible completely with the lid using tongs. The filter paper will slowly dry out and turn black, while releasing some smoke. Once the paper is completely charred (black), incline the crucible to one side, as in Figure 3.9, and apply the full heat of the Meker burner to the bottom of the crucible for a minimum of 20 min. Follow the guidelines given in Section 3.7.3 for this ignition step. If only one Meker burner is available, alternate with a Bunsen burner for a total of about 30 min (15 min per burner per sample). Allow to cool in place for 5 min, and using tongs, transfer to the desiccator to cool for an additional 15 to 20 min. Weigh the crucibles and lids and record the weights in your notebook.

14. Compute the weights of your sample and precipitates from the data in your notebook. The percentage of $SO_3$ in the unknown is calculated from these weights. Report the results to your instructor.

## Experiment 7: The Gravimetric Determination of Iron in a Commercial Unknown

*Remember safety glasses.*

### Session 1

1. Obtain a sample from your instructor. Record the unknown number and dry as discussed in Experiment 6 and in Section 3.7. Also, obtain and prepare two porcelain crucibles and lids, as done

in steps 2 to 4 of Experiment 6. After the sample has cooled, weigh by difference (Section 3.7.2) two samples, each around 0.2 g, into labeled 400-mL beakers. Store the beakers with your desiccator until session 2. (Note: session 1 here can coincide with session 3 for Experiment 6, if one is required.)

## Session 2

2. Add 5 mL of distilled water and 10 mL of concentrated HCl to each beaker, and place stirring rods in the beakers and watch glasses on them.

3. Place the beakers on a hot plate in a good fume hood and bring to boiling. Allow to boil about 2 min to ensure complete dissolution.

4. When the samples are completely dissolved, remove them from the hot plate and, while they are still in the fume hood, add 100 mL of water. Then remove from the hood and add 2 mL of concentrated $HNO_3$. (Be sure not to lose any yellow solution that might be adhering to the watch glass.) The purpose of the nitric acid is to ensure that all of the iron is present as $Fe^{3+}$ and not $Fe^{2+}$.

5. Bring the solutions to boiling again (use a hot plate). While waiting for the solutions to boil, fill half of a squeeze bottle with 7 $M$ $NH_4OH$ and label this bottle.

6. When the solutions begin to boil, remove them from the hot plate and add the $NH_4OH$ from the squeeze bottle with continuous stirring until the solutions become quite murky with the $Fe(OH)_3$ precipitate and the precipitate settles readily upon standing. Now test the solution with litmus paper (be sure to rinse the paper back into the beaker). If the paper does not turn a deep blue color, add more $NH_4OH$ until it does. Be careful not to add too much. If the solution becomes too basic, other metals will precipitate, causing an error.

7. Boil gently for 5 min to digest the precipitate. Then begin filtering immediately through Whatman #41 filter paper, prepared in the same manner as in Experiment 6, step 9. This filtering step is identical to that in Experiment 6, except that rinsing is done with a 1% solution of $NH_4NO_3$ at room temperature, rather than hot distilled water. The reason for using $NH_4NO_3$ is that distilled water can cause peptization of the precipitate. This means that the precipitate particles can become dispersed throughout the solution and become small enough to be lost through the filter paper. The use of a dissolved salt such as $NH_4NO_3$ prevents this, and in addition, it completely volatilizes during the ignition step, thus not adding any weight to the precipitate.

   Once all of the precipitate has been transferred to the funnel, rinse it three more times with the $NH_4NO_3$ solution. Test for chloride in the filtrate by acidifying a few milliliters of the most recent rinsings with dilute $HNO_3$ and adding $AgNO_3$ dropwise, as done in Experiment 6, step 11. Continue to rinse until all chloride is eliminated.

8. Remove the filter paper and precipitate from the funnel and carefully fold and place into the preweighed crucible. If you find the precipitate to be quite bulky and difficult to handle, place the folded filter paper "head first" into the crucible, so that if any precipitate oozes out, it will go into the crucible. Be careful not to tear the filter paper at this point. It is wet and quite fragile. You will need at least 60 min to finish the experiment. If you do not have this amount of time, store the crucibles outside of your desiccator until the next laboratory period.

## Session 3

9. Finish the experiment as directed in Experiment 6, steps 13 and 14. Slowly dry the filter paper using a Bunsen burner. Heat until black. Then incline the crucible at an angle, lid ajar, and direct the flame of a Meker burner to the bottom of the crucible, allowing air to get inside the crucible (Figure 3.9). Apply the full heat of the Meker burner for 20 min, or if only one Meker burner is available, alternate, as before, with a Bunsen burner for $^1/_2$ h. Cool and weigh. Calculate the percent Fe in the sample. The precipitate is $Fe_2O_3$.

# Questions and Problems

1. Define gravimetric analysis.
2. Compare Figure 3.1 with Figure 1.1. What is special about the gravimetric method of analysis?
3. What is the difference between weight and mass?
4. Why is a weighing device called a balance?
5. What is a single-pan balance?
6. What does it mean to say that a technician measures a weight to the nearest 0.01 g? What does it mean to say that the precision of a given balance is ±0.1 mg?
7. What is a top-loading balance?
8. What does it mean to say that a balance has a tare feature?
9. What is an analytical balance?
10. The analytical balance is a much more sensitive weighing device than an ordinary balance. Name three things that are important to remember when using the analytical balance that are not important when using an ordinary balance (Refer to Appendix 2).
11. What is a desiccator? What is Drierite? What is an indicating desiccant?
12. What does it mean to check the calibration of a balance?
13. When a weight measurement enters directly into the calculation of a quantitative analysis result that is to be reported to four significant figures, is an ordinary top-loading balance satisfactory for this measurement? Explain.
14. What is a physical separation of an analyte, as opposed to a chemical separation?
15. (a) Define loss on drying.
    (b) A soil sample, as received by a laboratory, weighed 5.6165 g. After drying in an oven, this same sample weighed 2.7749 g. What is the percent loss on drying?
16. In an experiment in which the percent loss on drying in a sample was determined, the following data were obtained:

| | |
|---|---|
| Weight of crucible + sample before drying | 11.9276 g |
| Weight of crucible + sample after drying | 10.7742 g |
| Weight of empty crucible | 7.6933 g |

What is the percent loss on drying in the sample?
17. What is meant by a step in a procedure that states "heat to constant weight"?
18. Define loss on ignition and residue on ignition.
19. A sample of grain was analyzed for loss on ignition. Initially, a sample of this grain in an evaporating dish weighed 29.6464 g. After heating in an oven at 500°C for 2 h, this same sample in the dish weighed 20.9601 g. If the evaporating dish alone weighed 11.6626 g, what is the percent loss on ignition and the percent residue on ignition?
20. An analyst performs an experiment to determine the percent loss on ignition of a grain sample. The grain is placed in a preweighed evaporating dish, the dish is weighed again, the volatiles are driven off by heating in a muffle furnace, and the dish is weighed a third time. Calculate the percent of volatile organics and the percent residue given the following data:

| | |
|---|---|
| Weight of empty evaporating dish | 28.3015 g |
| Weight of evaporating dish with the grain | 39.4183 g |
| Weight of dish with grain after heating | 33.1938 g |

21. What does insoluble matter in the general analysis report on a chemical bottle label refer to?
22. Define the following water and wastewater analysis terms: total solids, suspended solids, total dissolved solids, volatile solids, and settleable solids.

23. Consider the analysis of the water from a lake for suspended solid particles. A sample of the water was filtered through a preweighed filter to separate the suspended solids from the water. The following data were recorded:

| | |
|---|---|
| Volume of water used | 100.0 mL |
| Weight of empty filter | 11.6734 g |
| Weight of filter with solids | 11.7758 g |

What is the concentration in milligrams per liter of suspended solids in the water?

24. Consider the analysis of a 100.0-mL sample of wastewater for total solids and settleable solids. If the evaporating dish containing the dried sample weighs 38.1193 g, and the empty evaporating dish weighs 37.0209 g, what is the concentration of total solids in milligrams per liter in this sample? If, after the suspended solids settled, 100.0 mL of the resulting sample was again tested for solids, with the empty evaporating dish weighing 37.3884 g and the evaporating dish containing the dried sample weighing 37.8929 g, what is the concentration of settleable solids in milligrams per liter?

25. Consider the analysis of a salt–sand mixture. If the mixture contained in a beaker is treated with sufficient water to dissolve the salt, what are the percents of both the salt and sand in the mixture given the following data?

| | |
|---|---|
| Weight of mixture | 5.3502 g |
| Weight of sand isolated from mixture after filtering and drying | 4.2034 g |

26. The following data are obtained from a particle size analysis. What percent of the sample has a particle size greater than 2 mm in diameter?

| | |
|---|---|
| Weight of sample | 12.5 g |
| Weight of 2-mm sieve before sieving | 382.1 g |
| Weight of 2-mm sieve after sieving | 389.3 g |

27. What is the gravimetric factor, expressed to four significant figures, for each of the following gravimetric analysis examples?

| Substance Sought | Substance Weighed |
|---|---|
| (a) $Ag$ | $AgBr$ |
| (b) $SO_3$ | $BaSO_4$ |
| (c) $Ag_2O$ | $AgCl$ |
| (d) $Na_3PO_4$ | $Mg_2P_2O_7$ |
| (e) $Pb_3O_4$ | $PbCrO_4$ |
| (f) $SiF_4$ | $CaF_2$ |
| (g) $Co_3O_4$ | $Co_2O_3$ |
| (h) $Bi_2S_3$ | $Bi_2O_3$ |

28. What is the gravimetric factor
    (a) for obtaining the weight of $Ag_2CrO_4$ from the weight of $AgCl$?
    (b) if one is calculating the percent of $Na_2SO_4$ in a mixture when the weight of $Na_3PO_4$ is measured?
    (c) when converting the weight of Hg to the weight of $Hg_2Cl_2$?

29. What is the gravimetric factor that must be used in each of the following experiments?
    (a) The weight of $Mg_2P_2O_7$ is known and the weight of MgO is to be calculated.
    (b) The weight of $Fe_3O_4$ is to be converted to the weight of FeO.
    (c) The weight of $Mn_3O_4$ is to be determined from the weight of $Mn_2O_3$.

30. If a technician wishes to prepare a solution containing 55.3 mg of barium, how many grams of barium chloride dihydrate does he or she need to weigh?
31. What weight of $K_2SO_4$ is equivalent to 0.6603 g of $K_3PO_4$?
32. What weight of $P_2O_5$ is equivalent to 0.6603 g of P?
33. What is the percent of $K_2CrO_4$ in a sample that weighed 0.7193 g if the weight of the $Cr_2O_3$ precipitate derived from the sample was 0.1384 g?
34. The gravimetric factor for converting the weight of $BaCO_3$ to Ba is 0.6959. If the weight of $BaCO_3$ derived from a sample was 0.2644 g, what weight of Ba was in this sample?
35. If 0.9110 g of a sample of silver ore yielded 0.4162 g of AgCl in a gravimetric experiment, what is the percentage of Ag in the ore?
36. Given the following data, what is the percent S in the sample?

| | |
|---|---|
| Weight of weighing bottle before dispensing sample | 5.3403 g |
| Weight of weighing bottle after dispensing sample | 4.8661 g |
| Weight of crucible with $BaSO_4$ precipitate | 19.3428 g |
| Weight of crucible empty | 18.7155 g |

37. Given the following data, what is the percent Fe in the sample?

| | |
|---|---|
| Weight of weighing bottle before dispensing sample | 3.5719 g |
| Weight of weighing bottle after dispensing sample | 3.3110 g |
| Weight of crucible with $Fe_2O_3$ precipitate | 18.1636 g |
| Weight of empty crucible | 18.0021 g |

38. Nickel can be precipitated with dimethylglyoxime (DMG) according to the following reaction:

$$Ni^{2+} + HDMG \rightarrow Ni(DMG)_2 + 2H^+$$

If 2.0116 g of a nickel-containing substance is dissolved and the nickel precipitated as above so that the $Ni(DMG)_2$ precipitate weighs 2.6642 g, what is the percentage of nickel in the substance? The formula weight of $Ni(DMG)_2$ is 288.92 g/mol.

39. Imagine an experiment in which the percentage of manganese, Mn, in a manganese ore is to be determined by gravimetric analysis. If 0.8423 g of the ore yielded 0.3077 g of $Mn_3O_4$ precipitate, what is the percent Mn in the ore?
40. What is the advantage to weighing by difference, as opposed to weighing using a spatula and weighing paper?
41. Why is the barium sulfate precipitate in Experiment 6 digested?
42. During what period of the weighing-by-difference process must fingerprints on the item being weighed be avoided? Explain.
43. When testing the precipitate rinsings for chloride to determine the completeness of the rinsing in Experiments 6 and 7, why is the white precipitate formed discarded and not added to the filter paper?
44. In Experiments 4 to 7, why must heat-stable labels be placed on the crucibles before the first heating?
45. In Experiments 6 and 7, why is the desiccator not used to store the crucible and filter paper immediately after the filtration and before heating with a burner?
46. What does it mean to say that a filter paper is ashless? If the filter paper used in Experiments 6 and 7 were not ashless, what specific effect would it have on the calculated results?
47. In Experiment 6, if a student were to mark a crucible with the heat-stable marker after the first heating, rather than before, but before the first weighing, would his or her final results be higher or lower than the correct answer? Explain.

48. In Experiment 6, consider an error in which a student accidentally poked a hole in the filter paper with the stirring rod as he or she was filtering and some precipitate was observed with the filtrate. If the student takes no steps to correct the error, would the percent $SO_3$ calculated at the end be higher or lower than the true percent? Explain.

49. In step 11 of Experiment 6, two drops of silver nitrate solution are added to a test tube containing rinsings from the funnel containing the precipitate. Suppose you were first asked to prepare this silver nitrate solution by weighing a given quantity of solid silver nitrate and dissolving it in water to make a given volume of solution. Would an analytical balance be required for this weight, or would an ordinary balance suffice? Explain.

50. **Report**

    Find a real-world gravimetric analysis in a methods book (AOAC, USP, ASTM, etc.), in a journal, or on a website and report on the details of the procedure according to the following scheme:

    (a) Title

    (b) General information, including type of material examined, name of the analyte, sampling procedures, and sample preparation procedures

    (c) Specifics, including type of equipment used, details of exactly how the analyte is separated from its matrix (oven temperatures, filtering method and material, etc.), and, if chemical reaction or precipitation is required, the chemical equation for the reaction, what the precipitating agent is, and how prepared

    (d) Data handling and reporting

    (e) References

# 4

# Introduction to Titrimetric Analysis

## 4.1  Introduction

The gravimetric analysis methods introduced in Chapter 3 represent one of two major subcategories of wet chemical analysis. As we discussed, gravimetric analysis methods primarily utilize the measurement of weight, and that is why this subcategory is called gravimetric analysis. In this chapter, we introduce **titrimetric analysis,** the other major subcategory. Titrimetric analysis methods utilize solution chemistry heavily, and therefore it should not be surprising that volumes of solutions are prepared, measured, transferred, and analyzed frequently in this type of analysis. In fact, titrimetric analysis is often referred to as **volumetric analysis** for that reason. Solution reaction stoichiometry lies at the heart of the methods. Thus, methods of solution preparation, methods of measuring and transferring liquid volumes, and methods of utilizing solution reaction stoichiometry in analyzing solutions are all topics with which we will be concerned in our discussion of titrimetric analysis. The analytical strategy flow chart specific to titrimetric analysis is given in Figure 4.1. Our discussion in this chapter involves the specifics of the methods and data handling, as represented by the highlighted portions of Figure 4.1.

## 4.2  Terminology

Figure 4.1 mentions **standard solution**. A standard solution is a solution that has a concentration of solute known to some high degree of precision, such as a molarity known to four decimal places. For example, a solution of HCl with a concentration of $0.1025\ M$ is a standard solution of HCl. The concentration can be known, in some cases, directly through the preparation of this solution. It may become known by performing an experiment.

If the concentration of a standard solution is to be known directly through its preparation, then the amount of solute present must be accurately measured with a high degree of precision. In addition, a **volumetric flask** (Figure 4.2), which is an accurate, high-precision, very common type of flask, must be used as the container so that the total volume of solution can also be accurately measured with good precision. If the solute is a pure solid material capable of being weighed accurately, then an analytical balance must be used to weigh it. If the solute is already in solution, then the solution can be prepared by diluting this solution. However, the solution to be diluted must be a standard solution with a concentration known with equal or better precision (same or more significant figures) than the solution being prepared. In addition, the volume of the solution to be diluted must be precisely measured. A **volumetric transfer pipet** (Figure 4.2), a high-precision type of transfer pipet, is often the piece of glassware used for this latter measurement, although other devices are available. Flasks, pipets, and pipetting devices will be discussed in detail in Sections 4.8 and 4.9.

If the concentration of a standard solution is to become known by experiment, the experiment must be carried out with a high degree of accuracy and precision. Such an experiment is called **standardization**.

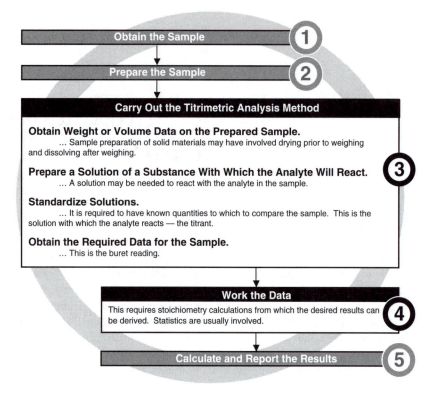

**FIGURE 4.1**    Analytical strategy flow chart for titrimetric analysis.

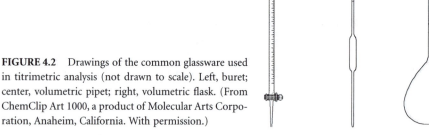

**FIGURE 4.2**    Drawings of the common glassware used in titrimetric analysis (not drawn to scale). Left, buret; center, volumetric pipet; right, volumetric flask. (From ChemClip Art 1000, a product of Molecular Arts Corporation, Anaheim, California. With permission.)

Similarly, to **standardize** a solution means to determine its concentration to three or more significant figures. For example, if a solution of NaOH is made up to be approximately 0.1 $M$, a standardization experiment may be performed and the concentration determined to be 0.1012 $M$. We will discuss the details of such an experiment in Section 4.6.

Another piece of glassware often used in titrimetric analysis (for both standardization experiments and sample analyses) is the buret. A traditional glass **buret** (Figure 4.2) is a long and narrow graduated cylinder with a dispensing valve at the bottom. The dispensing valve is called a **stopcock**. The solution in the buret is dispensed by turning the stopcock to the open position. The experiment used for titrimetric analysis and solution standardization is called a titration. A **titration** is an experiment in which a solution of a reactant is added to a reaction flask with the use of a buret. The solution being dispensed via the buret (or the substance dissolved in the solution dispensed via the buret) is called the **titrant**. The solution held in the reaction flask before the titration (or the substance dissolved in this solution) is called the

**substance titrated** . For example, a given titration experiment may have a solution of NaOH in the buret and a solution of HCl in the reaction flask. In that case, NaOH is the titrant, and HCl is the substance titrated.

In order to effectively utilize the stoichiometry of the reaction involved in a titration, both the titrant and the substance titrated need to be measured exactly. The reason is that one is the known quantity, and the other is the unknown quantity in the stoichiometry calculation. The buret is an accurate (if carefully calibrated) and relatively high-precision device because it is long and narrow. If a meniscus is read in a narrow graduated tube, it can be read with higher precision (more significant figures) than in a wider tube. Thus a buret provides the required precise measurement of the titrant.

The addition of titrant from the buret must be stopped at precisely the correct moment—the moment at which the last trace of substance titrated is consumed by a fraction of a drop of titrant added, so that the correct volume can be read on the buret. That exact moment is called the **equivalence point** of the titration. In order to detect the equivalence point, an indicator is often used. An **indicator** is a substance added to the reaction flask ahead of time in order to cause a color change at or near the equivalence point, i.e., to provide a visual indication of the equivalence point. For example, the use of a chemical named phenolphthalein as an indicator for a titration in which a strong base is used as the titrant and an acid as the substance titrated would give a color change of colorless to pink in the reaction flask near the equivalence point. The color change occurring near, not exactly at, the equivalence point is usually not a concern. The reason will become clear in a later discussion. The point of a titration at which an indicator changes color, the visual indication of the equivalence point, is called the **end point** of the titration. As we will see, equivalence points can be determined in other ways too.

Electronic devices such as **automatic titrators** and **digital burets** may be used in place of the traditional glass buret and manual titration. Such devices provide electronic control over the addition of titrant and thus, with proper calibration, are accurate, high-precision devices. These will be discussed in Section 4.9.

Solution preparation, standardization, and sample analysis activities all involve solution concentration. Let us review molarity and normality as methods of expressing solution concentration.

# 4.3 Review of Solution Concentration

## 4.3.1 Molarity

**Molarity** is defined as the number of moles of solute dissolved per liter of solution and may be calculated by dividing the number of moles dissolved by the number of liters of solution:

$$\text{molarity} = \frac{\text{moles of solute}}{\text{liters of solution}} \tag{4.1}$$

For example, solutions are referred to as being 2.0 molar, or 2.0 *M*. The *M* refers to molar, and the solution is said to have a molarity of 2.0, or 2.0 mol dissolved per liter of solution. It is important to recognize that, for molarity, it is the number of moles dissolved per liter of *solution* and not per liter of solvent.

### Example 4.1

What is the molarity of a solution that has 4.5 mol of solute dissolved in 300.0 mL of solution?

*Solution 4.1*

Equation (4.1) defines molarity. The number of milliliters given must first be converted to liters, and so 0.3000 L is used in the denominator:

$$\text{molarity} = \frac{4.5 \text{ mol}}{0.3000 \text{ L}} = 15 \, M$$

### Example 4.2

What is the molarity of a solution of NaOH that has 0.491 g dissolved in 400.0 mL of solution?

*Solution 4.2*

In order to use Equation (4.1), the grams of solute must be converted to moles by dividing the grams by the formula weight (FW). Once again, the milliliters are first converted to liters.

$$\text{molarity} = \frac{\text{grams}/\text{FW}}{\text{liters of solution}} = \frac{(0.491/39.997)\ \text{mol}}{0.4000\ \text{L}} = 0.0307\ M$$

When solutions are prepared such that their concentration is known directly through their preparation by weighing a pure solid into the container, as discussed in Section 4.2, the calculation performed in Example 4.2 is typical.

If a solution is prepared by diluting another solution whose concentration is precisely known (another possibility mentioned in Section 4.2), the following **dilution equation** may be used to calculate molarity:

$$C_B \times V_B = C_A \times V_A \tag{4.2}$$

Here C refers to concentration (any unit, but the same on both sides), V refers to volume (any unit, but the same on both sides), B refers to before dilution, and A refers to after dilution.

### Example 4.3

What is the molarity of a solution prepared by diluting 10.00 mL of a 4.281 $M$ solution to 50.00 mL?

*Solution 4.3*

Using Equation (4.2), we have the following:

$$4.281\ M \times 10.00\ \text{mL} = C_A \times 50.00\ \text{mL}$$

$$C_A = \frac{4.281\ M \times 10.00\ \text{mL}}{50.00\ \text{mL}} = 0.8562\ M$$

## 4.3.2   Normality

**Normality** is similar to molarity except that it utilizes a quantity of chemical called the **equivalent** rather than the mole; it is defined as **the number of equivalents per liter** rather than the number of moles per liter. It is dependent on the specific chemistry of the analysis. It may be calculated by dividing the equivalents of solute by the liters of solution in which that number of equivalents is dissolved:

$$\text{normality} = \frac{\text{equivalents of solute}}{\text{liters of solution}} \tag{4.3}$$

For example, if there are 2.0 equivalents dissolved per liter, a solution would be referred to as 2.0 normal, or 2.0 $N$. The equivalent is either the same as the mole or some fraction of the mole, depending on the reaction involved, and the **equivalent weight**, or **the weight of one equivalent**, is either the same as the formula weight or some fraction of the formula weight. Normality is either the same as molarity or some multiple of molarity. Let us illustrate with acids and bases in acid–base neutralization reactions.

The **equivalent weight of an acid in an acid–base neutralization reaction** is defined as the formula weight divided by the number of hydrogens lost per formula of the acid in the reaction. Acids may lose one or more hydrogens (per formula) when reacting with a base.

$$\text{HCl} + \text{NaOH} \rightarrow \text{NaCl} + \text{H}_2\text{O} \text{ (one hydrogen lost by HCl)} \tag{4.4}$$

$$\text{H}_2\text{SO}_4 + 2\,\text{NaOH} \rightarrow \text{Na}_2\text{SO}_4 + 2\,\text{H}_2\text{O} \text{ (two hydrogens lost by H}_2\text{SO}_4) \tag{4.5}$$

The equivalent weight of HCl is the same as its formula weight, and the equivalent weight of $H_2SO_4$ is half the formula weight. The equivalent weight is analogous to the formula weight in our molarity discussions. In the case of HCl, there is one hydrogen lost per formula; one equivalent is therefore the same as the mole, and the equivalent weight is the same as the formula weight. In the case of $H_2SO_4$ in the reaction in Equation (4.5), there are two hydrogens lost per formula—two equivalents per mole when used as in the above reaction; the equivalent weight is half the formula weight.

The **equivalent weight of a base in an acid–base neutralization reaction** is defined as the formula weight divided by the number of hydrogens accepted per formula of the base in the reaction. This definition is based on the Bronsted–Lowry concept of a base (a compound that accepts hydrogens when reacting with an acid). Thus all hydroxides, as well as carbonates, ammonia, etc., are Bronsted–Lowry bases and accept hydrogens when reacting with acids. Bases may accept one or more hydrogens (per formula) when reacting with an acid. In both reactions above (Equations (4.4) and (4.5)), sodium hydroxide is accepting one hydrogen per formula unit. Indeed, there is only one hydroxide group in one formula of NaOH to combine with one hydrogen to form water; therefore, sodium hydroxide can accept only one hydrogen per formula.

On the other hand, sodium carbonate is capable of accepting either one or two hydrogens per formula:

$$Na_2CO_3 + HCl \rightarrow NaCl + NaHCO_3 \quad \text{(one hydrogen accepted by } Na_2CO_3) \qquad (4.6)$$

$$Na_2CO_3 + 2\,HCl \rightarrow 2\,NaCl + H_2CO_3 \quad \text{(two hydrogens accepted by } Na_2CO_3) \qquad (4.7)$$

Thus, the equivalent weight of sodium carbonate may be equal to either the formula weight divided by 1 (105.99 g per equivalent, Equation (4.6)) or the formula weight divided by 2 (52.995 g per equivalent, Equation (4.7)), depending on which reaction is involved.

## Example 4.4

What is the normality of a solution of sulfuric acid if it is used as in Equation (4.5) and there are 0.248 mol dissolved in 250.0 mL of solution?

*Solution 4.4*

$$\text{normality} = \frac{\text{equivalents of solute}}{\text{liter of solution}}$$

$$= \frac{\text{moles of solute} \times \text{equivalents per mole}}{\text{liters of solution}}$$

$$= \frac{0.248\ \text{mol} \times 2\ \text{equivalents}/\text{mol}}{0.2500\ \text{L}}$$

$$= 1.98\ N$$

(Note that there are exactly two equivalents per mole and that the number 2 does not diminish the number of significant figures in the answer because it has an infinite number of significant figures. This is also true in other calculations in this section.)

## Example 4.5

What is the normality of a solution of oxalic acid dihydrate ($H_2C_2O_4 \cdot 2\,H_2O$; formula weight, 126.07 g per mole) if it is to be used as in the following reaction and 0.4920 g of it is dissolved in 250.0 mL of solution?

$$H_2C_2O_4 + 2\,NaOH \rightarrow Na_2C_2O_4 + 2\,H_2O$$

*Solution 4.5*

In the reaction given, two hydrogens are lost per molecule of oxalic acid. The equivalent weight is therefore the formula weight divided by 2, or 63.035 g per equivalent. The equivalents of oxalic acid used here are calculated by dividing the weight (of the hydrate) by the equivalent weight. Then normality is calculated in the same way as molarity was in Example 4.2, but using Equation (4.3):

$$\text{normality} = \frac{(0.4920/63.035) \text{ equivalents}}{0.2500 \text{ L}} = 0.03122 \, N$$

### Example 4.6

What is the normality of a solution of sodium carbonate if 0.6003 g of it is dissolved in 500.00 mL of solution and it is to be used as in Equation (4.7)?

### Solution 4.6

As stated above, the equivalent weight of sodium carbonate as used in Equation (4.7) is the formula weight divided by 2 because it accepted two hydrogens in the reaction. Thus, utilizing Equation (4.3), we have the following:

$$\text{normality} = \frac{(0.6003/52.9945) \text{ equivalents}}{0.5000 \text{ L}} = 0.02266 \, N$$

Normality applies mostly to acid–base neutralization, but the concentrations of other chemicals used in other kinds of reactions, such as in oxidation–reduction reactions, may also be expressed in normality.

## 4.4   Review of Solution Preparation

We will now consider the preparation of solutions. Solutions are prepared for a wide variety of reasons. We have already discussed the use of standard solutions in titrimetric analysis and that these solutions sometimes must be prepared with high precision and accuracy so that their concentrations may be known directly through the preparation process. Even if the need for good precision and accuracy through the preparation is not necessarily important, an analyst frequently prepares other solutions with concentrations known less precisely. Thus the familiarity with solution preparation schemes, highly precise or not, is very important.

### 4.4.1   Solid Solute and Molarity

To prepare a given volume of a solution of a given molarity of solute when the weight of the pure, solid solute is to be measured, it is necessary to calculate the grams of solute that are required. This number of grams can be calculated from the desired molarity and volume of solution desired. Grams can be calculated by multiplying moles by the formula weight:

$$\text{moles} \times \frac{\text{grams}}{\text{moles}} = \text{grams} \tag{4.8}$$

The moles that are required can be calculated from the volume (liters) and molarity (moles per liter):

$$\text{moles} = \text{liters} \times \frac{\text{moles}}{\text{liter}} \tag{4.9}$$

Combining Equations (4.8) and (4.9) gives us the formula needed:

$$\text{grams} = \text{liters} \times \frac{\text{moles}}{\text{liter}} \times \frac{\text{grams}}{\text{mole}} \tag{4.10}$$

The liters in Equation (4.10) are the liters of solution that are desired; the moles per liter is the molarity desired; and the grams per mole is the formula weight of the solute. Thus, we have

$$\text{grams to weigh} = L_D \times M_D \times FW_{SOL} \tag{4.11}$$

in which $L_D$ refers to the liters desired, $M_D$ to the desired molarity, and $FW_{SOL}$ to the formula weight of solute. The grams of solute thus calculated is weighed and placed in the container. Water is added to dissolve the solute and to dilute the solution to volume. Following this, the solution is shaken to make it homogeneous.

## Example 4.7

How would you prepare 500.0 mL of a 0.20 $M$ solution of NaOH from pure, solid NaOH?

### Solution 4.7
Using Equation (4.11) we have the following:

$$\text{grams to weigh} = L_D \times M_D \times FW_{SOL}$$

$$\text{grams to weigh} = 0.5000 \not{L} \times 0.20 \text{ mol}/\not{L} \times 39.997 \text{ g/mol} = 4.0 \text{ g}$$

The analyst would weigh 4.0 g of NaOH, place it in a container with a 500-mL calibration line, add water to dissolve the solid, and then dilute to volume and shake, stir, or swirl.

If the solute is a liquid (somewhat rare), the weight calculated from Equation (4.11) can, but rather inconveniently, be measured on a balance. However, this weight can also be converted to milliliters by using the density of the liquid. In this way, the volume of the liquid can be measured, rather than its weight, and it can be pipetted into the container.

## 4.4.2   Solid Solute and Normality

To determine the weight of a solid that is needed to prepare a solution of a given normality, we can derive an equation in a manner similar to that in Section 4.4.1 for molarity:

$$\text{grams to weigh} = \text{liters} \times \frac{\text{equivalents}}{\text{liter}} \times \frac{\text{grams}}{\text{equivalent}} \qquad (4.12)$$

$$\text{grams to weigh} = L_D \times N_D \times EW_{SOL} \qquad (4.13)$$

in which $L_D$ is the liters of solution desired, $N_D$ is the normality desired, and $EW_{SOL}$ is the equivalent weight of the solute.

As in Section 4.3, acid–base neutralization reactions will be illustrated here. In order to calculate the equivalent weight of an acid, the balanced equation representing the reaction in which the solution is to be used is needed so that the number of hydrogens lost per formula in the reaction can be determined. The equivalent weight of an acid is the formula weight of the acid divided by the number of hydrogens lost per molecule (see Section 4.3).

## Example 4.8

How many grams of $KH_2PO_4$ are needed to prepare 500.0 mL of a 0.200 $N$ solution if it is to be used as in the following reaction?

$$KH_2PO_4 + 2\,KOH \rightarrow K_3PO_4 + 2\,H_2O$$

### Solution 4.8
There are two hydrogens lost per formula of $KH_2PO_4$. Therefore the equivalent weight of $KH_2PO_4$ is the formula weight divided by 2. Utilizing Equation (4.13), we have the following:

$$\text{grams to weigh} = 0.5000 \not{L} \times 0.200 \text{ equivalents}/\not{L} \times \frac{136.085 \text{ g}}{2 \text{ equivalents}} = 6.80 \text{ g}$$

### 4.4.3 Solution Preparation by Dilution

If a solution is to be prepared by diluting another solution, whether high precision and accuracy are important or not, the **dilution equation** , Equation (4.2), is again used:

$$C_B \times V_B = C_A \times V_A \qquad (4.14)$$

C, V, B, and A have the same meanings as they do in Equation (4.2). For solution preparation, $V_B$, the volume of the solution to be diluted (the volume before dilution), is calculated so that the analyst can know how much of this more concentrated solution to measure out in order to prepare the less concentrated solution. Remember that $V_B$ and $V_A$ can have any volume unit, but the units for both $V_B$ and $V_A$ must be the same (for example, milliliters). The same is true of the concentration units.

**Example 4.9**

How many milliliters of 12 *M* hydrochloric acid are needed to prepare 500.0 mL of a 0.60 *M* solution?

*Solution 4.9*

Rearranging the dilution equation to calculate $V_B$, we have the following:

$$V_B = \frac{C_A \times V_A}{C_B}$$

$$V_B = \frac{0.60 \,\cancel{M} \times 500.0 \text{ mL}}{12.0 \,\cancel{M}} = 25 \text{ mL}$$

## 4.5   Stoichiometry of Titration Reactions

At the equivalence point of a titration, exact stoichiometric amounts of the reactants have reacted, i.e., the amount of titrant added is the exact amount required to consume the amount of substance titrated in the reaction flask. If the reaction is one-to-one in terms of moles (moles of titrant equals the moles of substance titrated, as is the case for the reaction represented by Equation (4.4), for example), then the moles of titrant added equals the moles of substance titrated consumed:

$$\text{moles}_T = \text{moles}_{ST} \qquad (4.15)$$

where $\text{moles}_T$ represents the moles of titrant added to get to the equivalence point and $\text{moles}_{ST}$ represents the moles of substance titrated in the reaction flask. If the reaction is two-to-one in terms of moles (as in Equation (4.5)), then the moles of titrant is twice the moles of substance titrated:

$$\text{moles}_T = \text{moles}_{ST} \times 2 \qquad (4.16)$$

and the moles of substance titrated is half the moles of titrant:

$$\text{moles}_{ST} = \text{moles}_T \times \tfrac{1}{2} \qquad (4.17)$$

or vice versa, depending on which substance is the titrant and which is the substance titrated. And so it goes for three-to-one, three-to-two, etc., reactions. In general terms, then, we have the following:

$$\text{moles}_{ST} = \text{moles}_T \times \text{mole ratio (ST/T)} \qquad (4.18)$$

where mole ratio (ST/T) is the ratio of moles of substance titrated (ST) to the moles of titrant (T) as gleaned from the chemical equation representing the reaction. When working with moles as in these equations, the molarity method of expressing concentration and formula weights is needed in the calculations. For example, to convert the buret reading for the titrant in liters ($L_T$) to moles of titrant (which can then be used in Equation (4.18)), we must multiply $L_T$ by the titrant molarity (moles per liter) as the following shows:

$$\text{liters} \times \frac{\text{moles}}{\text{liter}} = \text{moles} \tag{4.19}$$

And thus we have the following:

$$L_T \times M_T = \text{moles}_T \tag{4.20}$$

Combining Equations (4.18) and (4.20), we have the following:

$$\text{moles}_{ST} = L_T \times M_T \times \text{mole ratio (ST/T)} \tag{4.21}$$

This equation will be useful as we explore titration calculations in Sections 4.6 and 4.7.

Some analysts prefer to work with equivalents rather than moles. In that case, the normality method of expressing concentration is used and the equivalent weight is needed, rather than the formula weight. The equivalent weight of one substance reacts with the equivalent weight of the other substance. In other words, the reaction is *always* one-to-one: one equivalent of one substance *always* reacts with one equivalent of the other. Thus, we can write the following as a true statement at the equivalence point of the titration:

$$\text{equiv}_T = \text{equiv}_{ST} \tag{4.22}$$

where $\text{equiv}_T$ represents the equivalents of titrant used to get to the equivalence point and $\text{equiv}_{ST}$ represents the equivalents of substance titrated that reacted in the reaction flask. Multiplying the liters of titrant ($L_T$) by the normality of the titrant ($N_T$) gives the equivalents of titrant:

$$\text{liters} \times \frac{\text{equivalents}}{\text{liter}} = \text{equivalents} \tag{4.23}$$

And therefore we have the following:

$$L_T \times N_T = \text{equiv}_T \tag{4.24}$$

Combining this with Equation (4.22), we have

$$\text{equiv}_{ST} = L_T \times N_T \tag{4.25}$$

Like Equation (4.21), this equation will also be useful as we explore titration calculations in Sections 4.6 and 4.7.

## 4.6  Standardization

Standardization was defined in Section 4.2 as a titration experiment in which the concentration of a solution becomes known to a high degree of precision and accuracy. In a standardization experiment, the solution being standardized is compared to a known standard. This known standard can be either a solution that is already a standard solution or an accurately weighed solid material. In either case, the solute of the solution to be standardized reacts with the known standard in the titration vessel. If the solution to be standardized is the titrant, then the known standard is the substance titrated, and vice versa. We will now describe these two methods and the calculations involved.

### 4.6.1  Standardization Using a Standard Solution

In the case of standardization using a standard solution, it is a volume of the substance titrated that is measured into the reaction flask. Since volume in liters multiplied by molarity gives moles (see Equation (4.19)),

Equation (4.21) becomes

$$L_T \times M_T \times \text{mole ratio (ST/T)} = L_{ST} \times M_{ST} \qquad (4.26)$$

Also, since volume in liters multiplied by normality gives us equivalents (see Equation (4.23)), Equation (4.25) becomes

$$L_T \times N_T = L_{ST} \times N_{ST} \qquad (4.27)$$

As in the discussion accompanying Equation (4.2), the volume of titrant and the volume of substance titrated may be expressed in any volume unit as long as they are both the same unit. Thus we have

$$V_T \times M_T \times \text{mole ratio (ST/T)} = V_{ST} \times M_{ST} \qquad (4.28)$$

$$V_T \times N_T = V_{ST} \times N_{ST} \qquad (4.29)$$

where $V_T$ and $V_{ST}$ represent the volumes of titrant and substance titrated, respectively.

The experiment is performed by precisely measuring the volume of the solution of substance titrated (either the solution to be standardized or the known standard solution) into the reaction flask and then titrating it with the other solution. At the end point, $V_T$ and $V_{ST}$ are known and one of the two concentrations is known (the known standard). Thus the other concentration can then be calculated. See Workplace Scene 4.1.

### Example 4.10
Standardization of a solution of sulfuric acid required 29.03 mL of 0.06477 *M* NaOH when exactly 25.00 mL of $H_2SO_4$ was used. What is the molarity of $H_2SO_4$? Refer to Equation (4.5) for the reaction involved.

### *Solution 4.10*
Equation (4.28) is used since the question has to do with molarity.

$$V_T \times M_T \times \text{mole ratio (ST/T)} = V_{ST} \times M_{ST}$$

$$\text{milliliters}_{NaOH} \times M_{NaOH} \times \text{mole ratio (H}_2\text{SO}_4/\text{NaOH)} = \text{milliliters}_{H_2SO_4} \times M_{H_2SO_4}$$

$$29.03 \text{ mL} \times 0.06477 \ M \times 1/2 = 25.00 \text{ mL} \times M_{H_2SO_4}$$

$$M_{H_2SO_4} = 0.03761 \ M$$

### Example 4.11
Standardization of a solution of sulfuric acid required 28.50 mL of 0.1077 *N* NaOH when exactly 25.00 mL of $H_2SO_4$ was used. What is the normality of $H_2SO_4$? Refer to Equation (4.5) for the reaction involved.

### *Solution 4.11*
Equation (4.29) is used since the question concerns normality.

$$V_T \times N_T = V_{ST} \times N_{ST}$$

$$\text{milliliters}_{NaOH} \times N_{NaOH} = \text{milliliters}_{H_2SO_4} \times N_{H_2SO_4}$$

$$28.50 \text{ mL} \times 0.1077 \ N = 25.00 \text{ mL} \times N_{H_2SO_4}$$

$$N_{H_2SO_4} = \frac{28.50 \text{ mL} \times 0.1077 \ N}{25.00 \text{ mL}} = 0.1228 \ N$$

# WORKPLACE SCENE 4.1

SACHEM Inc. of Cleburne, Texas, manufactures various concentrations of tetramethylammonium hydroxide (TMAH) solutions to meet customer specifications. To ensure consistent performance, electronic industry requires very narrow concentration specifications for the solutions. In SACHEM's quality control laboratory, standardized acids such as HCl or $H_2SO_4$ are used as titrants for the TMAH solutions to check their concentrations. The performance of the assay titration is controlled by daily analysis of internal reference standards (IRSs). If the IRS results are within controlled limits, then the assay results of a product can be reported. If not, the results cannot be reported until the root cause is uncovered and eliminated. Safety glasses and gloves are worn while performing this work in the laboratory.

Bettina Pfeiffenberger of SACHEM Inc. uses an automatic titrator to titrate concentrations of TMAH to ensure that the products meet customers' specifications.

## 4.6.2  Standardization Using a Primary Standard

The alternative to the above is using an accurately weighed solid material as the known standard. Such a material is called a **primary standard** . Thus, a primary standard is a material that can be weighed accurately either for the purpose of preparing a standard solution (which then does not have to be standardized) or for comparison to a solution with which it reacts for the purpose of standardizing that solution. For standardization with a primary standard, Equation (4.21) becomes

$$L_T \times M_T \times \text{mole ratio (PS/T)} = \text{moles}_{PS} \tag{4.30}$$

where PS refers to primary standard. Knowing that grams divided by the formula weight gives moles, we have the following:

$$L_T \times M_T \times \text{mole ratio (PS/T)} = \text{grams}_{PS}/FW_{PS} \tag{4.31}$$

In addition, Equation (4.25) becomes

$$L_T \times N_T = \text{grams}_{PS}/EW_{PS} \tag{4.32}$$

The solution to be standardized is the titrant. In this case, the volume of the titrant must be expressed in liters, as shown.

The experiment consists of weighing the primary standard on an analytical balance into the titration flask, dissolving, and then titrating it with the solution to be standardized.

### Example 4.12

In a standardization experiment, 0.4920 g of primary standard sodium carbonate ($Na_2CO_3$) was exactly neutralized by 19.04 mL of hydrochloric acid solution. What is the molarity of the HCl solution? Refer to Equation (4.7) for the reaction involved.

### Solution 4.12

Since the problem concerns molarity, Equation (4.31) applies.

$$L_T \times M_T \times \text{mole ratio (PS/T)} = \text{grams}_{PS}/FW_{PS}$$

$$\text{liters}_{HCl} \times M_{HCl} \times \text{mole ratio (Na}_2\text{CO}_3/\text{HCl)} = \frac{\text{grams}_{Na_2CO_3}}{FW_{Na_2CO_3}}$$

$$0.01904\,L \times M_{HCl} \times 1/2 = \frac{0.4920\,g}{105.989\,g/mol}$$

$$M_{HCl} = 0.4876\,M$$

### Example 4.13

In a standardization experiment, 0.5067 g of primary standard sodium carbonate ($Na_2CO_3$) was exactly neutralized by 27.86 mL of a hydrochloric acid solution. What is the normality of the HCl solution? Refer to Equation (4.7) for the reaction involved.

### Solution 4.13

Equation (4.18) is used as follows:

$$\text{liters}_{HCl} \times N_{HCl} = \frac{\text{grams}_{Na_2CO_3}}{EW_{Na_2CO_3}}$$

$$0.02786\,L \times N_{HCl} = \frac{0.5067\,g}{52.9945\,g/\text{equivalent}}$$

$$N_{HCl} = \frac{0.5067\,g}{0.02786\,L \times 52.9945\,g/\text{equivalent}} = 0.3432\,N$$

Obviously the quality of the primary standard substance is ultimately the basis for a successful standardization. This means that it must meet some special requirements with respect to purity, etc.; these are enumerated below:

1. It must be 100% pure, or at least its purity must be known.
2. If it is impure, the impurity must be inert.
3. It should be stable at drying oven temperatures.
4. It should not be hygroscopic—it should not absorb water when exposed to laboratory air.
5. The reaction in which it takes part must be quantitative and preferably fast.
6. A high formula weight is desirable, so that the number of significant figures in the calculated result is not diminished.

Most substances used as primary standards can be purchased as a primary standard grade; this is usually appropriate and sufficient for standardization experiments. However, how a primary standard grade chemical, as sold by a chemical supply company, itself becomes a standard is a legitimate question. In the U.S., there is a federal agency that produces and certifies this ultimate standard. The agency is the National Institute of Standards and Technology (NIST), the same agency that provides standard weights for balance calibration (Chapter 3). NIST manufactures and sells standards (standard reference materials (SRMs)) for a wide variety of laboratory applications. Being the ultimate standards, such materials are expensive and are not used beyond essential standardizations and calibrations, and are used for that purpose most often only by the chemical supply companies. Standards distributed by the supply companies are then called certified reference materials (CRMs). CRMs are, in modern analytical chemistry language, **traceable** to SRMs.

### 4.6.3   Titer

The strength (concentration) of a titrant can also be expressed as its titer. The **titer** of a titrant is defined as the weight (in milligrams) of substance titrated that is consumed by 1 mL of the titrant. Thus it is specific to a particular substance titrated, meaning that it is expressed with respect to a specific substance titrated. For example, if the analyst is using an oxalic acid solution to titrate a solution of calcium oxide, the titer of the oxalic acid solution would be expressed as its CaO titer, or the weight of CaO that is consumed by 1 mL of the oxalic acid solution. Titer is typically used for repetitive routine work in which the same titrant is used repetitively to titrate a given analyte.

**Example 4.14**

What is the titer (expressed in milligrams per milliliter) of a solution of oxalic acid dihydrate with respect to calcium oxide if 21.49 mL of it was needed to titrate 0.2203 g of CaO?

*Solution 4.14*

$$\text{titer} = \frac{\text{milligrams of substance titrated}}{\text{milliliters of titrant}}$$

The titer of the oxalic acid solution with respect to calcium oxide is calculated as follows:

$$\text{titer of oxalic acid solution} = \frac{220.3 \text{ mg of CaO}}{21.49 \text{ mL of oxalic acid}} = 10.25 \text{ mg / mL}$$

## 4.7   Percent Analyte Calculations

The ultimate goal of any titrimetric analysis is to determine the amount of the analyte in a sample. This involves the stoichiometry calculation mentioned in the "Work the Data" section of the analytical strategy flow chart in Figure 4.1. This amount of analyte is often expressed as a percentage, as it was for the gravimetric analysis examples in Chapter 3. This percentage is calculated via the basic equation for percent used previously for the gravimetric analysis examples:

$$\% \text{ analyte} = \frac{\text{weight of analyte}}{\text{weight of sample}} \times 100 \tag{4.33}$$

As with gravimetric analysis, the weight of the sample (the denominator in Equation (4.33)) is determined by direct measurement in the laboratory or by weighing by difference. The weight of the analyte in the sample is determined from the titration data via a stoichiometry calculation. As discussed previously, we calculate moles of substance titrated (in this case, the analyte) as in Equation (4.21):

$$\text{moles}_{\text{analyte}} = L_T \times M_T \times \text{mole ratio (ST/T)} \tag{4.34}$$

The weight of the analyte is then calculated by multiplying by its formula weight (grams per mole):

$$grams_{analyte} = moles_{analyte} \times grams/mole \qquad (4.35)$$

Combining Equations (4.34) and (4.35), we have

$$grams_{analyte} = L_T \times M_T \times \text{mole ratio (ST/T)} \times FW_{analyte} \qquad (4.36)$$

Finally, combining Equations (4.33) and (4.36), we have

$$\% \text{ analyte} = \frac{L_T \times M_T \times \text{mole ratio (ST/T)} \times FW_{analyte}}{\text{weight of sample}} \times 100 \qquad (4.37)$$

If normality and equivalents are used, we calculate equivalents of analyte as in Equation (4.25) (again, the substance titrated is the analyte):

$$equivalents_{analyte} = L_T \times N_T \qquad (4.38)$$

The weight of the analyte (the numerator in Equation (4.33)) can then be calculated by multiplying the equivalents of substance titrated by the equivalent weight (grams per equivalent):

$$L_T \times N_T \times EW_{ST} = grams_{ST} \qquad (4.39)$$

Thus the percent calculation then becomes

$$\% \text{ analyte} = \frac{L_T \times N_T \times EW_{analyte}}{\text{weight of sample}} \times 100 \qquad (4.40)$$

## Example 4.15

In the analysis of a sample for $KH_2PO_4$ content, a sample weighing 0.3994 g required 18.28 mL of 0.1011 $M$ KOH for titration. The equation below represents the reaction involved. What is the percent $KH_2PO_4$ in this sample?

$$2\ KOH + KH_2PO_4 \rightarrow K_3PO_4 + 2\ H_2O$$

### Solution 4.15

Since the question involves molarity, Equation (4.37) applies and we have the following:

$$\% KH_2PO_4 = \frac{liters_{KOH} \times M_{KOH} \times \text{mole ratio }(KH_2PO_4/KOH) \times FW_{KH_2PO_4}}{\text{weight of sample}} \times 100$$

$$= \frac{0.01828\ L \times 0.1011\ mol/L \times 1/2 \times 136.085\ g/mol}{0.3994\ g} \times 100$$

$$= 31.48\%$$

## Example 4.16

In the analysis of a soda ash (impure $Na_2CO_3$) sample for sodium carbonate content, 0.5203 g of the soda ash required 36.42 mL of 0.1167 $N$ HCl for titration. What is the percent $Na_2CO_3$ in this sample?

$$Na_2CO_3 + 2\ HCl \rightarrow 2\ NaCl + H_2CO_3$$

### Solution 4.16

Since the question involves normality, Equation (4.40) applies and we have the following:

$$\% Na_2CO_3 = \frac{liters_{HCl} \times N_{HCl} \times EW_{Na_2CO_3}}{\text{sample weight}} \times 100$$

There are two hydrogens accepted by sodium carbonate, so the equivalent weight is the formula weight divided by two, or 52.995 g per equivalent. Thus we have

$$\% \, Na_2CO_3 = \frac{0.03642 \, L \times 0.1167 \, \text{equivalents}/L \times 52.9945 \, g/\text{equivalent}}{0.5203 \, g} \times 100 = 43.29\%$$

# 4.8 Volumetric Glassware

As indicated in Section 4.1 (and as should be apparent from the discussion thus far in this chapter), titrimetric analysis methods heavily utilize solution chemistry, and therefore volumes of solutions are prepared, measured, transferred, and analyzed with some degree of frequency in this type of analysis. It should not be surprising that analytical laboratory workers need to be well versed in the selection and proper use of the glassware and devices used for precise volume measurement.

The three volumetric glassware products we will discuss are the volumetric flask, the pipet, and the buret. Let us study the characteristics of each type individually.

## 4.8.1 The Volumetric Flask

The container that is typically used for precise solution preparation is the volumetric flask. This container has a single calibration line placed in a narrow diameter neck. See Figure 4.3. The reason a narrow diameter neck is desirable is that the volume of solution can be controlled very precisely. Even a fraction of a drop of solvent gives a noticeable change in the position of the meniscus when preparing the solution. If one were to use a beaker or Erlenmeyer flask for this, it may take as much as several milliliters of solvent to cause noticeable change in the position of the meniscus. Thus, volumes of solutions measured with a volumetric flask are precise to four significant figures.

Conveniently, the volumetric flask can be purchased in a variety of sizes, from 5 mL up to several liters. Figure 4.4 shows some of the various sizes.

The calibration line is affixed so that the indicated volume is *contained* rather than delivered. Accordingly, the legend TC is imprinted on the base of the flask, thus marking the flask as a vessel "to contain" the volume indicated, as opposed to "to deliver" (TD) the volume. The reason this imprint is important is that a contained volume is different from a delivered volume, since a small volume of solution remains adhering to the inside wall of the vessel and is not delivered when the vessel is drained. If a piece of

**FIGURE 4.3** Left, a 1000-mL volumetric flask. Right, a close-up view of the neck of the flask showing the single calibration line.

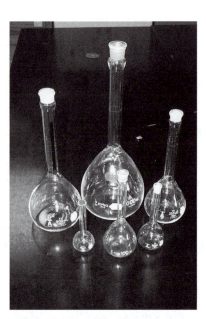

**FIGURE 4.4** Volumetric flasks come in a variety of sizes.

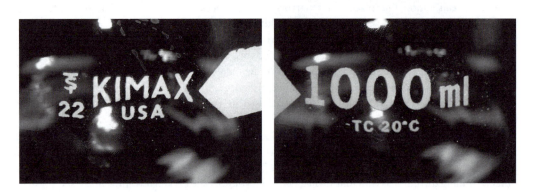

**FIGURE 4.5** Close-up views of the labels found on a 1000-mL volumetric flask. On the left, the stopper size (22) is evident, and on the right, besides the capacity of the flask (1000 mL), the TC designation and the 20°C specification are evident.

glassware is intended to deliver a specified volume, the calibration must obviously take this small volume into account in the sense that it *will not* be part of the *delivered* volume. On the other hand, if a piece of glassware is not intended to deliver a specified volume, but rather to contain the volume, the calibration must take this small volume into account in the sense that it *will* be part of the *contained* volume. Other pieces of glassware, namely most pipets and all burets, are TD vessels, meaning that they are calibrated to deliver. On the right in Figure 4.5, a close-up photograph of the base of a volumetric flask clearly shows the TC imprint (just below the flask's capacity, 1000 mL).

Notice the other markings on the base of the flask in Figure 4.5. The imprint 20°C indicates that the flask is calibrated to contain the indicated volume when the temperature is 20°C, which is a standard temperature of calibration. This marking is needed since the volume of liquids and liquid solutions changes slightly with temperature. For highest accuracy, the temperature of the contained fluid should be adjusted to 20°C. Notice the 22 marking visible in the left photograph in Figure 4.5. This refers to the size of the tapered top found on some flasks. The stopper that is used can be either a ground glass stopper, to match the flask opening, or a plastic tapered stopper, which can also be used in a ground glass opening. It has an identical numerical imprint on it (22 in this case), and such number designations on these two

**FIGURE 4.6**   A tapered ground glass stopper may be used with a flask with a ground glass opening. A tapered plastic stopper may also be used.

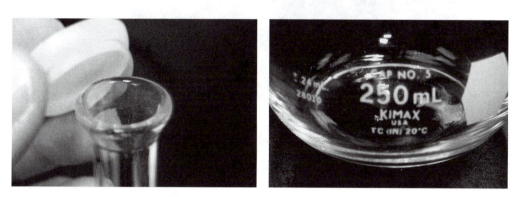

**FIGURE 4.7**   Left, the top of a volumetric flask designed for a snap cap. Right, the imprint on the base of the flask indicating the size of snap cap required (#5).

items must match to indicate that the stopper is the correct size. Figure 4.6 shows a ground glass stopper being inserted into the ground glass opening.

The opening may not necessarily be designed for a tapered stopper as in this example. It is fairly common for a flask to be designed for a snap cap rather than a ground glass stopper. In this case, neither the top of the flask is tapered, nor is it a ground glass opening. Rather, it has an unusually large lip around the opening, over which the snap cap is designed to seal. A number designation is usually used, as with the tapered opening, to indicate the size of the opening and the size of the cap required. This number is usually found on both the flask base and the cap, as in the taper design. The analyst should always be aware of the type and size of cap required for a given volumetric flask. If any amount of solution should leak out due to an improperly sealed cap while shaking before the solution is homogeneous, its concentration cannot be trusted. Figure 4.7 shows a volumetric flask with the large lip being fitted with a snap cap and the imprint on the flask indicating the size of the cap required.

Some volumetric glassware products have a large A imprint on the label. This designates the item as a class A item, meaning that more stringent calibration procedures were undertaken when it was manufactured. Class A glassware is thus more expensive, but it is most appropriate when highly precise work is important. This imprint may be found on both flasks and pipets.

# WORKPLACE SCENE 4.2

At Molex, Inc., Lincoln, Nebraska, various electronic connectors, such as telephone jacks, are gold-plated for optimum electrical conductivity. In the laboratory, not only must the gold concentration in the plating baths be monitored, but the amount of gold deposited on the connectors must be routinely determined as a quality check. In the case of the concentration of gold in a plating bath, an aliquot of each plating bath is diluted to the mark of a volumetric flask with water so as to have an appropriate concentration for the instrumental analysis technique used. In the case of the quality check, the gold from an electroplated specimen is dissolved in aqua regia and then also diluted to the mark of a volumetric flask. Needless to say, this laboratory utilizes volumetric flasks heavily in this work.

Clayton Allsman of Molex, Inc., is dwarfed by large volumetric flasks filled with diluted solutions of plating baths and dissolved electroplated specimens as he prepares to analyze these solutions for gold content.

One problem that exists with the volumetric flask, because of its unique shape, is the difficulty in making prepared solutions homogeneous. When the flask is inverted and shaken, the solution in the neck of the flask is not agitated. Only when the flask is set upright again is the solution drained from the neck and mixed. A good practice is to invert and shake at least a dozen times to ensure homogeneity.

Volumetric flasks should *not* be used to prepare solutions of reagents that can etch glass (such as sodium hydroxide and hydrofluoric acid), since if the glass is etched, its accurate calibration is lost. Volumetric flasks should *not* be used for storing solutions. Their purpose is to prepare solutions accurately. If they are used for storage, then they are not available for their intended purpose. Finally, volumetric flasks should *not* be used to contain solutions when heating or performing other tasks for which their accurate calibration serves no useful purpose. There are plenty of other glass vessels to perform these functions. See Workplace Scene 4.2.

## 4.8.2 The Pipet

As indicated previously, most pipets are pieces of glassware that are designed to deliver (TD) the indicated volume. Pipets come in a variety of sizes and shapes. The most common is probably the **volumetric pipet**, or **transfer pipet**, shown in Figure 4.8. This pipet, like the volumetric flask, has a single calibration

**TABLE 4.1** Steps Involved in Transferring a Solution with a Volumetric or Transfer Pipet

| | |
|---|---|
| Step 1 | If the pipet has been stored filled with water, invert and drain through the top. Wipe the outside dry, and with a rubber pipet bulb (Figure 4.9), blow out any water on the inside that does not drain out naturally. Again wipe the area around the tip; dry with a paper towel. You may have to blow and wipe again (or several times) to rid the inside of as much wetness as possible. This will help minimize the potential problem of wetness from the inside leaking into the original container of the solution being transferred. If the pipet has been in dry storage, this is not a problem. |
| Step 2 | Evacuate the pipet bulb (by squeezing), and while keeping it evacuated, seat the bulb opening over the top opening of the pipet. |
| Step 3 | If the pipet is wet on the inside, such as with storage water, immerse the tip of the pipet into the solution to be transferred while simultaneously releasing the squeezing pressure to immediately draw the solution in to about half the pipet's capacity. This immediate simultaneous release helps ensure that none of the wetness from the inside leaks into the original solution container. Empty by inverting and draining into a sink through the top. Repeat this rinsing step at least three times. If the inside of the pipet is dry, it should be rinsed with the solution; the danger of contaminating the solution in the original container is not a problem. |
| Step 4 | Fill the pipet to well past the calibration line by releasing the squeezing pressure, as in step 4. Reevacuate the bulb if necessary. |
| Step 5 | Quickly remove the bulb and seal the top of the pipet with your index finger. |
| Step 6 | Keeping your index finger in place, remove the tip from the solution and wipe with a towel. To avoid contamination from the towel, tilt the pipet to a 45° angle so that a small volume of air is drawn into the tip before wiping. |
| Step 7 | Slowly release your finger to adjust the meniscus to the calibration line. |
| Step 8 | When the meniscus is at the calibration line, stop it there and touch the tip to the *outside* of the receiving vessel to remove any solution that may be suspended there. |
| Step 9 | Touch the tip to the *inside* of the receiving vessel and completely release the finger. The tip should stay in contact with the inside wall of the receiving vessel until the end of the delivery. Be careful not to shake the pipet so that some solution is lost and an air bubble appears in the tip before releasing your finger. |
| Step 10 | When draining is complete, give the pipet a half-twist and remove the pipet from the receiving vessel. |

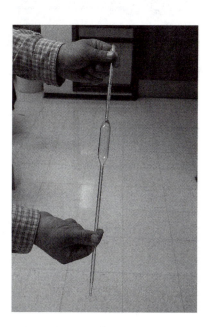

**FIGURE 4.8** A volumetric, or transfer, pipet.

line. It can thus be used in delivering only rather common volumes, meaning whole-number volumes. The correct use of a volumetric pipet is outlined in Table 4.1.

Summarizing the steps in the table, we can state the following. The bottom tip of the pipet is placed into the solution to be transferred. A rubber pipet bulb (Figure 4.9) is evacuated, sealed over the top of the pipet, and slowly released. The solution is drawn up into the pipet as a result of the vacuum. When the level of the

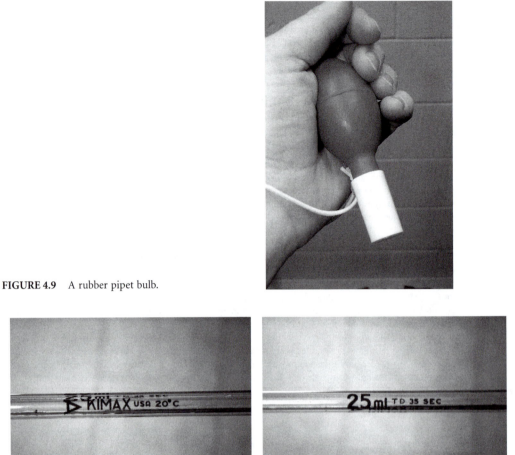

**FIGURE 4.9**    A rubber pipet bulb.

**FIGURE 4.10**    The top portion of a 25-mL volumetric pipet showing the labels indicating that it is class A (left) and that it has a run-down time of 35 sec (right).

solution has risen above the calibration line, the bulb is quickly removed from the pipet and replaced with the index finger.

With the index finger firmly in place to prevent the solution from draining out, the tip of the pipet is removed from the solution and wiped with a towel. The pressure exerted with the index finger is then used to adjust the bottom of the meniscus to coincide with the calibration line. The tip of the pipet is contacted to the inside wall of the receiving vessel, and the finger is released. The solution is thus drained into the receiving vessel. *Under no circumstances should this last drop in a volumetric pipet be blown out into the receiving vessel with the bulb.* The volumetric pipet is not calibrated for blowout. Also, the tip of the pipet should never contact the solution in the receiving vessel.

For precise work, volumetric pipets that are labeled class A have a certain time in seconds imprinted near the top (see Figure 4.10); this is the time that should be allowed to elapse from the time the finger is released until the pipet is completely drained. The reason for this is that the film of solution adhering to the inner walls will continue to slowly run down with time, and the length of time one waits to terminate the delivery thus becomes important. The intent with class A pipets, then, is to take this run-down time into account by terminating the delivery in the specified time. After this specified time has elapsed, the delivery is terminated.

Several additional styles of pipets other than the volumetric pipet are available. Two of these are shown in Figure 4.11. These are pipets that have many graduation lines, much like a buret, and are called **measuring pipets**. They are used whenever odd volumes are needed. There are two types of measuring pipets: the **Mohr pipet** (left in Figure 4.11) and the **serological pipet** (right). The difference is whether or not the calibration lines stop short of the tip (Mohr pipet) or go all the way to the tip (serological pipet). The serological pipet is better in the sense that the meniscus need be read only once, since the solution can be allowed to drain completely out. In this case, the last drop of solution is blown out with the pipet bulb. With the Mohr pipet, the meniscus must be read twice—once before the delivery (such as on the top graduation line) and again after the delivery is complete. The solution flow out of the pipet must be halted at the correct calibration line, and the error associated with reading a meniscus is doubled since the meniscus must be read twice. The delivery of 4.62 mL, for example, is done with a Mohr pipet in Figure 4.12(a) and with a serological pipet in Figure 4.12(b).

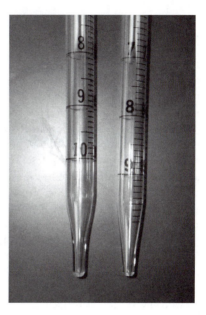

**FIGURE 4.11** The tips of two styles of measuring pipets. The Mohr pipet is shown on the left, and the serological pipet on the right. The graduation lines on the Mohr pipet stop short of the tip, but on the serological pipet, pass through the tip.

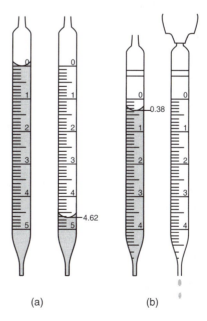

**FIGURE 4.12** The delivery of 4.62 mL of solution (a) with a Mohr pipet the meniscus is read twice and (b) with a serological pipet the meniscus can be read just once (at 0.38 mL) and the solution drained and blown out.

(a)  (b)

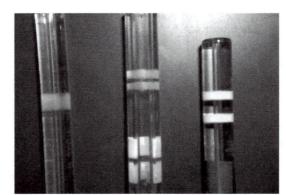

**FIGURE 4.13** The tops of three pipets calibrated for blowout showing either a single ring (left) or double ring (center and right) circumscribing the tops of the pipets.

It should be stressed that with the serological pipet, every last trace of solution capable of being blown out must end up inside the receiving vessel. Some analysts find that this is more difficult and that it perhaps introduces more error than reading the meniscus twice with a Mohr pipet. For these reasons, these analysts prefer to use a serological pipet as if it were a Mohr pipet. It is really a matter of personal preference. A double- or single-frosted ring circumscribing the top of a pipet (above the top graduation line) indicates that the pipet is calibrated for blowout (Figure 4.13).

**Disposable serological pipets** are available. They are termed disposable because the calibration lines are not necessarily permanently affixed to the outside wall of the pipet. The calibration process is thus less expensive, resulting in a less expensive product that can be discarded after use.

Some pipets are calibrated TC. Such pipets are used to transfer unusually viscous solutions such as syrups, blood, etc. With such solutions, the wetness remaining inside after delivery is a portion of the sample and would represent a significant nontransferred volume, which translates into a significant error by normal TD standards. With TC pipets, the calibration line is affixed at the factory so that every trace of solution contained within is transferred by flushing the solution out with a suitable solvent. Thus, the pipetted volume is contained within and then quantitatively flushed out. Such a procedure would actually be acceptable with any TC glassware, including the volumetric flask. Obviously, diluting the solution in the transfer process must not adversely affect the experiment.

### 4.8.3 The Buret

The buret has some unique attributes and uses. One could call it a specialized graduated cylinder, having graduation lines that increase from top to bottom, with a usual precision of ±0.01 mL, and a stopcock at the bottom for dispensing a solution. There are some variations in the type of stopcock that warrant some discussion. The stopcock itself, as well as the barrel into which it fits, can be made of either glass or Teflon™. Some burets have an all-glass arrangement, others have a Teflon stopcock and a glass barrel, and yet others have the entire system made of Teflon. These three types are pictured in Figure 4.14.

When the all-glass system is used, the stopcock needs to be lubricated so that it will turn with ease in the barrel. There are a number of greases on the market for this purpose. Of course, the grease must be inert to chemical attack by the solution to be dispensed. Also, the amount of grease used should be carefully limited so that excess grease does not pass through the stopcock and plug the tip of the buret. Any material stuck in a buret tip can usually be dislodged with a fine wire inserted from the bottom when the stopcock is open and the buret full of solution. The Teflon stopcocks are free of the lubrication problem. The only disadvantage is that the Teflon can become deformed, and this can cause leakage.

The correct way for a right-handed person to position his or her hands to turn the buret's stopcock during a titration is shown in Figure 4.15 The natural tendency with this positioning is to pull the stopcock in as it is turned. This will prevent the stopcock from being pulled out, causing the titrant to bypass the

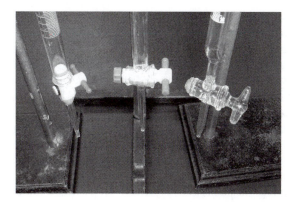

**FIGURE 4.14**   The three styles of buret stopcocks. Left, all Teflon; center, Teflon stopcock and glass barrel; right, all glass.

**FIGURE 4.15**   The correct way for a right-handed person to position his or her hands when performing a titration.

stopcock. The other hand is free to swirl the flask as shown. This may appear a bit clumsy at first, but it is the most efficient way to perform a titration, eliminating the novice's usual one-handed technique.

## 4.8.4   Cleaning and Storing Procedures

The use of clean glassware is of utmost importance when doing a chemical analysis. In addition to the obvious need of keeping the solution free of contaminants, the walls of the vessels, particularly the transfer vessels (burets and pipets), must be cleaned so that the solution will flow freely and not "bead up" on the wall as the transfer is performed. If the solution beads up, it is obvious that the pipet or buret is not delivering the volume of solution intended. It also means that there is a greasy film on the wall that could introduce contaminants. The analyst should examine, clean, and reexamine his or her glassware in advance so that the free flow of solution down the inside of the glassware can be observed. For the volumetric flask, at least the neck must be cleaned in this manner so as to ensure a well-formed meniscus.

Both hand washing and machine washing and drying procedures are in use in industrial analytical laboratories. The variety of available soaps include alkaline phosphate-based and phosphate-free soaps. While phosphate-based soaps are quite satisfactory for cleaning purposes because phosphate helps to soften hard tap water, concerns for the environment as well as for phosphate contamination from soap residues are important considerations. In any case, profuse rinsing with distilled water, often at an elevated temperature, is important. Sometimes, to neutralize the effect of alkaline cleaners, acid rinses are important. However, acid residue must also be removed by profuse rinsing with distilled water. Programmable washing machines are available that can cycle the glassware through all of the above procedures.

Hand washing typically involves the appropriate soap and a brush. For burets, a cylindrical brush with a long handle (buret brush) is used to scrub the inner wall. With flasks, a bottle or test tube brush is used to clean the neck. Also, there are special bent brushes available to contact and scrub the inside of the base of the flask.

Pipets pose a special problem. Brushes cannot be used because of the shape of some pipets and the narrowness of the openings. If soap is to be used, one must resort to soaking with a warm soapy water solution for a period of time proportional to the severity of the particular cleaning problem. Commercial soaking and washing units are available for this latter technique. Soap tablets are manufactured for such units and are easy to use.

In the past, chromic acid solutions have been used for cleaning. These solutions consist of concentrated sulfuric acid in which solid potassium dichromate has been dissolved. Because of safety concerns and concerns with chromium contamination in the environment, chromic acid has been essentially eliminated from use.

Once the glassware has been cleaned (by whatever method), one should also take steps to keep it clean. One technique is to rinse the items thoroughly with distilled water and then dry them either in the open air or in a dryer. Following this, they are cooled and stored in a drawer. For shorter time periods, it may be convenient to store them in a soaker under distilled water. This prevents their possible recontamination during dry storage. When attempting to use a soaker-stored pipet or buret, however, it must be remembered that a thin film of water is present on the inner walls; this must be removed by rinsing with the solution to be transferred, being careful not to contaminate the solution in the process.

# 4.9   Pipetters, Automatic Titrators, and Other Devices

## 4.9.1   Pipet Fillers

There are alternatives to the rubber bulb mentioned earlier (Figure 4.9 and Table 4.1) as the means for filling a glass pipet. Analysts, especially novices, often find the rubber bulb rather clumsy. One alternative is the plastic filler–dispenser shown in Figure 4.16. This device fits snugly over the top of the pipet, and the solution is drawn in by manually rotating a plastic wheel (shown next to the person's thumb in the figure). The solution may also be dispensed with this device by rotating the wheel in the direction opposite from filling. Various electrical and battery-operated pipet fillers also exist. With these, filling and dispensing are as easy as the push of a button. Some of these even draw in a preset volume at a preset rate.

## 4.9.2   Pipetters

Alternatives to the glass pipets themselves, called **pipetters**, are available and used rather extensively. Such devices are positive displacement devices, meaning that 100% of the liquid drawn in is forced out with a plunger, rather than allowed to flow out by gravity. Volumes transferred using such devices usually do not exceed 10 mL and are most often used for very small volumes, such as from 0 to 500 $\mu$L. Those used for microliter volumes are appropriately called **micropipets** or **micropipetters**. The typical pipetter employs a bulb concealed within a plastic fabricated body, a spring-loaded push button at the top, and a nozzle at the bottom for accepting a plastic disposable tip. They may be fabricated for either single or variable volumes. In the latter case, a ratchet-like device with a digital volume scale is used to "dial in" the desired volume. An example is shown in Figure 4.17.

A term often used in reference to such devices is **repeater pipetters** because of the ease with which a given volume can be repeatedly transferred. In addition, some devices use multiple tips for transferring a given volume up to eight times in just one stroke. These are especially useful in conjunction with well

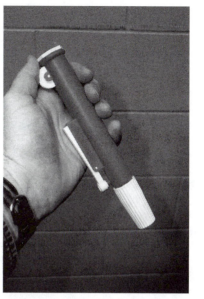

**FIGURE 4.16** A less clumsy alternative to the rubber pipet bulb.

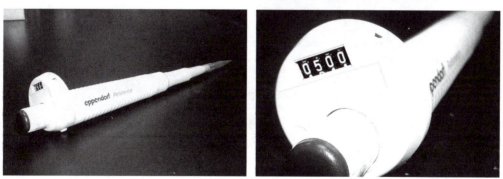

**FIGURE 4.17** A typical commercial pipetting device. The close-up on the right shows that it is "dialed in" to deliver 500 μL, or 0.5 mL.

plates, which often have 96 small depressions (wells) into each of which a given volume must be transferred.

### 4.9.3 Bottle-Top Dispensers

Bottle-top dispensers, often called **Repipets**™*, are quite popular. These are devices that fit on the top of reagent bottles threaded to receive screw caps. The dispensers themselves have screw caps that screw onto the bottles. The caps, however, are fitted with a hand pump with a plunger that draws liquid from the bottle in the upstroke and then dispenses a calibrated volume on the downstroke through a glass tip. Such devices are convenient and help prevent contamination of the reagent from the various pipets that the analyst might use for a transfer.

### 4.9.4 Digital Burets and Automatic Titrators

There are also various devices that are commonly used for titrations in place of the glass burets previously described. A digital buret, for example, is an electronically controlled bottle-top dispenser that delivers 0.01-mL increments from a reagent bottle containing the titrant. There are also automatic titrators, such

---

*"Repipet" is a trademark of Barnstead-Thermolyne Company.

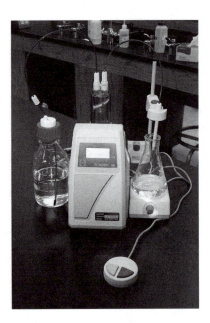

**FIGURE 4.18**    An automatic titrator. A pump draws the titrant from the reagent bottle on the left and fills the reservoir in the back of the unit. Pressing a key on the keypad in the foreground delivers titrant to the titration flask on the right as the solution is automatically stirred. The volume delivered is displayed on a digital readout.

as that shown in Figure 4.18. Such titrators draw the titrant from a reagent bottle and store it in a built-in reservoir fitted with a plunger. The titration is performed by pressing a key on the keypad. One tap of the key delivers 0.01 mL to the titration flask. See Workplace Scene 4.3.

# 4.10   Calibration of Glassware and Devices

It is important to remember that whatever volume measurement device is used, the analyst depends on it to deliver the indicated volume, i.e., it must be properly calibrated. Analytical chemists and technicians often depend on class A glassware to represent accurate calibration and hence do not usually feel the need for independent calibration. However, where non-class A glassware and alternative transfer devices are used, there is legitimate concern over their proper calibration.

The calibration of glassware and devices involves the measurement of the weight of a delivered volume, since laboratory workers can be quite confident in the accuracy of weight measurements in the laboratory because of the common use of calibrated analytical balances. The usual procedure is to weigh an empty glass weighing bottle, dispense the volume of water into the bottle, and then weigh it again. The weight of the water is then converted to volume using the known density of water at the temperature of the water. If the volume is not within the accepted precision of the device, the analyst concludes that the device is out of calibration and is discarded or returned to the manufacturer.

# 4.11   Analytical Technique

Previously, we discussed the need to know when a weight measurement does and does not need to be precise so that the appropriate balance is chosen for the measurement. We now repeat many of our previous comments, but for volume measurements rather than weight.

The question of which piece of glassware to use for a given procedure, such as for solution preparation (pipet or graduated cylinder, beaker or volumetric flask), remains. An analyst must be able to recognize when a high degree of precision is and is not important so that precise volumetric glassware is used. Pipets, burets, and volumetric flasks are used when precision is important; regular glassware—graduated cylinders and beakers—is used when precision is not important.

First, volumetric glassware should *not* be used for any volume measurement when the overall objective is strictly qualitative or when quantitative results are to be reported to two significant figures or less.

# WORKPLACE SCENE 4.3

In the laboratory for the electroplating facility at Molex, Inc., Lincoln, Nebraska, seventeen plating baths set up for tin and tin–lead electroplating must be tested three times daily for acid content. The procedure involves an acid–base titration using standard sodium hydroxide as the titrant. Because the volume of samples is so large, an automatic bottle-top buret is used with a 2-gal bottle filled with the standard sodium hydroxide solution.

Clayton Allsman, a laboratory technician for Molex, Inc., titrates a plating bath sample for acid content using an automatic bottle-top buret.

Whether the volume measurements are for preparing solutions to be used in such a procedure, for transferring an appropriate volume of solution or solvent for such an analysis, etc., they need not be accurate since the outcome will be either only qualitative or not necessarily accurate. Such volumes should, however, be measured with other marked glassware, such as graduated cylinders or marked beakers and flasks.

Second, if the results of a quantitative analysis are to be reported to three or more significant figures, then volume measurements that enter directly into the calculation of the results should be made with volumetric glassware so that the accuracy of the analysis is not diminished when the calculation is performed.

Third, if a volume measurement is only incidental to the overall result of an accurate quantitative analysis, then it need not be precise. This means that if the volume measurement to be performed has no bearing whatsoever on the quantity of the analyte tested, but is only needed to support the chemistry or other factor of the experiment, then it need not be precise.

Fourth, if a volume measurement does directly affect the numerical result of an accurate quantitative analysis (in a way other than entering directly into the calculation of the results), then volumetric glassware must be used.

These last two points require that the analyst carefully consider the purpose of the volume measurement and whether it will directly affect the quantity of analyte tested, the numerical value to be reported. It is often true that volume measurements taken during such a procedure need not be precise even though the analytical results are to be precisely reported.

# Experiments

## Experiment 8: Preparation and Standardization of HCl and NaOH Solutions

*Remember safety glasses.*

1. For this experiment you will need a minimum of 3 to 4 g of primary standard potassium hydrogen phthalate (KHP) for three titrations. Place at least 6 g of it in a weighing bottle and dry in a drying oven for 2 h. Your instructor may choose to dispense this to you.

2. Prepare $CO_2$-free water for a 0.10 $M$ NaOH solution by boiling 1000 mL of distilled water in a covered beaker on a hot plate. While waiting for this water to boil, go on to step 3.

3. Prepare 1 L of 0.10 $M$ HCl by diluting the appropriate volume of concentrated HCl (12.0 $M$). Use a 1-L glass bottle and fill half of it with water before adding the concentrated acid. Use a graduated cylinder to measure the acid. Add more water to have about 1 L, shake well, and label.

4. Once the water from step 2 has boiled, remove from the heat and cool so that it is only warm to the touch. This can be accomplished by immersing the beaker in cold water (such as in a stoppered sink).

5. Using this freshly boiled water, prepare 1000 mL of a 0.10 $M$ NaOH solution. Weigh out the appropriate number of grams of NaOH, and place it in a 1000-mL plastic bottle. Add the water to a level approximately equal to 1000 mL. Shake well to completely dissolve the solid. Allow to cool completely to room temperature before proceeding. Label the bottle.

6. Assemble the apparatus for a titration. The buret should be a 50-mL buret and should be washed thoroughly with a buret brush and soapy water. Clamp the buret to a ring stand with either a buret clamp or an ordinary ring stand clamp. The receiving flask should be a 250-mL Erlenmeyer flask. You should clean and prepare three such flasks. Place a piece of white paper (a page from your notebook will do) on the base of the ring stand. This will help you see the end point better.

7. Give both your acid and base solutions one final shake at this point to ensure their homogeneity. Rinse the buret with 5 to 10 mL of NaOH twice, and then fill it to the top. Open the stopcock wide open to force trapped air bubbles from the stopcock and tip. Allow this excess solution to drain into a waste flask. Bring the bottom of the meniscus to the 0.00-mL line. Using a clean 25-mL pipet (volumetric), carefully place 25.00 mL of the acid solution into each of the three flasks. Add three drops of phenolphthalein indicator to each of the three flasks.

8. Titrate each acid sample with the base, one at a time. For the first sample, you can very easily make a good estimate of how much titrant should be required. Since both solutions are the same molar concentration and since the reaction is one-to-one in terms of moles (see Equation (4.4)), the volume of NaOH required should be approximately the same as the volume of HCl pipetted, 25.00 mL. You should be able to open the stopcock and allow about 20 mL of the base to enter the flask without having to worry about overshooting the end point. From that point on, however, you should proceed with caution so that the indicator changes color (from colorless to pink) upon the addition of the smallest amount of titrant possible: a fraction of a drop. At the point that you think the indicator will change color in this manner, rinse down the walls of the flask with water from a squeeze bottle. This will ensure that all of the titrant has had a chance to react with all of the substance titrated. This step can be performed more than once—addition of more water has no effect on the location of the end point. The fraction of a drop can then be added, if necessary, by (1) a very rapid rotation of the stopcock, or (2) slowly opening the stopcock, allowing the fraction of a drop to hang on the tip of the buret, and then washing it into the flask with water from a squeeze bottle. At the end point, a faint pink color will persist in the flask for at least 20 sec. Read the buret to four significant figures, record the reading, and repeat with the next sample.

9. All three of your titrations should agree to within at least 0.05 mL in order to be acceptable. This would mean a parts per thousand (ppt) relative standard deviation of less than approximately 2.5 ppt. If they do not agree in this manner, more titrations must be performed until you have three good answers to rely on or until you have a precision satisfactory to your instructor.

10. After the 2-h drying period for KHP has expired, remove it from the drying oven and allow it to cool to room temperature in your desiccator.

11. Prepare three solutions of KHP for titration. To do this, again clean three 250-mL Erlenmeyer flasks and give each a final thorough rinse with distilled water. Now, weigh into each flask, by difference, on the analytical balance, a sample of KHP weighing between 0.7 and 0.9 g. Add approximately 50 mL of distilled water to each and swirl to dissolve.

12. Add three drops of phenolphthalein indicator to each flask and titrate as before.

13. Calculate the normality of the NaOH solution from each set of KHP data, and then calculate the mean. Calculate the standard deviation. If your three normalities do not agree to within 2.5 ppt relative standard deviation, repeat until you have three that do or until you have a precision that is satisfactory to your instructor.

14. Calculate the exact normality of the HCl solution using the three volume readings from step 8 and the average NaOH normality from step 13. Compute the average of these three results.

15. Record in your notebook at least a sample calculation for each of these standardizations and all results.

## Experiment 9: Relationship of Glassware Selection to Variability of Results

1. Prepare a stock solution of red food coloring by mixing 0.02 g of Food, Drug, and Cosmetic (FD&C) #33 dye in 250 mL of water solution.

2. Prepare five identical dilutions of the stock solution prepared in step 1 by diluting 10 mL of the solution to 100 mL. Use a clean 10-mL volumetric pipet and a clean 100-mL volumetric flask, and make the measurement and transfer as carefully as possible (see Table 4.1). Label each as 2/1, 2/2, etc., to indicate that they are solutions 1 to 5 prepared in step 2.

3. Repeat step 2, but use a clean class A 10-mL graduated cylinder instead of the 10-mL volumetric pipet to measure the 10 mL of stock solution. Again, make each measurement and transfer as carefully as possible. Label each as 3/1, 3/2, etc., to indicate that they are solutions 1 to 5 prepared in step 3.

4. Repeat step 2 again, but use a larger graduated cylinder (25 or 50 mL) to measure the 10 mL of stock solution. Label as 4/1, 4/2, etc., to indicate that they are solutions 1 to 5 prepared in step 4.

5. Prepare a spectrophotometer to measure absorbance at 490 nm. If you are not familiar with the spectrophotometer, your instructor can assist you. Use distilled water for a blank. Measure the absorbance at 490 nm of each of the solutions.

6. Calculate the mean and standard deviation for each of the data sets. Which method of preparing the solutions results in the least variability?

# Questions and Problems

1. How is titrimetric analysis different from gravimetric analysis?

2. Why is titrimetric analysis sometimes called volumetric analysis?

3. Compare Figure 3.1 with Figure 4.1 and tell how the analytical strategy for gravimetric analysis differs from that for titrimetric analysis.

4. Define standard solution, volumetric flask, volumetric pipet, standardization, buret, stopcock, titration, titrant, substance titrated, equivalence point, indicator, end point, and 9automatic titrator.

5. What is the molarity of the following?
   (a) 0.694 mol dissolved in 3.55 L of solution
   (b) 2.19 mol of NaCl dissolved in 700.0 mL of solution
   (c) 0.3882 g of KCl dissolved in 0.5000 L of solution
   (d) 1.003 g of $CuSO_4 \cdot 5\ H_2O$ dissolved in 250.0 mL of solution
   (e) 30.00 mL of 6.0 $M$ NaOH diluted to 100.0 mL of solution
   (f) 0.100 L of 12.0 $M$ HCl diluted to 500.0 mL of solution

6. What is the equivalent weight of both reactants in each of the following?
    (a) $NaOH + HCl \rightarrow NaCl + H_2O$
    (b) $2\ NaOH + H_2SO_4 \rightarrow Na_2SO_4 + 2\ H_2O$
    (c) $2\ HCl + Ba(OH)_2 \rightarrow BaCl_2 + 3\ H_2O$
    (d) $3\ NaOH + H_3PO_4 \rightarrow Na_3PO_4 + 3\ H_2O$
    (e) $2\ HCl + Mg(OH)_2 \rightarrow MgCl_2 + 2\ H_2O$
    (f) $2\ NaOH + H_3PO_4 \rightarrow Na_2HPO_4 + 2\ H_2O$
    (g) $NaOH + Na_2HPO_4 \rightarrow Na_3PO_4 + H_2O$
    (h) $NaOH + H_3PO_4 \rightarrow NaH_2PO_4 + H_2O$
    (i) $Na_2CO_3 + 2\ HCl \rightarrow 2\ NaCl + H_2CO_3$

7. Calculate the normality of the following solutions:
    (a) 0.238 equivalents of an acid dissolved in 1.500 L of solution.
    (b) 1.29 mol of sulfuric acid dissolved in 0.5000 L of solution used for the following reaction:

$$H_2SO_4 + 2\ NaOH \rightarrow Na_2SO_4 + 2\,H_2O$$

   (c) 0.904 mol of $H_3PO_4$ dissolved in 250.0 mL of solution used for the following reaction:

$$H_3PO_4 + 3\ KOH \rightarrow K_3PO_4 + 3\,H_2O$$

   (d) 0.827 mol of $Al(OH)_3$ dissolved in 0.2500 L of solution used for the following reaction:

$$3\ HCl + Al(OH)_3 \rightarrow AlCl_3 + 3\,H_2O$$

   (e) 1.38 g of KOH dissolved in 500.0 mL of solution used for the chemical reaction in part (c) above.
   (f) 2.18 g of $NaH_2PO_4$ dissolved in 1.500 L used for the following reaction:

$$NaH_2PO_4 + Al(OH)_3 \rightarrow Al(NaHPO_4)_3 + H_2O$$

   (g) 0.728 g of $KH_2PO_4$ dissolved in 250.0 mL of solution used for the following reaction:

$$KH_2PO_4 + Ba(OH)_2 \rightarrow KBaPO_4 + 2\,H_2O$$

8. An $H_3PO_4$ solution is to be used to titrate an NaOH solution as in the equation in the following reaction. If the normality of the $H_3PO_4$ solution is 0.2411 $N$, what is its molarity?

$$H_3PO_4 + 2\ KOH \rightarrow KH_2PO_4 + 2\,H_2O$$

9. Tell how you would prepare each of the following:
    (a) 500.0 mL of a 0.10 $M$ solution of KOH from pure solid KOH
    (b) 250.0 mL of a 0.15 $M$ solution of NaCl from pure solid NaCl
    (c) 100.0 mL of a 2.0 $M$ solution of glucose from pure solid glucose $(C_6H_{12}O_6)$
    (d) 500.0 mL of a 0.10 $M$ solution of HCl from concentrated HCl, which is 12.0 $M$
    (e) 100.0 mL of a 0.25 $M$ solution of NaOH from a solution of NaOH that is 2.0 $M$
    (f) 2.0 L of a 0.50 $M$ solution of sulfuric acid from concentrated sulfuric acid, which is 18.0 $M$
10. Tell how you would prepare each of the following:
    (a) 500.0 mL of 0.20 $N$ $KH_2PO_4$ used for the following reaction:

$$KH_2PO_4 + 2\ NaOH \rightarrow KNa_2PO_4 + 2\ H_2O$$

   (b) 500.0 mL of 0.11 $N$ $H_2SO_4$ from concentrated $H_2SO_4$ (18.0 $M$) used for the following reaction:

$$H_2SO_4 + Ca(OH)_2 \rightarrow CaSO_4 + 2\ H_2O$$

   (c) 750.0 mL of 0.11 $N$ $Ba(OH)_2$ from pure solid $Ba(OH)_2$ given the following reaction:

$$2\ Na_2HPO_4 + Ba(OH)_2 \rightarrow Ba(Na_2PO_4)_2 + 2\ H_2O$$

(d)  200.0 mL of a 0.15 $N$ solution of the base in the following reaction:

$$2\ HBr + Na_2CO_3 \rightarrow 2\ NaBr + H_2O + CO_2$$

(e)  700.0 mL of a 0.25 $N$ solution of the acid in the following reaction:

$$2\ NaHCO_3 + Mg(OH)_2 \rightarrow Mg(NaCO_3)_2 + 2\ H_2O$$

(f)  700.0 mL of a 0.30 $N$ solution of $Ba(OH)_2$ from a 15.0 $N$ solution of $Ba(OH)_2$ used for the following reaction:

$$2\ H_3PO_4 + Ba(OH)_2 \rightarrow Ba(H_2PO_4)_2 + 2\ H_2O$$

(g)  300.0 mL of 0.15 $N$ solution of $H_3PO_4$ from concentrated $H_3PO_4$ (15 $M$) used for the following reaction:

$$H_3PO_4 + Al(OH)_3 \rightarrow AlPO_4 + 3\ H_2O$$

11.  How would you prepare 250.0 mL of a 0.35 $M$ solution of NaOH using
   (a)  A bottle of pure, solid NaOH?
   (b)  A solution of NaOH that is 6.0 $M$?
12.  How many milliliters of a KOH solution, prepared by dissolving 60.0 g of KOH in 100.0 mL of solution, are needed to prepare 450.0 mL of a 0.70 $M$ solution?
13.  How many milliliters of a KCl solution, prepared by dissolving 45 g of KCl in 500.0 mL of solution, are needed to prepare 750.0 mL of a 0.15 $M$ solution?
14.  A solution of $KNO_3$, 500.0 mL with 0.35 $M$ concentration, is needed. Tell how you would prepare this solution
   (a)  From pure, solid $KNO_3$.
   (b)  From a solution of $KNO_3$ that is 4.5 $M$.
15.  To prepare a certain solution, it is determined that acetic acid must be present at 0.17 $M$ and that sodium acetate must be present at 0.29 $M$, both in the same solution. If the sodium acetate to be used to prepare this solution is a pure solid chemical and the acetic acid to be used is a concentrated solution (17 $M$), how would you prepare 500.0 mL of this buffer solution?
16.  How many milliliters of a $NaH_2PO_4$ solution, prepared by dissolving 0.384 g in 500.0 mL of solution, are needed to prepare 1.000 L of a 0.00200 $N$ solution given the following reaction?

$$NaH_2PO_4 + Ca(OH)_2 \rightarrow CaNaPO_4 + 2\ H_2O$$

17.  Suppose 0.7114 g of KHP was used to standardize a $Mg(OH)_2$ solution, as in the following reaction:

$$Mg(OH)_2 + 2\ KHC_8H_4O_4 \rightarrow Mg(KC_8H_4O_4)_2 + 2\ H_2O$$

If 31.18 mL of $Mg(OH)_2$ was needed, what is the molarity of $Mg(OH)_2$?

18.  A NaOH solution was standardized against a $H_3PO_4$ solution, as in the following reaction:

$$H_3PO_4 + 3NaOH \rightarrow Na_3PO_4 + 3\ H_2O$$

If 25.00 mL of 0.1427 $M$ $H_3PO_4$ required 40.07 mL of NaOH, what is the molarity of NaOH?

19.  A solution of KOH is standardized with primary standard KHP ($KHC_8H_4O_4$). If 0.5480 g of KHP exactly reacted with 25.41 mL of the KOH solution, what is the molarity of KOH?

$$KOH + KHP \rightarrow K_2P + H_2O$$

20.  What is the normality of a solution of HCl, 35.12 mL of which was required to titrate 0.4188 g of primary standard $Na_2CO_3$?

$$2\ HCl + Na_2CO_3 \rightarrow 2\ NaCl + CO_2 + H_2O$$

21. What is the normality of a solution of sulfuric acid that was used to titrate a 0.1022 *N* solution of KOH, as in the following reaction, if 25.00 mL of the base was exactly neutralized by 29.04 mL of the acid?

$$H_2SO_4 + 2\ KOH \rightarrow K_2SO_4 + 2\ H_2O$$

22. Primary standard tris-(hydroxymethyl)amino methane, also known as THAM or TRIS (FW = 121.14 g/mol), is used to standardize a hydrochloric acid solution. If 0.4922 g of THAM is used and 23.45 mL of HCl is needed, what is the normality of HCl?

$$(HOCH_2)_3CNH_2 + HCl \rightarrow (HOCH_2)_3CNH_3{}^+\ Cl^-$$

23. Suppose a sulfuric acid solution, rather than the hydrochloric acid solution in question 22, is standardized with primary standard THAM. Does the calculation change in any way? Explain.
24. What is a primary standard?
25. What are three requirements of a primary standard?
26. What is an SRM? What is a CRM? What is NIST an acronym for?
27. Define titer.
28. What is the titer (expressed in milligrams per milliliter) of a solution of disodium dihydrogen ethylenediaminetetraacetate (EDTA) with respect to calcium carbonate if 17.29 mL of it was needed to titrate 0.0384 g of calcium carbonate?
29. What is the percent of $K_2HPO_4$ in a sample when 46.79 mL of 0.04223 *M* $Ca(OH)_2$ exactly neutralizes 0.9073 g of the sample according to the following equation?

$$2\ K_2HPO_4 + Ca(OH)_2 \rightarrow Ca(K_2PO_4)_2 + 2\ H_2O$$

30. What is the percent of $Al(OH)_3$ in a sample when 0.3792 g of the sample is exactly neutralized by 23.45 mL of 0.1320 *M* $H_3PO_4$ according to the following equation?

$$3\ H_3PO_4 + Al(OH)_3 \rightarrow Al(H_2PO_4)_3 + 3\ H_2O$$

31. What is the percent of $NaH_2PO_4$ in a sample if 24.18 mL of 0.1032 *N* NaOH was used to titrate 0.3902 g of the sample according to the following?

$$NaH_2PO_4 + 2\ NaOH \rightarrow Na_3PO_4 + 2\ H_2O$$

32. A 0.1057 *N* HCl solution was used to titrate a sample containing $Ba(OH)_2$. If 35.78 mL of HCl was required to exactly react with 0.8772 g of the sample, what is the percent of $Ba(OH)_2$ in the sample?

$$2\ HCl + Ba(OH)_2 \rightarrow BaCl_2 + 2\ H_2O$$

33. Tell which statements are true and which are false.
    (a) TC means "to control."
    (b) The volumetric pipet has graduation lines on it, much like a buret.
    (c) The volumetric flask is never used for delivering an accurate volume of solution to another vessel.
    (d) Volumetric pipets are not calibrated for blowout.
    (e) Two frosted rings found near the top of a pipet indicate that it is a Mohr pipet.
    (f) The volumetric flask has TD imprinted on it.
    (g) The volumetric pipet is a type of measuring pipet.
    (h) The serological pipet is a type of volumetric pipet.
34. Complete the following:
    (a) The kind of pipet that has graduations on it much like a buret is called the
        _____ pipet.
    (b) To say that a given pipet is not calibrated for blowout means that
        _____.

(c) Burets and pipets that are not dry should be rinsed first with the solution to be used because _____.

35. Which is more accurate, a Mohr pipet or a volumetric pipet? Why?
36. Should a 100-mL volumetric flask be used to measure out 100 mL of solution to be added to another vessel? Why or why not?
37. What does a frosted ring near the top of a pipet indicate?
38. How does a volumetric flask differ from an Erlenmeyer flask?
39. Explain the reasons for rinsing a pipet as directed in step 3 of Table 4.1.
40. With a Mohr pipet, the meniscus must be read twice, but with a serological pipet, the meniscus may be read only once, if desired. Explain this.
41. A student is observed using a 50-mL volumetric flask to "accurately" transfer 50 mL of a solution from one container to another. What would you tell the student (a) to explain his or her error? (b) to help him or her do the "accurate" transfer correctly?
42. Why must a pipet that has been stored in distilled water be thoroughly rinsed with the solution to be transferred before use?
43. A technician is directed to prepare accurately to four significant figures 100 mL of a solution of $Na_2CO_3$ that is around 0.25 $M$. The laboratory supervisor provides a solution of 4.021 $M$ $Na_2CO_3$ and some pure, solid $Na_2CO_3$ and tells the technician to proceed by whichever method is easier. Give specific details as to how you would prepare the solution by two different methods, one by dilution and one by weighing the pure chemical, including how many grams or milliliters to measure (show calculation), what type of glassware is used (include the size and kind of pipets and flasks, if applicable), and how the glassware is used.
44. For one of the methods in problem 43, you needed a pipet. Is the pipet you would choose calibrated for blowout? How can you tell by looking at the markings on the pipet? What is the name of the pipet you would choose? Give specific details as to how the volume is delivered using this pipet.
45. The concentration of solutions can be known accurately to four significant figures either directly through their preparation or by standardization.
    (a) If the solute is a pure solid, give specific instructions that would ensure such accuracy directly through its preparation.
    (b) What is meant by standardization?
46. Compare a volumetric pipet with a serological measuring pipet in terms of:
    (a) The number of graduation lines on the pipet
    (b) Whether it is calibrated for blowout
    (c) Which one to select if you need to deliver 3.72 mL
    (d) Whether it is calibrated TC or TD
47. Some pipets are calibrated TC. Why would one ever want a pipet calibrated TC rather than TD?
48. Consider an experiment in which a pure solid chemical is weighed on an analytical balance into an Erlenmeyer flask. The nature of the experiment is such that this substance must be dissolved before proceeding. The experiment calls for you to dissolve it in 75 mL of water. In the subsequent calculation of the results, which are to be reported to four significant figures, the weight measurement is needed but the 75-mL measurement is not. Should the 75 mL of water be measured with a pipet, or is a graduated cylinder good enough? Explain.

# 5
# Applications of Titrimetric Analysis

## 5.1  Introduction

The standardization and percent analyte examples in Chapter 4 (Examples 4.10 to 4.13, 4.15, and 4.16 and Experiment 8) involved acid–base reactions. In this chapter, we discuss the chemistry of these and others.

In order for a titrimetric analysis to be successful, the equivalence point must be easily and accurately detected, the reaction involved must be fast, and the reaction must be quantitative. If an equivalence point cannot be detected (i.e., if there is no acceptable indicator or other detection method), then the correct volume of titrant cannot be determined. If the reaction involved is not fast, the end point cannot be detected immediately upon adding the last fraction of a drop of titrant and there would be some doubt as to whether the end point had been reached. If the reaction is not quantitative—every trace of reactant in the titration flask is not consumed by the titrant at the end point—then again the correct volume of titrant cannot be determined. This latter point means that equilibrium reactions that do not go essentially to completion immediately are not acceptable reactions for this type of analysis. Thus, not all reactions are acceptable reactions.

In this chapter, we investigate individual types of reactions that meet all the requirements. We will also discuss back titrations, in which some of the limitations that we may encounter are solved. Our discussions in Chapter 4 involved acid–base reactions. These reactions as well as other applicable reactions will also be discussed here.

## 5.2  Acid–Base Titrations and Titration Curves

Various acid–base titration reactions are discussed in this section, including a number of scenarios of base in the buret, acid in the reaction flask, and vice versa, as well as various monoprotic and polyprotic acids titrated with strong bases and various weak monobasic and polybasic bases titrated with strong acids. A **monoprotic acid** is an acid that has only one hydrogen ion (or proton) to donate per fomula. Examples are hydrochloric acid, HCl, a strong acid, and acetic acid, $HC_2H_3O_2$, a weak acid. A **polyprotic acid** is an acid that has two or more hydrogen ions to donate per formula. Examples include sulfuric acid, $H_2SO_4$, a **diprotic acid**, and phosphoric acid, $H_3PO_4$, a **triprotic** acid.

A **monobasic base** is one that will accept just one hydrogen ion per formula. Examples include sodium hydroxide, NaOH, a strong base; ammonium hydroxide, $NH_4OH$, a weak base; and sodium bicarbonate, $NaHCO_3$, a weak base. A **polybasic base** is one that will accept two or more hydrogen ions per formula. Examples include sodium carbonate, $Na_2CO_3$, a **dibasic base**, and sodium phosphate, $Na_3PO_4$, a **tribasic base**.

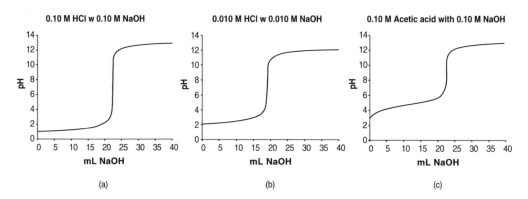

**FIGURE 5.1**  Acid–base titration curves: (a) 0.10 *M* HCl (strong acid) titrated with 0.10 *M* NaOH (strong base), (b) 0.010 *M* HCl titrated with 0.010 *M* NaOH, and (c) 0.10 *M* acetic acid (weak acid) titrated with 0.10 *M* NaOH.

## 5.2.1   Titration of Hydrochloric Acid

A graphic picture of what happens during an acid–base titration is easily produced in the laboratory. Consider again what is happening as a titration proceeds. Consider, specifically, NaOH as the titrant and HCl as the substance titrated. In the titration flask, the following reaction occurs when titrant is added:

$$H^+ + OH^- \rightarrow H_2O$$

As $H^+$ ions are consumed in the reaction flask by the $OH^-$ added from the buret, the pH of the solution in the flask will change, since pH = –log [$H^+$]. In fact, the pH should increase as the titration proceeds, since the number of $H^+$ ions decrease due to the reaction with $OH^-$. The lower the [$H^+$], the higher the pH. This increase in pH can be monitored with the use of a pH meter. Thus, if we were to measure the pH in this manner after each addition of NaOH and graph the pH vs. milliliters of NaOH added, we would have a graphical display of the experiment. Figure 5.1(a) shows the results of such an experiment for the case in which 0.10 *M* HCl is titrated with 0.10 *M* NaOH. The graph is called a **titration curve**. The sharp increase in the pH at the center of the graph occurs at the equivalence point of the titration, or the point at which all the acid in the flask has been neutralized by the added base. The point at which the slope of a titration curve is a maximum (a sharp change in pH, whether an increase or decrease) is called an **inflection point**.

Initial responses to this curve might be: Why the strange shape? Why the steady increase in the beginning? Why is there an inflection point? Why the steady increase at the end? An acid–base titration curve is not without theoretical foundation. The entire curve can also be recorded independent of the pH meter experiment by calculating the [$H^+$] and the pH after each addition and plotting the results. Such calculations and plotting are beyond our scope here, but they can be done. The calculations would indicate a steady increase in pH over a wide range of milliliters of NaOH added in the beginning. This would be followed by the inflection point, which in turn is followed by a steady increase again over a broad range—exactly what Figure 5.1(a) shows.

We should emphasize that the concentrations of the acid and base are important to consider. These concentrations in the discussions thus far have been 0.10 *M*: 0.10 *M* NaOH in the buret and 0.10 *M* HCl in the reaction flask. The lower the concentration of the acid, the fewer $H^+$ ions there are present, the higher the initial pH, and the higher the pH level of the initial steady increase. The lower the concentration of the base, the lower the level of the pH after the inflection point. See Figure 5.1(b).

## 5.2.2   Titration of Weak Monoprotic Acids

It is interesting to compare the case of a strong acid titrated with a strong base (in the last section) with the titration of a *weak* acid, such as acetic acid, with a strong base, such as NaOH. The difference between the weak acid–strong base case and the strong acid–strong base case just discussed is that for the same

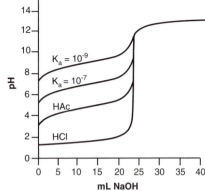

**FIGURE 5.2** A family of acid–base titration curves for a 0.10 $M$ strong acid (HCl) and three weak acids, as indicated (0.10 $M$ each), titrated with 0.10 $M$ NaOH (strong base). HAc is a representation of acetic acid.

0.10 $M$ concentrations, the pH starts and continues to the equivalence point at a higher pH level, in a manner similar to that in the lower concentration of HCl. See Figure 5.1(c). This should not be unexpected, since a solution of a weak acid is being measured, and once again there are fewer $H^+$ ions in the solution. Still weaker acids start at even higher pH values.

The weak acid curves can also be calculated. This involves the use of the equilibrium constant expression for a weak monoprotic acid ionization:

$$HA \rightleftharpoons H^+ + A^- \tag{5.1}$$

$$K_a = \frac{[H^+][A^-]}{[HA]} \tag{5.2}$$

In these equations, HA symbolizes a weak acid and $A^-$ symbolizes the anion of the weak acid. The calculations are beyond our scope. However, we can correlate the value of the equilibrium constant for a weak acid ionization, $K_a$, with the position of the titration curve. The weaker the acid, the smaller the $K_a$ and the higher the level of the initial steady increase. Figure 5.2 shows a family of curves representing several acids at a concentration of 0.10 $M$ titrated with a strong base. The curves for HCl and acetic acid (represented as HAc) are shown, as well as two curves for two acids even weaker than acetic acid. (The $K_a$'s are indicated.)

## 5.2.3 Titration of Monobasic Strong and Weak Bases

Now consider an alternative in which the acid is in the buret and the base is in the reaction flask. If the titration of 0.10 $M$ NaOH with 0.10 $M$ HCl (a strong base titrated with strong acid) is considered, we have a curve that starts at a high pH value (a solution of a base has a high pH) and ends at a low pH—just the opposite of that observed when titrating an acid with a base. See Figure 5.3(a). Likewise, the curves for 0.10 $M$ weak bases titrated with a strong acid, such as ammonium hydroxide titrated with HCl, start out at a lower pH than those for the strong base, just as those for 0.10 $M$ weak acids started out at a higher pH than those for the strong acid. See Figure 5.3(b). A family of curves for the titration of bases with acids is given in Figure 5.4. These curves, too, have a theoretical foundation and can be calculated.

## 5.2.4 Equivalence Point Detection

Before continuing with other examples, it is important to consider how the equivalence point in an acid–base titration is found and what relationship this has with titration curves. As we have said, the inflection point at the center of these curves occurs at the equivalence point, the point at which all of the substance titrated has been exactly consumed by the titrant. The exact position for this in the case

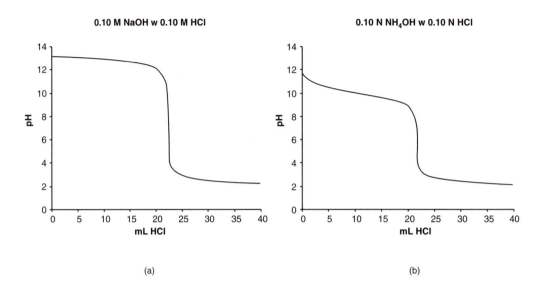

FIGURE 5.3   The titration curves for (a) 0.10 *N* NaOH titrated with 0.10 *N* HCl and (b) 0.10 *N* NH₄OH titrated with 0.10 *N* HCl.

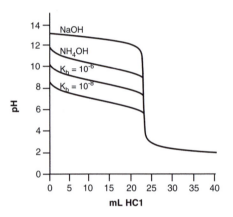

FIGURE 5.4   The family of titration curves for various bases titrated with a strong acid.

of strong acid titrated with strong base, and also in the case of strong base titrated with strong acid, is at pH = 7, exactly in the middle of the inflection point. For the case of weak acid with strong base, the equivalence point is again in the middle of the inflection point, but this occurs at a pH higher than 7 (refer to Figure 5.2). In fact, the weaker the acid, the higher the pH value that corresponds to the equivalence point. The opposite is observed in the case of weak base with strong acid (see Figure 5.4). The equivalence point occurs at progressively lower pH values, the weaker the base. The exact pH at these equivalent points can be calculated, as indicated previously.

The problem the analyst has is to choose indicators that change color close enough to an equivalence point so that the accuracy of the experiment is not diminished, which really means at any point during the inflection point. (Refer to Section 4.2 for the definitions of equivalence point and end point.) It almost seems like an impossible task, since there must be an indicator for each possible acid or base to be titrated. Fortunately, there are a large number of indicators available, and there is at least one available for all acids and bases, with the exception of only the extremely weak acids and bases. Figure 5.5 lists some of these indicators and shows the pH ranges over which they change color.

Thus, referring to Figure 5.5, in the case of HCl titrated with NaOH (Figure 5.1(a)), the following indicators would work: phenolphthalein, thymolphthalein, methyl red, methyl orange, etc.—virtually

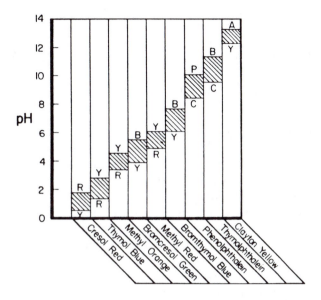

**FIGURE 5.5**   Some acid–base indicators and their color change ranges. R = red, Y = yellow, B = blue, P = pink, C = colorless, and A = amber.

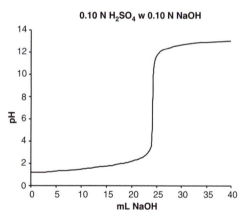

**FIGURE 5.6**   Titration curve of 0.10 $N$ $H_2SO_4$ titrated with 0.10 $N$ NaOH.

all of them in the list except for thymol blue and cresol red. As the acid becomes weaker and weaker, the pH range available for the indicator to change color becomes narrower and narrower, and a smaller number of indicators are useful. The color change must take place at the inflection point. The same observation is made when titrating bases with acids. For the titration of acetic acid, phenolphthalein, thymolphthalein, and perhaps bromthymol blue are useful, while for ammonium hydroxide (see Figure 5.4), methyl red, bromcresol green, etc., are useful.

## 5.2.5   Titration of Polyprotic Acids: Sulfuric Acid and Phosphoric Acid

Titrations curves for polyprotic acids have an inflection point for each hydrogen in the formula if the dissociation constant ($K_a$) for each hydrogen is very different from the others and if any dissociation constant is not too small. The titration curves of the polyprotic acids $H_2SO_4$ and $H_3PO_4$ are shown in Figures 5.6 and 5.7. Sulfuric acid has essentially one inflection point (like hydrochloric acid—compare with Figure 5.1(a)), while phosphoric acid has two apparent inflection points. Both hydrogens on the

# EQUIVALENCE POINT DETECTION USING DERIVATIVES

T he volume of titrant added at the equivalence point of a titration can be accurately determined by plotting the first and second derivatives of the titration curve. A first derivative is a plot of the rate of change of the pH, $\Delta$pH, vs. milliliters of titrant, and the second derivative is a plot of the rate of change of the first derivative, $\Delta(\Delta$pH$)$, vs. milliliters of titrant. The plot in the center is the first derivative of the titration curve on the left, and the plot on the right is the second derivative. The rate of change of the curve on the left is a maximum at the midpoint of the inflection point, so the maximum on the first derivative coincides with this point, which is the equivalence point of the titration. Similarly, the rate of change is zero at the maximum of the curve in the center, so the equivalence point is also the point where the second derivative crosses zero. Thus, the equivalence point is the milliliters of titrant at the peak of the first derivative and the milliliters of titrant at the point where the line crosses zero for the second derivative. The second derivative provides the most precise measurement of the equivalence point.

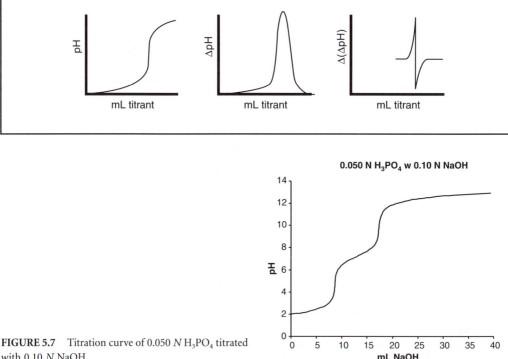

**FIGURE 5.7**   Titration curve of 0.050 $N$ H$_3$PO$_4$ titrated with 0.10 $N$ NaOH.

sulfuric acid molecule are strongly acidic (the dissociation of each in water solution is very nearly complete) and are neutralized simultaneously; therefore, there is only one inflection point. The dissociation constants for the three hydrogens on the phosphoric acid molecule are very different from each other; therefore, we would expect that they are neutralized one at a time. However, the third dissociation constant is very small, hence just two inflection points.

Thus, for sulfuric acid there is essentially one reaction along the way to the lone inflection point:

$$H_2SO_4 + 2\ OH^- \rightarrow SO_4^{2-} + 2\ H_2O \tag{5.3}$$

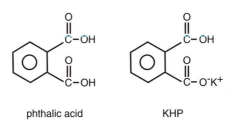

phthalic acid                    KHP

**FIGURE 5.8**   The molecular structures of phthalic acid and potassium hydrogen phthalate (KHP).

For phosphoric acid there are three reactions along the way:

$$H_3PO_4 + OH^- \rightarrow H_2PO_4^- + H_2O \tag{5.4}$$

$$H_2PO_4^- + OH^- \rightarrow HPO_4^{2-} + H_2O \tag{5.5}$$

$$HPO_4^{2-} + OH^- \rightarrow PO_4^{3-} + H_2O \tag{5.6}$$

The dissociation constants for the three hydrogens are as follows:

$$H_3PO_4 \rightleftharpoons H_2PO_4^- + H^+ \qquad K_{a1} = 7.11 \times 10^{-3} \tag{5.7}$$

$$H_2PO_4^- \rightleftharpoons HPO_4^{2-} + H^+ \qquad K_{a2} = 6.32 \times 10^{-8} \tag{5.8}$$

$$HPO_4^{2-} \rightleftharpoons PO_4^{3-} + H^+ \qquad K_{a3} = 7.1 \times 10^{-13} \tag{5.9}$$

## 5.2.6   Titration of Potassium Biphthalate

For standardizing a base solution, primary standard grade potassium biphthalate is a popular choice. Also called potassium hydrogen phthalate, potassium acid phthalate, or simply KHP, it is the salt representing partially neutralized phthalic acid and is a monoprotic weak acid. The true formula is $KHC_8H_4O_4$. Figure 5.8 shows the chemical structure of phthalic acid and KHP. The reaction with a base is as follows:

$$KHC_8H_4O_4 + OH^- \rightarrow K_2C_8H_4O_4 + H_2O \tag{5.10}$$

KHP is a white crystalline substance that has a high formula weight (204.23), is stable at oven drying temperatures, and is available in a very pure form. It is a weak acid with a titration curve similar to acetic acid. See Figure 5.9 and compare with acetic acid, Figure 5.1(c). KHP was used in Experiment 8 (Chapter 4) to standardize a 0.1 $M$ NaOH solution. In Experiment 10 in this chapter, KHP is the analyte in an experiment in which the sample is impure KHP. In Experiment 8, the equivalence point is detected with the use of phenolphthalein, while in Experiment 10, the equivalence point is detected by actually measuring the pH with a pH meter. It is only a slightly weaker acid than acetic acid, and so the titration curve for a 0.10 $M$ solution is similar to that of acetic acid, shown in Figure 5.1(c).

## 5.2.7   Titration of Tris-(hydroxymethyl)amino Methane

For standardizing acid solutions, primary standard tris-(hydroxymethyl)amino methane, THAM (also referred to as TRIS), can be used. Its formula is

$$(HOCH_2)_3CNH_2$$

It has a high formula weight (121.14), is stable at moderate drying oven temperatures, and is available in pure form. The reaction with an acid involves the acceptance of one hydrogen by the amine group, $-NH_2$:

$$(HOCH_2)_3CNH_2 + H^+ \rightarrow (HOCH_2)_3CNH_3^+ \tag{5.11}$$

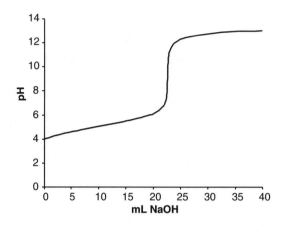

**FIGURE 5.9**   Titration curve for the titration of KHP with NaOH.

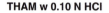

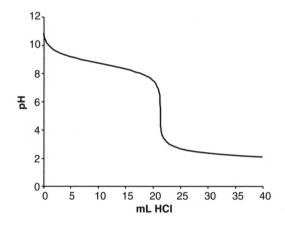

**FIGURE 5.10**   Titration curve for the titration of THAM with 0.10 *M* HCl.

It is a weak base, and the end point occurs at a pH between 4.5 and 5. It is a slightly stronger base than ammonium hydroxide, so its titration curve would appear similar to that of ammonium hydroxide in Figure 5.3(b). See Figure 5.10.

## 5.2.8   Titration of Sodium Carbonate

Primary standard sodium carbonate may also be used to standardize acid solutions. Sodium carbonate also possesses all the qualities of a good primary standard, like KHP and THAM. When titrating sodium carbonate, carbonic acid, $H_2CO_3$, is one of the products and must be decomposed with heat to push the equilibria below to completion to the right:

$$Na_2CO_3 + 2\ HCl \rightleftharpoons 2\ NaCl + H_2CO_3 \qquad (5.12)$$

$$H_2CO_3 \rightleftharpoons CO_2 + H_2O \qquad (5.13)$$

Heat also will eliminate $CO_2$ from the solution, which further aids in this completion push.

The $H_2CO_3 \leftrightarrows CO_2$ equilibrium (Equation (5.13)) can be a problem in all base solutions because $CO_2$ from the air dissolves in the solution, forming carbonic acid due to the reverse reaction represented by

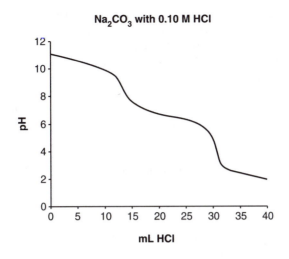

**FIGURE 5.11** Titration curve for $Na_2CO_3$ titrated with 0.10 $N$ HCl.

Equation (5.13), resulting in the formation of carbonate since carbonic acid is a weak acid:

$$H_2CO_3 \rightleftharpoons 2\ H^+ + CO_3^{2-} \tag{5.14}$$

The carbonic acid would then also react with whatever acid may be used in titrimetric procedures, thus creating an error. The water used to prepare such solutions is boiled in advance (or freshly distilled) so that all $CO_2$ is eliminated such that there is no extra carbonate to be neutralized. Such solutions may have to be restandardized, however, due to $CO_2$ from the air redissolving in the solution over time.

Figure 5.11 represents the titration curve of sodium carbonate titrated with a strong acid. Notice that there are two inflection points. This is because sodium carbonate is a dibasic base—there are two hydrogen ions accepted by the carbonate. On the way to the first inflection point, hydrogen ions are accepted by the carbonate to form bicarbonate:

$$CO_3^{2-} + H^+ \rightarrow HCO_3^- \tag{5.15}$$

This reaction is complete at the first inflection point. On the way to the second inflection point, the bicarbonate from the first reaction reacts with more hydrogen ions from the buret to form carbonic acid:

$$HCO_3^- + H^+ \rightarrow H_2CO_3 \tag{5.16}$$

The titration of $Na_2CO_3$ (see Experiment 11) usually uses the second inflection point for the end point. At that pH range, either bromcresol green or methyl red is usually used (refer to Figure 5.5). The end point, however, is not sharp (the drop in pH is not sharp) because of a buffering effect due to the presence of a large concentration of $H_2CO_3$, the product of the titration to the second inflection point (Equation (5.7)). If the solution is heated (boiled) close to the end point, however, the $H_2CO_3 \rightleftharpoons CO_2$ equilibrium (Equation (5.13)) is pushed to the right (recall previous discussion), consuming the $H_2CO_3$ and thus greatly diminishing this buffering effect. This end point then becomes sharp and very satisfactory at a pH of approximately 4.5. Boiling at the end point is therefore what is done in Experiment 11 and also in acid standardization experiments in which sodium carbonate is the primary standard.

## 5.2.9 Alkalinity

One important application of acid–base titrations is the determination of the alkalinity of various kinds of samples. It is an especially important measurement for the proper treatment of municipal water and wastewater. **Alkalinity** of a water sample is defined as its acid-neutralizing capacity. It is determined by titrating the water sample with standard acid until a particular pH is achieved. The alkalinity value depends on the pH used for the end point. **Total alkalinity** of a water sample is determined by titrating the sample usually to a pH of 4.5. The hydroxides, carbonates, bicarbonates, and other bases in the water are all neutralized when a pH of 4.5 is reached during the titration, and thus this pH is considered to be the equivalence point. Note that it is the same pH as that of the second inflection point in the titration of carbonate discussed in Section 5.2.8.

Alkalinity is usually expressed as the millimoles of $H^+$ required to titrate 1 L of water. Millimoles is calculated by multiplying the molarity of the acid (millimoles per milliliter) by buret reading in milliliters, as shown in the solution to Example 5.1 below.

### Example 5.1

What is the alkalinity of a water sample if 100.0 mL of the water required 5.29 mL of 0.1028 $M$ HCl solution to reach a pH of 4.5?

### Solution 5.1

$$\text{Alkalinity} = \frac{\text{mmol of } H^+}{\text{liters of water}} = \frac{0.1028 \text{ mmol/mL} \times 5.29 \text{ mL}}{0.1000 \text{ L}}$$

$$= 5.44 \text{ mmol/L}$$

## 5.2.10  Back Titrations

Sometimes a special kind of titrimetric procedure known as a **back titration** is required. In this procedure, the analyte is consumed using excess titrant (the end point is intentionally overshot), and the end point is then determined by titrating the excess with a second titrant. Thus, two titrants are used, and the exact concentration of each is needed for the calculation.

Figure 5.12 shows a graphic representation of a back titration. The long vertical block on the left (down arrow) represents the equivalents of the first titrant added. This amount actually exceeds the equivalents of the substance titrated present in the reaction flask. The short vertical block on the lower right (up arrow) represents the amount of the second titrant (the so-called back titrant) used to come back to the end point, to titrate the excess amount of the initial titrant. The difference between the total equivalents of the first titrant and the total equivalents of the second titrant is the number of equivalents that actually reacted with the analyte. It is this number of equivalents that is needed for the calculation. The calculation therefore uses the following equation, derived from Equation (4.40) in Chapter 4:

$$\% \text{ analyte} = \frac{(L_T \times N_T - L_{BT} \times N_{BT}) \times EW_{analyte}}{\text{sample weight}} \times 100 \tag{5.17}$$

in which BT symbolizes the back titrant.

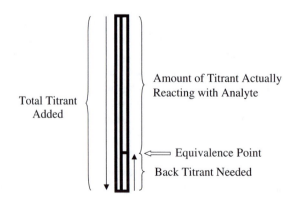

**FIGURE 5.12**  A representation of a back titration.

It may seem strange that we would ever want to perform an experiment of this kind. First of all it would be used in the event of a slow reaction taking place in the reaction flask. Perhaps the sample is not dissolved completely, and addition of the titrant causes dissolution to take place over a period of minutes or hours. Adding an excess of the titrant and back-titrating it later would seem an appropriate course of action in a case of this kind. An example would be the determination of the calcium carbonate in an antacid tablet (Experiment 13).

**Example 5.2**

What is the percent of $CaCO_3$ in an antacid given that a tablet that weighed 1.3198 g reacted with 50.00 mL of 0.4486 *N* HCl that subsequently required 3.72 mL of 0.1277 *N* NaOH for back titration? Also report the milligrams of $CaCO_3$ in the tablet.

$$2\ HCl + CaCO_3 \rightarrow CaCl_2 + CO_2 + H_2O$$

*Solution 5.2*

Equation (5.17) is used as follows:

$$\% \ CaCO_3 = \frac{(liters_{HCl} \times N_{HCl} - liters_{NaOH} \times N_{NaOH}) \times EW_{CaCO_3}}{sample\ weight} \times 100$$

$$= \frac{(0.05000\ L \times 0.4486\ N - 0.00372\ L \times 0.1277\ N) \times 100.09/2}{1.3198\ g} \times 100$$

$$= 83.25\%$$

The grams of $CaCO_3$ is the numerator above, or 1.0987 g.

It is also possible that the analyte must be calculated from a gaseous product of another reaction. In this case, one would want the gas to react with the titrant as soon as it is formed, since it could escape into the air because of a high vapor pressure. Thus an excess of the titrant would be present in a solution through which the gas is bubbled. After the gas-forming reaction has stopped, the excess titrant in this "bubble flask" could be titrated with the back titrant and the results calculated. This latter experiment is one form of the Kjeldahl titration.

## 5.2.11 The Kjeldahl Method for Protein

A titrimetric method that has been used for many years for the determination of nitrogen or protein in a sample is the Kjeldahl method. Examples of samples include grain, protein supplements for animal feed, fertilizers, and food products. It is a method that often makes use of the back titration concept mentioned above. We will now describe this technique in detail.

The method consists of three parts: digestion, distillation, and back titration. The digestion step is in essence the dissolving step. The sample is weighed and placed in a Kjeldahl flask, which is a round bottom flask with a long neck, similar in appearance to a volumetric flask, except for the round bottom and the lack of a calibration line. A fairly small volume of concentrated sulfuric acid along with a quantity of $K_2SO_4$ (to raise the boiling point of the sulfuric acid) and a catalyst (typically an amount of $CuSO_4$, selenium, or a selenium compound) is added, and the flask is placed in a heating mantle and heated. The sulfuric acid boils and the sample digests for a period of time until it is evident that the sample is dissolved, and a clear solution is contained in the flask. The digestion must be carried out in a fume hood, since thick $SO_3$ fumes evolve from the flask until the sample is dissolved. At this point, the contents of the flask are diluted with water and an amount of fairly concentrated sodium hydroxide is added to neutralize the acid. Immediately upon neutralization, the nitrogen originally present in the sample is converted to ammonia. At this point the distillation step is begun.

**FIGURE 5.13**   A photograph of a commercial Kjeldahl distillation unit. The heating chamber (with heating element) is in the left center, the baffle in the top center, the condenser in the right center, and the beaker (receiving vessel) for holding the standard acid solution in the lower right.

A laboratory that runs Kjeldahl analyses routinely would likely have a special apparatus set up for the distillation. One variation of this apparatus commercially available is shown in Figure 5.13. A baffle is placed on the top of the Kjeldahl flask and subsequently connected to a condenser, which in turn guides the distillate into a receiving vessel, as shown. The ammonia is then distilled into the receiving vessel. The receiving vessel contains an acid for reaction with the ammonia.

The acid in the receiving vessel can be either a dilute (perhaps 0.10 $N$) standardized solution of a strong acid, such as sulfuric acid, or a solution of boric acid. If it is the former, it is an example of a back titration. If it is the latter, it is an example of an **indirect titration**.

In the back titration method, an excess, but carefully measured, amount of the standardized acid is contained in the receiving vessel such that after the ammonia bubbles through it and is consumed, an excess remains. The flask is then removed from the apparatus and the excess acid titrated with a standardized NaOH solution. The analyte in this procedure is the nitrogen (in the form of ammonia as it enters the flask), and thus the amount of acid consumed is the important measurement. The amount of acid consumed is the difference between the total amount present and the amount that was in excess. It is a back titration because the amount of acid in the flask is in excess and in essence goes beyond the equivalence point for the reaction with the ammonia. Thus, the analyst must come *back* to the equivalence point with the sodium hydroxide. Equation (5.17) is then used to calculate the results. The equivalent weight of the analyte is the atomic weight of nitrogen. The percent protein may also be calculated, in which case the equivalent weight of the protein is substituted for the atomic weight of nitrogen, or a gravimetric factor is used to convert the weight of the nitrogen to the weight of the protein.

In the indirect method using boric acid ($H_3BO_3$), the ammonia reacts with the boric acid, producing a partially neutralized salt of boric acid ($H_2BO_3^-$):

$$NH_3 + H_3BO_3 \rightarrow H_2BO_3^- + NH_4^+ \tag{5.18}$$

which can then be titrated with a standardized acid:

$$H_2BO_3^- + H^+ \rightarrow H_3BO_3 \tag{5.19}$$

The amount of standardized acid needed is proportional to the amount of ammonia that bubbled through. It is an indirect method because the ammonia is determined but not titrated. It is determined indirectly by titration of $H_2BO_3^-$. In a direct titration, the analyte would be reacted directly with the titrant, as per the discussion in Section 4.6. The concentration of the boric acid in the receiving vessel does not enter into the calculation and need not be known. Equation (4.40) is used for the calculation.

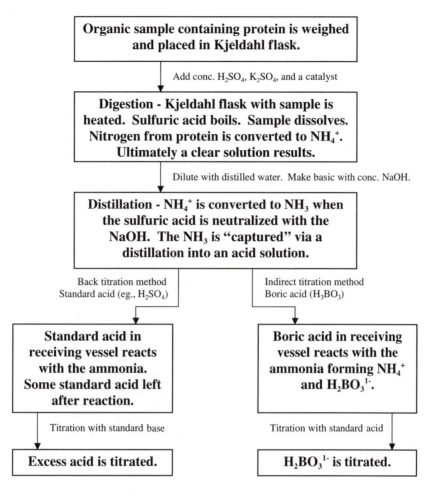

**FIGURE 5.14**   A flow chart of the Kjeldahl titration procedures.

See Figure 5.14 for a summary flow chart of the Kjeldahl titration. Experiment 12 in this chapter is a Kjeldahl titration experiment, a back titration experiment as written. See also Workplace Scene 5.1.

### Example 5.3

In a Kjeldahl analysis, a flour sample weighing 0.9857 g was digested in concentrated $H_2SO_4$ for 45 min. A concentrated solution of NaOH was added such that all of the nitrogen was converted to $NH_3$. Following this, the $NH_3$ was distilled into a flask containing 50.00 mL of 0.1011 N $H_2SO_4$. The excess required 5.12 mL of 0.1266 N NaOH for titration. What is the percent of nitrogen in the sample?

### Solution 5.3

% nitrogen

$$= \frac{(\text{liters}_{H_2SO_4} \times N_{H_2SO_4} - \text{liters}_{NaOH} \times N_{NaOH}) \times 14.00}{\text{sample weight}} \times 100$$

$$= \frac{(0.050000 \, \cancel{L} \times 0.1011 \, \cancel{\text{equivalents}/L} - 0.00512 \, \cancel{L} \times 0.1266 \, \cancel{\text{equivalents}/L}) \times 14.00 \, \cancel{g}/\cancel{\text{equivalent}}}{0.9857 \, \cancel{g}} \times 100$$

$$= 6.259\% \text{ nitrogen}$$

# WORKPLACE SCENE 5.1

At the Nebraska State Agriculture Laboratory in Lincoln, Nebraska, the analysis of animal feed samples for protein content by the Kjeldahl procedure is routine. In midwestern agriculture communities, the protein content of feeds and grains given to cattle and hogs is critical to the animals' growth and development. The state guards against fraud by routinely and randomly testing to see if the protein content matches the specifications stated on the product label and in the feed companies' literature. In this particular laboratory, the boric acid indirect titration method is used. A large fume hood is dedicated to the digestion of the relatively large number of samples. Full safety precautions are taken, including face shields, wrist-length gloves, and fume hood barriers.

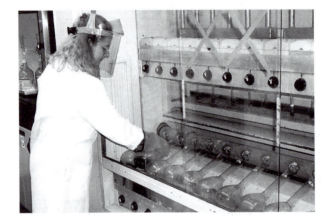

Cindy Wagner-Wiebeck sets a Kjeldahl flask in place in a fume hood dedicated to Kjeldahl sample digestion.

## Example 5.4

In a Kjeldahl analysis, a grain sample weighing 1.1033 g was digested in concentrated $H_2SO_4$ for 40 min. A concentrated solution of NaOH was added such that all of the nitrogen was converted to $NH_3$. The $NH_3$ was then distilled into a flask containing a solution of boric acid. Following this, the solution in the receiving flask was titrated with 0.1011 $N$ HCl, requiring 24.61 mL. What is the percent nitrogen in the sample?

### Solution 5.4

$$\% \text{ nitrogen} = \frac{\text{liters}_{HCl} \times N_{HCl} \times 14.00}{\text{sample weight}} \times 100$$

$$= \frac{0.02461\,\cancel{L} \times 0.1011\,\cancel{\text{equivalents}/L} \times 14.00\,\cancel{g}/\cancel{\text{equivalent}}}{1.1033\,\cancel{g}} \times 100$$

$$= 3.157\%$$

## 5.2.12 Buffering Effects and Buffer Solutions

Titration curves help us to understand a very important concept in many areas of chemistry and biochemistry. This concept is that of buffering and buffer solutions. A **buffer solution** is a solution that resists changes in pH even when a strong acid or base is added or when the solution is diluted with water. The typical composition of a buffer solution includes a weak acid or base of a particular concentration and its conjugate base or acid of a particular concentration. A **conjugate base** is the product of an acid neutralization that can gain back a hydrogen that it lost during the neutralization. Conversely, a **conjugate acid** is the product of a base neutralization that can lose the hydrogen that it gained during the neutralization. A **conjugate acid–base pair** would consist of the acid and its conjugate base or the base and its conjugate acid. An important example of a conjugate base is the acetate ion formed during the neutralization of acetic acid ($HC_2H_3O_2$) with hydroxide:

$$HC_2H_3O_2 + OH^- \rightarrow C_2H_3O_2^- + H_2O \tag{5.20}$$

Acetate is a conjugate base because it can gain a hydrogen and become acetic acid again (the reverse of Equation (5.20)). Acetic acid and acetate ion (such as from sodium acetate) would constitute a conjugate acid–base pair.

An important example of a conjugate acid is the ammonium ion formed by the neutralization of ammonia with an acid:

$$H^+ + NH_3 \rightarrow NH_4^+ \tag{5.21}$$

The ammonium ion is a conjugate acid because it can lose a hydrogen and become ammonia again (the reverse of Equation (5.21)), and thus ammonia (or ammonium hydroxide) and ammonium ion (such as from ammonium chloride) together constitute another conjugate acid–base pair.

In the process of a weak acid or weak base neutralization titration, a mixture of a conjugate acid–base pair exists in the reaction flask in the time period of the experiment leading up to the inflection point. For example, during the titration of acetic acid with sodium hydroxide, a mixture of acetic acid and acetate ion exists in the reaction flask prior to the inflection point. In that portion of the titration curve, the pH of the solution does not change appreciably, even upon the addition of more sodium hydroxide. Thus this solution is a buffer solution, as we defined it at the beginning of this section.

Buffer solutions can become very effective when the concentrations of the conjugate acid–base pair are higher. When the concentrations are higher, the portion of the curve leading up to the inflection point is extended; thus the **buffer capacity**, or the ability to resist neutralization, is extended. Such a mixture would require even more strong acid or base before it is rendered neutral. The region of a titration curve leading up to the inflection point is often called the **buffer region**.

Buffer solutions can be prepared quite simply by appropriate combination of conjugate acid–base pairs, as in the above examples. Although commercially prepared buffer solutions are available, these are most often utilized solely for pH meter calibration and not for adjusting or maintaining a chemical reaction system at a given pH. It is not surprising, therefore, that the analyst often needs to prepare his or her own solutions for this purpose. It then becomes a question of what proportions of the conjugate acid and base should be mixed to give the desired pH.

The answer is in the expression for the ionization constant, $K_a$ or $K_b$, where the ratio of the conjugate acid and base concentrations is found. In the case of a weak monoprotic acid, HA, we have the following:

$$HA \rightleftharpoons H^+ + A^- \tag{5.22}$$

$$K_a = \frac{[H^+][A^-]}{[HA]} \tag{5.23}$$

in which [HA] approximates the weak acid concentration and [A$^-$] is the conjugate base concentration. In the case of a weak base, B, we have the following:

$$B + H_2O \rightleftharpoons BH^+ + OH^- \tag{5.24}$$

$$K_b = \frac{[BH^+][OH^-]}{[B]} \tag{5.25}$$

in which [B] is the weak base concentration and [BH$^+$] is the conjugate acid concentration.

Knowing the value of $K_a$ or $K_b$ for a given weak acid or base, and knowing the desired pH value, one can calculate the ratio of conjugate base (or acid) concentration to acid (or base) concentration that will produce the given pH. Rearranging Equation (5.23), for example, would give the following:

$$[H^+] = K_a \times \frac{[HA]}{[A^-]} \tag{5.26}$$

Taking the negative logarithm of both sides would give

$$pH = pK_a - \log\frac{[HA]}{[A^-]} \tag{5.27}$$

or

$$pH = pK_a + \log\frac{[A^-]}{[HA]} \tag{5.28}$$

In these equations, $pK_a$ is defined as the negative logarithm of $K_a$. The equivalent expression, derived from Equation (5.25), for a weak base would be

$$pOH = pK_b + \log\frac{[BH^+]}{[B]} \tag{5.29}$$

It may be more convenient in the case of a weak base to use Equation (5.28): think of the conjugate acid as the weak acid and use its $pK_a$. In this case, we could write the following:

$$pH = pK_a + \log\frac{[B]}{[BH^+]} \tag{5.30}$$

where [BH$^+$] is the concentration of the conjugate acid of the base, and $pK_a$ is the negative logarithm of $K_a$ for this acid.

Thus, for a weak acid with a given $K_a$ (or $pK_a$) and a given ratio of conjugate base concentration to acid concentration, the pH may be calculated. Or, given the desired pH and $K_a$ ($pK_a$), the ratio of salt concentration to acid concentration can be calculated and the buffer subsequently prepared. Equations (5.26) to (5.30) are each a form of the **Henderson–Hasselbalch equation** for dealing with buffer solutions.

### Example 5.5

What is the pH of an acetic acid–sodium acetate solution if the acid concentration is 0.10 $M$ and the acetate concentration is 0.20 $M$? (The $K_a$ of acetic acid is $1.8 \times 10^{-5}$.)

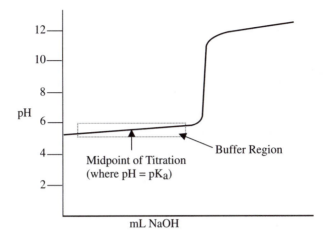

**FIGURE 5.15** A drawing of a titration curve for a weak acid pointing out the buffer region and midpoint.

**Solution 5.5**
Utilizing Equation (5.28), we have

$$pH = -\log(1.8 \times 10^{-5}) + \log \frac{0.20}{0.10} = 5.04$$

## Example 5.6

What concentration of THAM hydrochloride is required to have a buffer solution of pH = 7.98 if the THAM concentration is 0.20 $M$? (The $K_a$ for THAM hydrochloride is $8.4 \times 10^{-9}$.)

**Solution 5.6**
Utilizing Equation (5.30), we have

$$7.98 = -\log(8.41 \times 10^{-9}) + \log \frac{0.20}{[\text{THAM hydrochloride}]}$$

$$7.98 = 8.075 + \log \frac{0.20}{[\text{THAM hydrochloride}]}$$

$$-0.095 = \log \frac{0.20}{[\text{THAM hydrochloride}]}$$

$$[\text{THAM hydrochloride}] = 0.25 \ M$$

It should be stressed that since $K_a$ (or $K_b$) enters into the calculation, how weak the acid or base is dictates what is a workable pH range for that acid or base. This range can be seen graphically on the titration curve for the weak acid or base—it is the pH range covered by the shallow increase (in the case of a weak acid) or decrease (in the case of a weak base) leading up to the inflection point (see Figure 5.15). The pH value at the midpoint of the pH range (the so-called **midpoint of a titration**) equals the $pK_a$ value, as can be seen in the following derivation starting with Equation (5.28):

$$pH = pK_a + \log \frac{[A^-]}{[HA]} \tag{5.31}$$

**TABLE 5.1** Commonly Used Examples of Conjugate Acid–Base Combinations and Corresponding Useful pH Ranges

| Combination | pH Range |
|---|---|
| Chloroacetate–chloroacetic acid ($K_a = 1.36 \times 10^{-3}$) | 1.8–3.8 |
| Acetate–acetic acid ($K_a = 1.8 \times 10^{-5}$) | 3.7–5.7 |
| Monohydrogen phosphate–dihydrogen phosphate ($K_a = 6.23 \times 10^{-8}$) | 6.1–8.1 |
| THAM–THAM hydrochloride ($K_a = 8.4 \times 10^{-9}$) | 7.5–9.5 |
| Ammonium hydroxide–ammonium ion ($K_a = 5.7 \times 10^{-10}$) | 8.3–10.3 |

At the midpoint of the titration, the weak acid concentration equals the conjugate base concentration (half of the acid has been converted to the conjugate base):

$$[A^-] = [HA] \tag{5.32}$$

Thus the ratio of $[A^-]$ to $[HA]$ is equal to 1:

$$\frac{[A^-]}{[HA]} = 1 \tag{5.33}$$

As a result, we have the following:

$$pH = pK_a + \log(1) \tag{5.34}$$

Because the logarithm of 1 is zero, we have

$$pH = pK_a \tag{5.35}$$

See Figure 5.15.

Table 5.1 gives commonly used examples of conjugate acid–base pair combinations and the pH range for which each is useful. This range corresponds to the pH range defined by the buffer region in the titration curve for each, and the middle of the range corresponds to the midpoint of each titration.

## Example 5.7

Suppose a buffer solution of pH = 2.00 is needed. Suggest a conjugate acid–base pair for this solution, and calculate the ratio of the concentration of conjugate base to acid that is needed to prepare it.

### Solution 5.7

From Table 5.1, we can see that a particular chloroacetic acid–chloroacetate combination would give a pH of 2.00. Rearranging Equation (5.28), we have

$$\log \frac{[A^-]}{[HA]} = pH - pK_a = 2.00 - 2.87 = -0.87$$

$$\frac{[A^-]}{[HA]} = 0.13$$

It should be stressed that the pH value of an actual buffer solution prepared by mixing quantities of the weak acid or base and its conjugate base or acid based on the calculated ratio will likely be different from what was calculated. The reason for this is the use of approximations in the calculations. For example, the molar concentration expressions found in Equations (5.23) to (5.30), e.g., $[H^+]$, are approximations. To be thermodynamically correct, the **activity** of the chemical should be used rather than the concentration. Activity is directly proportional to concentration, the **activity coefficient** being the proportionality constant:

$$a = \gamma C \tag{5.36}$$

**TABLE 5.2**   Recipes for Some of the More Popular Buffer Solutions

| | |
|---|---|
| pH = 4.0 phthalate buffer | Dissolve 10.12 g of potassium hydrogen phthalate (KHP) in 1 L of solution |
| pH = 6.9 phosphate buffer | Dissolve 3.39 g of potassium dihydrogen phosphate and 3.53 g of dried sodium monohydrogen phosphate in 1 L of solution |
| pH = 10.0 ammonia buffer | Dissolve 70.0 g of ammonium chloride and 570 mL of concentrated ammonium hydroxide (ammonia) in 1 L of solution |

where a is the activity, $\gamma$ is the activity coefficient, and C is the molar concentration. However, for most applications, especially at small concentrations, the activity coefficient is assumed to equal 1; therefore, the activity is equal to the concentration. In addition, the values for the un-ionized acid and base concentrations in the denominators in Equations (5.28) and (5.29) are actually approximations. For example, $C_{HA}$, the total concentration of acid, is often substituted into Equation (5.28) rather than [HA], the un-ionized acid concentration. This was the case in Examples 5.5 and 5.6. A better method of accurate buffer solution preparation would be to prepare a solution (the concentration is not important) of the conjugate base (or acid) and then add a solution of a strong acid (or base) until the pH, as measured by a pH meter, is the desired pH. At that point, the required conjugate acid–base pair combination required is in the container.

For example, to prepare a pH = 9 buffer solution, one would prepare a solution of ammonium chloride (refer to Table 5.1), and then add a solution of sodium hydroxide while stirring and monitoring the pH with a pH meter. The preparation is complete when the pH reaches 9. The required conjugate acid–base pair would be $NH_3 - NH_4^+$. Recipes for standard buffer solutions can be useful. Table 5.2 gives specific directions for preparing some popular buffer solutions.

Many applications for buffer solutions are found in the analytical laboratory. It is frequently required to have solutions that do not change pH during the course of an experiment. An example is cited in Section 5.3.3 and used in Experiments 11 and 12.

# 5.3   Complex Ion Formation Reactions

## 5.3.1   Introduction

Acid–base (neutralization) reactions are only one type of many that are applicable to titrimetric analysis. There are reactions that involve the formation of a precipitate. There are reactions that involve the transfer of electrons. There are reactions, among still others, that involve the formation of a complex ion. This latter type typically involves transition metals and is often used for the qualitative and quantitative colorimetric analysis (Chapters 8 and 9) of transition metal ions, since the complex ion that forms can be analyzed according to the depth of a color that it imparts to a solution. In this section, however, we are concerned with a titrimetric analysis method in which a complex ion-forming reaction is used.

## 5.3.2   Complex Ion Terminology

A **complex ion** is a polyatomic charged aggregate consisting of a positively charged metal ion combined with either a neutral molecule or negative ion. The neutral molecule or negative ion in this aggregate is called a **ligand**. The ligand can consist of a monatomic, negative ion, such as $F^-$, $Cl^-$, etc., or a polyatomic molecule or ion, such as $H_2O$, $CN^-$, $CNS^-$, $NH_3$, $CN^-$, etc. Some simple examples of the complex ion-forming reaction are presented in Table 5.3. Notice that complex ions can be either positively or negatively charged.

Ligands can be classified according to the number of bonding sites that are available for forming a coordinate covalent bond to the metal ion. A **coordinate covalent bond** is one in which the shared

**TABLE 5.3**   Examples of Reactions between
Metal Ions and Ligands to Form Complex Ions

| Metal Ion | | Ligand | | Complex Ion |
|---|---|---|---|---|
| $Cu^{2+}$ | + | $4NH_3$ | $\rightleftharpoons$ | $Cu(NH_3)_4^{2+}$ |
| $Fe^{3+}$ | + | $CNS^-$ | $\rightleftharpoons$ | $Fe(CNS)^{2+}$ |
| $Co^{2+}$ | + | $4Cl^-$ | $\rightleftharpoons$ | $CoCl_4^{2-}$ |

**FIGURE 5.16**   The electron dot structure of a nitrogen atom (left) and an ammonia molecule (right). The pair of electrons above the nitrogen is the nonbonding pair available for coordinate covalent bonding.

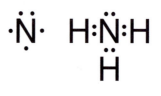

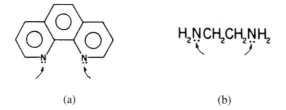

(a)                                                                    (b)

**FIGURE 5.17**   Two bidentate ligands: (a) 1,10 phenanthroline, and (b) ethylenediamine. The arrows point out the bonding sites.

electrons are contributed to the bond by only one of the two atoms involved. In the case of complex ion-forming reactions, both electrons are always donated by the ligand. If only one such pair of electrons is available per molecule or ion, we say that the ligand is **monodentate**. If two are available per molecule or ion, it is described as **bidentate**, etc. Thus we have the following terminology:

| No. of Sites | Descriptive Term |
|---|---|
| 1 | Monodentate |
| 2 | Bidentate |
| 3 | Tridentate |
| 6 | Hexadentate |

Examples of monodentate ligands are all of those given in Table 5.3 ($Cl^-$, $CN^-$, $NH_3$, etc.). Nitrogen-containing ligands are especially evident in reactions of this kind. This is true because of the pair of electrons occupying the nonbonding orbital found on the nitrogen in many nitrogen-containing compounds. This pair of electrons is present in the Lewis electron dot structure for the nitrogen atom shown in Figure 5.16. It is easy to see that nitrogen would form three covalent bonds, since it has three unpaired electrons, thus often leaving the nonbonded pair of electrons available for coordinate covalent bonding. An example is ammonia, $NH_3$, as shown in Figure 5.16.

A good example of a bidentate ligand is the 1,10-phenanthroline molecule, which, since it forms a stable complex ion with $Fe^{2+}$ ions that is deep orange color, is used in the colorimetric analysis of iron(II) ions (see Experiment 19 in Chapter 7). This ligand is shown in Figure 5.17(a). The two bonding sites are

# WORKPLACE SCENE 5.2

During the bleaching of wood pulp in paper mills, certain metal ions present in the wood can cause an increase in the degradation of the hydrogen peroxide contained in the bleach liquor. It is best to remove as many of these metal ions as possible before the bleach stage, or to efficiently control them during the bleach stage. Removing or controlling these metal ions with chelating agents is the answer in many paper mills. In the laboratory, different chelating agents at varying levels are added to the pulp. This is best when carried out in an aqueous environment. After removal of the excess liquid, the wood pulp is dried, weighed, and ashed. Next the samples are dissolved in concentrated acid and diluted to a predetermined volume. A metal analysis is then performed using an inductively coupled plasma (ICP) spectrometer (Chapter 9). The results help the paper mills to decide which chelating agent to use and how much is needed.

Pamela Slavings of the Dow Chemical Company in Midland, Michigan, performs the metal analysis required following the removal of metals with chelating agents.

pointed out with the arrows. Another bidentate ligand is ethylenediamine, shown in Figure 5.17(b). These ligands, although bidentate, will form a complex ion with only one metal ion. It is also common for more than one such ligand to combine with a single metal ion.

Complex ions that involve ligands with two or more bonding sites (bidentate, tridentate, etc.) are also called **chelates** and the ligands **chelating agents**. Thus, the ligands shown in Figure 5.17 are examples of chelating agents, and the complex ions formed are examples of chelates. See Workplace Scene 5.2. Another term associated with complex ion chemistry is **masking**. Masking refers to the use of ligands and complex ion formation reactions for the purpose of avoiding interferences. When the complex ion formation equilibrium lies far to the right, such that the equilibrium constant (more often called **formation constant** in the case of complex ion formation) is very large, the complex ion formed is very stable. This has the effect of "tying up," or **masking**, the metal ion such that the interference does not occur. The ligand used in this application is called the **masking agent**. A good example is in the water hardness analysis to be studied in the next section. Dissolved iron ions can interfere with this analysis. This interference can be removed if the iron ions are reacted with the cyanide ligand ($CN^-$). The formation constant of the $Fe(CN)_6^{3-}$

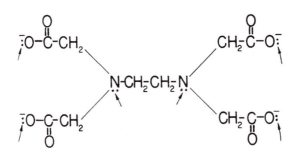

**FIGURE 5.18**    The structure of the EDTA ligand. The bonding sites are pointed out with arrows.

complex ion is large, meaning that the iron ions are effectively removed from the solution, or masked, cyanide being the masking agent. The reaction involved and the formation constant, $K_f$, are as follows:

$$Fe^{3+} + 6\ CN^- \rightleftharpoons Fe(CN)_6^{3-} \tag{5.37}$$

$$K_f = \frac{[Fe(CN)_6^{3-}]}{[Fe^{2+}][CN^-]^6} = 10^{31} \tag{5.38}$$

A very important ligand (or chelating agent) for titrimetric analysis is the ethylenediaminetetraacetate (EDTA) ligand. It is especially useful in reacting with calcium and magnesium ions in hard water such that water hardness can be determined. The next section is devoted to this subject.

### 5.3.3   EDTA and Water Hardness

EDTA is a hexadentate ligand. Its structure is shown in Figure 5.18 (the bonding sites are pointed out by arrows). In addition to the four charged sites at the carboxyl group oxygens, each of the two nitrogens has an unshared pair of electrons, making six electron pairs available to form coordinate covalent bonds. In forming a complex ion with calcium ions, for example, all six bonding sites bond to calcium, forming a large aggregate consisting of a single EDTA ligand wrapped around the calcium ion, as shown in Figure 5.19.

Thus a one-to-one reaction is involved, and in fact, all reactions of EDTA with metal ions (most metals do react with EDTA) are one-to-one. Therefore, we will not be concerned with a formal scheme for determining equivalent weights, as we were with acid–base reactions.

The usual source of EDTA for use in metal analysis is the disodium dihydrogen salt of ethylenediaminetetraacetic acid (Figure 5.20). This is the partially neutralized salt of ethylenediaminetetraacetic acid.

Ethylenediaminetetraacetic acid is frequently symbolized $H_4Y$, and the disodium salt, $Na_2H_2Y$. The hydrogens in these formulas are the acidic hydrogens associated with the carboxyl groups, as in any weak organic carboxylic acid; they dissociate from the EDTA ion in a series of equilibrium steps:

Step 1                              $$H_4Y \rightleftharpoons H^+ + H_3Y^- \tag{5.39}$$

Step 2                              $$H_3Y^- \rightleftharpoons H^+ + H_2Y^{2-} \tag{5.40}$$

Step 3                              $$H_2Y^{2-} \rightleftharpoons H^+ + HY^{3-} \tag{5.41}$$

Step 4                              $$HY^{3-} \rightleftharpoons H^+ + Y^{4-} \tag{5.42}$$

The exact position of this equilibrium is established by controlling the pH of the solution. With extremely basic pH values, $Y^{4-}$ will predominate, while in extremely acidic pH values, $H_4Y$ will predominate. The intermediate species will predominate at intermediate pH values. See Figure 5.21.

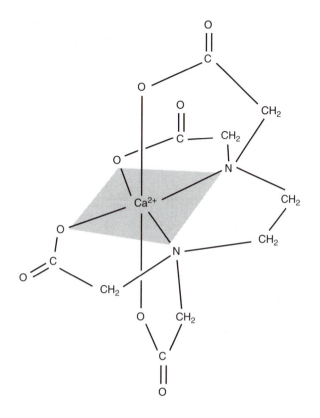

**FIGURE 5.19** The calcium–EDTA complex ion.

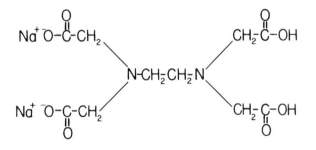

**FIGURE 5.20** The disodium dihydrogen salt of ethylenediaminetetraacetic acid.

When $Na_2H_2Y$ is dissolved in water and mixed with a solution of a metal ion, such as calcium, at a basic pH, the metal ion will react with the predominant EDTA species such that the following equilibrium is established:

$$Ca^{2+} + Y^{4-} \rightleftharpoons CaY^{2-} \tag{5.43}$$

A problem exists with this procedure, however, in that at basic pH values, many metal ions precipitate as the hydroxide, e.g., $Mg(OH)_2$, and thus would be lost to the analysis. This occurs with the magnesium in the water hardness procedure alluded to earlier. Luckily, a happy medium exists. At pH = 10, the reaction of the metal ion with the predominant $HY^{3-}$ and $Y^{4-}$ species (Figure 5.21) is shifted sufficiently to the right for the quantitative requirement to be fulfilled, while at the same time the solution is not basic enough for the magnesium ions to precipitate appreciably. Thus, all solutions in the reaction flask in the water hardness determination are buffered at pH = 10, meaning that a conjugate acid–base pair

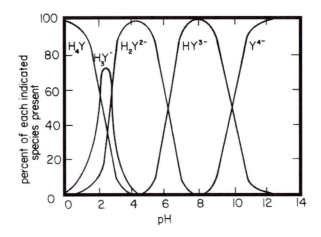

**FIGURE 5.21**    The predominance of EDTA species as a function of pH.

is added to each solution so that the pH = 10. In this case the buffer is the ammonia–ammonium ion system mentioned previously (see Tables 5.1 and 5.2).

EDTA titrations are routinely used to determine water hardness in a laboratory. Raw well water samples can have a significant quantity of dissolved minerals that contribute to a variety of problems associated with the use of such water. These minerals consist chiefly of calcium and magnesium carbonates, sulfates, etc. The problems that arise are mostly a result of heating or boiling the water over a period of time such that the water is evaporated, and the calcium and magnesium salts become concentrated and precipitate in the form of a "scale" on the walls of the container, hence the term hardness. This kind of problem is evident in boilers, domestic and commercial water heaters, humidifiers, tea kettles, and the like.

Consumers and industrial companies can install water softeners to eliminate the problem. Water softeners work on an ion exchange principle (see Chapter 11) and remove the metal ions that cause the hardness properties, replacing them with sodium ions.

A second problem with hard water is that these metals react with soap molecules and form a scum to which bathtub rings, etc., are attributed. Hard water is therefore not the best water to use for efficient soapy water cleaning processes, since the metal–soap precipitation reaction competes with the cleaning action. Water softeners assist with solving this problem too.

Water from different sources can have very different hardness values. Water samples can vary from being extremely soft to extremely hard. While this description of hardness is anything but quantitative, a quantitative description based on an EDTA titration can be given.

The EDTA determination of water hardness results from the reaction of the EDTA ligand with the metal ions involved, calcium and magnesium. An interesting question is "How are the results reported?" Hardness is not usually reported precisely as so much calcium, plus so much magnesium, etc. There is no distinction made between the metals involved. All species reacting with the EDTA are considered as one, and the results are reported as an amount of one species, calcium carbonate, $CaCO_3$. That is, in the calculation, when a formula weight is used to convert moles to grams, the formula weight of calcium carbonate is used, and thus a quantity of $CaCO_3$ equivalent to the sum total of all contributors to the hardness is reported. Hardness is most often expressed as either parts per million $CaCO_3$ or grains per gallon $CaCO_3$.

The indicator that is most often used is called **Eriochrome Black T** (EBT). EBT is actually a ligand that also reacts with the metal ions, like EDTA. In the free uncombined form, it imparts a sky-blue color to the solution, but if it is part of a complex ion with either calcium or magnesium ions, it is a wine-red color. Thus, before adding any EDTA from a buret, the hard water sample containing the pH = 10 ammonia buffer and several drops of EBT indicator will be wine red. As the EDTA solution is added, the EDTA ligand reacts with the free metal ions and then actually reacts with the metal–EBT complex ion, complexing the metal and resulting in the free EBT ligand, which, as mentioned earlier, gives a

**TABLE 5.4** Parts Per Million Unit for Solid Samples and Liquid Samples and Solutions

| For Solid Samples | For Liquid Samples and Solutions |
|---|---|
| 1 ppm = 1 mg/kg<br>= 1 $\mu$g/g | 1 ppm = 1 mg/L<br>= 1 $\mu$g/mL |

sky-blue color to the solution. The color change, then, is the total conversion of the wine-red color to the sky-blue color, with every trace of red disappearing at the end point.

It is known that this color change is quite sharp when magnesium ions are present. In cases in which magnesium ions are not present in the water samples, the end point will not be sharp. Because of this, a small amount of magnesium chloride is added to the EDTA as it is prepared, and thus a sharp end point is assured.

## 5.3.4 Expressing Concentration Using Parts Per Million

As indicated in the last section, the amount of hardness in a water sample is often reported as parts per million (ppm) $CaCO_3$. **Parts per million** is a unit of concentration similar to molarity, percent, etc. A solution that has a solute content of 1 ppm has 1 part of solute dissolved for every million ($10^6$) parts of solution. A "part" can be a weight measurement, a volume measurement, or both. It if is a weight measurement, it might mean l mg/$10^6$ mg, or 1 mg/kg. A conversion of units here indicates that it might also mean 1 $\mu$g/$10^6$ $\mu$g, or 1 $\mu$g/g. If it is a volume measurement, it might mean 1 mL/$10^6$ mL, 1 mL/1000 L, or 1 $\mu$L/L. If the intent is to represent a concentration in a water solution, and this is frequently the case, we can recognize that 1 g of water (and dilute water solutions) is 1 mL of the water, and that 1 L of water weighs 1 kg. Thus, 1 ppm might also mean 1 mg/L.

Most commonly, if the sample to be analyzed is a solid material (for example, soil), parts per million is expressed as milligram per kilogram (or microgram per gram), since the weight (as opposed to volume) of the material is most easily measured. However, if the volume is most easily measured (such as with a liquid sample like water or water solutions), parts per million is expressed as milligrams per liter (or micrograms per milliliter). Thus if a certain solution is described as 10 ppm iron, this means 10 mg of iron dissolved per liter of solution. If, however, an analysis of soil for potassium is reported as 250 ppm K, this means 250 mg of potassium per kilogram of soil. See Table 5.4. Some solution preparation procedures a technician may encounter involving the parts per million unit are discussed below.

### 5.3.4.1 Solution Preparation

If you wanted to prepare 1 L of a 10 ppm solution of zinc, Zn, you would weigh out 10 mg of zinc metal, place it in a 1-L flask, dissolve it, and dilute to the 1-L mark with distilled water. If you need to prepare a volume other than 1 L, the parts per million (ppm) concentration is multiplied by the volume in liters:

$$\text{ppm}_{\text{desired}} \times \text{liters}_{\text{desired}} = \text{milligrams to be weighed} \tag{5.44}$$

$$\frac{\text{milligrams}}{\text{liters}} \times \text{liters} = \text{milligrams to be weighed} \tag{5.45}$$

**Example 5.8**

Tell how you would prepare 500 mL of a 25 ppm copper solution from pure copper metal.

*Solution 5.8*

$$\frac{25.0 \text{ mg}}{L} \times 0.500 \text{ } L = 12.5 \text{ mg}$$

Weigh 12.5 mg of copper into a 500-mL flask, dissolve (in dilute nitric acid), and dilute to the mark with water.

Another example of the preparation of parts per million solutions is by dilution. It is a very common practice to purchase solutions of metals that are fairly concentrated (1000 ppm) and then dilute them to obtain the desired concentration. This is done to save solution preparation time in the laboratory. As per the discussion in Chapter 4, (see Equation (4.14) and the accompanying discussion), the concentration is multiplied by the volume both before and after dilution. Using the parts per million unit, we have

$$ppm_B \times V_B = ppm_A \times V_A \tag{5.46}$$

where A and B refer to before dilution and after dilution, respectively, as in Chapter 4.

### Example 5.9

How would you prepare 100 mL of a 20 ppm iron solution from a 1000 ppm solution?

*Solution 5.9*

$$1000\ ppm \times V_B = 20\ ppm \times 100\ mL$$

$$V_B = \frac{20\ ppm \times 100\ mL}{1000\ ppm} = 2.0\ mL$$

Dilute 2.0 mL of the 1000 ppm solution to 100 mL.

There is a third type of solution preparation problem that could be encountered with the parts per million unit. This is the case in which the solution of a metal is to be prepared by weighing a metal salt, rather than the pure metal, when only the parts per million of the metal in the solution is given. In this case, the weight of the metal needs to be converted to the weight of the metal salt via a gravimetric factor (see Section 3.6.3) so that the weight of the metal salt is known.

### Example 5.10

How would you prepare 250.0 mL of a 50.0 ppm solution of nickel from pure solid nickel chloride hexahydrate, $NiCl_2 \cdot 6\ H_2O$, FW = 237.71 g/mol?

*Solution 5.10*

$$ppm_D \times L_D \times gravimetric\ factor = milligrams\ to\ be\ weighed$$

$$\frac{50.0\ mg}{L} \times 0.2500\ L \times \frac{NiCl_2 \cdot 6\ H_2O}{Ni} = milligrams\ to\ be\ weighed$$

$$\frac{50.0\ mg}{L} \times 0.2500\ L \times \frac{237.71}{58.71} = 50.6\ mg\ of\ NiCl_2 \cdot 6\ H_2O$$

Weigh 50.6 mg of $NiCl_2 \cdot 6\ H_2O$, place in a 250-mL volumetric flask, dissolve with water, and dilute to the mark.

## 5.3.5  Water Hardness Calculations

As mentioned earlier, the hardness of a water sample is often reported as parts per million of calcium carbonate. Using our definition of parts per million for liquid solutions (Table 5.4), we have

$$ppm_{CaCO_3} = \frac{weight\ of\ CaCO_3\ (in\ milligrams)}{volume\ of\ water\ (in\ liters)} \tag{5.47}$$

At the end point of the titration, the moles of $CaCO_3$ reacted are the same as the moles of EDTA added (it is a one-to-one reaction; see Equation (5.43)):

$$\text{moles}_{CaCO_3} = \text{moles}_{EDTA} \tag{5.48}$$

The moles of EDTA, and therefore the moles of $CaCO_3$, are computed from the titration data:

$$\text{moles}_{CaCO_3} = \text{moles}_{EDTA} = \text{liters}_{EDTA} \times M_{EDTA} \tag{5.49}$$

The grams of $CaCO_3$ can be computed by multiplying by the formula weight of $CaCO_3$ (100.03 g/mol) and converted to milligrams by multiplying by 1000 mg/g:

$$\text{liters}_{EDTA} \times M_{EDTA} \times FW_{CaCO_3} \times 1000 = \text{milligrams}_{CaCO_3} \tag{5.50}$$

The dimensional analysis of this equation is as follows, using 100.09 g/mol as the formula weight of $CaCO_3$:

$$\cancel{\text{liters}} \times \frac{\cancel{\text{moles}}}{\cancel{\text{liters}}} \times \frac{100.03\,\cancel{g}}{\cancel{\text{mol}}} \times \frac{1000\text{ mg}}{\cancel{g}} = \text{milligrams} \tag{5.51}$$

Parts per million is then calculated by dividing by the liters of water used in the titration. Combining Equations (5.47) and (5.50), we have

$$\text{ppm}_{CaCO_3} = \frac{\text{liters}_{EDTA} \times M_{EDTA} \times FW_{CaCO_3} \times 1000}{\text{liters}_{\text{water used}}} \tag{5.52}$$

## Example 5.11

What is the hardness in a water sample, 100 mL of which required 27.95 mL of a 0.01266 *M* EDTA solution for titration?

### Solution 5.11

Substituting into Equation (5.52), we have the following:

$$\text{ppm}_{CaCO_3} = \frac{\text{liters}_{EDTA} \times M_{EDTA} \times FW_{CaCO_3} \times 1000}{\text{liters}_{\text{water used}}}$$

$$\text{ppm}_{CaCO_3} = \frac{0.02795\,\cancel{L} \times 0.01266\,\cancel{\text{mol}}/\cancel{L} \times 100.09\,\cancel{g}/\cancel{\text{mol}} \times 1000\text{ mg}/\cancel{g}}{0.1000\text{ L}} = 354.2\text{ ppm } CaCO_3$$

Multiplying by the formula weight of $CaCO_3$ in the numerator has the effect of reporting all metals that reacted with EDTA as $CaCO_3$.

The molarity of the EDTA solution, $M_{EDTA}$ in Equation (5.52), can be known directly through its preparation with the use of an analytical balance and a volumetric flask. That is, one can purchase pure disodium dihydrogen EDTA and use it as a primary standard. In that case, the solution is prepared and the concentration calculated according to the discussion in Chapter 4 (see Section 4.3, especially Example 4.2, and Section 4.4.1).

## Example 5.12

The following EDTA solution is needed in a water hardness laboratory: 500.0 mL of a 0.01000 *M* solution of disodium dihydrogen EDTA dihydrate (FW = 372.23 g/mol). How many grams would be needed if you were to prepare it to be exactly 0.01000 *M*? How would you prepare this solution?

*Solution 5.12*

The grams of solute required is calculated according to Equation (4.11) in Chapter 4, in which $L_D$ is the liters of solution desired, $M_D$ is the molarity desired, and $FW_{SOL}$ is the formula weight of the solute.

$$\text{grams to weigh} = L_D \times M_D \times FW_{SOL}$$

$$= 0.5000 \, \cancel{L} \times 0.01000 \, \cancel{mol/L} \times 372.23 \, g/\cancel{mol}$$

$$= 1.861 \, g$$

The solution is prepared by weighing 1.861 g of dried disodium dihydrogen EDTA dihydrate on an analytical balance, transferring to a clean 500-mL volumetric flask, adding water to dissolve, diluting to the mark with water, and making homogeneous by shaking.

## Example 5.13

Suppose you followed the procedure in Example 5.12, but did not weigh *exactly* 1.861 g, but came close with 1.9202 g. What is the molarity of the EDTA solution prepared with this weight of solute?

*Solution 5.13*

The exact molarity can be calculated using Equation (4.1) in Chapter 4 as follows:

$$\text{molarity} = \frac{\text{moles of solute}}{\text{liters of solution}} = \frac{\text{grams of solute/FW of solute}}{\text{liters of solution}}$$

$$= \frac{(1.9202/372.23) \, \text{mol}}{0.5000 \, \text{L}} = 0.01032 \, M$$

Note: Standardization with a primary standard may still be important as a cross-check. See discussion and examples below.

   The molarity of the EDTA solution may also become known via standardization. In this case, the solution was prepared with an ordinary balance and a vessel other than a volumetric flask such that its molarity is not known to more than two significant figures. Just as with acids and bases (see Section 4.5), the standardization can take place by either weighing an appropriate primary standard accurately into the reaction flask or accurately pipetting a standard solution of a metal into the flask. The calculations are similar to those discussed in Chapter 4, and we can derive equations similar to Equations (4.17) and (4.18), but using molarity instead of normality. Starting with Equation (5.51), we have the following at the end point of a titration using calcium carbonate:

$$\text{moles}_{EDTA} = \text{moles}_{CaCO_3} \tag{5.53}$$

In the case of standardization with a solution of calcium carbonate, we have

$$V_{EDTA} \times M_{EDTA} = V_{CaCO_3} \times M_{CaCO_3} \tag{5.54}$$

In the case of standardization with a weighed quantity of calcium carbonate, we have

$$\text{liters}_{EDTA} \times M_{EDTA} = \frac{\text{grams}_{CaCO_3}}{FW_{CaCO_3}} \tag{5.55}$$

   It is also possible to prepare a standard solution of calcium carbonate accurately using an analytical balance and a volumetric flask, as was suggested previously for EDTA, and pipetting an **aliquot** (a portion of a larger volume) of this solution into the reaction flask in preparation for the standardization of the EDTA solution. In this case, the concentration of the calcium carbonate solution is first calculated from the weight–volume preparation data (refer back to Example 4.2) and then the molarity of the EDTA solution is determined using Equation (5.54).

## Example 5.14

A $CaCO_3$ solution is prepared by weighing 0.5047 g of $CaCO_3$ into a 500-mL volumetric flask, dissolving with HCl, and diluting to the mark. If a 25.00-mL aliquot of this solution required 28.12 mL of an EDTA solution, what is the molarity of the EDTA?

### Solution 5.14

The molarity of the $CaCO_3$ solution is first calculated from the weight and volume data given:

$$M_{CaCO_3} = \frac{(grams_{CaCO_3} / FW_{CaCO_3})\ mol}{liters_{solution}}$$

$$= \frac{0.5047\ g / 100.09\ g/mol}{0.5000\ L}$$

$$= 0.01008\ M$$

Next, using Equation (5.54), we have

$$28.12\ mL \times M_{EDTA} = 25.00\ mL \times 0.01008\ mol/L$$

$$M_{EDTA} = \frac{25.00\ \cancel{mL} \times 0.01008\ mol/L}{28.12\ \cancel{mL}} = 0.008962\ M$$

# 5.4 Oxidation–Reduction Reactions

Oxidation–reduction reactions represent yet another type of reaction that titrimetric analysis can utilize. In other words, a solution of an oxidizing agent can be in the buret, and a solution of a reducing agent can be in the reaction flask (and vice versa). In this section, we review the fundamentals of oxidization–reduction chemistry and discuss the titrimetric analysis applications.

## 5.4.1 Review of Basic Concepts and Terminology

Many chemical species have a tendency to either give up or take on electrons. This tendency is based on the premise that a greater stability, or lower energy state, is achieved as a result of this electron donation or acceptance. A hypothetical example is the reaction between a sodium atom and a chlorine atom:

$$\text{sodium atom (Na)} + \text{chlorine atom (Cl)} \rightarrow \text{sodium chloride formula unit (NaCl)} \quad (5.56)$$

A sodium atom with only one electron in its outermost energy level would achieve a lower energy state if this electron were released to the chlorine atom, which would also achieve a lower energy state as a result. Both atoms would become ions ($Na^+$ and $Cl^-$), and each would have a stable, filled outermost energy level identical with those of the noble gases, neon in the case of sodium and argon in the case of chlorine. Thus, the "electron transfer" does take place and sodium chloride, NaCl, is formed.

The reactions that this sodium–chlorine case typifies are called **oxidation–reduction reactions**. The term **oxidation** refers to the loss of electrons, while the term **reduction** refers to the gain of electrons. A number of oxidation–reduction reactions (nicknamed **redox reactions**) are useful in titrimetric analysis, and many are encountered in other analysis methods.

It may appear strange that the term reduction is associated with a gaining process. Actually, the term reduction was coined as a result of what happens to the oxidation number of the element when the electron transfer takes place. The **oxidation number** of an element is a number representing the state of the element with respect to the number of electrons the element has given up, taken on, or contributed to a covalent bond. For example, pure sodium metal has neither given up, taken on, nor shared electrons, and thus its oxidation number is zero. In sodium chloride, however, the sodium has given up an electron and becomes a +1 charge; thus its oxidation number is +1. The chlorine in NaCl has taken on an electron

**TABLE 5.5**    Rules for Assigning Oxidation Numbers

| Element | Oxidation Number |
|---|---|
| 1. Any uncombined element | 0 |
| 2. Combined alkali metal (group I) | +1 |
| 3. Combined alkaline earth metal (group II) | +2 |
| 4. Combined aluminum | +3 |
| 5. Halogen (group VIIA) in binary compound with a metal or hydrogen | −1 |
| 6. Hydrogen combined with nonmetal | +1 |
| 7. Hydrogen combined with metal (hydrides) | −1 |
| 8. Combined oxygen (except peroxides) | −2 |
| 9. Oxygen in peroxides | −1 |
| 10. All others—oxidation numbers are determined from the fact that the sum of all oxidation numbers within the molecule or polyatomic ion must add up to either zero (in the case of a neutral molecule) or the charge on the ion (in the case of a polyatomic ion) | |

and becomes a −1 charge; thus its oxidation number is −1. Chlorine also had an oxidation number of zero prior to the reaction. Thus, while it is true that the chlorine gained an electron, it is also true that its oxidation number became lower (from 0 to −1), hence the term reduction.

As we shall see, it is useful to be able to determine the oxidation number of a given element in a compound. Since most elements can exist in a variety of oxidation states, it is necessary to adopt a set of rules or guidelines for this determination. These are listed in Table 5.5.

A discussion of several examples should clarify the general scheme. First, rules 1 through 9 cover many situations in which, with few exceptions, the oxidation number is a set number. For example, in $BaCl_2$, barium is +2 (rule 3) and Cl is −1 (rule 5). In $K_2O$, potassium is +1 (rule 2) and O is −2 (rule 8). In $CaH_2$, calcium is +2 (rule 3) and H is −1 (rule 7).

Next, rule 10 is used for determining all oxidation numbers of all elements that rules 1 through 8 do not cover (this must be all but one element in a formula), and then assigning the remaining elements oxidation numbers, knowing that the total must add up to either zero or the ionic charge.

## Example 5.15

What is the oxidation number of sulfur in $H_2SO_4$?

*Solution 5.15*

$$\text{Oxygen is } -2 \text{ and } 4 \times (-2) = -8$$

$$\text{Hydrogen is } +1 \text{ and } 2 \times (+1) = +2$$

$$\text{Sum} = -8 + 2 = -6$$

Therefore, sulfur must be +6 in order for the total to be zero.

## Example 5.16

What is the oxidation number of manganese in $KMnO_4$?

*Solution 5.16*

$$\text{Potassium is } +1 \text{ and } 1 \times (+1) = +1$$

$$\text{Oxygen is } -2 \text{ and } 4 \times (-2) = -8$$

$$\text{Sum} = -8 + 1 = -7$$

Manganese must be +7 in $KMnO_4$ in order for the total to be zero.

## Example 5.17

What is the oxidation number of nitrogen in the nitrate ion $NO_3^{-1}$ ?

### Solution 5.17

$$Oxygen \text{ is } -2 \text{ and } 3 \times (-2) = -6$$

$$Sum = -6 + 5 = -1$$

Nitrogen must be +5 in $NO_3^-$ in order for the net charge to be −1.

The usefulness of determining the oxidation number in analytical chemistry is twofold. First, it will help determine if there was a change in oxidation number of a given element in a reaction. This always signals the occurrence of an oxidation–reduction reaction. Thus, it helps tell us whether a reaction is a redox reaction or some other reaction. Second, it will lead to the determination of the number of electrons involved, which will aid in balancing the equation. These latter points will be discussed in later sections.

## Example 5.18

Which of the following is a redox reaction?

(a)                                    $Cl_2 + 2\ NaBr \rightarrow Br_2 + 2\ NaCl$

(b)                                    $BaCl_2 + K_2SO_4 \rightarrow BaSO_4 + 2\ KCl$

### Solution 5.18

The oxidation numbers of Cl and Br in (a) have changed. Cl has changed from 0 to −1, while Br has changed from −1 to 0. (In diatomic molecules, the elements are considered to be in the uncombined state; thus rule 1 applies.) There is no change in any oxidation number of any element in (b). Thus, (a) is a redox reaction.

The terms oxidation and reduction, as should be obvious from the discussion to this point, can be defined in two different ways—according to the gain or loss of electrons, or according to the increase or decrease in oxidation number:

Oxidation—the loss of electrons or the increase in oxidation number

Reduction—the gain of electrons or the decrease in oxidation number

To say that a substance has been oxidized means that the substance has lost electrons. To say that a substance has been reduced means that the substance has gained electrons. To say that substance A has oxidized substance B means that substance A has caused B to lose electrons. Similarly, to say that substance A has reduced substance B means that substance A has caused B to gain electrons. When substance A causes substance B to be oxidized, substance A is called the **oxidizing agent**. When substance A causes substance B to be reduced, substance A is a **reducing agent**. To say that substance A "causes" oxidation or reduction means that substance A either removes electrons from or donates electrons to substance B. Thus, oxidation is always accompanied by reduction and vice versa. Also, every redox reaction has an oxidizing agent (which is the substance reduced) and a reducing agent (which is the substance oxidized).

## Example 5.19

Tell what is oxidized and what is reduced and what the oxidizing agent and the reducing agent are in the following:

$$Zn + 2HCl \rightarrow ZnCl_2 + H_2$$

### Solution 5.19

$$oxidation: Zn \rightarrow Zn^{2+}$$

(The oxidation number of Zn has increased from 0 to +2.)

reduction: $2 H^+ \rightarrow H_2$

(The oxidation number of H has decreased from +1 to 0.)

Zn has been oxidized; therefore, it is the reducing agent. $H^+$ has been reduced; therefore, it is the oxidizing agent.

## 5.4.2    The Ion-Electron Method for Balancing Equations

We have seen how analytical calculations in titrimetric analysis involve stoichiometry (Sections 4.5 and 4.6). We know that a balanced chemical equation is needed for basic stoichiometry. With redox reactions, balancing equations by "inspection" can be quite challenging, if not impossible. Thus, several special schemes have been derived for balancing redox equations. The **ion-electron method** for balancing redox equations takes into account the electrons that are transferred, since these must also be balanced. That is, the electrons given up must be equal to the electrons taken on. A review of the ion-electron method of balancing equations will therefore present a simple means of balancing redox equations.

The method makes use of only those species, dissolved or otherwise, that actually take part in the reaction. So-called **spectator ions**, or ions that are present but play no role in the chemistry, are not included in the balancing procedure. Solubility rules are involved here, since spectator ions result only when an ionic compound dissolves and ionizes. Also, the scheme is slightly different for acid and base conditions. Our purpose, however, is to discuss the basic procedure; thus spectator ions will be absent from all examples from the start, and acidic conditions will be the only conditions considered. The step-wise procedure we will follow is below:

Step 1: Look at the equation to be balanced and determine what is oxidized and what is reduced. This involves checking the oxidation numbers and discovering which have changed.

Step 2: Write a **half-reaction** for both the oxidation and reduction processes and label "oxidation" and "reduction." These half-reactions show only the species being oxidized (or the species being reduced) on the left side, with only the product of the oxidation (or reduction) on the right side.

Step 3: If oxygen appears in any formula on either side in either equation, it is balanced by writing $H_2O$ on the opposite side. This is possible since the reaction mixture is a water solution. The hydrogen in the water is then balanced on the other side by writing $H^+$, since we are dealing with acid solutions. Now balance both half-reactions for all elements by inspection.

Step 4: Balance the charges on both sides of the equation by adding the appropriate number of electrons ($e^-$) to whichever side is deficient in negative charges. The charge balancing is accomplished as if the electron is like any chemical species—place the appropriate multiplying coefficient in front of $e^-$.

Step 5: Multiply through both equations by appropriate coefficients, so that the number of electrons involved in both half-reactions is the same. This has the effect of making the total charge loss equal to the total charge gain and thus eliminates electrons from the final balanced equation, as you will see in step 6.

Step 6: Add the two equations together. The number of electrons, being the same on both sides, cancels out and thus does not appear in the final result. One can also cancel out some $H^+$ and $H_2O$, if they appear on both sides at this point.

Step 7: Make a final check to see that the equation is balanced.

### Example 5.20

Balance the following equation by the ion-electron method:

$$MnO_4^- + Fe^{2+} \rightarrow Fe^{3+} + Mn^{2+}$$

*Solution 5.20*

Step 1:    $Fe^{2+}$ oxidized from $+2 \rightarrow +3$ (loss of electrons)

$MnO_4^-$ reduced from $+7 \rightarrow +2$ (gain of electrons)

Step 2:     Oxidation: $Fe^{2+} \rightarrow Fe^{3+}$
            Reduction: $MnO_4^- \rightarrow Mn^{2+}$
Step 3:     Oxidation: $Fe^{2+} \rightarrow Fe^{3+}$
            Reduction: $8\ H^+ + MnO_4^- \rightarrow Mn^{2+} + 4\ H_2O$
Step 4:     Oxidation: $Fe^{2+} \rightarrow Fe^{3+} + 1\ e^-$
            Reduction: $5\ e^- + 8\ H^+ + MnO_4^- \rightarrow Mn^{2+} + 4\ H_2O$
Step 5:     Oxidation: $5(Fe^{2+} \rightarrow Fe^{+3} + 1\ e^-)$
            Reduction: $1(5\ e^- + 8\ H^+ + MnO_4^- \rightarrow Mn^{2+} + 4\ H_2O)$
Step 6:     Oxidation: $5(Fe^{2+} \rightarrow Fe^{3+} + 1e^-)$
            Reduction: $1(5\ e^- + 8\ H^+ + MnO_4^- \rightarrow Mn^{2+} + 4\ H_2O)$

$$5\ Fe^{2+} + 8\ H^+ + MnO_4^- \rightarrow 5\ Fe^{3+} + Mn^{2+} + 4\ H_2O$$

Step 7:     5 Fe on each side
            8 H on each side
            1 Mn on each side
            4 O on each side

The equation is balanced.

## 5.4.3   Analytical Calculations

Following are some examples of analytical calculations involving redox titrations.

**Example 5.21**

What is the molarity of a solution of $K_2Cr_2O_7$ if 0.6729 g of ferrous ammonium sulfate hexahydrate (sometimes referred to as FAS; FW = 392.14 g/mol) is exactly consumed by 24.92 mL of the solution as in the following equation?

$$Fe^{2+} + Cr_2O_7^{2-} \rightarrow Fe^{3+} + Cr^{3+}$$

*Solution 5.21*

Balancing the equation using the ion-electron method results in the following (check it):

$$6\ Fe^{2+} + 14\ H^+ + Cr_2O_7^{2-} \rightarrow 6\ Fe^{3+} + 2\ Cr^{3+} + 7\ H_2O$$

This is an example of a standardization with a primary standard. Calculating the molarity involves Equation (4.31) from Chapter 4:

$$L_T \times M_T \times \text{mole ratio (PS/T)} = \frac{\text{grams}_{PS}}{FW_{PS}}$$

$$\text{liters}_{K_2Cr_2O_7} \times M_{K_2Cr_2O_7} \times \text{mole ratio } (Fe^{2+}/Cr_2O_7^{2-}) = \frac{\text{grams}_{FAS}}{FW_{FAS}}$$

$$0.12492\ L \times M_{K_2Cr_2O_7} \times 6/1 = \frac{0.6729\ g}{392.14\ g\ mol^{-1}}$$

$$M_{K_2Cr_2O_7} = 0.01148\ M$$

**Example 5.22**

What is the percent of hydrogen peroxide ($H_2O_2$) in a sample if 23.01 mL of 0.04739 $M$ KMnO$_4$ was needed to titrate 10.0232 g of a hydrogen peroxide solution according to the following?

$$H_2O_2 + MnO_4^- \rightarrow O_2 + Mn^{2+}$$

*Solution 5.22*

Balancing the equation using the ion-electron method results in the following (check it):

$$5\ H_2O_2 + 6\ H^+\ 2\ MnO_4^- \rightarrow 5\ O_2 + 2\ Mn^{2+} + 8\ H_2O$$

Calculating the percent hydrogen peroxide utilizes Equation (4.37) from Chapter 4:

$$\%\ \text{analyte} = \frac{L_T \times M_T \times \text{mole ratio (ST/T)} \times FW_{\text{analyte}}}{\text{weight of sample}} \times 100$$

$$\%\ H_2O_2 = \frac{\text{liters}_{KMnO_4} \times M_{KMnO_4} \times \text{mole ratio } (H_2O_2/KMnO_4) \times FW_{H_2O_2}}{\text{weight of sample}} \times 100$$

$$\%\ H_2O_2 = \frac{0.02301\ L \times 0.04739\ mol\ L^{-1} \times 5/2 \times 34.01\ g\ mol^{-1}}{10.0232\ g} \times 100$$

$$= 0.9250\%$$

## 5.4.4   Applications

### 5.4.4.1   Potassium Permanganate

An oxidizing agent that has significant application in redox titrimetry is potassium permanganate, $KMnO_4$. As discovered earlier, the manganese in $KMnO_4$ has a +7 oxidation number. It is as if the manganese has contributed all seven of its outermost electrons ($4s^2$, $3d^5$) to a bonding situation—an observation that implies significant instability. Manganese atoms are in lower energy states if they are found with +4 or +2 oxidation numbers. Thus, like sodium and chlorine in the zero state, manganese in the +7 state is very unstable and will take electrons, given the opportunity, and be reduced. Hence, it is a strong (relatively speaking) oxidizing agent. Some notable examples of its oxidizing powers include $Fe^{2+}$ to $Fe^{3+}$, $H_2C_2O_4$ to $CO_2$, As (III) to As (V), and $H_2O_2$ to $O_2$. Organic compounds could also be included in this list, and in fact, potassium permanganate solutions, which are deep purple, are used to test qualitatively for alkenes, the reaction being the reduction of $MnO_4^-$ to $MnO_2$. The purple color disappears in this test, and the brown precipitate, $MnO_2$, forms and indicates a positive test.

Working with $KMnO_4$ presents some special problems because of its significant oxidizing properties. The keys to successful redox titrimetry using $KMnO_4$ are: 1) to prepare solutions well in advance so that any oxidizable impurities (usually organic in nature) in the distilled water used are completely oxidized, and 2) to protect the standardized solution from additional oxidizable materials (such as lint, fingerprints, rubber, etc.) so that its concentration remains constant until the solution is no longer needed. One additional problem is that once some $MnO_2$ has formed through the oxidation of such organic substances as those listed, it can catalyze further decomposition and thereby cause further changes in the $KMnO_4$ concentration. It is obvious that if the solutions are not carefully protected, the concentration of the standardized solution cannot be trusted. Even ordinary light can catalyze the reaction, and for this reason, solutions must be protected from light. The $MnO_2$ present from the initial reactions (prior to standardization) is filtered out, so that the catalytic reactions are minimized.

There are three major points to be made concerning the actual titrations using $KMnO_4$. First, since the solutions are unstable when first prepared due to the presence of oxidizable materials in the distilled water, distilled water cannot be used as a primary standard. Thus, $KMnO_4$ solutions must always be standardized prior to use. Second, all titrations are carried out in acid solution. Thus, $Mn^{2+}$ is the product, rather than $MnO_2$, and no brown precipitate forms in the reaction flask. Third, the fact that the color of the $KMnO_4$ solution is a deep purple means that the solution may serve as its own indicator. At the point when the substance titrated is exactly consumed, the $KMnO_4$ will no longer react, and since it is

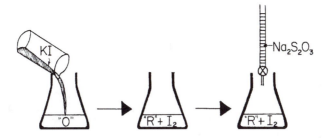

**FIGURE 5.22** In iodometry, a solution of KI is added to a solution of the analyte (O). The products are R and iodine ($I_2$). The amount of iodine, which is proportional to the amount of O, is titrated with sodium thiosulfate, $Na_2S_2O_3$.

a highly colored chemical species, the slightest amount of it in the reaction flask is easily seen. Thus, the end point is the first detectable pink color due to unreacted permanganate.

$KMnO_4$ solutions are standardized using primary standard grade reducing agents. Typical reducing agents include sodium oxalate, $Na_2C_2O_4$, and ferrous ammonium sulfate hexahydrate, $Fe(NH_4)_2(SO_4)_2 \cdot 6 H_2O$, also known as Mohr's salt.

### 5.4.4.2 Iodometry: An Indirect Method

Another important reactant in redox titrimetry is potassium iodide, KI. KI is a reducing agent ($2\ I^- \rightarrow I_2 + 2\ e^-$) that is useful in analyzing for oxidizing agents. The interesting aspect of the iodide–iodine chemistry is that it is most often used as an *indirect* method (recall the indirect Kjeldahl titration involving boric acid discussed previously). This means that the oxidizing agent analyte is not measured directly by a titration with KI, but is measured indirectly by the titration of the iodine that forms in the reaction. The KI is actually added in excess, since it need not be measured at all. The experiment is called iodometry. Figure 5.22 shows the sequence of events. Thus, the percent of the oxidizing agent (O in Figure 5.22) is calculated indirectly from the amount of titrant since the titrant actually reacts with $I_2$ and not O. This titrant is normally sodium thiosulfate ($Na_2S_2O_3$).

The sodium thiosulfate solution must be standardized. Several primary standard oxidizing agents are useful for this. Probably the most common one is potassium dichromate, $K_2Cr_2O_7$. Primary standard potassium bromate, $KBrO_3$, or potassium iodate, $KIO_3$, can also be used. Even primary standard iodine, $I_2$, can be used (but because solid iodine releases corrosive fumes, it should not be weighed on an analytical balance). Usually in the standardization procedures, KI is again added to the substance to be titrated ($Cr_2O_7^{2-}$, etc.) and the liberated iodine titrated with thiosulfate. If $I_2$ is the primary standard, it is titrated directly. The end point is usually detected with the use of a starch solution as the indicator. Starch, in the presence of iodine, is a deep blue color. It is not added, however, until near the end point, after the color of the solution changes from mahogany to straw yellow. Upon adding starch, the color changes to the deep blue. The addition of the thiosulfate is then continued until one drop changes the solution color from blue to colorless. Some important precautions concerning the starch, however, are to be considered. The starch solution should be fresh, should not be added until the end point is near, cannot be used in strong acid solutions, and cannot be used with solution temperatures above about 40°C.

An important application of iodometry can be found in many wastewater treatment plant laboratories. Chlorine, $Cl_2$, is used in a final treatment process prior to allowing the wastewater effluent to flow into a nearby river. Of course, the chlorine in both the free and combined forms can be just as harmful environmentally as many components in the raw wastewater. Thus, an important measurement for the laboratory to make is the amount of residual chlorine remaining unreacted in the effluent. Such chlorine, which is an oxidizing agent, can be determined by iodometry. It is the O in Figure 5.22. See also Workplace Scene 5.3.

# WORKPLACE SCENE 5.3

When bleaching wood pulp in a paper mill, the amount of leftover or residual hydrogen peroxide present in the spent bleach liquor can be important. Sometimes the bleach liquor is used in a second bleach or recycled back to the first bleach. Also, the residual peroxide level can indicate whether the bleach liquor makeup needs adjusting. Laboratory scale bleach tests are performed on the wood pulp, and the residual peroxide levels are checked by titration of the spent bleach liquors using potassium iodide and sodium thiosulfate. By monitoring the residual peroxide levels of these bleaches, the bleach liquor makeup can be adjusted, giving the best level, as needed by each paper mill.

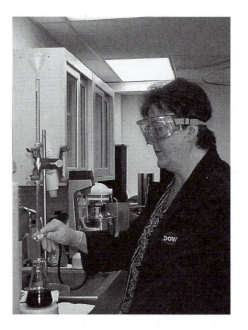

Pamela Slavings of the Dow Chemical Company in Midland, Michigan, performs the peroxide titration.

### 5.4.4.3  Prereduction and Preoxidation

Perhaps the most important application of redox chemicals in the modern laboratory is in oxidation or reduction reactions that are required as part of a preparation scheme. Such preoxidation or prereduction is also frequently required for certain instrumental procedures for which a specific oxidation state is essential in order to measure whatever property is measured by the instrument. An example in this textbook can be found in Experiment 19 (the hydroxylamine hydrochloride keeps the iron in the +2 state). Also in wastewater treatment plants, it is important to measure dissolved oxygen (DO). In this procedure, $Mn(OH)_2$ reacts with the oxygen in basic solution to form $Mn(OH)_3$. When acidified and in the presence of KI, iodine is liberated and titrated. This method is called the Winkler method.

## 5.5  Other Examples

Precipitation reactions are used for some determinations. These involve principally reactions utilizing the highly insoluble nature of silver compounds. Two example reactions are the Volhard method for silver $(Ag^+ + CNS^- \rightarrow AgCNS)$ and the Mohr method for chloride (also called the chloride method for silver; $Ag^+ + Cl^- \rightarrow AgCl$). End points in these can be detected in any one of several ways. A color change resulting from the first excess of the titrant may be utilized. In this case, the color change is due to the titrant reacting with another added component, the product being colored. With the precipitate also present in the solution, this color may be difficult to see. In these cases, a blank correction is needed.

End points may also be detected using so-called ion-selective electrodes. Such titrations are called potentiometric titrations and will be discussed in Chapter 14.

# Experiments

## Experiment 10: Titrimetric Analysis of a Commercial KHP Unknown for KHP

Note: This experiment calls for you to use a pH meter and a combination pH electrode (Chapter 14) to detect the end point of a titration. Your instructor may choose to have you use an indicator instead. Safety glasses are required.

1. Obtain an unknown KHP sample from your instructor and dry, as before, for 2 h. Allow to cool, and store in your desiccator.
2. Prepare three 250-mL beakers and weigh, by difference, three samples of the unknown KHP, each weighing between 1.0 and 1.3 g, into the beakers. Dissolve in 75 mL of distilled water. Place watch glasses on the beakers.
3. Prepare and standardize a pH meter with a combination probe for pH measurement. Your instructor may provide special instructions for the pH meter you are using. You will use an automatic stirrer with a magnetic stirring bar to stir the solution in the beaker while you are titrating. Mount the electrode in a ringstand clamp on a ringstand so that it is just immersed in the solution in one of the beakers. The beaker should be positioned on the stirrer with the stirring bar in the center of the beaker and the pH probe off to one side so as to not contact the stirring bar. The stirring speed should be slow enough so as not to splash the solution, but fast enough to thoroughly mix the added titrant quickly.
4. The titrant is the standardized 0.10 *M* NaOH from Experiment 8. The procedure is to monitor the pH as the titrant is added. A sharp increase in the pH will signal the end point. You will want to determine the midpoint of the sharp increase as closely as possible, since that will be the end point. The titrant is added very slowly when the pH begins to rise, since only a very small volume will be required at that point to reach the end point. You can add the titrant rapidly at first, but slow to a fraction of a drop when the pH begins to rise. The pH at the end point will be in the range of 8 to 10.
5. Record the buret readings at the end points for all three beakers. Calculate the percent KHP in the sample for all three titrations and include at least a sample calculation in your notebook, along with all results and the average. Calculate the ppt relative standard deviation. If your three results do not agree to within 10.0 ppt relative standard deviation, repeat until you have three that do or until you have a precision satisfactory to you instructor.

## Experiment 11: Titrimetric Analysis of a Commercial Soda Ash Unknown for Sodium Carbonate

Note: Safety glasses are required.

1. Obtain a soda ash sample from your instructor and dry, as before, for 2 h. Allow to cool and store in your desiccator.
2. Prepare the flasks and weigh, by difference, as before, three samples of the soda ash, each weighing between 0.4 and 0.5 g. Dissolve in 75 mL of water.
3. Perform this on each of your samples one at a time, starting with the one of least weight. Add three drops of bromcresol green indicator and titrate with your standard HCl solution from Experiment 8. The color will change slowly from blue to a light green. When the light green color is apparent, stop the titration, place a watch glass on the flask, and bring to a boil on a hot plate (see explanation in Section 5.2.8). Boil for 2 min. The color of the solution will turn back to blue. Cool to room temperature (you can use a cold water bath) and resume the titration (do not refill the buret). The color change now should be sharp, from blue to greenish yellow. Record the buret reading next to the corresponding weight.

4. Calculate the percentage of $Na_2CO_3$ in the sample, and record at least a sample calculation in your notebook, along with all results and the average. Calculate the ppt relative standard deviation. If your three results do not agree to within 10.0 ppt relative standard deviation, repeat until you have three that do or until you have a precision satisfactory to you instructor.

## Experiment 12: Determination of Protein in Macaroni by the Kjeldahl Method

### Part A: Preparation of Distillation Apparatus

1. The Labconco digestion and rapid distillation units are used for this experiment. Both are placed in a fume hood. Prepare the distillation unit ahead of time according to the manufacturer's instructions so that it is ready for use in Part D. Be sure that a beaker is in place to capture the distilled water as it drips from the condenser. The water in the flask used to generate the steam for the steam distillation should be boiling, and the flow rate of the replacement water and the rate of distillation should be equalized so that it is ready to accept samples immediately following the digestion step. The beaker capturing the distilled water will need to be emptied periodically.

### Part B: Solution Standardization

2. A 0.05 $N$ $H_2SO_4$ solution is the titrant for the ammonia to be distilled into the receiving vessel of the Kjeldahl distillation apparatus. A 0.01 $N$ NaOH solution is used for the back titrant to titrate the excess 0.05 $N$ $H_2SO_4$ after the distillation is completed. All students will use the same standard 0.05 $N$ $H_2SO_4$ and 0.01 $N$ NaOH solutions. Each student should perform just one standardization titration for each solution. When standardizing the NaOH with the KHP, since the concentration of the NaOH is one tenth what it was in Experiment 8, use one tenth as much of the KHP. For the standardization of the $H_2SO_4$, pipet only 5 mL into the titration flask rather than the 25 mL indicated in Experiment 8, since the $H_2SO_4$ is five times more concentrated than the NaOH. Pool your results with those of other students and perform the same precision tests you did in Experiment 8.

### Part C: Digestion

3. Weigh between 0.10 and 0.13 g of the macaroni sample on the analytical balance using weighing paper. Add this sample to a clean, dry 100-mL Kjeldahl flask to avoid having sample particles stick to the neck of the flask. Set aside for a few minutes (inside a beaker so that it stays upright) while you do step 4.

4. Pipet 5.00 mL of the 0.05 $N$ $H_2SO_4$ into a clean 250-mL beaker. Add 40 mL (use a graduated cylinder) of water and six drops of methyl red indicator to the beaker. This is the receiving vessel for the distillation. Place a watch glass on this beaker and set it aside while you finish this step and step 5. Take a 50-mL beaker on a top-loading balance and weigh 3.2 to 3.5 g of solid NaOH into the beaker. Dissolve in 10ml of water. USE CAUTION: this is a hazardous solution. Set these aside while you do step 5.

5. Add 2.5 mL of concentrated $H_2SO_4$ (CAUTION: hazardous solution) and 2.0 g of the Kjeldahl catalyst to the Kjeldahl flask from step 3. Digest by heating on the Kjeldahl digestion apparatus in the fume hood. The contents of this flask will turn black as the digestion proceeds, but eventually the blackness will go away while the $H_2SO_4$ boils. Ultimately, the flask contents will be a clear solution. Let cool for about 10 min, keeping upright as before. Be aware that the flask contains very hot, concentrated $H_2SO_4$, which is quite hazardous. Add a few milliliters of distilled water very slowly from a squeeze bottle and swirl to dissolve any crystallized material.

### Part D: Distillation

6. Set the 250-mL beaker in place on the platform of the distillation apparatus in the fume hood such that the glass tip of the condenser is under the surface of the solution in the beaker. Make sure that the stopcock on the upper left of the distillation unit (the addition stopcock) is closed. Also make sure that the other stopcock (the siphon stopcock) is closed. Add the contents of the Kjeldahl flask to the funnel above the addition stopcock. Open this stopcock to deliver the solution fairly

slowly to the mixing chamber below. Rinse the digestion flask with several small amounts of distilled water (total volume of rinsing not to exceed 10 mL), adding each rinse to the funnel and letting it drain into the mixing chamber each time. Close the addition stopcock after each addition. Similarly, rinse the funnel twice with a few milliliters of water from a squeeze bottle and add it to the mixing chamber. Close the stopcock, leaving a plug of distilled water in the funnel stem.

7. Carefully add the NaOH solution prepared in step 4 to the funnel. Slowly and intermittently, add this solution to the mixing chamber by opening the addition stopcock. The NaOH neutralizes the $H_2SO_4$ and causes the ammonia to form. The contents of the mixing chamber will get very dark at the point of neutralization. Again, leave a liquid plug in the stem. The distillation should be complete in 3 to 5 min. After this period, lower the receiving flask and allow the distillation to continue for another minute. This allows any solution from the receiving flask that was drawn into the condenser to rinse back into the receiving flask. Then set the receiving vessel aside for step 9. Place the distilled water beaker back under the condenser.

8. Open the siphon stopcock. The contents of the mixing chamber should siphon out. Close the siphon stopcock. Rinse the mixing chamber several times by adding small amounts of distilled water through the addition stopcock and subsequently siphoning through the siphon stopcock. This prepares the apparatus for the next student.

## Part E: Back Titration and Calculation

9. Titrate the solution in the receiving vessel with the standardized 0.01 $N$ NaOH until there is a color change from pink to light yellow. Calculate the percent protein in the sample as follows. A factor (F) is used to convert the weight of nitrogen to the weight of protein.

$$\% \text{ protein} = \frac{[\text{milliliters}_{H_2SO_4} \times N_{H_2SO_4} - \text{milliliters}_{NaOH} \times N_{NaOH}] \times 14.007 \times F}{\text{sample weight (milligrams)}} \times 100$$

F = 6.25 for macaroni

# Experiment 13: Analysis of Antacid Tablets[*]

## Introduction

In this experiment the neutralizing power of various antacids will be determined. Antacids contain basic compounds that will neutralize stomach acid (stomach acid is HCl). The amount of base in the antacid tablets will be determined by an acid–base titration. It is a back titration method. This method is used because most antacids produce carbon dioxide gas, which can interfere with the titration. By initially adding an excess of acid, one can drive off the $CO_2$ by boiling the solution before titrating the excess acid. There are many brands of commercial antacids with various ingredients. A few of the common ones are listed below:

| Brand | Active Ingredient |
|---|---|
| Mylanta[TM] | $Mg(OH)_2$ and $Al(OH)_3$ |
| Milk of Magnesia[TM] | $Mg(OH)_2$ |
| Rolaids[TM] | $NaAl(OH)_2CO_3$ |
| Tums[TM] | $CaCO_3$ |
| Alka-Seltzer[TM] II | $NaHCO_3$ and $KHCO_3$ |
| Maalox[TM] | $Mg(OH)_2$ and $Al(OH)_3$ |

[*]This experiment was provided by Susan Marine of Miami University Middletown, Middletown, Ohio.

These bases react with acid as shown below:

$$Mg(OH)_2 + 2\ H^+ \rightarrow Mg^{2+} + 2\ H_2O$$

$$Al(OH)_3 + 3\ H^+ \rightarrow Al^{3+} + 3\ H_2O$$

$$NaAl(OH)_2CO_3 + 4\ H^+ \rightarrow NaAl^{4+} + CO_2 + 3\ H_2O$$

$$CaCO_3 + 2\ H^+ \rightarrow Ca^{2+} + CO_2 + H_2O$$

$$NaHCO_3 + H^+ \rightarrow Na^+ + CO_2 + H_2O$$

*Remember safety glasses.*

**Procedure**

1. Prepare a solution of HCl that is approximately 0.2 *M* and standardize it. Prepare a solution of NaOH that is between 0.10 and 0.15 *M* and standardize it.
2. Record the name of the antacid you are testing. Describe the bottle; include price, net weight, number of tablets, active ingredients, and other information from the label.
3. Weigh three individual antacid tablets (of the same brand) to the nearest 0.001 g and place them in separate 250-mL Erlenmeyer flasks. Crush the tablets thoroughly with a glass stirring rod.
4. Using a volumetric pipet, add 50.00 mL of 0.2 *M* HCl to each flask and dissolve the tablets. Using a graduated cylinder, add 50 mL of distilled water to each flask. Heat the flasks on a hot plate and boil the solutions gently for approximately 10 min.
5. After the flasks have cooled, swirl the solutions and add four or five drops of bromcresol green or methyl red indicator mix. If the solution is green, all the base has not been neutralized. In this case, add 25.00 mL of 0.2 *M* HCl with a volumetric pipet and boil again for 10 min. Repeat this process until the solution remains red.
6. When the solution is red, there is an excess of HCl. At this point, titrate the solution with the standardized NaOH to a green end point. Record the readings of the buret.
7. Calculate the average moles of acid neutralized per tablet, the average moles of acid neutralized per gram of tablet, and the average percent base (calculated as $CaCO_3$) in the antacid. Compare your results with others in the class for the different brands analyzed.

## Experiment 14: Determination of Water Hardness

Note: All Erlenmeyer and volumetric flasks used in this experiment must be rinsed thoroughly with distilled water prior to use. Ordinary tap water contains hardness minerals that will contaminate. The pH = 10 ammonia buffer required can be prepared by dissolving 35 g of $NH_4Cl$ and 285 mL of concentrated ammonium hydroxide in water and diluting to 500 mL. The EBT indicator should be fresh and prepared by dissolving 200 mg in a mixture of 15 mL of triethanolamine and 5 mL of ethyl alcohol.

*Remember safety glasses.*

1. Weigh 4.0 g of disodium dihydrogen EDTA dihydrate and 0.10 g of magnesium chloride hexahydrate into a 1-L glass bottle. Add one pellet of NaOH and fill to approximately 1 L with water. Shake well. These ingredients will require some time to dissolve, so it is recommended that this solution be prepared one lab session ahead of its intended use.
2. Dry a quantity (at least 0.5 g) of primary standard $CaCO_3$ for 1 h. Your instructor may choose to dispense this to you. While drying, continue with the procedure below.
3. Obtain a water sample. It will be contained in a 1-L volumetric flask. Dilute to the mark with distilled water. Shake well.

4. Pipet 100.0-mL aliquots of the water sample (an **aliquot** is a portion of a solution) into each of three clean 500-mL Erlenmeyer flasks. Add 5.0 mL (graduated cylinder) of the pH = 10 buffer and three drops of EBT indicator to each flask.

5. Give your EDTA solution one final shake to ensure its homogeneity. Then, clean a 50-mL buret, rinse it several times with your EDTA solution (the titrant), and fill, as usual, with your titrant. Be sure to eliminate the air bubbles from the stopcock and tip. Titrate the solutions in your flask, one at a time, until the last trace of red disappears in each with a fraction of a drop. This final color change will be from a violet color to a deep sky blue, but should be a sharp change. All three titrations should agree to within 0.05 mL of each other. If they do not, repeat until you have three that do. Record all readings in your notebook.

6. With this step, we begin the EDTA standardization process. Weigh accurately 0.3 to 0.4 g of the dried and cooled $CaCO_3$, by difference, into a clean, dry, short-stemmed funnel in the mouth of a 500-mL volumetric flask. Tap the funnel gently to force the $CaCO_3$ into the flask. Wash any remaining $CaCO_3$ into the flask with a squeeze bottle. Add a small amount of concentrated HCl (less than 2.0 mL total) to the flask through the funnel such that the funnel is rinsed thoroughly in the process. Rinse the funnel into the flask one more time with distilled water and remove the funnel. Rinse the neck of the flask with distilled water. Swirl the flask until all the $CaCO_3$ is dissolved and effervescence has ceased. Dilute to the mark with distilled water and shake well. If the solution is warm at this point, cool in a cold water bath and then add more distilled water to bring the meniscus back to the mark. Shake well again.

7. Pipet a 25.00-mL aliquot of this standard calcium solution into each of three 250-mL Erlenmeyer flasks, add the buffer solution, and titrate each to the same end point as before, but use only two drops of EBT. All buret readings should agree to within 0.05 mL. If they do not, repeat until you have three that do.

8. Calculate the molarity of your EDTA for each of the three titrations and calculate the average. Record in your notebook. Also calculate the parts per million $CaCO_3$ in the sample and compute the average. Record, as usual, in your notebook.

## Questions and Problems

1. What are three attributes of a successful titrimetric analysis?
2. Define monoprotic acid, polyprotic acid, monobasic base, polybasic base, titration curve, and inflection point.
3. Compare the titration curves for 0.10 *M* hydrochloric acid and 0.10 *M* acetic acid each titrated with 0.10 *M* sodium hydroxide. What parts of the titrations curves are the same and what parts are different? Why? Compare the inflections points for the two curves and tell what impact the differences have on indicator selection.
4. Repeat problem 3, but compare the titration curves of 0.10 *M* sodium hydroxide with 0.10 *M* ammonium hydroxide titrated with 0.10 *M* hydrochloric acid.
5. Repeat problem 3, but compare the titration curves of 0.10 *M* sulfuric acid and 0.10 *M* phosphoric acid titrated with 0.10 *M* sodium hydroxide.
6. Phenolphthalein indicator changes color in the pH range of 8 to 10. Methyl orange changes color in the pH range of 3 to 4.5. Roughly sketch two titration curves as follows:
   (a) One that represents a titration in which phenolphthalein would be useful but methyl orange would not
   (b) One that represents a titration in which methyl orange would be useful but phenolphthalein would not
7. Roughly sketch the following three titration curves:
   (a) A weak acid titrated with a strong base
   (b) A strong base titrated with a strong acid
   (c) A weak base titrated with a strong acid

8. Bromcresol green was used as the indicator for the $Na_2CO_3$ titration in Experiment 11. Would phenolphthalein also work? Explain.

9. Would bromcresol green be an appropriate indicator for an acetic acid titration? Explain.

10. Look at Figures 5.5 and 5.7 and tell what indicator you would recommend for the titration of phosphoric acid at the second inflection point. Explain.

11. What is the P in KHP? Draw the structure of KHP. Tell why it is useful as a primary standard chemical.

12. What is the name of the chemical often referred to as THAM or TRIS? What is its structure? It does not contain $OH^-$ ions, yet it is a base. Explain. Tell why it is useful as a primary standard chemical.

13. Why does the titration curve of sodium carbonate have two inflection points? Why does this titration require that the solution be boiled as you approach the second equivalence point? Why can bromcresol green be used as the indicator and not phenolphthalein?

14. Define total alkalinity.

15. What is the alkalinity of a water sample if 50.00 mL of the water sample required 3.09 mL of 0.09928 *M* HCl to reach a pH of 4.5?

16. Describe in your own words what a back titration is.

17. Why is a back titration useful in the analysis of an antacid tablet containing calcium carbonate as the active ingredient?

18. What is the percent of $CaCO_3$ in an antacid given that a tablet that weighed 1.2918 g reacted with 50.00 mL of 0.4501 *N* HCl that subsequently required 3.56 mL of 0.1196 *N* NaOH for back titration? Also report the milligrams of $CaCO_3$ in the tablet.

$$2\ HCl + CaCO_3 \rightarrow CaCl_2 + CO_2 + H_2O$$

19. In the Kjeldahl titration
    (a) What acid is used to dissolve or digest the sample?
    (b) What substance (be specific) is added to neutralize this acid?
    (c) What substance (be specific) is distilled into the receiving flask?
    (d) What kind of substance (acid or base) is in the receiving flask before the distillation?
    (e) What kind of substance (acid or base) is the titrant?
    (f) What kind of substance (acid or base) is the back titrant?

20. Concerning the Kjeldahl method for nitrogen,
    (a) What is concentrated sulfuric acid used for?
    (b) What might boric acid be used for?

21. In the calculation of the percent analyte when using a back titration, the following appears in the numerator:

$$(L_T \times N_T - L_{BT} \times N_{BT})$$

Why is it necessary to do this subtraction?

22. A grain sample was analyzed for nitrogen content by the Kjeldahl method. If 1.2880 g of the grain was used, and 50.00 mL of 0.1009 *N* HCl was used in the receiving flask, what is the percent *N* in the sample when 5.49 mL of 0.1096 *N* NaOH was required for back titration?

23. A flour sample was analyzed for nitrogen content by the Kjeldahl method. If 0.9819 g of the flour was used, and 35.10 mL of 0.1009 *N* HCl was used to titrate the boric acid solution in the receiving flask, what is the percent nitrogen in the sample?

24. Define buffer solution, conjugate acid, conjugate base, conjugate acid–base pair, buffer capacity, and buffer region.

25. Under what circumstances can the acetate ion be thought of as a conjugate base? Under what circumstances can the ammonium ion be thought of as a conjugate acid?

26. What is the Henderson–Hasselbalch equation? Tell how it is useful in the preparation of buffer solutions.

27. What is the pH of a solution of chloroacetic acid (0.25 $M$) and sodium chloroacetate (0.20 $M$)? The $K_a$ of chloroacetic acid is $1.36 \times 10^{-3}$.

28. What is the pH of a solution that is 0.30 $M$ in iodoacetic acid and 0.40 $M$ in sodium iodoacetate? The $K_a$ of iodoacetic acid is $7.5 \times 10^{-4}$.

29. What is the pH of a solution of THAM and THAM hydrochloride if the THAM concentration is 0.15 $M$ and the THAM hydrochloride concentration is 0.40 $M$? The $K_a$ of THAM hydrochloride is $8.41 \times 10^{-9}$.

30. What ratio of the concentrations of acetic acid to sodium acetate is needed to prepare a buffer solution of pH = 4.00? The $K_a$ of acetic acid is $1.76 \times 10^{-5}$.

31. A buffer solution of pH 3.00 is needed. From Table 5.1, select a weak acid–conjugate base combination that would give that pH and calculate the ratio of acid concentration to conjugate base concentration that would give that pH.

32. To prepare a certain buffer solution, it is determined that acetic acid must be present at 0.17 $M$ and that sodium acetate must be present at 0.29 $M$, both in the same solution. If the sodium acetate ($NaC_2H_3O_2$) is a pure solid chemical and the acetic acid to be measured out is a concentrated solution (17 $M$), how would you prepare 500 mL of this buffer solution?

33. Tell how you would prepare 500 mL of the buffer solution in question 29. Both THAM and THAM hydrochloride are pure solid chemicals. The formula weight of THAM is 121.14 g/mol. The formula weight of THAM hydrochloride is 157.60 g/mol.

34. In Experiment 14, a recipe for a pH = 10 buffer solution is given. This recipe calls for dissolving 35 g of $NH_4Cl$ and 285 mL of concentrated ammonium hydroxide (15 $M$) in water in the same container and diluting with water to 500 mL. Calculate the pH of the buffer using the data given, and confirm that it really is 10. The $K_a$ for the ammonium ion is $5.70 \times 10^{-10}$.

35. Define monodentate, bidentate, hexadentate, ligand, complex ion, chelate, chelating agent, masking, masking agent, formation constant, coordinate covalent bond, water hardness, aliquot.

36. Define ligand and complex ion. Give an example of each.

37. Given the following reaction, tell which of the three species is the ligand and which is the complex ion:

$$Co^{2+} + 4\ Cl^- \rightleftharpoons CoCl_4^{2-}$$

38. Give one example each (either structure or name) of a monodentate ligand and a hexadentate ligand. Explain what is meant by monodentate.

39. Is the ligand ethylenediamine, $H_2NCH_2CH_2NH_2$, monodentate, bidentate, tridentate, or what? Explain your answer.

40. Consider the reaction shown in Figure 5.23.
    (a) Which chemical species is a ligand?
    (b) Which chemical species is a complex ion?
    (c) Is the ligand monodentate, bidentate, or what? Explain your answer.
    (d) Is the complex ion a chelate? Explain.

41. Concerning the EDTA ligand,
    (a) How many bonding sites on this molecule bond to a metal ion when a complex ion is formed?
    (b) How many EDTA molecules will bond to a single metal ion?
    (c) What is the word describing the property pointed out in (a)?

42. Explain why water samples titrated with EDTA need to be buffered at pH = 10 and not at pH = 12 or pH = 8.

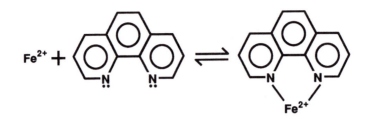

**FIGURE 5.23**   The reaction for problem 40.

43. In the water hardness titration,
    (a) What chemical species is the wine-red color at the beginning of the titration due to?
    (b) What chemical species is the sky-blue color at the end point due to?
    (c) What does it mean to say that cyanide is a masking agent?
44. Explain why the pH = 10 ammonia buffer is required in EDTA titrations for water hardness.
45. How would you prepare each of the following?
    (a) 250.0 mL of a 25 ppm solution of magnesium from pure magnesium metal
    (b) 750.0 mL of a solution that is 30.0 ppm silver using pure silver metal
    (c) 600.0 mL of a solution that is 40.0 ppm aluminum using pure Al metal
    (d) 500.0 mL of a 15 ppm Mg solution using pure magnesium for the solute
    (e) 250.0 mL of a 30.0 ppm solution of iron using pure iron metal for the solute
    (f) 100.0 mL of a solution that is 125 ppm copper using pure copper metal
46. How many milliliters of a 1000.0 ppm solution of the metal is needed to prepare each of the following?
    (a) 100.0 mL of a 125 ppm solution of copper
    (b) 600.0 mL of a 15 ppm solution of zinc
    (c) 250.0 mL of a 25 ppm solution of sodium
47. How would you prepare 500.0 mL of a 50.0 ppm Na solution
    (a) From pure, solid NaCl?
    (b) From a solution that is 1000.0 ppm Na?
48. How would you prepare 250.0 mL of a 30.0 ppm solution of iron
    (a) Using pure iron wire as the solute?
    (b) Using solid $Fe(NO_3)_3 \cdot 9 H_2O$ (FW = 404.02) as the solute?
    (c) If you needed to dilute a 1000.0 ppm solution of iron?
49. Tell how you would prepare 100.0 mL of a 50.0 ppm solution of copper
    (a) Using pure, solid copper sulfate pentahydrate, $CuSO_4 \cdot 5 H_2O$
    (b) Using pure copper metal
    (c) By diluting 1000.0 ppm copper solution?
50. A technician wishes to prepare 500 mL of a 25.0 ppm solution of barium.
    (a) How many mL of 1000.0 ppm barium would be required if he or she were to prepare this by dilution?
    (b) How many grams would be required if he or she were able to prepare this from pure barium metal?
    (c) How many grams would be required if he or she were to prepare this from pure, solid $BaCl_2 \cdot 2 H_2O$?
51. How many milligrams of the solute is needed to prepare the following volumes of solution?
    (a) Solute is KBr, 600.0 mL of a 40.0 ppm bromide solution is needed
    (b) Solute is $K_2HPO_4$, 450.0 mL of a 10.0 ppm phosphorus solution is needed
    (c) Solute is $KNO_3$, 100.0 mL of a solution that is 50.0 ppm nitrogen is needed

52. Tell how you would prepare 500.0 mL of a 0.0250 *M* solution of the solid disodium dihydrogen EDTA dihydrate for use as a standard solution without having to standardize it.

53. How many grams of disodium dihydrogen EDTA dihydrate is required to prepare 1000 mL of a 0.010 *M* EDTA solution?

54. What is the molarity of an EDTA solution given the following standardization data?
    (a) If 10.0 mg of Mg required 40.08 mL of the EDTA
    (b) If 0.0236 g of solid $CaCO_3$ was dissolved and exactly consumed by 12.01 mL of the EDTA solution
    (c) If 30.67 mL of it reacts exactly with 45.33 mg of calcium metal
    (d) If 34.29 mL of it is required to react with 0.1879 g of $MgCl_2$
    (e) If a 100.0-mL aliquot of a zinc solution required 34.62 mL of it (zinc solution was prepared by dissolving 0.0877 g of zinc in 500.0 mL of solution)
    (f) If a solution of primary standard $CaCO_3$ was prepared by dissolving 0.5622 g of $CaCO_3$ in 1000 mL of solution; a 25.00-mL aliquot of it required 21.88 mL of the EDTA
    (g) If 25.00 mL of a solution prepared by dissolving 0.4534 g of $CaCO_3$ in 500.0 mL of solution reacts with 34.43 mL of the EDTA solution
    (h) If a solution has 0.4970 g of $CaCO_3$ dissolved in 500.0 mL and 25.00 mL of it reacts exactly with 29.55 mL of the EDTA solution
    (i) If 25.00 mL of a $CaCO_3$ solution reacts with 30.13 mL of the EDTA solution and there is 0.5652 g of $CaCO_3$ per 500.0 mL of the solution

55. What is the hardness of the water sample in parts per million $CaCO_3$ in each of the following situations?
    (a) If a 100.0-mL aliquot of the water required 27.62 mL of 0.01462 *M* EDTA for titration
    (b) If 25.00 mL of the water sample required 11.68 mL of 0.01147 *M* EDTA
    (c) If 12.42 mL of a 0.01093 *M* EDTA solution was needed to titrate 50.00 mL of the water sample
    (d) If, in the experiment for determining water hardness, 75.00 mL of the water sample required 13.03 mL of a 0.009242 *M* EDTA solution
    (e) If the EDTA solution used for the titrant was 0.01011 *M*, a 150.0-mL sample of water was used, and 16.34 mL of the titrant was needed
    (f) If 14.20 mL of an EDTA solution, prepared by dissolving 4.1198 g of $Na_2H_2EDTA \cdot 2\,H_2O$ in 500.0 mL of solution, was needed to titrate 100.0 mL of a water sample
    (g) When 100.0 mL of the water required 13.73 mL of an EDTA solution prepared by dissolving 3.8401 g of $Na_2H_2EDTA \cdot 2\,H_2O$ in 500.0 mL of solution

56. Define oxidation, reduction, oxidation number, oxidizing agent, and reducing agent.

57. What is the oxidation number of each of the following?
    (a) P in $H_3PO_4$
    (b) Cl in $NaClO_2$
    (c) Cr in $CrO_4^{2-}$
    (d) Br in $KBrO_3$
    (e) I in $IO_4^-$
    (f) N in $N_2O$
    (g) S in $H_2SO_3$
    (h) S in $H_2SO_4$
    (i) N in $NO_2^-$
    (j) P in $PO_3^{3-}$

58. What is the oxidation number of bromine (Br) in each of the following?
    (a) HBrO
    (b) NaBr
    (c) $BrO_3^-$
    (d) $Br_2$

(e) $Mg(BrO_2)_2$

(f) $BrO_3^-$

59. What is the oxidation number of chromium (Cr) in each of the following?

    (a) $CrBr_3$

    (b) $Cr$

    (c) $CrO_3$

    (d) $CrO_4^{2-}$

    (e) $K_2Cr_2O_7$

60. What is the oxidation number of iodine (I) in each of the following?

    (a) $HIO_4$

    (b) $CaI_2$

    (c) $I_2$

    (d) $IO_2^-$

    (e) $Mg(IO)_2$

61. What is the oxidation number of sulfur (S) in each of the following?

    (a) $SO_2$

    (b) $H_2S$

    (c) $S$

    (d) $SO_4^{2-}$

    (e) $K_2SO_3$

    (f) $K_2S$

    (g) $H_2SO_4$

    (h) $SO_3^{2-}$

    (i) $SO_2$

    (j) $SF_6$

62. What is the oxidation number of phosphorus (P) in each of the following?

    (a) $P_2O_5$

    (b) $Na_3PO_3$

    (c) $H_3PO_4$

    (d) $PCl_3$

    (e) $HPO_3^{2-}$

    (f) $PO_4^{3-}$

    (g) $P^{3-}$

    (h) $Mg_2P_2O_7$

    (i) $P$

    (j) $NaH_2PO_4$

63. In the following redox reactions, tell what has been oxidized and what has been reduced, and explain your answer:

    (a) $3\ CuO + 2\ NH_3 \rightarrow 3\ Cu + N_2 + 3\ H_2O$

    (b) $Cl_2 + 2\ KBr \rightarrow Br_2 + 2\ KCl$

64. In each of the following reactions, tell what is the oxidizing agent and what is the reducing agent, and explain your answers.

    (a) $Mg + 2\ HBr \rightarrow MgBr_2 + H_2$

    (b) $4\ Fe + 3\ O_2 \rightarrow 2\ Fe_2O_3$

65. Which of the following unbalanced equations represent redox reactions? Explain your answers.

    (a) $H_2SO_4 + NaOH \rightarrow Na_2SO_4 + H_2O$

    (b) $H_2S + HNO_3 \rightarrow S + NO + H_2O$

    (c) $Na + H_2O \rightarrow NaOH + H_2$

    (d) $H_2SO_4 + Ba(OH)_2 \rightarrow BaSO_4 + H_2O$

    (e) $K_2CrO_4 + Pb(NO_3)_2 \rightarrow 2\ KNO_3 + PbCrO_4$

(f) $K + Br_2 \rightarrow 2\ KBr$

(g) $2\ KClO_3 \rightarrow 2\ KCl + 3\ O_2$

(h) $KOH + HCl \rightarrow KCl + H_2O$

(i) $BaCl_2 + Na_3PO_4 \rightarrow Ba_3(PO_4)_2 + NaCl$

(j) $Mg + HCl \rightarrow MgCl_2 + H_2$

66. Consider the following two unbalanced equations:

(a) $KOH + HCl \rightarrow KCl + H_2O$

(b) $Cu + HNO_3 \rightarrow Cu(NO_3)_2 + NO + H_2O$

Which represents a redox reaction, (a) or (b)? In the redox reaction, what has been oxidized and what has been reduced? What is the oxidizing agent and what is the reducing agent?

67. One of the following unbalanced equations represents a redox reaction and one represents a reaction that is not a redox reaction. Select the equation that is a redox reaction and answer the questions that follow.

(1) $$Pb(NO_3)_2 + K_2CrO_4 \rightarrow PbCrO_4 + 2\ KNO_3$$

(2) $$Zn + HCl \rightarrow ZnCl_2 + H_2$$

(a) Which one is redox, (1) or (2)?

(b) What is the oxidizing agent?

(c) What has been oxidized?

(d) Did the reducing agent lose or gain electrons?

68. Balance the following equations by the ion-electron method:

(a) $Cl^- + NO_3^- \rightarrow ClO_3^- + N^{3-}$

(b) $Cl^- + NO_3^- \rightarrow ClO_2^- + N_2O$

(c) $ClO^- + NO_3^- \rightarrow ClO_{3-} + NO_2^-$

(d) $ClO^- + NO_3^- \rightarrow ClO_2^- + NO$

(e) $ClO_3^- + SO_4^{2-} \rightarrow ClO_4^- + S^{2-}$

(f) $BrO_3^- + SO_4^{2-} \rightarrow BrO_4^- + SO_3^{2-}$

(g) $IO_3^- + SO_3 \rightarrow IO_4^- + S^{2-}$

(h) $Cl^- + SO_4^{2-} \rightarrow ClO^- + SO_2$

(i) $Cl^- + SO_4^{2-} \rightarrow ClO_4^- + S$

(j) $Br^- + SO_3 \rightarrow Br_2 + S$

(k) $I^- + NO_3^- \rightarrow IO_2^- + N_2$

(l) $P + IO_4^- \rightarrow PO_4^{3-} + I^-$

(m) $SO_2 + BrO_3^- \rightarrow SO_4^{2-} + Br^-$

(n) $Fe + P_2O_5 \rightarrow Fe^{3+} + P$

(o) $Cr + PO_4^{3-} \rightarrow Cr^{3+} + PO_3^{3-}$

(p) $Ni + PO_4^{3-} \rightarrow Ni^{2+} + P$

(q) $MnO_4^- + H_2C_2O_4 \rightarrow Mn^{2+} + CO_2$

(r) $I^- + Cr_2O_7^{2-} \rightarrow I_2 + Cr^{3+}$

(s) $Cl_2 + NO_2^- \rightarrow Cl^- + NO_{3-}$

(t) $S^{2-} + NO_3^- \rightarrow S + NO$

(u) $SO_3^{2-} + NO_3^- \rightarrow SO_4^{2-} + NO_2^-$

69. (a) Balance the following equation:

$$S_2O_3^{2-} + Cr_2O_7^{2-} \rightarrow S_4O_6^{2-} + Cr^{3+}$$

(b) If 0.5334 g of $K_2Cr_2O_7$ was titrated with 24.31 mL of the $Na_2S_2O_3$ solution, what is the molarity of the $Na_2S_2O_3$?

70. Consider the reaction of $Fe^{2+}$ with $K_2Cr_2O_7$ according to the following:

$$Fe^{2+} + Cr_2O_7^{2-} \rightarrow Fe^{3+} + Cr^{3+}$$

What is the exact molarity of a solution of $K_2Cr_2O_7$ if 1.7976 g of Mohr's salt—an $Fe^{2+}$ compound, $Fe(NH_4)_2(SO_4)_2 \cdot 6\, H_2O$—was exactly reacted with 22.22 mL of the solution?

71. Consider the standardization of a solution of $KIO_4$ with Mohr's salt (see problem 70) according to the following:

$$Fe^{2+} + IO_4^- \rightarrow Fe^{3+} + I^-$$

What is the exact molarity of the solution if 1.8976 g of Mohr's salt was exactly reacted with 24.22 mL of the solution?

72. Consider the standardization of a solution of $K_2Cr_2O_7$ with iron metal according to the following:

$$Fe + Cr_2O_7^{2-} \rightarrow Fe^{3+} + Cr^{3+}$$

What is the exact molarity of the solution if 0.1276 g of iron metal was exactly reacted with 48.56 mL of the solution?

73. What is the percent $SO_3$ in a sample if 45.69 mL of a 0.2011 $M$ $KIO_3$ solution is needed to consume 0.9308 g of sample according to equation (g) of problem 68?

74. What is the percent of $K_2SO_4$ in a sample if 35.01 mL of a 0.09123 $M$ $KBrO_3$ solution is needed to consume 0.7910 g of sample according to equation (f) of problem 68?

75. What is the percent of Fe in a sample titrated with $K_2Cr_2O_7$ according to the following equation if 2.6426 g of the sample required 40.12 mL of 0.1096 $M$ $K_2Cr_2O_7$?

$$Fe^{2+} + Cr_2O_7^{2-} \rightarrow Fe^{3+} + Cr^{3+}$$

76. What is the percent of Sn in a sample of ore if 4.2099 g of the ore was dissolved and titrated with 36.12 mL of 0.1653 $M$ $KMnO_4$?

$$MnO_4^- + Sn^{2+} \rightarrow Sn^{4+} + Mn^{2+}$$

77. What does it mean to say that potassium permanganate is its own indicator?
78. (a) Why is potassium permanganate, $KMnO_4$, termed an oxidizing agent?
    (b) Why must a standardized potassium permanganate solution be carefully protected from oxidizable substances if you expect it to remain standardized?
79. Explain the difference between an indirect titration and a back titration.
80. One redox method we have discussed is called iodometry. What is iodometry and why is it called an indirect method?
81. Briefly explain the use of the following substances in iodometry:
    (a) KI
    (b) $Na_2S_2O_3$
    (c) $K_2Cr_2O_7$
82. **Report**
    Find a real-world titrimetric analysis in a methods book (AOAC, USP, ASTM, etc.) or journal or on a website and report on the details of the procedure according to the following scheme:
    (a) Title
    (b) General information, including type of material examined, the name of the analyte, sampling procedures, and sample preparation procedures

(c) Specifics, including what titrant is used and how it is standardized (including what primary standards are used); what solutions are needed and how they are prepared; what glassware is needed and for what; what end point detection method is used for both standardization and analysis steps; what reactions (write balanced equations) are involved in both the standardization and analysis steps; whether it is a direct, indirect, or back titration (both standardization and analysis steps); and any special procedures, potential problems, etc.

(d) Data handling and reporting

(e) References

# Introduction to Instrumental Analysis

## 6.1  Review of the Analytical Strategy

In our introduction to analytical chemistry in Chapter 1, we presented a flow chart outlining the general strategy of analytical chemistry. This flow chart is reproduced in Figure 6.1. As shown in the figure, the analytical strategy consists of five parts: 1) sampling, 2) sample preparation, 3) the analytical method, 4) data handling, and 5) the calculation and reporting of results. Following is a brief summary of what is involved: 1) the sample is obtained from the bulk system under consideration, 2) this sample is appropriately prepared for analysis, 3) the chosen method is completed, 4) the data obtained from the method are analyzed, and 5) the results are calculated and reported.

As stated previously, and as shown in Figure 6.1, all analytical schemes begin with proper sampling and sample preparation procedures. These procedures were studied in detail in Chapter 2. However, we mention them again in this brief summary to emphasize their importance. In terms of sampling, we indicated that it is important for the sample to represent the entire bulk system for which an analytical result is to be reported. We also indicated that it is important to maintain the integrity of the sample at all times, so that the sample is not inappropriately altered prior to the analysis, and to document the chain of custody of all samples so that any doubts about sample integrity can be answered. The sampling procedures are central to the success of a chemical analysis, and the importance of proper sampling procedures should never be discounted. An analytical result is only as good as the sample used.

In terms of sample preparation, we indicated that in nearly every case, preanalysis procedures must be performed to get the sample into a form that can be utilized by the chosen analytical method. Common sample preparation schemes can involve any number of physical or chemical processes, such as drying, dissolving, extracting, or chemical alteration.

Because the basics of sampling and sample preparation were covered in detail in earlier chapters, they will appear in this and the coming chapters only if there is a need to discuss the state of a sample for the particular method under consideration or if there is some special relationship between a sampling or sample preparation procedure and the method. This fact should not imply that the importance of these topics is diminished to any degree. *Sampling and sample preparation procedures are very important to the success of all analytical work.*

The analysis methods we have covered to this point (Chapters 3 to 5) are the wet chemical analysis methods of gravimetric analysis and titrimetric analysis. The data handling procedures for these were also discussed. For gravimetric analysis and titrimetric analysis, these were summarized in flow charts that are reproduced here as Figures 6.2 and 6.3. For gravimetric analysis (Figure 6.2), either physical or chemical separation of the analyte is part of the method. Physical separation can result in a weight loss or gain that is directly entered into the calculation of the results. Examples include moisture in a soil sample (weight loss upon drying), water in a hydrate (weight loss upon heating), and the suspended solids in a wastewater sample (weight gain of a filtering crucible). Chemical separations require the use

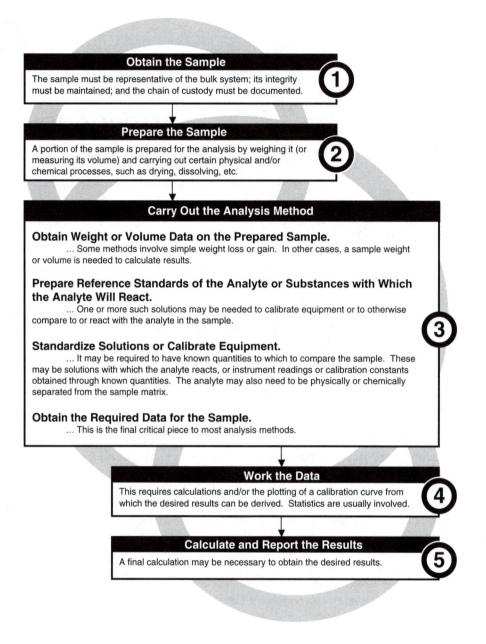

**FIGURE 6.1**    The general analytical strategy.

of a solution of a reagent that reacts with the analyte, usually forming a precipitate. In these cases, the precipitate is weighed and this weight is the critical datum from which the results are calculated.

In titrimetric analysis (Figure 6.3), a weight or volume of the prepared sample is needed and the sample may need to be dissolved or otherwise pretreated following the weight or volume measurement. The most important part, however, is that a standard solution (the titrant) needs to be prepared and standardized. This solution is used to titrate the sample, and the buret reading is the datum needed for calculating the results.

With the wet chemical methods and methods involving physical properties now behind us, we begin a thorough discussion of the methods, data handling, and calculation and reporting of results relating to instrumental analysis.

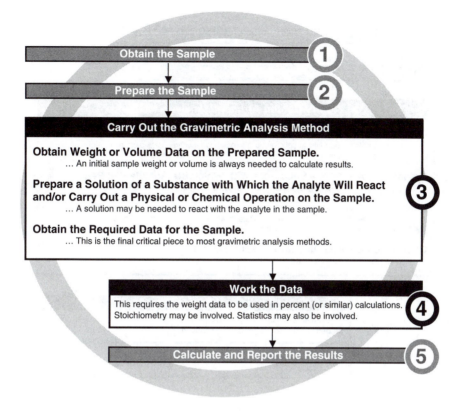

**FIGURE 6.2** The analytical strategy for gravimetric analysis.

# 6.2 Instrumental Analysis Methods

Quantitative instrumental analysis mostly involves sophisticated electronic instrumentation that generates electrical signals that are related to some property of the analyte and proportional to the analyte's concentration in a solution. In other words, as depicted in Figure 6.4, the standards and samples are provided to the instrument, the instrument measures the property, the electrical signal is generated, and the signal, or often the numerical value of the property itself, is displayed on a readout and subsequently related to concentration. A simple example with which you may be familiar is the pH meter. In this case, the pH electrode is immersed into the solution, an electrical signal proportional to the pH of the solution is generated, electronic circuitry converts the signal to a pH value, and the pH is then displayed.

Most instrumental analysis methods can be classified in one of three general categories: spectroscopy, which uses instruments generally known as a spectrometers; chromatography, which uses instruments generally known as chromatographs; and electroanalytical chemistry, which uses electrodes dipped into the analyte solution. These are the three categories that are emphasized in the remainder of this book. Spectroscopic methods (Chapters 7 to 10) involve the use of light and measure the amount of either light absorbed (absorbance) or light emitted by solutions of the analyte under certain conditions. Chromatographic methods (Chapters 11 to 13) involve more complex samples in which the analyte is separated from interfering substances using specific instrument components and electronically detected, with the electrical signal generated by any one of a number of detection devices. Electroanalytical methods (Chapter 14) involve the measurement of a voltage or current resulting from electrodes immersed into the solution. The pH meter mentioned earlier is an example of an electroanalytical instrument.

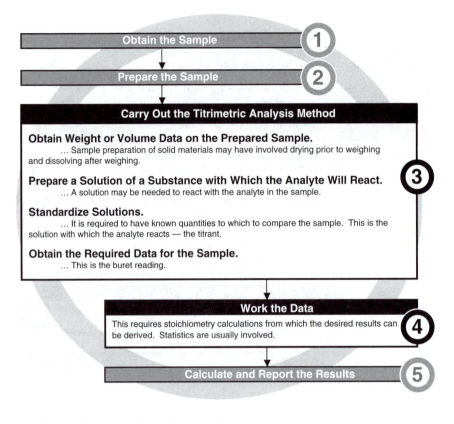

**FIGURE 6.3** The analytical strategy for titrimetric analysis.

**FIGURE 6.4** The general principle of analysis with electronic instrumentation.

A flow chart for instrumental methods is presented in Figure 6.5. Along with the determination of the weight or volume of the prepared sample, one or several standard solutions must be prepared. Such solutions are needed to calibrate the instrument. In general terms, **calibration** is a procedure by which any instrument or measuring device is tested with a standard in order to determine its response for an analyte in a sample for which the true response is already known or needs to be established. In cases in which the true response is already known, such as the use of a known weight when calibrating a balance (Chapter 3) or the pH of a buffer solution when calibrating a pH meter, either the device may be removed from service if the response is not correct (balance) or the electronics of the instrument may be tweaked to display the correct response (pH meter). In other cases, the true response may need to be established; this is done by measuring the responses of the standards (the known quantities). If a single standard is used, the calibration often results in a calibration constant to be used in subsequent calculations. If a series of standards are used, a **calibration curve** (or **standard curve**) is usually plotted. This is a plot of the instrument response vs. the concentration, or other known quantity, of the standard. The response of the sample is then applied to the standard curve, and the concentration, or other desired quantity, is determined. Details of calibration and the calibration curve are presented in Section 6.4.

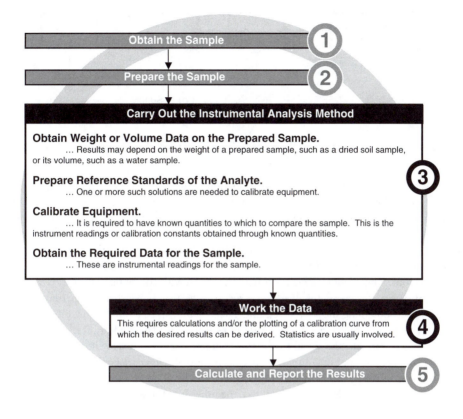

**FIGURE 6.5** The general analytical strategy for instrumental analysis.

## 6.3 Basics of Instrumental Measurement

Let us dwell on Figure 6.4 for a moment. The standards and sample solutions are introduced to the instrument in a variety of ways. In the case of a pH meter and other electroanalytical instruments, the tips of one or two probes are immersed in the solution. In the case of an automatic digital Abbe refractometer (Chapter 15), a small quantity of the solution is placed on a prism at the bottom of a sample well inside the instrument. In an ordinary spectrophotometer (Chapters 7 and 8), the solution is held in a round (like a test tube) or square container called a cuvette, which fits in a holder inside the instrument. In an atomic absorption spectrophotometer (Chapter 9), or in instruments utilizing an auto-sampler, the solution is sucked or aspirated into the instrument from an external container. In a chromatograph (Chapters 12 and 13), the solution is injected into the instrument with the use of a small-volume syringe. Once inside, or otherwise in contact with the instrument, the instrument is designed to act on the solution. We now address the processes that occur inside the instrument in order to produce the electrical signal that is seen at the readout.

### 6.3.1 Sensors, Signal Processors, Readouts, and Power Supplies

In general, an instrument consists of four components: a **sensor** that converts a property of the solution into a weak electrical signal, a **signal processor** that amplifies or scales the signal and converts it to a useable form, a **readout device** that displays the signal for the analyst to see, and a **power supply** to provide the power to run these three components. The information flow within the instrument occurs with the movement of electrons, or electrical current.

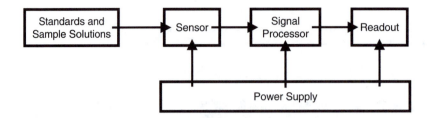

**FIGURE 6.6**   The four instrument components for translating a property of a solution to a readout for the analyst to observe.

A sensor is a kind of translator. It receives specific information about the system under investigation and transmits this information in the form of an electrical signal. Sensors are specific to a given property of the system under investigation. Some sensors are sensitive to temperature, others are sensitive to light, still others to pH, pressure, etc.

The signal processor is also measurement specific. A different mathematical treatment, such as a logarithmic conversion, is required for data from each kind of sensor, depending on what the operator desires as a readout. Some data treatment is often conducted with computer software.

The readout device is a translator like the sensor. It translates the electrical signal produced by the signal processor to something the analyst can understand. This can be a number on a digital display, the position of a needle on a meter, a computer monitor display, etc. The readout device is not specific to the measurement. It can take the signal from any signal processor and display it.

The power supply is also not measurement specific. Almost all sensors, signal processors, and readouts operate from similar voltages. The power supply converts the alternating current (ac) line power (or battery power) to voltage levels needed to operate the other functional elements of the instrument. See Figure 6.6 and Workplace Scene 6.1.

## 6.3.2   Some Basic Principles of Electronics

Electronics might be described as the science of moving and counting electrons for a useful purpose. As stated in the last section, the information flow within an instrument occurs with the movement of electrons, or electrical current. The current can only flow through a **complete circuit**, which means that the path of the flow must be continuous with no breaks (like a circle). A sensor releases or pushes on electrons with a force (voltage) that depends on the properties being sensed. For most chemical systems, this voltage is usually very small and must be processed and amplified (see Section 6.3.3). Thus, this voltage is often the input for the signal processor. **Input** refers to the voltage level (electrical signal) fed into a component of an electronic circuit, such as an amplifier. The amplifier circuit within the signal processor controls the flow of a larger group of electrons by monitoring the presence of electrons from the sensor at its input. Data processing circuits release or attract electrons in some proportion to the number of electrons present at their inputs. And finally, readout devices present information appropriately proportional to the original voltage produced by the chemical system.

**Voltage** is a measure of the force with which electrons are being pushed from one place to another, like water pressure is a measure of the force with which water is being pumped through a pipe. A battery or power supply acts as an electron pump, pulling electrons into one terminal (the positive (+) terminal) and pushing them out the other terminal (the negative (−) terminal). Voltage is symbolized by the letter V (sometimes E) and is expressed in units of volts, abbreviated V. An ordinary flashlight battery pulls and pushes electrons with about 1.5 V. All of the common flashlight battery sizes—AA, AAA, C, and D—pull and push with the same amount of force. Higher-voltage batteries can supply more electrons per second.

Batteries or voltage sources are symbolized with a series of short and longer bars (Figure 6.7), with the longer bars representing the positive (+) pole of the battery, into which electrons are pulled. Electrons are pushed out the negative (−) pole of the battery. In general, low-voltage batteries such as flashlight batteries are symbolized with one long and one short bar, while higher-voltage batteries (9-V radio batteries, for

# WORKPLACE SCENE 6.1

A unique application of sensor technology is used in an instrument called a **total organic carbon** (TOC) analyzer. Such an analyzer uses a combination of phosphoric acid, ammonium persulfate, and ultraviolet (UV) light to oxidize the organic compounds to carbon dioxide. Carbon dioxide gas, the product of the oxidation, is sensed electronically with the use of membrane-based conductometric sensors. A membrane that is permeable only to the carbon dioxide and not the other solution components is used. Once the gas passes through the membrane, it is converted to ionic form and the resulting conductance give rise to the electrical signal.

An example of its application is in the analysis of the water used in producing pharmaceutical formulations that must be free of organic contaminants. Such water can be monitored routinely in the company laboratory for the concentration of organic compounds using the TOC analyzer described here.

Jessica Weinland, a laboratory technician for a pharmaceutical company, operates a TOC analyzer to monitor the organic contaminants in the water used in preparing pharmaceutical formulations.

**FIGURE 6.7**   Symbols for batteries.

example) are symbolized with more bars. In most batteries, the small metal post on the end is the positive terminal, and the case of the battery (exposed at the other end) is the negative terminal.

**Resistance** to electron flow is measured in ohm units and is symbolized by the letter R. The ohm unit is symbolized by the Greek letter omega ($\Omega$). A resistance slows the movement of electrons in a circuit much like a smaller-diameter pipe would slow the movement of water. While all electronic circuit components have a certain resistance, components known as **resistors**—components that have a certain defined resistance in ohms, kilohms (k$\Omega$, or $10^3$ $\Omega$), or megaohms (M$\Omega$, or $10^6$ $\Omega$) and are inserted into circuits for their resistance values—are available. These usually are small, cylindrical epoxy or plastic packages with leads (short wires) protruding from each end and small granules of carbon or resistance wire inside to slow the electrons. Resistors are manufactured with different degrees of accuracy. One percent accuracy resistors have their value printed on the side. Five and ten percent accuracy resistors use a color code to indicate the value of the resistance (see Experiment 15). The design of electrical circuits often enables low-accuracy resistors to have no impact on the final readout. Thus, low-cost resistors can be used and deterioration of the resistors does not effect instrument accuracy.

**Current** is a measure of electron flow rate in an electrical circuit, analogous to water flow rate through a pipe, and is symbolized by I. Current is measured in amperes (amps), symbolized as A; milliamperes (milliamps), symbolized as mA; or microamperes (microamps), symbolized as $\mu A$. An ampere is an electron flow of $6.23 \times 10^{18}$ electrons per second passing through the circuit.

Just as a narrower-diameter pipe reduces the water pressure in the pipe, a resistor reduces the voltage from the battery. We say that there is a voltage drop across a resistor. The amount by which a voltage is reduced can be calculated by Ohm's law. **Ohm's law** is the mathematical relationship between this voltage drop (V), the value of the resistance (R), and the current (I) through the resistor:

$$V = IR$$

With a current of 1 A, the voltage drop across a resistance of 1 $\Omega$ is 1 V. This voltage is also represented by a current of 1 mA, through a resistance of 1 k$\Omega$, and a current of 1 $\mu A$, through a resistance of 1 M$\Omega$. Thus, volts are calculated when the current is in amperes and the resistance in ohms, but also when the current is in milliamperes and the resistance in kilohms, etc. A circuit can be drawn in electronic symbols, as illustrated in Figure 6.8. The symbol for a resistor in such a drawing is a sawtooth line.

A **voltmeter** is a device for measuring volts. Voltmeters have two input terminals. The voltage drop across a resistor is measured by connecting these inputs to each end of the resistor, as shown in Figure 6.9. An **ammeter** is a device for measuring current. It is shown in Figure 6.8 as a circle with an arrow. An **ohmeter**

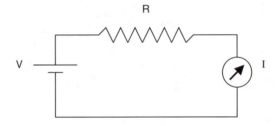

**FIGURE 6.8**   A simple circuit drawn using the symbols for voltage and resistance. The measurement of current is illustrated with the arrow inside the circle.

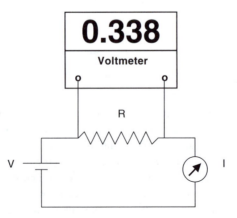

**FIGURE 6.9**   A voltmeter measures the voltage drop across a resistor. A voltage of 0.338 V would result from a current of 0.000338 A through a resistor of 1 k$\Omega$.

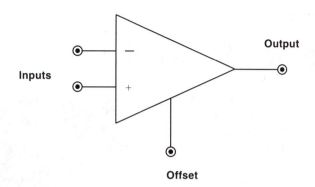

**FIGURE 6.10** The symbol of an amplifier.

is a device for measuring resistance. It is possible for one device to serve all three purposes. In this case, the device is often called a **multimeter**.

**Example 6.1**

What is the voltage drop across a resistor if the resistance value is 1.2 kΩ and the current is 2.5 mA?

*Solution 6.1*

As indicated above, using Ohm's law, multiplying current in milliamperes by resistance in kilohms gives the voltage in volts:

$$\text{volts (V)} = I \text{ (mA)} \times R \text{ (k}\Omega)$$

Thus we have the following:

$$V = 2.5 \times 1.2 = 3.0 \text{ V}$$

### 6.3.3 Signal Amplification

Some sensors produce a large enough output signal to be read directly with a voltmeter. However, most sensors do not. In that case, integrated circuit amplifiers are used to increase (amplify) the weak sensor signal to useable values. The amplification factor, the amount by which the signal is multiplied, is called the **gain** of the amplifier. An **integrated circuit** is a complex circuit contained on a single, tiny semiconductor chip. The most common of these are the instrumentation amplifier and the operational amplifier. Although these are complex versatile circuits, they are easy to understand and use. An **instrumentation amplifier** provides precise voltage amplification for sensors that produce a voltage output. There is usually an offset control for expanded scale measurements. It is a **difference amplifier**, meaning that it amplifies the difference between two input signals. An **operational amplifier** converts current output from sensors to a voltage and performs mathematical operations, such as comparison, summing, and logarithm calculation. It may also serve as a constant current source. The symbol of an amplifier is shown in Figure 6.10. See Workplace Scene 6.2.

## 6.4  Details of Calibration

There are many examples in our everyday life in which the calibration of a measuring device is important. When we step on the bathroom scale, we want to be assured that the scale reads our correct weight. When we measure the body temperature of our sick child, we want the temperature that is displayed to be correct. When we fill our car's gas tank at the gas pump, we want the pump to display the correct number of gallons and therefore the correct cost. In other words, we want these devices to be properly

# WORKPLACE SCENE 6.2

At the City of Lincoln, Nebraska, Wastewater Treatment Plant Laboratory, **flow injection analysis** (FIA) is used to analyze for various ions dissolved in wastewater. This instrument uses small-diameter tubing to carry and mix reagents and samples in a continuous flow system. Following this mixing, the test samples are channeled to a flow cell where a fiber optic light source and light sensor are used to measure the concentration of the ions as the solution flows through. The flow cell and fiber optic and sensor electronics are housed in a detector module with inlet and outlet pathways for the flowing solution. A problem recently arose when the baseline sensor signal was too high and could not be zeroed. New reagents were prepared to try to solve the problem, but to no avail. Finally, the detector was disassembled to view the solution pathway and fiber optic–sensor system. It was discovered that the solution was leaking from a joint in the flow path and chemicals had crystallized on and corroded the tip of the fiber optic, thus partially blocking the light. It was determined that the module had to be replaced.

Steve Kruse of the City of Lincoln, Nebraska, Wastewater Treatment Plant Laboratory examines a faulty detector module taken from the flow injection analysis system in use in the laboratory.

calibrated. In the instrumental analysis laboratory, all instruments and devices used must be properly calibrated and have their calibrations checked on a regular basis to ensure accurate data.

## 6.4.1   Thermocouples: An Example of a Calibration

Consider temperature as an example. Temperature measurement is needed in a variety of laboratory applications and, in the modern laboratory, is done with a temperature sensor, such as a **thermocouple**. A thermocouple is a junction of two metals that produces a voltage proportional to temperature that can be measured via electrical connections to the two metals. The voltage difference between the two connections can be amplified by the difference amplifier discussed in Section 6.3.3.

The most common reference points for temperature sensor calibration are the freezing and boiling points of water. The freezing point is a function of the purity of the water–ice system. When both pure ice and pure liquid water are present, the ice is melting and the temperature of a well-stirred mixture is, by definition, 0°C. It is the most accurate calibration point for that reason. If it is possible to use such a quantity in a calibration, then the calibration is true and without question.

The boiling point of water, however, depends on both the purity of the water and the prevailing atmospheric pressure. Thus a boiling water bath is not as accurate a reference point as the water–ice system. When calibrating a temperature sensor, the upper level calibration point is commonly a hot water bath, the temperature of which is measured with a regular thermometer. The accuracy of the calibration then depends on the accuracy of this thermometer.

This latter point begs the question "Who can verify the accuracy of a reference point if its value may vary?" In other words, who is the ultimate source of calibration materials, such as a thermometer? In the U.S., it is the National Institute of Standards and Technology (NIST). This is the same organization that we cited as the source of accurate standardization materials in Chapters 3 and 4. Experiments 16 and 17 in this chapter are exercises in the calibration of a temperature sensor and how such a calibrated sensor can be used.

## 6.4.2 Calibration of an Analytical Instrument

We mentioned in Section 6.2 that the response of an instrument used for chemical analysis is proportional to the concentration of the analyte in a solution. This proportionality can be expressed as follows:

$$R = KC \tag{6.1}$$

where $R$ is the instrument readout, C is the concentration of the analyte, and K is the proportionality constant. (The most common example is **Beer's law**, which will be introduced in Chapter 7.) If the object of the experiment is the concentration (C) of the analyte in the sample solution, then the R and the K for this solution must be known so that C can be calculated:

$$C = \frac{R}{K} \tag{6.2}$$

We also indicated in Section 6.2 that the calibration of an instrument for quantitative analysis can utilize a single standard, resulting in a **calibration constant**, or a series of standards, resulting in a **standard curve**. If a single standard is used, the value of K is the calibration constant. It is found by determining the instrument response for a standard solution of the analyte and then calculating K:

$$R_S = KC_S \tag{6.3}$$

$$K = \frac{R_s}{C_s} \tag{6.4}$$

The concentration of the unknown sample solution is then calculated from its instrument response and the K determined from Equation (6.4):

$$C_u = \frac{R_u}{K} \tag{6.5}$$

In the above equations, $R_S$ is the readout for the standard solution, $C_S$ is the concentration of the analyte in the standard solution, $R_U$ is the readout for the unknown sample solution, $C_U$ is the concentration of the analyte in the unknown solution, and the proportionality constant (K) is the calibration constant.

There are two limitations with the above process: 1) the analyst is "putting all the eggs in one basket" by comparing the sample to just one standard (not very statistically sound), and 2) the calibration constant, K, must truly be constant at the two concentration levels, $C_S$ and $C_U$ (possible, but not guaranteed). Because of these limitations, the concept of the standard curve is used most of the time.

The concept of a series of standards refers to an experiment in which a series of standard solutions is prepared covering a concentration range within which the unknown concentration is expected to fall. For example, if an unknown concentration is thought to be around 4 parts per million (ppm), then a series of standards bracketing this value, such as 1, 3, 5, and 7 ppm, are prepared. The readout for each of these is then measured. The standard curve is a plot of the readout vs. concentration. The unknown concentration is determined from this plot.

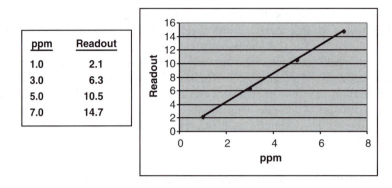

| ppm | Readout |
|-----|---------|
| 1.0 | 2.1 |
| 3.0 | 6.3 |
| 5.0 | 10.5 |
| 7.0 | 14.7 |

**FIGURE 6.11**   Sample data and standard curve showing the desired linearity. The value of K is constant at all concentrations at 2.1.

The standard curve procedure is free from the limitations involved in comparing an unknown to a single standard because: 1) if the unknown concentration is compared to more than one standard, the results are more statistically sound, and 2) if the curve is a straight line, the value of K is constant through the concentration range used. If the standard curve is a straight line free of bends or curves, the value of K is the slope of the line and is constant. See Figure 6.11. If a portion of the standard curve is linear, this linear portion may be used to determine the unknown concentration.

We should indicate at this point that, while the limitations of the single standard are eliminated, there is still considerable opportunity for experimental error in this procedure (error in solution preparation and instrument readout), and this error may make it appear that the data points are "scattered" and not linear. This problem is addressed in the next two subsections.

This series of standard solutions method is commonplace in an instrumental analysis laboratory for the vast majority of all quantitative instrumental procedures. Examples of this abound for many spectrophotometric, chromatographic, and other techniques.

### 6.4.3   Mathematics of Linear Relationships

The mathematical equation representing a straight-line relationship between variables x and y is as follows:

$$y = mx + b \qquad (6.6)$$

If the variables x and y bear a linear relationship to one another, then m and b are constants defining the slope (m) of the line (how steep it is) and the y-intercept (b), the value of y when x is zero. The slope of a line can be determined very simply if two or more points, $(x_1, y_1)$ and $(x_2, y_2)$, that make up the line are known.

$$m = \frac{\Delta y}{\Delta x} = \frac{(y_2 - y_1)}{(x_2 - x_1)} \qquad (6.7)$$

The y-intercept can be determined by observing what the y value is on the graph where the line crosses the y-axis, or if the slope and one point are known, it may be calculated using Equation (6.6).

The important observation for instrumental analysis is that the K in Equation (6.1) is the same as the m in Equation (6.6). Thus, in a graph of R vs. C, the slope of the line is the K value, the proportionality constant.

It would seem that the y-intercept would always be zero, since if the concentration is zero, the instrument readout would logically be zero, especially since a blank is often used. The blank, a solution prepared so that all sample components are in it except the analyte (see Section 6.6), is a solution of zero concentration. Such a solution is often used to zero the readout, meaning that the instrument is manually set to read zero when the blank is being read by the instrument. This represents a calibration step in which the result is known and the instrument readout can be tweaked. However, real data sometimes do not fit a straight-line graph all the way to the y-intercept, and thus the y-intercept is *not* always zero and the (0, 0) point is usually not included in the standard curve.

Another problem with real data is that due to random indeterminate errors (Chapter 1), the analyst cannot expect the measured points to fit a straight-line graph exactly. Thus it is often true that we draw the *best* straight line that can be drawn through a set of data points and the unknown is determined from *this* line. A linear regression, or least squares, procedure is then done to obtain the correct position of the line and therefore the correct slope, etc.

## 6.4.4  Method of Least Squares

The linearity of instrument readout vs. concentration data must be determined. As stated in Section 6.4.2, random indeterminate errors during the solution preparation and during the measurement of R will likely cause the resulting points to appear scattered to some extent. If the data appear to be linear despite the scatter, a method must be adopted that will fit a straight line to the data as well as possible. It may happen that some (or even all) of the points may not fall exactly on the line because of these random errors, but a straight line is still drawn, since the random errors are indeed random and cannot be compensated for directly. Thus, the best straight line possible is drawn through the points. See Figure 6.12. If only a portion of the curve appears to be linear, then this portion is plotted and the best straight line is drawn through these points.

In the modern laboratory, standard curves are generated with computer programs utilizing the method of least squares, or linear regression, which determines the placement of the straight line mathematically. By this method, the best straight-line fit is obtained when the sum of the squares of the individual y-axis value deviations (deviations between the plotted y values and the values on the proposed line) are at a minimum. This proposed line is actually calculated from the given data; a slope and y-intercept (m and b, respectively, in Equation (6.6)) are then obtained and the deviations ($y_{point} - y_{line}$) for each given x are calculated. Finding the values of the slope and the y-intercept that minimize the sum of the squares of the deviations involves calculations with which we will not concern ourselves.[*] Computers and programmable calculators, however, handle this routinely in the modern laboratory (see Experiment 18 for a procedure using Excel[TM] spreadsheet), and the results are very important to a given analysis since the line that is determined is statistically the most correct line that can be drawn with the data obtained. The concentrations of unknown samples are also readily obtainable on the calculator or computer since the equation of the straight line, including the slope and y-intercept, becomes known as a result of the

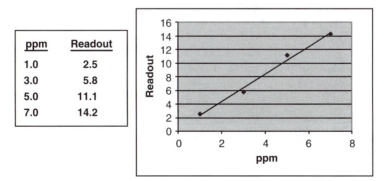

| ppm | Readout |
|-----|---------|
| 1.0 | 2.5 |
| 3.0 | 5.8 |
| 5.0 | 11.1 |
| 7.0 | 14.2 |

**FIGURE 6.12**   An example of a standard curve in which, due to random errors, the data points do not lie on the line.

[*]See the following reference for details: Harvey, David, *Modern Analytical Chemistry*, McGraw-Hill (2000), page 119.

least squares procedure. Other statistically important parameters are readily obtainable as well, including the correlation coefficient.

## Example 6.2

The equation of the straight line determined from a least squares fit procedure for an experiment in which the instrument readout, R, was plotted on the y-axis and the concentration, C, in parts per million was plotted on the x-axis is

$$R = 1.37 \ C + 0.029 \qquad (6.8)$$

What is the analyte concentration in the unknown if the instrument readout for the unknown is 0.481?

*Solution 6.2*
Solving Equation (6.8) for C, we get the following:

$$C = \frac{(R - 0.029)}{1.37} = 0.330 \text{ ppm}$$

## 6.4.5   The Correlation Coefficient

The correlation coefficient is one measure of how well the straight line fits the analyst's data—how well a change in one variable correlates with a change in another. A correlation coefficient of exactly one indicates perfectly linear data. This, however, rarely occurs in practice. It occurs if all instrument readings increase by exactly the same factor from one concentration to the next, as in the data in Figure 6.11. Due to the indeterminate errors that are present in all analyses, such data are quite rare. Data that approach such linearity will show a correlation coefficient of less than one, but very nearly equal to one. Numbers such as 0.9997 or 0.9996 are considered excellent and attainable correlation coefficients for many instrumental techniques. Good pipetting and weighing techniques when preparing standards and well-maintained and calibrated instruments can minimize random errors and produce excellent correlation coefficients, and therefore accurate results. The analyst usually strives for *at least* two nines, possibly three, in his or her correlation coefficients, depending on the particular instrumental method used. Again, these coefficients can be determined on programmable calculators and laboratory computers by using the appropriate software. A step-by-step procedure using Microsoft Excel spreadsheet software is presented in Experiment 18. The usual symbol for the correlation coefficient is r. The square of the correlation coefficient, $r^2$, may also be reported.

# 6.5   Preparation of Standards

The series of standard solutions of the analyte can be prepared in several ways. Probably the most common is through several **dilutions of a stock standard solution**. The stock standard may be prepared by the analyst or purchased from a chemical vendor. It may be prepared by weighing a quantity of the primary standard analyte chemical (using an analytical balance) into a volumetric flask and diluting to the mark (as previously discussed in Chapters 4 and 5). It may also be a secondary standard prepared by diluting a primary stock that was purchased from a chemical vendor or prepared in the lab. The series of standards are then prepared by diluting various volumes of this stock (as determined by dilution calculations discussed in Chapter 4) to volume in a series of volumetric flasks. These dilutions require diligent solution transfer technique, such as with one of the various pipetting devices discussed in the previous chapters, to minimize the indeterminate errors.

Another is a so-called **serial dilution** procedure. In this procedure, the second solution is prepared by diluting the first, the third by diluting the second, the fourth by diluting the third, etc. Once again, volumetric flasks and suitable solution transfer devices are required.

## Example 6.3

How would you prepare 50.00 mL each of a series of standards that have analyte concentrations of 1.0, 3.0, 5.0, and 7.0 ppm from a stock standard that is 100.0 ppm?

**Solution 6.3**

We use the dilution equation (Equation (4.2)) for each solution to be prepared. The dilution equation is

$$C_B \times V_B = C_A \times V_A$$

where $C_B$ is the concentration before dilution, $V_B$ is the volume to be diluted, $C_A$ is the concentration after dilution, and $V_A$ is the final solution volume.

1.0 ppm solution: $100.0$ ppm $\times V_B = 1.0$ ppm $\times 50.00$ mL; $V_B = 0.50$ mL

3.0 ppm solution: $100.0$ ppm $\times V_B = 3.0$ ppm $\times 50.00$ mL; $V_B = 1.5$ mL

5.0 ppm solution: $100.0$ ppm $\times V_B = 5.0$ ppm $\times 50.00$ mL; $V_B = 2.5$ mL

7.0 ppm solution: $100.0$ ppm $\times V_B = 7.0$ ppm $\times 50.00$ mL; $V_B = 3.5$ mL

The solutions are prepared by pipetting the calculated volumes of the 100.0 ppm solutions into separate 50.00-mL volumetric flasks. Each is then diluted to the mark and made homogeneous by thorough mixing.

## 6.6 Blanks and Controls

In addition to the series of standard solutions needed for an instrumental analysis, there are often other solutions needed for the procedure. We have already briefly mentioned the need for and use of a blank (Section 6.4.3). As stated previously, the **blank** is a solution that contains all the substances present in the standards, and the unknown if possible, except for the analyte. The readout for such a solution should be zero, and as we have indicated, the readout is often manually adjusted to read zero when this blank is being measured. Thus, the blank is useful as a sort of precalibration step for the instrument.

### 6.6.1 Reagent Blanks

A blank such as the one described above is appropriately called a **reagent blank**. Although an analyst may be tempted to use distilled water or other pure sample solvent as a blank, this may not be desirable because other chemicals (reagents) may have been added to enhance the readout for the analyte in the standards and unknowns, and these chemicals may independently affect the readout in some small way. The blank must take into account the effects of all the reagents (and any contaminants in the reagents) used in the analysis. Thus if the reagents or their contaminants contribute to an instrument's readout when the analyte is being measured, such effects would be canceled out as a result of the zeroing step. The value of a reagent blank is thus obvious.

### 6.6.2 Sample Blanks

While a reagent blank is frequently prepared and used as described above, a sample blank is sometimes appropriate instead. A **sample blank** takes into account any chemical changes that may take place as the sample is taken or prepared. For example, the analysis of an air sample for a component in the particulates in the air involves drawing the air sample through a filter in order to capture the particulates so that they can be dissolved. This dissolution step involves not just the particulates but also the filter itself. Thus, the dissolved filter changes the chemistry of the sample and may itself contribute to the instrument readout, and this would not be taken into account by a simple reagent blank. The answer to this problem is to take a clean filter and run it through the dissolution procedure alongside the sample and to use the resulting solution as the blank. Such a blank is called a sample blank.

Of course, anytime the dissolving of a sample includes a heating step, or any other step that may produce chemical changes in the solution, the blank should undergo the same steps. The reagent blank then becomes a sample blank, and the resulting solution would represent a matrix as close to the composition of the sample solution as possible, thus enhancing accuracy when reading the samples.

### 6.6.3   Controls

A **control** is a standard solution of the analyte prepared independently, often by other laboratory personnel, for the purpose of cross-checking the analyst's work. If the concentration found for such a solution agrees with the concentration it is known to have (within acceptable limits based on statistics), then this increases the confidence a laboratory has in the answers found for the real samples. If, however, the answer found differs significantly from the concentration it is known to have, then this signals a problem that would not have otherwise been detected. The analyst then knows to scrutinize his or her work for the purpose of discovering an error.

The results of the analysis of a control are often plotted on a control chart (Chapter 1) in order to visualize the history of the analysis in the laboratory so that a date and time can be identified as to when the problem was first detected. Thus, the problem can be traced to a bad reagent, instrument, or other component of the procedure if such a component was first put into use the day the problem was first detected. Your instructor may want you to use controls in various experiments in this text.

## 6.7   Effects of Sample Pretreatment on Calculations

The concentration obtained from the standard curve is rarely the final answer in a real-world instrumental analysis. In most procedures, the sample has undergone some form of preanalysis treatment prior to the actual measurement. In some cases, the sample must be diluted prior to the measurement, as mentioned in Workplace Scene 6.3. In other cases, a chemical must be added prior to the measurement, possibly changing the analyte's concentration. In still other cases, the sample is a solid and must be dissolved or extracted prior to the measurement.

The instrument measurement is the measurement of the solution tested, and the concentration found is the concentration in that solution. What the concentration is in the original, untreated solution or sample must then be calculated based on what the pretreatment involved. Often this is merely a dilution factor. It may also be a calculation of the grams of the constituent from the molar concentration of the solution, or the calculation of the parts per million in a solid material based on the weight of the solid taken and the volume of extraction solution used and whether or not the extract was diluted to the mark of a volumetric flask. Some examples of this follow. Remember that parts per million for a solute in dilute water solutions is in milligrams per liter, while for an analyte in solid samples it is in milligrams per kilogram. Review Chapter 5 for more information about the parts per million unit.

In order to set up a calculation for a given real-world analysis, it may be useful to first decide how the answer is to be reported and then work backwards from this to the nature of the data obtained. For example, if the parts per million of a constituent in a solid material sample is to be reported, the final step will involve dividing the milligrams of constituent by the kilograms of sample. Thus the sample weight in kilograms will be in the denominator. The milligrams of constituent in the numerator will likely need to be calculated from the concentration (such as parts per million in milligrams per liter) of the solution tested, as discovered from the plot of the standard curve. The first step is frequently then the conversion of milligrams per liter to milligrams, which is done by multiplying the parts per million found by the volume of the solution tested (in liters). If this is the total milligrams in the sample (if there is no additional dilution or pretreatment), then dividing by the kilograms of sample weighed out would give the answer.

### Example 6.4

A water sample was tested for iron content, but was diluted prior to obtaining the instrument reading. This dilution involved taking 10.00 mL of the sample and diluting it to 100.00 mL. If the instrument reading gave a concentration of 0.891 ppm for this diluted sample, what is the concentration in the undiluted sample?

# WORKPLACE SCENE 6.3

In order to determine chemical elements in soil, samples of the soil must undergo a solid–liquid extraction. Sometimes the extracts resulting from this procedure have analyte concentrations that are too high to be measured accurately by the chosen method. Therefore, they must be diluted. At the Natural Resources Conservation Service (NRCS) Soil Survey Laboratory in Lincoln, Nebraska, an automated diluting device is used. Using this device, the analyst accurately transfers aliquots of the extract and a certain volume of extraction solution to the same container. This dilutor may also be used to pipet standards and prepare serial dilutions.

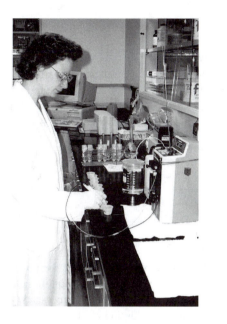

Patty Jones, a physical science technician at the NRCS Soil Survey Laboratory, prepares extracts for phosphorus analysis using a diluting device.

**Solution 6.4**

$$\frac{0.891\text{mg}}{\text{L}} \times 0.10000 \text{ L} = 0.0891 \text{ mg of Fe in flask}$$

This is also the milligrams of Fe in the 10.00 mL of undiluted sample. Thus,

$$\frac{0.891\text{mg of Fe}}{0.01000 \text{ L of water}} = 8.91 \text{ ppm Fe in original water}$$

**Alternate Solutions**
The dilution factor is

$$\frac{\text{dilution volume}}{\text{sample volume}} = \frac{0.10000 \text{ L}}{0.01000 \text{ L}} \text{ or } \frac{100.00 \text{ mL}}{10.00 \text{ mL}} \text{ or } 10.00$$

Therefore, 0.891 ppm × 10.00 = 8.91 ppm Fe.

The dilution equation (see solution to Example 6.3) may also be used to solve this problem. In that case, all variables are known except $C_B$.

$$C_B \times V_B = C_A \times V_A$$

$$C_B \times 10.00 \text{ mL} = 0.891 \text{ ppm} \times 100.00 \text{ mL}$$

$$C_B = 8.91 \text{ ppm}$$

## Example 6.5

A 0.5693-g sample of insulation was analyzed for formaldehyde residue by extraction with 25.00 mL of extracting solution. After extraction, the sample was filtered and the filtrate analyzed without further dilution. The concentration of formaldehyde in this extract was determined to be 4.20 ppm. What is the concentration of formaldehyde in the insulation?

*Solution 6.5*

$$\frac{\frac{4.20 \text{ mg}}{\cancel{L}} \times 0.025 \cancel{L}}{0.0005693 \text{ kg}} = 1.8 \times 10^2 \text{ ppm}$$

## Example 6.6

A 2.000-g soil sample was analyzed for potassium content by extracting the potassium using 10.00 mL of aqueous ammonium acetate solution. Following the extraction, the soil was filtered and rinsed. The filtrate with rinsings was diluted to exactly 50.00 mL. Then 1.00 mL of this solution was diluted to 25.00 mL, and this dilution was tested with an instrument. The concentration in this 25.00 mL was found to be 3.18 ppm. What is the concentration of the potassium in the soil in parts per million?

*Solution 6.6*

One fiftieth of the potassium in the soil sample is in the final 25.00-mL dilution; therefore, there is 50.00 times more K in the extract than in the solution tested, and it all originated from the 2.000 g of soil. Thus,

$$\text{ppm K} = \frac{\text{milligrams of K}}{\text{kilograms of soil}} = \frac{3.18 \text{ mg}/\cancel{L} \times 0.02500 \cancel{L} \times 50.00}{0.002000 \text{ kg}} = 1.99 \times 10^3 \text{ ppm K in soil}$$

# 6.8   Laboratory Data Acquisition and Information Management

In this chapter, we have advocated the use of a computer for plotting the standard curve and performing the least squares fit procedure. Indeed, computers play a central role in the analytical laboratory for acquiring and manipulating data generated by instruments and for information management.

## 6.8.1   Data Acquisition

It is very commonplace today for the analyst to use a computer for acquiring data. The term data acquisition refers to the fact that the data generated by an instrument can be fed directly into and acquired by a computer via an electronic connection between the two. Often, the computer monitor then serves as the readout device in real time. For example, a continuously changing readout can be monitored as a function of time on the screen, with the signal tracing across the screen as this signal is being generated by the instrument's signal processor. Such data are thereby stored in the short term in the computer's memory and in the long term on magnetic storage media, such as disks. Hard copies can be generated through connection to a printer.

The actual mode of connection between instrument and computer varies depending on the type of signal generated and the design of the instrument. The connection can be made via a serial port, a parallel port, or a USB port. The electronic circuitry required is built into the instrument's internal readout electronics or into an external box used for conditioning the instrument's output signal. In some cases, the instrument will not operate without the computer connection and is switched on and off as the computer is switched on and off. In other cases, the entire computer is built into the instrument.

When in actual use, a software program is run on the computer that establishes the sampling time interval and other parameters at the discretion of the operator and begins the data acquisition at the touch of a key on the computer keyboard or at the click of the mouse.

## 6.8.2 Laboratory Information Management

Modern laboratories are complex multifaceted units with vast amounts of information passing to and from instruments and computers and to and from analysts and clients daily. The development of high-speed, high-performance computers has provided laboratory personnel with the means to handle the situation with relative ease. Software written for this purpose has meant that ordinary personal computers can handle the chores. The hardware and software system required has come to be known as the **laboratory information management system** (LIMS).

LIMSs vary in sophistication. However, all systems perform basic information management as it relates to sample labeling and tracking, lab tests, personnel assignments, analytical results, report writing, and client communication. More sophisticated versions include verifications, validations, and approvals. They produce reports on work in progress, maintain a backlog, incorporate data for productivity and resources, and provide the means for conducting an audit, computing costs, and maintaining an archive.

The LIMS computer is located on the site, and several terminals may be provided for entry of data from notebooks and instrument readouts and for the retrieval of information. Bar coding for sample tracking and access codes for laboratory personnel are part of the system. Instruments may be interfaced directly with the LIMS computer to allow direct data entry without help from the analyst. The LIMS may also incorporate statistical methods and procedures, including statistical control and control chart maintenance. See Workplace Scene 6.4.

# Experiments

## Experiment 15: Voltage, Current, and Resistance

This experiment uses a socket board available from Radio Shack (catalog number 276-175). The socket board has the layout shown in Figure 6.13, each black square representing a socket (or hole) with metal lining into which a wire can be inserted for electrical contact. The sockets in each set of five sockets in the top and bottom horizontal rows are connected internally, as indicated with the horizontal lines. The sockets in each set of five sockets A to E and F to J in each vertical column are also connected internally,

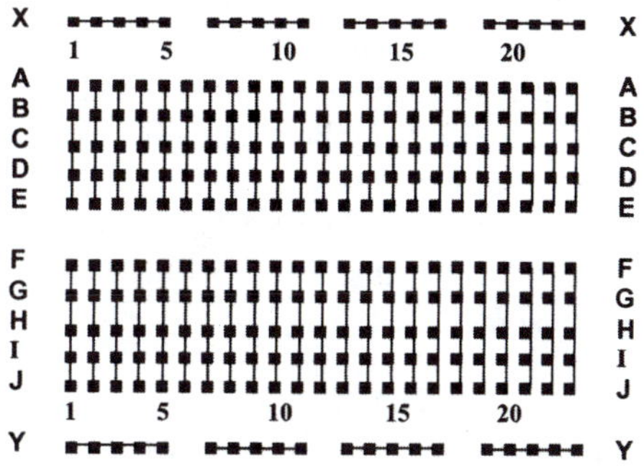

**FIGURE 6.13**  The layout of the socket board for Experiment 15.

# WORKPLACE SCENE 6.4

At the National Soil Survey Laboratory in Lincoln, Nebraska, tens of thousands of analyses are performed each year on an average of 7500 soil samples from all over the world. A LIMS is crucial to the organization and management of the work. It is very important to correctly label and identify samples as they come into the laboratory for analysis in order to tie each sample to the field information accompanying it and to the results of the analyses performed. Each sample is unique and has its own sample ID that is carried with it through its life in the laboratory. A group of samples from a site are made into a project. After the projects have been logged into the LIMS, they are assigned the analyses to be done and the results are stored in the LIMS so that they can be made available to customers in a form that fits their needs.

Larry Arnold of the National Soil Survey Laboratory assigns unique identification labels to soil samples in preparation to logging the samples and their unique field information into a LIMS.

as indicated with the vertical lines. The board is useful for holding resistors in place while measuring voltages, resistances, and currents.

This experiment also uses a device known as a digital multimeter. A multimeter is capable of measuring all three Ohm's law parameters—voltage, current, and resistance—hence the use of the prefix multi-. A selector switch on the face of the meter is used to switch to the parameter one wishes to measure. Each parameter has several positions representing several ranges over which the parameter can be measured in each position. For example, positions labeled with a V and showing values of 600, 200, 20, and 2 are for measuring voltages, and the positions can display voltages up to maximums of 600, 200, 20, and 2 V, respectively. Most multimeters are capable of measuring both alternating current and direct current (dc) voltages and will have separate sets of positions for both. The usual symbol for ac voltages is ∼, and the usual symbol for dc voltages is —.

Multimeters have sockets into which test leads are inserted. Test leads are insulated wires that have "banana plugs" on one end, for insertion into the sockets, and metal tips on the other end, for contacting the points in the circuit that are being measured.

## Procedure

1. Obtain a socket board and multimeter with test leads from your instructor. If not already done, insert the banana plugs on the one end of the test leads into the sockets on the multimeter for volt and ohm measurement.

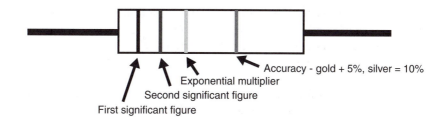

Accuracy - gold + 5%, silver = 10%
Exponential multiplier
Second significant figure
First significant figure

**FIGURE 6.14** Interpretation of colored stripes found on some resistors.

2. Check out the socket board to verify the statements above indicating which sockets are connected internally. To do this, set the multimeter to measure resistance (set to the 2-KΩ range) and touch the contact tips of the test leads to any two sockets on the board. If there is internal contact, the resistance measurement will be zero (no resistance). If there is no internal connection, it is an open circuit and the resistance will be infinite. Record the results for each pair of sockets tested.

3. Obtain a battery (such as AA, C, or D size) and a battery holder (with leads) from your instructor. Insert the battery into the holder according to the drawing on the holder. Now insert the leads from the battery holder into sockets J1 and J5 (see Figure 6.13). Set the selector switch to measure voltage (set to the 2-V range) and then measure the voltage of the battery by contacting the multimeter lead tips to sockets I1 and I5, red to I1 and black to I5. Record the reading and comment on how it compares with the expected value. Note the sign of the reading. Change the polarity by contacting the black lead tip to I1 and red to I5. Record the results and discuss briefly what changing the polarity means.

4. Obtain six colored-coded resistors from your instructor and prepare a three-column table in your notebook for recording the color code, the resistance as ascertained from the color code, and the resistance as ascertained by measurement with the multimeter. The color code consists of colored stripes that encircle the cylindrical shape, as shown in Figure 6.14. Also obtain from your instructor a resistor color code wheel (e.g., Radio Shack catalog number 271-1210) and determine the value of all your resistances as ascertained from the color code. For example, if the colors of the stripes left to right as viewed above are red, violet, yellow, and gold, the resistance is 270,000 Ω. Record the values in your table.

5. Now measure one of your resistances with the multimeter as follows. Insert the wire ends into two sockets on the socket board so that they are not connected internally, such as in sockets F1 and F5. (You will have to bend the end wires at about a 90° angle.) Measure the resistance with your multimeter by setting the selector switch to measure resistance and then contacting the lead tips to sockets that make contact with F1 and F5, such as G1 and G5. You may have to adjust the selector switch to the proper range for the resistor being measured. Record the value in your table. Measure all the other resistors in the same way. Determine if the accuracy of each, as indicated by either the gold or silver stripes, is correct based on your data and comment on this.

6. Next, measure your resistors connected two at a time in series. This means that two of the resistors are connected end to end. When two resistors are connected in this fashion, the total resistance is the sum of the two. To do this, use the socket board and insert the wire ends of one resistor into sockets F1 and F5 (as in step 5), for example, so that the ends are not connected internally. Then insert the wire ends of the other resistor into sockets G5 and G9, for example. Since G5 and F5 are connected internally, this connects both resistors in series. Now measure the total resistance by touching the lead tips of the multimeter to sockets H1 and H9. Record the individual values of the two resistors, the sum of the two, and the measured value of the sum in another table (four columns) in your notebook. Repeat with several different combinations. Comment on the agreement (or lack thereof) between the calculated and measured values.

7. Construct a complete circuit with a battery and two resistors in series. To do this, choose the two resistors you want to use and connect them in series on the socket board, as in step 6. Then,

connect the 1.5-V battery by inserting the leads from the battery holder into the appropriate sockets on the circuit board. If the F1/F5 and G5/G9 sockets are used, as suggested in step 6, then the battery leads can be inserted into J1 and J9. You now have a complete circuit and a current is flowing. To conserve battery power, disconnect one battery lead (pull it out of its socket) while you do step 8.

8. Using Ohm's law, compute the current through the circuit. *V* is the voltage of the battery (measured in step 3) and *R* is the sum of the resistances. Record the calculation and answer in your notebook. Convert the current to milliamperes and record this answer in your notebook too.

9. Now reconnect the battery and measure the voltage drop across each of the two resistors with the multimeter in the volts mode. To do this, touch the multimeter lead tips to sockets H1 and H5 (to measure the voltage drop across the resistor in sockets F1 and F5) and H5 and H9 (to measure the voltage drop across the resistor in sockets G5 and G9). Disconnect the battery again to conserve battery power. The sum of the two voltage drops should equal the voltage of the battery. Record the voltage values, the calculation of the sum, and the comparison of the sum with the battery voltage.

10. Calculate the voltage drops across the two resistors based on the current value computed in step 8 and the individual resistance values. Comment on whether these voltage drops agree with what was measured in step 9.

## Experiment 16: Checking the Calibration of a Temperature Sensor

This experiment utilizes the following electronic components available from MicroLab, Inc., Bozeman, Montana (www.microlabinfo.com):

National Semiconductor LM-35 temperature sensor with a category 5 connector
Battery module, catalog number 050-061-110, with two 9-V batteries
Digital voltmeter or digital multimeter, like the one used in Experiment 15

The LM-35 temperature sensor operates in the 0 to 110°C range. With no special signal processing, it will measure temperatures with a precision of ±0.1°C. Its calibration can be checked against a water–ice mixture (0.0°C) at other temperatures. These other temperatures must be compared to the reading on a regular thermometer. At 0°C, its output is 0.000 V, and at 110°C, its output is 1.100 V. Thus, multiplying the voltage output by 100 will give the Celcius temperature to the nearest 0.1°C. The sensor is a shrink-wrapped rod about 6 in. long and a quarter inch in diameter, and as indicated above, it has a category 5 connector for connection to a power source and output display.

The battery module is a circuit board with the power source (two 9-V batteries) mounted on it and a receptacle for receiving the category 5 connector. It also has the output display terminals that the sensor requires for connection to a voltmeter to display the output voltage. The voltmeter or multimeter connects to these display terminals. The module has additional electronic components and connections with which we will not concern ourselves in this experiment.

### Procedure

1. Plug the LM-35's category 5 connector into the battery module at the receptacle labeled "sensor input." Also plug the digital multimeter into the output terminals using the same leads as in Experiment 15. The contact tips plug into the receptacles labeled GND and V on the battery module, either on the top of the receptacle or on the short side. Connect the COM port on the multimeter to the GND terminal (the black one) and the voltage terminal to the V terminal (red). Set the selector switch to the 2-V range.

2. Prepare an ice–water mixture by adding a quantity of crushed ice to a Styrofoam™ cup (Styrofoam so that it will be slow in melting) with a small amount of water—enough for a good slushy mix. Now check the calibration of the sensor by dipping the tip into the slush, stirring (with the sensor itself), and observing the reading on the multimeter. Allow the reading to become stable. The reading should be 0.000 V, corresponding to 0.0°C.

3. Prepare a three-column table in your notebook: one column for describing the sample whose temperature is being measured, one for recording the actual temperature (known either because it is the freezing point of water or because it was measured with a regular thermometer), and one for the voltage reading. Record the information for the measurement in step 2.

4. Next measure a series of water samples, first at room temperature and then at temperatures above room temperature. Place a beaker of water on a hot plate for these latter measurements. Be sure to measure the temperature of the water with a regular thermometer and enter all measurements into your table.

5. Comment in your notebook on how well the temperature sensor is calibrated.

## Experiment 17: Working with an Instrumentation Amplifier

This experiment utilizes the following electronic components available from MicroLab, Inc., Bozeman, Montana (www.microlabinfo.com).

Sensor amplifier module, catalog number 050-061-120 (see Figure 6.15)

National Semiconductor LM-35 temperature sensor with a category 5 connector, like the one used in Experiment 16

Digital voltmeter or digital multimeter, like the one used in Experiments 15 and 16

Battery module, catalog number 050-061-110, with two 9-V batteries, like the one used in Experiment 16

In this experiment, you will become familiar with an instrumentation amplifier circuit that is used to amplify weak sensor output signals.

### Procedure

1. Obtain the sensor amplifier module board from your instructor, and examine it closely as follows. Notice the outlined horizontal box in the top center of the board—it is labeled "instrumentation amplifier." Notice the symbol of an amplifier in this box (refer to Figure 6.10). Also, notice the inputs to the amplifier (labeled P14 and P15). These are sockets for the wires that will bring the input signals to the amplifier. It is the difference between these two signals that is amplified. The output of the amplifier (the amplified signal) is connected to sockets P19 and P20. Wires inserted into these sockets allow us to observe the amplified signal with a voltmeter.

2. Next, notice the outlined vertical box on the right side of the amplifier module board. This experiment involves two items in this box. The first is the voltmeter connection (labeled DVM for digital voltmeter) with switch (SW3). When the switch is down, the DVM measures the voltage at socket P45. When the switch is up, the DVM measures the voltage at socket P44. The second item with which this experiment is concerned is the signal/power connector, a category 5 connector. It is at this connector where the battery module will be connected.

3. At the bottom center of the module is another outlined box. This experiment involves the two potentiometers labeled "coarse" and "fine." These regulate the power from the battery module in order to have a specific voltage as a power supply to power the amplifier. This voltage is available at sockets P42 and P43. A connection can be made to these sockets wherever the regulated power is needed on the module board.

4. The vertical box on the left side of the amplifier module board is for connection to the sensor. The temperature sensor we will use has a category 5 connection and will be connected to either J3 or J4.

5. Connect the battery module to the amplifier module by running a category 5 cable from J2 on the battery module to the signal/power connector (J5) on the amplifier module. Also connect the digital multimeter as follows. The black color represents ground, the TP2 terminal on the amplifier module. This must connect to the COM port on the multimeter. Use the black lead for this connection. Connect the banana plug to the V socket on the multimeter. Set the multimeter selector switch to the 20-V range.

**FIGURE 6.15** The sensor amplifier module board used in Experiment 17.

6. Observe how the voltage output of the regulated power supply can be controlled. To do this, touch the red multimeter lead to the output of the regulated power supply, either P42 or P43, and rotate the knobs of the potentiometers, first the coarse, then the fine. You should observe that this output can vary between −5 and +5 V. Record your observations and your comments in your notebook.

7. Plug the temperature sensor into the category 5 socket labeled J3. Touch the red multimeter lead to the V socket of the sensor input (P8). The voltage reading at P8 is proportional to the temperature sensed by the temperature sensor. Record this reading. Warm the tip of the sensor with your fingers and notice how the reading changes. Record your observations. As in Experiment 16, you should observe that the sensor produces 10 mV per Celcius degree. A room temperature of 22°C should produce a voltage of 220 mV, or 0.22 V.

8. Since the instrumentation amplifier is a difference amplifier, it amplifies the difference between the inputs at P14 and P15 by a factor set by the operator. This factor is set at the switch box labeled SW1. Notice that the amplification factor can be ×5, ×10, ×50, or ×100. There are four switches, one for each of the factors. If all the switches are set to the left (off), then the amplification factor is ×1. When checking out each amplification factor, set all switches to off except for the factor being tested.

9. To check out the amplification, we will use as input a voltage of 0.50 V from the regulated power supply. To adjust the voltage from the regulated power to 0.50 V, connect the output of the power supply (P43) to the input to the DVM (P45). Use a short insulated wire for this connection and insert the wire into the sockets. With switch SW3 in the down position, read the voltage on the DVM. Adjust the coarse and fine potentiometers so that this voltage is 0.50 V. Now feed this voltage into the amplifier by connecting the output of the power supply (P42) to the input of the amplifier (P15), again with a short insulated wire. Connect the other input of the amplifier to ground by connecting P14 to P35. The difference between the two inputs is now 0.50 V, and it is this voltage that will be amplified.

10. Connect the output of the amplifier (P19) to the input to the DVM (P44). Move the switch SW3 to the up position so that the DVM measures the P44 point. Record the voltage readings for ×1, ×5, and ×10, and comment on how well they agree with the expected readings. Also check out ×50 and ×100. These, however, will not give the expected result because the limit of the amplifier has been reached.

11. Reverse the inputs to the amplifier by contacting P42 to P14 and P35 to P15. This means that the input to the amplifier is now −0.50 V instead of +0.50 V. Comment on the output of the amplifier in this case.

12. Next we will use the temperature sensor to input a voltage to the amplifier to amplify its signal and make the sensor more precise than it was in Experiment 16. Make sure the temperature sensor is connected to the J3 connector. Connect the temperature sensor voltage (P8) to the positive (+) input of the amplifier (P15) and the output of the regulated power (P42) to the negative (−) input of the amplifier. The output of the regulated power supply should be set to 0.00 V. You can adjust it to this setting in the same way you set it to 0.50 V in step 9 by connecting P43 to P45 (if not already done), changing switch SW3 to the down position, and rotating the coarse and fine potentiometers.

13. After adjusting the negative (−) input to 0.00 V, change switch SW3 to the up position again and read the output of the amplifier on the DVM. If the amplifier is set to ×1 amplification, the output should be the same as if you did not use the amplifier (as in Experiment 16). Next, increase the voltage at P45 to this output and set the switch SW1 to ×10. This results in a temperature measurement that is now ten times as sensitive as it was in Experiment 16. Measure and record a few temperatures as you did in Experiment 16 (comparing with thermometer readings) to confirm. Comment on the results.

## Experiment 18: Use of a Computer in Laboratory Analysis

### Part A: Plotting a Standard Curve Using Excel® Spreadsheet Software

Note: Use hypothetical data to practice plotting a standard curve with the Excel spreadsheet software procedure below. Later experiments will refer you back to this procedure to plot standard curves for real experiments. Launch Excel (or click "file" then "new" if already launched) to begin.

1. Type in the data for the standard curve in the A and B spreadsheet columns. Use the A column for the concentrations and the B column for the corresponding instrument readout values. For the unknowns and control, type in the instrument readout values in the B column cells, but leave the concentration cells blank. When finished, the A and B columns should appear as in Table 6.1, in which there are four standards with concentrations of 1, 2, 3, and 4 ppm, two unknowns, and one control.

2. Place the cursor in the A1 cell. Under "tools" select "data analysis," highlight "regression," and click "OK." (Note: If "data analysis" is not active in your Excel program, go to "help" for instructions on how to load it.)

3. Place the cursor in the "input Y range" slot. Enter B1:B3 if there are three standards, B1:B4 if there are four standards, etc.

4. Place the cursor in the "input x range" slot. Enter A1:A3 if there are three standards, A1:A4 if there are four standards, etc.

5. Select "output range." Place cursor in "output range" slot. Enter G3. Click OK. The y-intercept is now found in Cell H19 and the slope in Cell H20. The correlation coefficient, identified as "multiple R," is found in H6. Move these three values to more convenient cells as follows. Highlight cells G6 and H6 and copy and paste to cell D1. Highlight cells G19, G20, H19, and H20 and copy and paste to cell D2. The y-intercept is now in cell E2 and the slope in E3.

6. To calculate the concentrations of the unknowns and the control, use the equation of a straight line as follows, in which b is the y-intercept and m is the slope:

$$y = mx + b$$

$$x = \frac{(y - b)}{m}$$

7. Place the cursor in the A cell corresponding to the first unknown concentration (A5 in the example in step 1). Click "=" in upper margin. Type in the above formula for x using the cell number for the instrument readout (y) that corresponds to the A cell highlighted. For the example in step 1, for the first unknown, this would be (B5–E2)/E3. Click "OK." The concentration of the unknown is now found in the A cell in which you placed the cursor (A5 for the example).

8. Place the cursor in the A cell for the next unknown and repeat. Repeat step 6 for the other unknowns and for the control sample. Once completed for the example in step 1, the A and B columns should appear as they do in Table 6.2.

9. Highlight the standard curve data (A1 to B4 in the example). Click "insert" (top tool bar), then "chart." Then click "XY (scatter)." Click "next" twice.

TABLE 6.1  Example of the A and B Columns
in the Spreadsheet in Experiment 18, Step 1

|   | A | B |
|---|---|---|
| 1 | 1 | 3 |
| 2 | 2 | 5.9 |
| 3 | 3 | 9.1 |
| 4 | 4 | 12 |
| 5 |   | 7.2 |
| 6 |   | 8.4 |
| 7 |   | 4.5 |

**TABLE 6.2** Example of the A and B Columns in the Spreadsheet in Experiment 18, Step 7

|   | A | B |
|---|---|---|
| 1 | 1 | 3 |
| 2 | 2 | 5.9 |
| 3 | 3 | 9.1 |
| 4 | 4 | 12 |
| 5 | 2.400662 | 7.2 |
| 6 | 2.798013 | 8.4 |
| 7 | 1.506623 | 4.5 |

10. Type in a chart title and labels for the x- and y-axes. Click "finish." With either the graph or the entire graph box highlighted, click "chart" (upper tool bar), then "add trendline," then "OK."

11. Click (on the white area surrounding the graph to highlight the entire graph box), and drag the graph box to directly below the data. Highlight the data (cells A through F) and graph box and copy and paste into a new spreadsheet. Click "file" and "print." Type in the number of copies and click "OK."

12. Click on "file," then "close," and then "no" to delete your spreadsheet (without saving) for the next user.

### Part B: Demonstration of Data Acquisition with a Microcomputer

#### Introduction

This experiment is an acid–base titration similar to those performed in Chapters 4 and 5. It is the titration of 0.10 *M* HCl with 0.10 *M* NaOH, as in Experiment 8, but a combination pH probe will be used to monitor the pH during the titration, as in Experiment 10. The pH meter to be used has an RS232 output to interface with a microcomputer. We will use this special feature of the pH meter to feed the pH data directly to a microcomputer in an example of data acquisition by computer, and observe the titration curve traced on the screen in real time.

*Remember safety glasses.*

1. Prepare 100 mL of a 0.1 *M* HCl solution and 100 mL of a 0.10 *M* NaOH solution.
2. Measure 25 mL of the HCl into a 250-mL beaker. Fill a 50-mL buret with the NaOH.
3. Place a magnetic stirring bar in the beaker and place the beaker on a magnetic stirrer.
4. Prepare the pH meter and a combination pH electrode. Place the electrode in the beaker such that the tip of the electrode is immersed, but suspended with a clamp or pointed to the side of the beaker to avoid contact with the stirring bar. You may have to add some distilled water in order for the tip (including the contact to the reference electrode) to be completely immersed.
5. Turn on the stirrer and measure the pH. The pH should be between 1 and 2. Load and run the data acquisition program; your instructor will guide you in getting this ready.
6. Begin the titration at the same time that the data acquisition begins. Your instructor will suggest an appropriate titration rate. Titrate until all 50 mL have been added. At that point, stop the data acquisition and store the data on disk.
7. Using the computer, plot the titration curve and display it on the computer screen. If possible, print out a hard copy.
8. If directed by your instructor, also obtain weak acid–strong base, strong base–strong acid, and weak base–strong acid titration curves.

## Questions and Problems

1. What are the five parts of the general analytical strategy?
2. Distinguish between wet methods of analysis and instrumental methods of analysis. What do the analytical strategies for the wet chemical methods and instrumental methods have in common?

3. Why are sampling and sample preparation activities important no matter what analytical method is chosen?

4. What is different about the analytical strategy for instrumental analysis, compared to wet chemical analysis?

5. Explain the general principle of analysis with electronic instrumentation and cite an example.

6. What are three major instrumental analysis classifications? What are the general names of the instruments?

7. Distinguish between spectroscopic methods and chromatographic methods.

8. Define calibration.

9. Briefly explain the differences between calibrating an analytical balance and calibrating a pH meter and what happens if the measured standard does not give the correct result.

10. What is a standard curve?

11. Briefly explain how the plotting of a standard curve is a calibration process.

12. Describe the various methods by which a sample is introduced into an instrument.

13. What is a sensor?

14. What are the four instrument components for translating a property of a solution to an instrument readout?

15. Define the following: complete circuit, input, voltage, resistance, current, Ohm's law, voltmeter, ammeter, and ohmmeter.

16. If the voltage drop across a resistor is measured to be 1.293 V and the value of the resistance is 3 k$\Omega$, what is the current through the resistor?

17. What is the difference between an instrumentation amplifier and an operational amplifier?

18. What is the symbol for an integrated circuit amplifier?

19. What is a thermocouple? Explain how a thermocouple might be calibrated.

20. What are the advantages of calibrating an instrument with a single standard solution rather than with a series of standard solutions? Which is preferred and why?

21. The proportionality constant between an instrument readout and concentration is 54.2. Assuming a linear relationship between the readout and concentration, what is the numerical value of the concentration of a solution when the instrument readout is 0.922?

22. What is the numerical value of the concentration in a solution that gave an instrument readout of 53.9 when the proportionality constant is 104.8?

23. An instrument reading for a standard solution whose concentration is 8.0 ppm is 0.651. This same instrument gave a reading of 0.597 for an unknown. What is the concentration of the unknown?

24. Calculate the concentration for the unknown given the following data:

| R | C (ppm) |
|------|---------|
| 72.0 | 0.693 |
| 68.1 | $C_u$ |

25. Using spreadsheet software, plot the following data and give the concentration of the unknown solution:

| R | C (ppm) |
|------|---------|
| 8.2 | 2.00 |
| 17.0 | 4.00 |
| 24.9 | 6.00 |
| 31.9 | 8.00 |
| 40.5 | 10.00 |
| 26.7 | $C_u$ |

26. Why must the instrument readout for an unknown fall within the range of of readouts for a series of standard solutions in order to be accurate?

27. Calculate the slope of the straight line defined by the following two points: (0.20, 0.439) and (0.50, 0.993).

28. Given the following data, calculate the proportionality constant, K, between A and C. (Assume that the y-intercept is zero.)

| A | C |
|---|---|
| 0.419 | 3.00 |
| 0.837 | 6.00 |

29. Given the data and results from problem 28, what is the concentration in a solution that gave an A value of 0.677?

30. When plotting the results of the measurement of a series of standard solutions, why do we draw the best straight line possible through the points rather than just connect the points?

31. What is the method of least squares and why is it useful in instrumental analysis?

32. What is it about the calculations involved in the method of least squares that gives this method its name?

33. What is meant by linear regression analysis?

34. Give four parameters that are readily obtainable as a result of the method of least squares treatment of a set of data.

35. What is meant by perfectly linear data? What value of what parameter would indicate that a data set is perfectly linear?

36. What are some realistic values of a correlation coefficient that would indicate to a laboratory worker that the error associated with his or her data is probably minimal?

37. What is meant by serial dilution?

38. The series of standard solutions method works satisfactorily most of the time. When does it *not* work well?

39. A stock standard solution of copper (1000.0 ppm) is available. Tell how you would prepare 50.00 mL each of five standard solutions, giving the amount of the 1000.0 ppm required for each one and how you would proceed with the preparation, including the kind of pipet needed. The concentrations of the standards should be 1.00, 2.00, 3.00, 4.00, and 5.00 ppm.

40. How many milliliters of a 100.0 ppm stock are needed to prepare 25.00 mL each of four standards with concentrations 2, 4, 6, and 8 ppm?

41. What is a blank? What is the difference between a reagent blank and a sample blank?

42. What is a control sample and how is it useful as an accuracy check?

43. Explain why the concentration of an unknown obtained from a graph of a series of standard solutions may not be the final answer in an instrumental analysis.

44. An unknown solution of riboflavin, contained in a 25-mL volumetric flask, was determined to have a concentration of 0.525 ppm. How many milligrams of riboflavin are in the flask?

45. After analysis for iron content, a water sample was found to have a concentration of 4.62 ppm iron. How many milligrams of iron are contained in 100.0 mL of the water?

46. A 2.000-g sample of a soil is found to yield 3.73 mg of phosphorus after extraction. What is the concentration of phosphorus in the soil in parts per million?

47. After a 5.000-g sample of concrete was dissolved, the resulting solution was found to contain 0.229 g of manganese. What is the concentration of manganese in the concrete in parts per million?

48. A certain water sample was diluted from 1 to 50 mL with distilled water. After an analysis for zinc was performed, this diluted sample was found to contain 10.7 ppm zinc. What is the zinc concentration in the undiluted sample?

49. A rag that a farmer was using was analyzed for pesticide residue. The rag weighed 49.22 g and yielded 25.00 mL of an extract solution that was determined to have a pesticide concentration of 102.5 ppm. How many grams of pesticide were in the rag and what is the concentration of pesticide in the rag in parts per million?

50. The soil around an old gasoline tank buried in the ground is analyzed for benzene. If 10.00 g of soil shows the concentration of benzene in 100.0 mL of soil extract to be 75.0 ppm, what is the concentration of benzene in the soil in parts per million?

51. Suppose 4.272 g of a soil sample undergoes an extraction with 50 mL of extracting solvent to remove the potassium. This 50 mL is then diluted to 250 mL and tested with an instrument. The concentration of potassium in this diluted extract is found to be 35.7 ppm. What is the potassium concentration, in parts per million, in the untreated soil sample?

52. A city's water supply is found to be contaminated with carbon tetrachloride. The chemical analysis procedure involves the extraction of the carbon tetrachloride from the water with hexane. A 4.00-L sample of water is extracted with 10 mL of hexane. If this hexane solution is diluted from 1 to 25 mL and the concentration of carbon tetrachloride in the diluted solution found to be 10.4 ppm, what is the concentration in the original water?

53. What is meant by data acquisition by computer?

54. Why has computerized data storage developed into an important function of computers in the modern laboratory?

55. Name one way that spreadsheet computer software is useful in the modern laboratory.

56. What is LIMS an acronym for? Explain how LIMS is useful in the modern laboratory.

# 7

# Introduction to Spectrochemical Methods

## 7.1 Introduction

More than half of all instrumental methods of analysis involve the absorption or emission of light. Such instrumental methods can be referred to as **spectrochemical methods**. The science that deals with light and its absorption and emission by solutions and other material substances is called **spectroscopy** or **spectrometry**. The broad term for the instrument used is **spectrometer**, while a slightly more specific term (when a light sensor known as a phototube is used) is **spectrophotometer**. In spectrochemical analysis procedures, the degree to which light is absorbed, or the intensity of light that is emitted, is related to the amount of an analyte present in the sample tested. Thus the degree of light absorption and the intensity of light emission are the critical measurements. The electrical signal readout referred to in Figure 6.4 is, in the case of spectroscopy, an electrical signal that is related to the degree of light absorption or the intensity of light emission. The instrument readings mentioned in Figure 6.5 are the readings generated by the instrument as a result of this absorption or emission. See Figures 7.1 and 7.2.

In this chapter we expand the above brief summary so that all aspects of spectroscopy as an analytical method can be clearly understood and practiced. This includes a full discussion of the nature of light, including energy, wavelength, frequency, and wavenumber. It includes exactly what is meant by light absorption and emission. It includes the spectral differences between atoms (**atomic spectroscopy**) and molecules (**molecular spectroscopy**). It includes the different effects caused by ultraviolet light, visible light, and infrared light. It includes instrument design and what exactly gives rise to the electrical signals that are generated. It includes the fine points of instrument and experiment design so that experimental results can be optimized. And it includes the instrument readings and how they are related to the amount of analyte present.

## 7.2 Characterizing Light

The modern characterization of light is that it has a dual nature. Some qualities of light are best explained if we describe it as consisting of moving particles, often called **photons** or **quanta** (called the **particle theory** of light). Other qualities are best explained if we describe light as consisting of moving electromagnetic disturbances, referred to as **electromagnetic waves** (called the **wave theory of light**). Such a dual nature is not unlike the modern description of electrons, a description that you likely encountered in your previous studies of chemistry. Electrons may be described as particles in order to explain some aspects of their behavior. However, in order to be fully in tune with all scientific observations, electrons must be described as entities of energy and not particles.

**FIGURE 7.1**    Principle of analysis with spectrochemical instrumentation (compare with Figure 6.4).

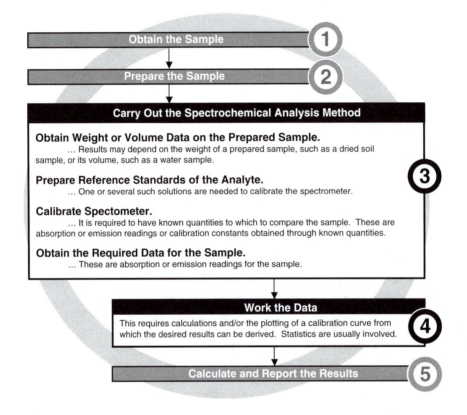

**FIGURE 7.2**    The general analytical strategy for spectrochemical analysis (compare with Figure 6.5).

The wave theory of light states that light travels in a fashion similar to that of a series of repeating waves of water, as in a wave pool at an amusement park. However, water waves are mechanical waves and, as such, require matter, such as water, to exist. Light waves are **electromagnetic waves**, meaning that they are wave disturbances that have an electrical component and a magnetic component and do not require matter to exist. They therefore can (and do) travel through a vacuum, such as outer space, where no matter is present.

## 7.2.1    Wavelength, Speed, Frequency, Energy, and Wavenumber

The water waves in a wave pool at an amusement park can be long or short, depending on the kind of motion at their origin. The same is true of electromagnetic waves. The length of an electromagnetic wave is called its **wavelength**, and its symbol is the lowercase Greek letter lambda, $\lambda$. In a set of repeating waves, wavelength is the physical distance from a point on one wave, such as the crest of the wave, to the same point (the crest) on the next wave. See Figure 7.3. In science, it is measured in metric system units.

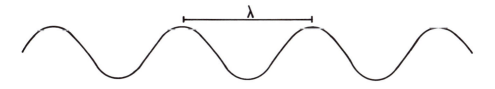

**FIGURE 7.3**    The definition of wavelength.

These can be meters, centimeters, nanometers, etc. Wavelengths of electromagnetic waves vary from as short as atomic diameters to as long as several miles.

The speed with which electromagnetic waves move is called the **speed of light** and is given the symbol c. The speed of light in a vacuum is approximately $3.00 \times 10^{10}$ cm/sec (cm sec$^{-1}$). This extraordinarily fast speed accounts for the seemingly instantaneous manner with which light fills a room when the light switch is switched on. It is also one huge difference between mechanical water waves, which travel almost infinitely slower, and electromagnetic waves. It is important to know that all electromagnetic waves travel at the same speed in a vacuum regardless of their wavelength.

The number of the moving electromagnetic waves that pass a fixed point in 1 sec is called the **frequency** of the light. Its symbol is the lowercase Greek letter nu, $\nu$. It is expressed in waves (most often called cycles) per second, or hertz (Hz). In mathematical formulas, the unit is reciprocal seconds (sec$^{-1}$).

Mathematically, wavelength, speed, and frequency are related by the formula

$$c = \lambda \nu \tag{7.1}$$

Multiplying the wavelength expressed in centimeters (cm) by the frequency in reciprocal seconds (sec$^{-1}$) gives the unit of speed, centimeters per second (cm sec$^{-1}$):

$$cm \times sec^{-1} = cm\ sec^{-1} \tag{7.2}$$

With the speed of all light in a vacuum being the same (approximately equal to $3.00 \times 10^{10}$ cm sec$^{-1}$, as stated previously), it is clear that wavelength and frequency are inversely proportional. This means that as one increases the other decreases such that when one is multiplied by the other, the same number, the speed of light, always results. It is also clear that a given wavelength is always associated with a particular frequency, and given one, the other can be calculated by rearranging Equation (7.1):

$$\nu = \frac{c}{\lambda} \tag{7.3}$$

$$\lambda = \frac{c}{\nu} \tag{7.4}$$

**Example 7.1**

What is the frequency of a light that has a wavelength of 537 nm?

***Solution 7.1***

$$\nu = \frac{c}{\lambda}$$

If $3.00 \times 10^{10}$ cm sec$^{-1}$ is to be used for the speed of light, then the wavelength must be converted to centimeters. There are $10^{-7}$ cm in 1 nm.

$$\nu = \frac{3.00 \times 10^{10}\ \cancel{cm}\ sec^{-1}}{537\ \cancel{nm} \times 10^{-7}\ \cancel{cm}/\cancel{nm}} = 5.59 \times 10^{14}\ sec^{-1}$$

**Example 7.2**

What is the wavelength of light that has a frequency of $7.89 \times 10^{14}$ sec$^{-1}$? Express the answer in both centimeters and nanometers.

*Solution 7.2*

$$\lambda = \frac{c}{v}$$

$$\lambda = \frac{3.00 \times 10^{10} \text{ cm sec}^{-1}}{7.89 \times 10^{14} \text{ sec}^{-1}}$$

$$= 0.0000380 \text{ cm} = 380 \text{ nm}$$

Light is a form of **energy**, and each wavelength or frequency has a certain amount of energy associated with it. This energy is considered to be the energy associated with a single photon of the light. Thus, the particle theory and the wave theory are linked via energy. The relationship between energy and frequency is as follows:

$$E = hv \tag{7.5}$$

where E is energy, $v$ is the frequency, and h is a proportionality constant known as **Planck's** constant, after the famous physicist Max Planck. The value of h depends on what unit of energy is used. If the energy is to be expressed in joules (J), a unit of energy in the metric system, Planck's constant is $6.63 \times 10^{-34}$ J sec.

Combining Equations (7.1) and (7.3), the relationship between energy and wavelength is as follows:

$$E = \frac{hc}{\lambda} \tag{7.6}$$

**Example 7.3**

What is the energy associated with a frequency of $6.18 \times 10^{12}$ sec$^{-1}$?

*Solution 7.3*

$$E = hv$$

$$= 6.63 \times 10^{-34} \text{ J sec} \times 6.18 \times 10^{12} \text{ sec}^{-1}$$

$$= 4.10 \times 10^{-21} \text{ J}$$

**Example 7.4**

What is the energy associated with a wavelength of 497 nm?

*Solution 7.4*

$$E = hc/\lambda$$

$$= \frac{6.63 \times 10^{-34} \text{ J sec} \times 3.00 \times 10^{10} \text{ cm sec}^{-1}}{497 \text{ nm} \times 10^{-7} \text{ cm} / \text{nm}}$$

$$= 4.00 \times 10^{-19} \text{ J}$$

**Example 7.5**

What is the wavelength of light with an energy of $5.92 \times 10^{-12}$ J? Express your answer in both centimeters and nanometers.

**Solution 7.5**

Rearranging Equation (7.6), we have the following:

$$\lambda = \frac{hc}{E}$$

$$= \frac{6.63 \times 10^{-34} \text{ J sec} \times 3.00 \times 10^{10} \text{ cm sec}^{-1}}{5.92 \times 10^{-12} \text{ J}}$$

$$= 3.36 \times 10^{-12} \text{ cm} = 3.36 \times 10^{-5} \text{ nm}$$

If the wavelength is expressed in centimeters, waves of light are sometimes characterized by the reciprocal of this wavelength. This parameter is known as the **wavenumber** and is given the symbol $\bar{\nu}$ (Greek letter nu with a bar).

$$\bar{\nu} = \frac{1}{\lambda}(\text{cm}) \qquad (7.7)$$

It has reciprocal centimeters units ($\text{cm}^{-1}$) and is used especially in conjunction with infrared light. The relationships between wavenumber and frequency and between wavenumber and energy are as follows:

$$\nu = c\bar{\nu} \qquad (7.8)$$

$$E = hc\bar{\nu} \qquad (7.9)$$

## Example 7.6

What is the wavenumber of light with a wavelength of 3030 nm?

**Solution 7.6**

$$\bar{\nu} = \frac{1}{\lambda(\text{cm})}$$

$$= \frac{1}{3030 \text{ nm} \times 10^{-7} \text{cm} / \text{nm}}$$

$$= 3300 \text{ cm}^{-1}$$

## Example 7.7

What is the energy of light with a wavenumber of 1720 $\text{cm}^{-1}$?

**Solution 7.7**

$$E = hc\bar{\nu}$$

$$= 6.63 \times 10^{-34} \text{ J sec} \times 3.00 \times 10^{10} \text{ cm sec}^{-1} \times 1720 \text{ cm}^{-1}$$

$$= 3.42 \times 10^{-20} \text{ J}$$

Frequency, energy, and wavenumber are directly proportional, as can be seen in Equations (7.5), (7.8), and (7.9). This means that as frequency increases, energy and wavenumber also increase. If we are dealing with a very high energy wave, then we are dealing with a wave of high frequency and high wavenumber. Wavelength, on the other hand, is inversely proportional to frequency, energy, and wavenumber, as can be seen from Equations (7.3), (7.6), and (7.7). If we are dealing with a very short wavelength, then we are dealing with a high frequency, high energy, and high wavenumber.

## 7.3   The Electromagnetic Spectrum

Wavelengths can vary in distance from as little as fractions of atomic diameters to as long as several miles. This suggests the existence of an extremely broad spectrum of wavelengths. This **electromagnetic spectrum** of light is so broad that we break it down into regions. Most of us are familiar with at least a few of the regions. The region of wavelengths that we see with our eyes is called the **visible region**. It is located approximately in the middle of the electromagnetic spectrum and has wavelengths that vary from approximately 350 nm to approximately 750 nm—a very narrow region compared to the entire spectrum and to the other regions. Others include the ultraviolet region, the infrared region, the x-ray region, and the radio and television wave region. These regions are considerably broader, as you can see from the representations in Figure 7.4. In this figure, wavelength increases left to right and energy, frequency, and wavenumber decrease left to right. The approximate borders of the various regions in nanometers are shown. In the ultraviolet, visible, and infrared regions, which are the regions that are emphasized in this chapter, the nanometer and the micrometer (or micron, μ) are the most commonly used units of wavelength.

The wavelength pictured in Figure 7.3 is several centimeters long. Most wavelengths that we deal with, however, are either much shorter or much longer than that. Radio and television waves, for example, are on the long end of the electromagnetic spectrum, on the order of kilometers long. These are very low energy waves (remember: long wavelength = low energy) that do no harm as far as our health and safety are concerned. That is a good thing, because the atmosphere in which we live is full of these wavelengths transporting the radio and television sound and pictures from every studio on Earth ultimately to our personal radios and televisions. Microwaves have wavelengths on the order of a centimeter and are also of low energy, but they can be dangerous because their absorption causes the generation of much internal heat. Microwave ovens are used in our kitchens, but they are also used in an analytical laboratory for drying samples.

In another familiar region, the infrared region, the wavelengths are extremely short by comparison. Wavelengths in this range are so short that we cannot represent them on paper or measure them with common measuring tools. While the wavelengths are shorter and have higher energy than radio and television waves or microwaves, they also cause us no harm. Indeed, the remote controls we use to control our stereos and VCRs utilize infrared light.

When we consider the ultraviolet region, the wavelengths are shorter still, meaning more energy. They are known to cause harm, such as sunburn. X-rays are extremely short wavelengths of extremely high energy, penetrating our skin and tissue and causing harm, hence the reason for the lead aprons used by doctors and dentists when our bodies are x-rayed. Gamma rays have wavelengths on the order of atomic diameters and cause extreme damage to the human body due to their extremely high energy.

The wavelengths within the visible region of the electromagnetic spectrum are associated with the colors we see. Consider a rainbow. When the light from the sun moves through the Earth's atmosphere after a rainstorm, a rainbow may appear in the sky. The reason for this is that the different wavelengths of visible light present in white light travel through atmospheric water at different speeds (see Chapter 4 for the related discussion of refraction of light). The result is the visible region of the electromagnetic spectrum—the violet, blue, green, yellow, orange, red sequence—displayed for all to see. See Figure 7.5.

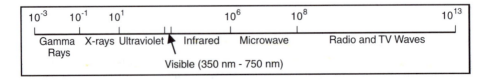

**FIGURE 7.4**   The electromagnetic spectrum. The exponential numbers across the top are approximate wavelength values. Energy increases right to left.

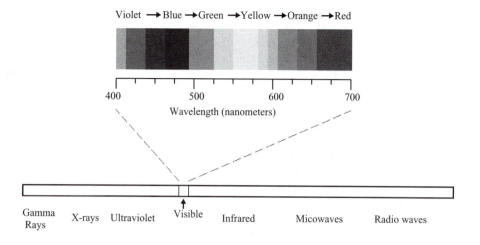

**FIGURE 7.5** The visible region of the electromagnetic spectrum. (From Kenkel, J., Kelter, P., and Hage, D., *Chemistry: An Industry-Based Introduction with CD-ROM*, CRC Press, Boca Raton, FL, 2001. With permission.)

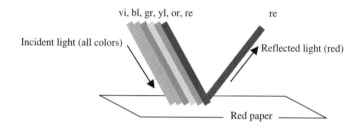

**FIGURE 7.6** A representation of the absorption of visible light by a sheet of paper, resulting in the paper having a red color. The following abbreviations are used: vi = violet, bl = blue, gr = green, yl = yellow, or = orange, and re = red.

# 7.4 Absorption and Emission of Light

As stated previously, spectrochemical methods of analysis involve the absorption and emission of light. Absorption is considered in Sections 7.4.1 to 7.4.3. Emission is considered in Section 7.4.4.

## 7.4.1 Brief Summary

The visible region of the spectrum provides a good starting point to understanding the process of light absorption. Visual evidence of light absorption abounds in our world because of our ability to see the colors that are the visible wavelengths. Objects display a color because some visible light wavelengths are absorbed. Why does a red sheet of paper appear red? All the wavelengths of visible light from the sun and the light bulbs on in the room are absorbed except for the red wavelengths, and these are reflected to our eye. See Figure 7.6. Why does a solution of potassium permanganate appear to be a deep purple color? The wavelengths of visible light that are incident on the solution from the light in the room are all absorbed, except for those in the violet and red ends of the visible region. The result is an intense purple color. Thus, those wavelengths of light that are not absorbed by a sample of matter are the wavelengths that are reflected, or transmitted, to our eye and give the sample its color.

What is the nature of the interaction between light and matter that causes certain wavelengths of light to be absorbed? The answer lies in the structure of the atoms and molecules of which matter is composed. First consider atoms. The modern theory of the atom states that electrons exist in energy levels around the nucleus. Electrons can be moved from one energy level to a higher one if conditions are right. For example, the outermost electron of a sodium atom (which has the electron configuration $1s^2 2s^2 2p^6 3s^1$) can be moved from the 3s level to the vacant 3p level if conditions are right. These conditions consist of: 1) the addition of a specific amount of energy to the electron (in the case of sodium, this is the energy difference between the 3s and 3p levels), and 2) a vacancy for the electron with this greater energy in a certain higher energy level (in the case of sodium, the 3p level is vacant). In other words, if an electron absorbs the energy required for it to be promoted (or elevated) to a higher vacant energy level, then it is promoted (or elevated) to that level.

Atoms in which no electrons are in the higher levels are said to be in the **ground state**. This state is designated in energy level diagrams as $E_0$. Atoms in which there *is* an electron in the higher level are said to be in an **excited state**. Excited states are designated in energy level diagrams as $E_1$, $E_2$, $E_3$, etc. An **energy level diagram** consists of short horizontal lines representing the levels or states, with each line labeled as $E_0$, $E_1$, etc. Often, an energy level diagram shows the movement of electrons between levels with longer vertical arrows. The movement of an electron between electron energy levels is called an **electronic energy transition**. See Figure 7.7.

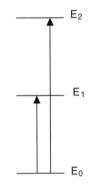

**FIGURE 7.7** An example of an energy level diagram for an atom depicting energy levels $E_0$, $E_1$, and $E_2$ and using vertical arrows to show electrons in the ground state ($E_0$) being promoted to excited states ($E_1$ and $E_2$)—electronic energy transitions.

Where can an electron obtain the required energy to be promoted to a higher level? One way is for it to come into contact with light of the same energy. When the light of this energy "strikes" an atom and causes an electron to be promoted to a higher energy level, the energy of the light becomes part of the electron's energy and therefore "disappears." It is absorbed. It is important to keep in mind that the light coming in must be exactly the same energy as the energy difference between the two electronic levels; otherwise, it will not be absorbed at all. See Figure 7.8. If it is not absorbed and is in the visible region, we see it. It becomes part of the light that is reflected and therefore detected by our eyes. The absorption of light by atoms consists of the absorption of only a few very specific wavelengths because the energy difference between two levels is very specific.

We cited sodium previously. Gaseous sodium atoms absorb in the visible region of the spectrum. The few very specific wavelengths for sodium are 589.0 and 589.6 nm. Both of these represent transitions from the 3s level to the 3p level. The 3p level is actually split narrowly into two levels due to the effect of two possible spin states for the electron in this level, hence the observation of two transitions that

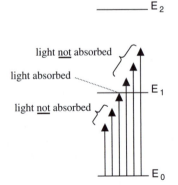

**FIGURE 7.8** An energy level diagram of an atom showing the fact that some wavelengths possess too much or too little energy to be absorbed, while another possesses the exact energy required and is therefore absorbed.

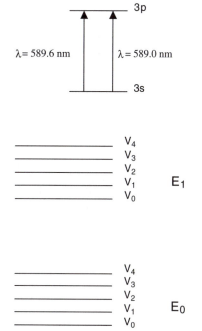

**FIGURE 7.9** The electronic energy transitions for gaseous sodium atoms. There are two slightly different transitions between the same two levels because of the effect of two spin states that differ slightly in energy.

**FIGURE 7.10** An energy level diagram for a molecule or complex ion showing the vibrational energy states superimposed on the electronic states.

are nearly equal in energy. The energy level diagram for gaseous sodium is illustrated in Figure 7.9. The energy associated with electronic energy transitions in atoms is equivalent to the energy of visible and ultraviolet light, and thus atoms absorb light in these regions.

What about molecules and complex ions? Absorption of light by molecules and complex ions results in the promotion of electrons to higher energy states in the same way as in atoms. However, it is more complicated because molecules and complex ions have energy states that atoms do not.

## 7.4.2 Atoms vs. Molecules and Complex Ions

Molecules and complex ions exist in vibrational and rotational energy states as well as electronic states. A vibrational energy state represents a particular state of covalent bond vibration that a molecule can have. This concept will be discussed in Chapter 8. A rotational energy state represents a particular state of rotation of a molecule. The vibrational energy states exist in each electronic state, and rotational energy states exist in each vibrational state. Energy level diagrams for molecules show additional horizontal lines, or levels, within (superimposed on) each electronic level to represent these vibrational states. The vibrational states can be labeled $V_0$, $V_1$, $V_2$, etc. See Figure 7.10. The rotational states are superimposed on the vibrational states in the same way. However, unless we are dealing with the rotational states directly, they are most often not depicted on energy level diagrams in order to keep the diagrams relatively simple. The rotational levels are not considered in our discussion here.

For molecules and complex ions, an electronic transition can refer to a transition from any vibrational level in one electronic level to any vibrational level in another electronic level. As with atoms, the amount of energy required for such a transition is found in either the visible or ultraviolet regions of the electromagnetic spectrum, and thus involves either visible or ultraviolet light. A vibrational transition refers to a transition from the lowest vibrational level within a certain electronic level to another vibrational level in the same electronic level. Since such a transition does not involve another electronic level, it requires much less energy and involves the infrared region of the electromagnetic spectrum. See Figure 7.11.

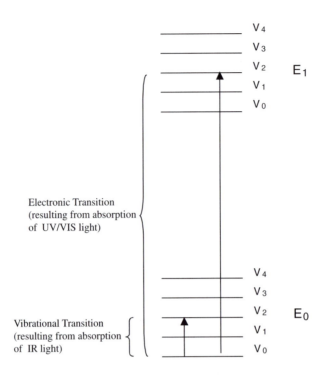

**FIGURE 7.11**    An energy level diagram for a molecule or complex ion expanded to show an electronic energy transition and a vibrational energy transition. The electronic transition involves visible or ultraviolet light while the vibrational transition involves infrared light.

## 7.4.3  Absorption Spectra

Many instruments that are used to measure the absorption of light by atoms, molecules, and complex ions are constructed to measure the ultraviolet and visible regions of the electromagnetic spectrum. Others are constructed to measure the infrared region of the spectrum. All of these instruments have a light source from which a beam of light is formed. The sample being measured is contained such that the light beam is passed through it and the absorption of the wavelengths present in the light beam is measured by a light sensor and signal processor. See Figure 7.12 for the case of ultraviolet and visible light absorption. There are sufficient differences in the mechanics of measuring molecules by ultraviolet and visible wavelengths, as opposed to infrared wavelengths, to warrant separate discussions of each. Also, there are sufficient differences in the mechanics of measuring atoms, as opposed to molecules and complex ions, in the ultraviolet and visible regions so as to also warrant separate discussions. For example, the atoms must be in the gas phase and there are a number of different methods used for converting metal ions in solution to atoms. If molecules are being measured in the ultraviolet and visible regions, the technique is referred to as **ultraviolet-visible (UV-VIS) spectrophotometry**. This technique is developed fully in Chapter 8. If molecules are being measured in the infrared region, the technique is referred to as **infrared (IR) spectrometry**. This technique is also developed fully in Chapter 8. If atoms are being measured, the technique is referred to as **atomic spectroscopy**; this technique, including a number of subtechniques, is developed fully in Chapter 9. In this section, we follow up the fundamentals discussed in Section 7.4.2 to describe one particular property important for both molecules and atoms: the absorption spectrum.

An **absorption spectrum** is a plot of the amount of light absorbed by a sample vs. the wavelength of the light. The amount of light absorbed is called the **absorbance**. It is symbolized as **A** and will be clearly defined in Section 7.5. An absorption spectrum is obtained by using a spectrometer to scan a particular wavelength region and to observe the amount of light absorbed by the sample along the way. Consider a

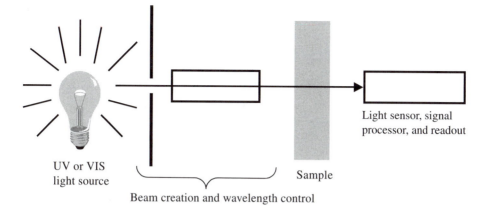

FIGURE 7.12 Instruments manufactured to measure ultraviolet or visible light absorption by a test sample have a light source from which a beam of light is formed. The intensities of the wavelengths of light present in the light beam are measured at the light sensor and converted to a readout so that the amount of light absorbed by the sample at selected wavelengths becomes known.

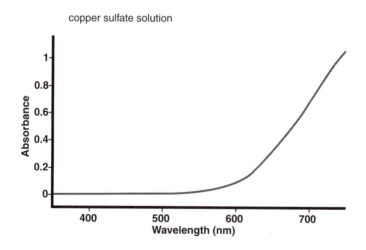

FIGURE 7.13 The absorption spectrum, visible region, of a copper sulfate solution.

solution of copper sulfate. Such a solution is an example of a complex ion solution, $Cu(H_2O)_4^{2+}$ (see Section 5.3). It displays a blue color, which means that the blue portion of the visible region is not absorbed, but transmitted to our eyes, while the red portion is absorbed. The absorption spectrum of this solution in the visible region is shown in Figure 7.13. This spectrum clearly shows that wavelengths in the blue and violet regions (350 to 500 nm) are not absorbed, while wavelengths in the red region (650 to 750 nm) are absorbed.

Notice also that this absorption spectrum is a **continuous spectrum**, meaning that the spectrum is an unbroken pattern, left to right. It does not display any breaks or sharp peaks of absorption at particular wavelengths, but rather shows that a smooth band of wavelengths in a given region, such as the red region, is absorbed.

Compare this with the absorption spectrum of gaseous copper atoms, Figure 7.14. First, the wavelength region shown (320 to 330 nm) is in the ultraviolet region. Gaseous copper atoms do not absorb in the visible region. Second, sharp lines of absorption are observed, one at 324.8 nm and one at 327.4 nm. It is a **line spectrum**, meaning that individual absorption *lines* are observed, rather than a continuous, unbroken band, like that observed for the copper sulfate solution.

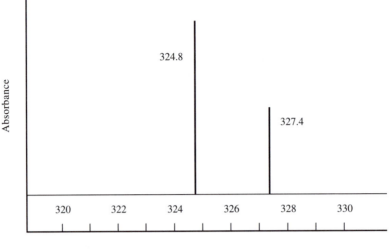

FIGURE 7.14    The absorption spectrum, in a narrow portion of the ultraviolet region, of gaseous copper atoms.

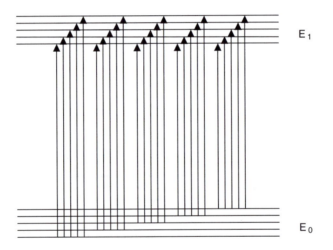

FIGURE 7.15    The energy transitions possible between two electronic states, each with five vibrational levels superimposed.

These observations are explained by relating them to energy level diagrams. In the case of atoms, only specific energies (wavelengths), represented by the precisely defined energy differences between two atomic electronic levels, can get absorbed. Thus, Figure 7.7 translates into Figure 7.14—two absorption lines corresponding to two electronic transitions. In the case of molecules or complex ions, an entire range of wavelengths can get absorbed due to the presence of the vibrational and rotational levels superimposed on the electronic levels. The transitions possible between electronic levels $E_0$ and $E_1$, for example, each with five vibrational levels, are shown in Figure 7.15. Figure 7.15 translates into Figure 7.13—many wavelengths absorbed, each corresponding to a particular transition. When the higher electronic levels are considered, and when the rotational levels superimposed on each vibrational level in each electronic level are considered, virtually all wavelengths may be absorbed and the result is the continuous absorption pattern of Figure 7.13.

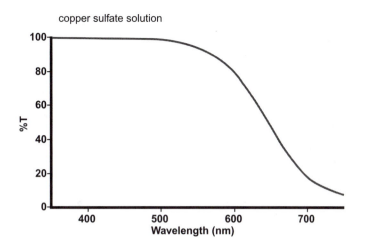

**FIGURE 7.16**    The transmission spectrum of a copper sulfate solution.

The absorption vs. wavelength information for molecules and complex ions may also be displayed as a **transmission spectrum**, rather than an absorption spectrum, by plotting the amount of light transmitted by a sample rather than the light absorbed. In this case, the parameter that is plotted on the y-axis is the **transmittance**, or **percent transmittance**, rather than the absorbance. Transmittance is symbolized as T. Transmittance will be specifically defined in Section 7.5. For molecules and complex ions, the transmission spectrum is similar to the corresponding absorption spectrum, but upside down, since a high absorbance corresponds to a low transmittance and vice versa. For example, the transmission spectrum of the copper sulfate solution is shown in Figure 7.16. Compare this with Figure 7.13.

A molecule or complex ion of one compound is different from a molecule or complex ion of another compound in terms of the energy differences between the ground state and the various excited states, and therefore the absorption pattern differs from compound to compound and from complex ion to complex ion. This results in a unique absorption spectrum for each compound or ion, and thus the absorption spectrum is a molecular fingerprint. Similarly, the atom of one element is different from the atom of every other element; the result is a unique line spectrum for that element, and thus the line spectrum for a given element is also a fingerprint for that element. This means that absorption and transmission spectra are useful for identification and for detecting impurities or other sample components. Additional examples of continuous absorption and transmission spectra for both the ultraviolet and visible regions are shown in Figure 7.17.

## 7.4.4   Light Emission

Under certain conditions, molecular and ionic analytes present in samples of matter will emit light, and this light can be useful for qualitative and quantitative analysis. For example, most processes used to obtain free ground state atoms in the gas phase from liquid phase solutions of their ions (Chapter 9) result in a small percentage of the atoms being elevated to the excited state, even if no light beam is used. Whether a light beam is used or not, excited atoms return to the ground state because the ground state is the lowest energy state. The energy loss that occurs when the atoms return to the ground state may be dissipated as heat, but it may also involve the emission of light because the difference in energy between the ground state and the excited state is equivalent to light energy. Energy level diagrams can be used to depict such a process, and an **emission spectrum**, the plot of emission intensity vs. wavelength, may be plotted. See Figure 7.18. The process is the reverse of that of light absorption by atoms, so downward-pointing arrows are used to indicate the return to the ground state; the wavelengths emitted are the same as those that are absorbed. Thus the lines in an emission spectrum of a metal are often at the same wavelengths as the lines in the absorption spectrum for the same metal. More details are given in Chapter 9.

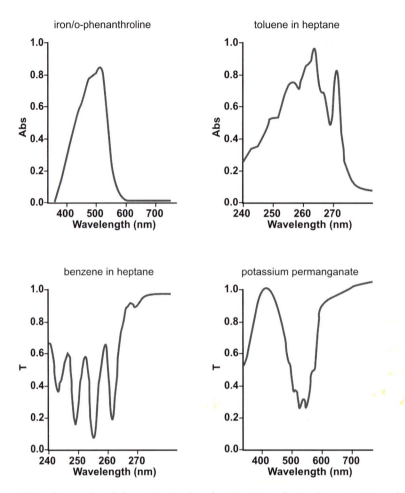

**FIGURE 7.17**  Additional examples of absorption (top) and transmission (bottom) spectra. Upper left, iron–o-phenanthroline; upper right, toluene in heptane; lower left, benzene in heptane; lower right, potassium permanganate.

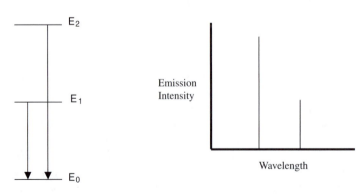

**FIGURE 7.18**  Left, an energy level diagram with arrows pointing downward, indicating the loss of energy and the return to the ground state. Right, a line emission spectrum (emission intensity vs. wavelength) showing two emission lines that may result from the process on the left.

Similarly, molecules and complex ions may also emit light under certain conditions, a phenomenon known as **fluorescence**. Specifically, it is when the absorption of light in the UV region is followed by the emission of light in the visible region. As with atomic emission, it involves the loss in energy from an excited state to a lower state, and this loss corresponds to the energy of light. More details of this phenomenon are presented in Chapter 8.

## 7.5   Absorbance, Transmittance, and Beer's Law

Let us return to the absorption phenomenon again and precisely define absorbance and transmittance, two terms previously used in this chapter. The intensity of light striking the light sensor (more often called a detector) in Figure 7.12 when a blank solution (no analyte species) is held in the path of the light is given the symbol $I_0$. The blank, as discussed in Section 6.6, is a solution that contains all chemical species that are present in the standards and samples to be measured (at equal concentration levels) except for the analyte species. Such a solution does not display any absorption due to the analyte, and thus $I_0$ represents the maximum intensity that can strike the detector at any time. When the blank is replaced with a solution of the analyte, a less intense light beam will be detected because it is absorbed by the analyte. The intensity of the light for this solution is given the symbol I. See Figure 7.19. The fraction of light transmitted is thus $I/I_0$. This fraction is defined as the transmittance (T):

$$T = \frac{I}{I_0} \tag{7.10}$$

The percent transmittance (%T) is similarly defined:

$$\%T = T \times 100 \tag{7.11}$$

Transmittance and percent transmittance are two parameters that can be displayed on the readout.

An important fact to recognize concerning transmittance is that it is not linear with concentration. In other words, a plot of transmittance or percent transmittance vs. concentration would not be an acceptable standard curve because it is not a straight line. Rather, it is a logarithmic relationship. Because it is logarithmic, the logarithm of the transmittance can be expected to be linear with concentration. Thus, we define the parameter of absorbance as the negative logarithm of the transmittance and give it the symbol A:

$$A = -\log T \tag{7.12}$$

Absorbance then is a parameter that increases linearly with concentration and is important for quantitative analysis. If the analyst measures transmittance, he or she must convert it to absorbance via Equation (7.12)

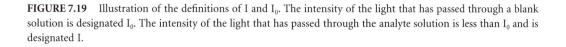

**FIGURE 7.19**   Illustration of the definitions of I and $I_0$. The intensity of the light that has passed through a blank solution is designated $I_0$. The intensity of the light that has passed through the analyte solution is less than $I_0$ and is designated I.

before plotting the standard curve. Instrument readouts can display either transmittance or absorbance, and either absorption spectra or transmission spectra can be displayed. The appropriate signal processor is often activated with a press of a key on the instrument keypad or with a click of a mouse.

## Example 7.8

What is the absorbance (A) of a sample if the transmittance (T) is 0.347?

*Solution 7.8*

$$A = -\log T$$

$$A = -\log (0.347)$$

$$A = 0.460$$

## Example 7.9

What is the absorbance (A) of a sample that has a percent transmittance (%T) of 49.6%?

*Solution 7.9*

$$\%T = T \times 100$$

$$T = \% \frac{T}{100} = \frac{49.6}{100} = 0.496$$

$$A = -\log T$$

$$A = -\log (0.496) = 0.305$$

The conversion of absorbance (A) to transmittance (T) or percent transmittance (%T), the reverse of the above, may also be important.

## Example 7.10

What is the percent transmittance (%T) if the absorbance (A) = 0.774?

*Solution 7.10*

$$A = -\log T$$

$$0.774 = -\log T$$

$$T = 0.168$$

$$\%T = T \times 100$$

$$\%T = 16.8\%$$

The equation of the straight-line A vs. C standard curve is known as the the **Beer-Lambert law**, or simply as **Beer's law**. A statement of Beer's law is

$$A = abc \tag{7.13}$$

in which A is the absorbance; a is the **absorptivity**, or **extinction coefficient**; b is the **pathlength**; and c is the concentration. Absorbance was defined previously in this section. Pathlength is the distance the light travels through the measured solution. It is the inside diameter of the sample container placed in the light path. Pathlength is measured in units of length, usually centimeters or millimeters. Absorptivity is the inherent ability of a chemical species to absorb light and is constant at a given wavelength, pathlength, and concentration. It is a characteristic of the absorbing species itself. Concentration can be expressed in any concentration unit with which you are familiar. Usually, however, it is expressed in molarity, parts per million, or grams per 100 mL. The value and units of absorptivity depend on the units of these other parameters, since absorbance is a dimensionless quantity. When the concentration is in molarity and the pathlength is in centimeters, the units of absorptivity must be liters per mole per centimeter ($L\ mol^{-1}\ cm^{-1}$). Under these specific conditions, the absorptivity is called the molar absorptivity, or the molar extinction coefficient, and is given a special symbol, the Greek letter epsilon, $\varepsilon$. Beer's law is therefore sometimes given as

$$A = \varepsilon bc \qquad (7.14)$$

## Example 7.11

The measured absorbance of a solution with a pathlength of 1.00 cm is 0.544. If the concentration is $1.40 \times 10^{-3}\ M$, what is the molar absorptivity for this analyte?

*Solution 7.11*

$$A = \varepsilon bc \text{ (b in centimeters, c in molarity)}$$

$$\varepsilon = \frac{A}{bc}$$

$$= \frac{0.544}{1.00\ cm \times 1.40 \times 10^{-3}\ mol/L}$$

$$= 389\ L\ mol^{-1}\ cm^{-1}$$

The nature of the container that is used for the sample varies according to the specific method. In UV-VIS molecular spectrophotometry, it is typically a small test tube or a square tube with an inside diameter (pathlength) of 1 cm. Such a container is called a **cuvette.** Cuvettes with different pathlengths, such as 1 mm, are available. More information on cuvette selection and handling is given in Chapter 8. In IR molecular spectrophotometry, the container is called the **IR liquid sampling cell**. In this cell, the liquid sample is contained in a space between two salt plates, created with a thin spacer between the plates, and the pathlength is the thickness of the spacer. More information is given in Chapter 8. In atomic absorption spectroscopy, the pathlength is defined by the device used to convert metal ions in solution to atoms. The traditional device used is a burner with a **flame.** In this case, the sample is contained in the flame and the width of the flame where the light beam passes is the pathlength. More information is given in Chapter 9. See Figure 7.20.

Defining the molar absorptivity parameter presents analytical chemists with a standardized method of comparing one spectrochemical method with another. The larger the molar absorptivity, the more sensitive the method. (It is not unusual for molar absorptivity values to be as large as 10,000 $L/mol^{-1}\ cm^{-1}$ and higher.) In addition, it was stated above that the absorptivity is constant at a given wavelength, implying that it changes as the wavelength changes. The greatest analytical **sensitivity** (the ability to detect and measure small concentrations accurately) occurs at the wavelength at which the absorptivity is a maximum. This is the same wavelength at which the absorbance is a maximum in the molecular absorption spectrum for that species. This wavelength is often referred to as the **wavelength of maximum absorbance**. The molecular absorption spectrum is sometimes shown as a plot of absorptivity vs. wavelength, rather than absorbance vs. wavelength. For a given absorbing species, such a plot would display the same characteristic shape and the same wavelength of maximum absorbance. Refer to Figure 7.17 for examples. For iron-o-phenanthroline, the wavelength of maximum absorbance is approximately 510 nm. For toluene in heptane, it is approximately 263 nm; for benzene in heptane, it is approximately 255 nm; and for potassium permanganate, it is approximately 520 nm.

Most quantitative analyses by Beer's law involve preparing a series of standard solutions, measuring the absorbance of each in identical containers (or the same container), and plotting the measured

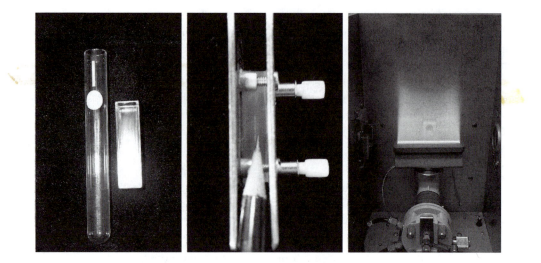

**FIGURE 7.20**    Examples of sample containers. Left, cuvettes used in UV-VIS spectrophotometry (pathlength is the inner diameter—approximately 1 cm); center, liquid sampling cell used in IR spectrometry (pathlength is thickness of spacer to the left of the pencil tip—approximately 0.1 mm); right, atomic absorption flame (pathlength is the width of the flame—approximately 4 in.).

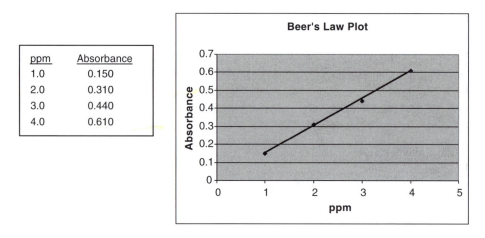

| ppm | Absorbance |
|-----|------------|
| 1.0 | 0.150 |
| 2.0 | 0.310 |
| 3.0 | 0.440 |
| 4.0 | 0.610 |

**FIGURE 7.21**    An example of data and standard curve for a quantitative spectrophotometric analysis. The standard curve in this case is also known as a Beer's law plot.

absorbance vs. concentration, thus creating a standard curve as defined in Chapter 6 (Section 6.4.2). The absorbance of an unknown solution is then measured and its concentration determined from the standard data. The standard curve in this case is sometimes called a **Beer's law plot**. See Figure 7.21 and Workplace Scene 7.1.

## 7.6    Effect of Concentration on Spectra

From the discussion in Section 7.5, it is clear that the absorbance at the wavelength of maximum absorbance increases as the concentration increases and decreases as the concentration decreases. This is true not just at the wavelength of maximum absorbance but at *all* wavelengths. This means that an entire absorption spectrum is affected by concentration changes. It is important to note, however, that the *pattern* of the absorption—the characteristic of the spectrum that makes it useful for *qualitative*

# WORKPLACE SCENE 7.1

A t Los Alamos National Laboratory, the silicon content of plutonium samples is determined spectrophotometrically using a silica-molybdenum blue complex ion, which has a blue color. The sample is dissolved in a mixture of hydrochloric acid and hydrofluoric acid. After the excess hydrofluoric acid is removed, the absorbance is measured at 825 nm. All operations are carried out in a silica-free apparatus.

Technician Velma Montoya, of the Actinide Analytical Chemistry Group at Los Alamos, works in a silica-free glove box to determine silicon content in plutonium samples using visible (near infrared) spectrophotometry.

analysis—does *not* change. The complete absorption spectrum for a given absorbing species for a series of different concentration levels shows that the absorption pattern is the same for all concentrations, but the level of the absorption is different.

## Experiments

### Experiment 19: Colorimetric Analysis of Prepared and Real Water Samples for Iron

*Remember safety glasses.*

1. Prepare 100 mL of a 100 ppm Fe stock solution from the available 1000 ppm stock solution. Use a clean 100-mL volumetric flask and a clean volumetric pipet. Shake well.
2. Prepare calibration standards of 1.0, 2.0, 3.0, and 4.0 ppm iron from the 100 ppm stock in 50-mL volumetric flasks, but do not dilute to the mark until step 6. Also prepare a flask for the blank (no iron). If real water samples are being analyzed and you expect the iron content to be low, prepare two additional standards that are 0.1 and 0.5 ppm.

3. Prepare the real water samples by pipetting 25.00 mL of each into 50-mL volumetric flasks. As with the standards, do not dilute to the mark until step 6. If an instructor-prepared unknown is provided, do not dilute it to the mark until directed in step 6 for all flasks. Your instructor may also provide you with a control sample. Prepare the control as if it were a real water sample by pipetting 25.00 mL of it into a 50-mL flask.

4. To each of the flasks (standards, samples, blank, and control) add:
   (a) 0.5 mL of 10% hydroxylamine hydrochloride
   (b) 2.5 mL of 0.1% o-phenanthroline solution
   (c) 4.0 mL of 10% sodium acetate

   These reagents are required for proper color development. Hydroxylamine hydrochloride is a reducing agent, which is required to keep the iron in the +2 state. The o-phenanthroline is a ligand that reacts with $Fe^{2+}$ to form an orange-colored complex ion. This ion is the absorbing species. In addition, since the reaction is pH dependent, sodium acetate is needed for buffering at the optimum pH.

5. Dilute each to the mark with distilled water and shake well.

6. Using one of your standards and the Spectronic 20, or other single-beam visible spectrophotometer, obtain an absorption (or transmittance) spectrum of the Fe-o-phenanthroline complex ion (instructor will demonstrate use of instrument). Determine the wavelength of maximum absorbance from the spectrum and use this wavelength to obtain absorbance readings of all the solutions. (Use the blank for the 100% transmittance setting.)

7. Plot the standard curve using the spreadsheet procedure used in Experiment 18 and obtain the correlation coefficient and the concentrations of the unknowns and control.

8. If real water samples and a control were analyzed, remember that these samples were diluted by a factor of 2. Multiply the concentrations found in step 7 by 2 to get the final answers in that case.

9. Maintain the logbook for the instrument used in this experiment. Record the date, your name(s), the experiment name or number, the correlation coefficient, and the results for the control sample. Also plot the control sample results on a control chart for this experiment posted in the laboratory.

## Experiment 20: Designing an Experiment: Determining the Wavelength at which a Beer's Law Plot Becomes Nonlinear

You should have noticed that the Beer's law plot constructed in Experiment 19 is linear between concentrations of 1 and 4 ppm Fe. At some point beyond 4 ppm, there will be a deviation from Beer's law. Design and conduct an experiment that will precisely determine the concentration at which this occurs.

## Experiment 21: The Determination of Phosphorus in Environmental Water

Note: In this experiment, the phosphorus in water samples is determined by visible spectrometry following a reaction of the phosphates in the sample with potassium antimonyl tartrate, ammonium molybdate, and ascorbic acid. All glassware should be washed in phosphate-free detergent prior to use to avoid phosphate contamination.

*Remember safety glasses.*

### Preparation of Reagent Solutions

1. Potassium antimonyl tartrate solution. In a 250-mL Erlenmeyer flask, dissolve 0.45 g of potassium antimonyl tartrate in about 50 mL of distilled water. Swirl gently until dissolved, and then dilute to 100 mL and swirl to make homogeneous. Prepare just prior to the run.

2. Ammonium molybdate solution. Dissolve 6 g of ammonium molybdate in distilled water in a 250-mL Erlenmeyer flask, dilute to 100 mL with distilled water, and swirl to make homogeneous. Prepare just prior to the run.

3. Ascorbic acid solution. Weigh 2.7 g of ascorbic acid into a 250-mL Erlenmeyer flask, dilute to 100 mL with distilled water, and swirl to make homogeneous. Prepare just prior to the run.

4. Dilute sulfuric acid. Prepare 100 mL of 2.52 $M$ $H_2SO_4$ from concentrated sulfuric acid.
5. Combined reagent. To prepare 100 mL of the color-producing reagent, mix the following portions in the following order, swirling gently after each addition: 50 mL of the dilute sulfuric acid, 5 mL of the potassium antimonyl tartrate solution, 15 mL of the ammonium molybdate solution, and 30 mL of the ascorbic acid solution. Use a 125-mL Erlenmeyer flask. If this solution turns blue after several minutes, phosphate contamination is indicated and all solutions used will need to be reprepared. This combined reagent is stable for a few hours.

### Preparation of Standard Solutions

6. Your instructor has prepared a 50 ppm phosphorus stock solution by dissolving 0.228 g of dried $KH_2PO_4$ in 1000 mL of solution.
7. Prepare four calibration standards that are 0.1, 0.5, 1.0, and 1.5 ppm P from the stock standard prepared in step 6. Use 100-mL volumetric flasks.

### Absorbance Measurements and Results

8. For each the calibration standard, unknown, control sample, and distilled water for the blank, pipet 50.00 mL into a clean, dry 125 mL Erlenmeyer flask. Add 8.0 mL of the combined reagent and mix thoroughly. After at least 10 min, but no more than 30 min, measure the absorbance of each at 880 nm.
9. Create the standard curve, using the procedure practiced in Experiment 18, by plotting absorbance vs. concentration and obtain the concentration of the unknowns and control.
10. Maintain the logbook for the instrument used in this experiment. Record the date, your name(s), the experiment name or number, the correlation coefficient, and the results for the control sample. Also plot the control sample results on a control chart for this experiment posted in the laboratory.

# Questions and Problems

1. Define: spectrochemical methods, spectroscopy, spectrometry, spectrometer, and spectrophotometer.
2. What is specific about the general analytical strategy for spectrochemical analysis methods compared to instrumental analysis in general?
3. What is meant by the dual nature of light?
4. Light waves are electromagnetic and not mechanical. What does this have to do with the fact that light can travel through outer space?
5. Define the following: wavelength, frequency, speed of light, energy of light, and wavenumber.
6. Express each of the following in centimeters:
   (a) 831 nm
   (b) 749 nm
   (c) 4927 A
   (d) 3826 A
7. Express each of the following in nanometers:
   (a) 0.0000000317 cm
   (b) $5.11 \times 10^{-4}$ cm
8. What is the wavelength of light that has a frequency of
   (a) $6.26 \times 10^{11}$ sec$^{-1}$?
   (b) $7.82 \times 10^{12}$ sec$^{-1}$?
   (c) $3.94 \times 10^{13}$ sec$^{-1}$?
9. What is the frequency of light that has a wavelength of
   (a) $4.26 \times 10^{-4}$ cm?
   (b) $7.27 \times 10^{2}$ nm?
   (c) 654 nm?

10. What is the energy of light that has a frequency of
    (a) $3.02 \times 10^{13}$ sec$^{-1}$?
    (b) $3.72 \times 10^{11}$ sec$^{-1}$?
    (c) $7.65 \times 10^{10}$ sec$^{-1}$?
11. What is the frequency of light that has an energy of
    (a) $6.88 \times 10^{-23}$ J?
    (b) $2.72 \times 10^{-25}$ J?
12. What is the energy of light with a wavelength of
    (a) 462 cm?
    (b) 46.9 nm?
13. What is the wavelength of light with energy of
    (a) $4.29 \times 10^{-24}$ J?
    (b) $9.03 \times 10^{-22}$ J?
14. What is the wavenumber of light with a frequency of
    (a) $5.28 \times 10^{12}$ sec$^{-1}$?
    (b) $3.17 \times 10^{13}$ sec$^{-1}$?
15. What is the frequency of light with a wavenumber of
    (a) $5.67 \times 10^{4}$ cm$^{-1}$?
    (b) $3.15 \times 10^{5}$ cm$^{-1}$?
16. What is the wavenumber of light that has a wavelength of
    (a) 792 nm?
    (b) 335 nm?
17. What is the wavelength of light that has a wavenumber of
    (a) $4.99 \times 10^{7}$ cm$^{-1}$?
    (b) $6.18 \times 10^{5}$ cm$^{-1}$?
18. A wavelength of 254 nm was formerly used in certain analyses for organic compounds because it is the most intense wavelength emitted by a mercury lamp. Today, however, mercury lamps are no longer used in those instruments.
    (a) What is this wavelength in centimeters?
    (b) What is the frequency corresponding to this wavelength?
    (c) What is the energy of light of this wavelength?
    (d) What is the wavenumber corresponding to this wavelength?
19. Which has a greater frequency, light of wavelength 627 A or light of wavelength 462 nm?
20. Which has the longer wavelength, light with a frequency of $7.84 \times 10^{13}$ sec$^{-1}$ or light with an energy of $5.13 \times 10^{-13}$ J?
21. Which has the greater energy, light of wavelength 591 nm or light with a frequency of $5.42 \times 10^{12}$ sec$^{-1}$?
22. Which has the greater wavenumber, light with a frequency of $7.34 \times 10^{13}$ sec$^{-1}$ or light with an energy of $5.23 \times 10^{-14}$ J?
23. Which has the lower energy, light with a frequency of $7.14 \times 10^{13}$ sec$^{-1}$ or light with a wavenumber of $1.91 \times 10^{4}$ cm$^{-1}$?
24. A certain light, A, has a greater frequency than a second light, B.
    (a) Which light has the greater energy?
    (b) Which light has the shorter wavelength?
    (c) Which light has the higher wavenumber?
25. If the wavelength used in an instrument is changed from 460 to 560 nm,
    (a) Has the energy been increased or decreased?
    (b) Has the frequency been increased or decreased?
    (c) Has the wavenumber been increased or decreased?
26. Compare IR light with UV light in terms of wavelength, frequency, and wavenumber.

27. List the following in order of increasing energy, frequency, and wavelength: x-rays, infrared light, visible light, radio waves, and ultraviolet light.
28. What are the upper and lower wavelength limits of visible light?
29. Compare infrared and ultraviolet radiation
    (a) In terms of energy
    (b) In terms of the type of disturbance they cause within molecules
30. Why does a yellow sweatshirt appear yellow and not some other color?
31. What is an energy level diagram? What is meant by ground state? What is meant by excited state?
32. Explain briefly the phenomenon of light absorption in terms of the energy associated with light and in terms of electrons and the energy levels in atoms and molecules.
33. What is meant by electronic transition, vibrational transition, rotational transition? Which of these transitions requires the most energy? Which requires the least energy?
34. Describe the basic instrument for measuring light absorption.
35. Distinguish between UV-VIS spectrophotometry, IR spectrometry, and atomic spectroscopy.
36. What is an absorption spectrum? What is the difference between a molecular absorption spectrum and an atomic absorption spectrum and why does this difference exist?
37. What is a line spectrum? Why is an atomic absorption spectrum a line spectrum?
38. Why is it that with atoms, only specific wavelengths get absorbed (resulting in a line spectrum), while with molecules, broad bands of wavelengths get absorbed (resulting in a continuous spectrum)?
39. Why is a molecular absorption spectrum called a molecular fingerprint?
40. What is a transmission spectrum? What is an emission spectrum?
41. Differentiate between an energy level diagram used to depict atomic absorption and one used to depict atomic emission.
42. What is the mathematical definition of transmittance, T? Define the parameters that are found in this mathematical definition.
43. What absorbance corresponds to a transmittance of
    (a) 0.821?
    (b) 0.492?
    (c) 0.244?
44. What absorbance corresponds to a percent transmittance of
    (a) 46.7% T?
    (b) 28.9% T?
    (c) 68.2% T?
45. What transmittance corresponds to an absorbance of
    (a) 0.622?
    (b) 0.333?
    (c) 0.502?
46. What is the percent transmittance given that the absorbance is
    (a) 0.391?
    (b) 0.883?
    (c) 0.490?
47. What is Beer's law? Using only one word for each, what are the parameters?
48. What is the absorbance given that the absorptivity is $2.30 \times 10^4$ L mol$^{-1}$ cm$^{-1}$, the pathlength is 1.00 cm, and the concentration is 0.0000453 $M$?
49. A sample in a 1-cm cuvette gives an absorbance reading of 0.558. If the absorptivity for this sample is 15,000 L mol$^{-1}$ cm$^{-1}$, what is the molar concentration?
50. If the transmittance for a sample having all the same characteristics as those in problem 49 is measured as 72.6%, what is the concentration?
51. The transmittance of a solution measured at 590 nm in a 1.5-cm cuvette was 76.2%.

(a)  What is the corresponding absorbance?

(b)  If the concentration is 0.0802 $M$, what is the absorptivity of this species at this wavelength?

(c)  If the absorptivity is 10,000 L mol$^{-1}$ cm$^{-1}$, what is the concentration?

52.  Calculate the transmittance of a solution in a 1.00-cm cuvette given that the absorbance is 0.398. What additional information, if any, would you need to calculate the molar absorptivity of this analyte?

53.  What is the molar absorptivity given that the absorbance is 0.619, the pathlength is 1.0 cm, and the concentration is $4.23 \times 10^{-6}$ $M$?

54.  Calculate the concentration of an analyte in a solution given that the measured absorbance is 0.592, the pathlength is 1.00 cm, and the absorptivity is $3.22 \times 10^{4}$ L mol$^{-1}$ cm$^{-1}$.

55.  What is the concentration of an analyte given that the percent transmittance is 70.3%, the pathlength is 1.0 cm, and the molar absorptivity is 8382 L mol$^{-1}$ cm$^{-1}$?

56.  What is the pathlength in centimeters when the molar absorptivity for a given absorbing species is $1.32 \times 10^{3}$ L mol$^{-1}$ cm$^{-1}$, the concentration is 0.000923 $M$, and the absorbance is 0.493?

57.  What is the pathlength in centimeters when the transmittance is 0.692, the molar absorptivity is $7.39 \times 10^{4}$ L mol$^{-1}$ cm$^{-1}$, and the concentration is 0.0000923 $M$?

58.  What is the transmittance when the molar absorptivity for a given absorbing species is $2.81 \times 10^{2}$ L mol$^{-1}$ cm$^{-1}$, the pathlength is 1.00 cm, and the concentration is 0.000187 $M$?

59.  What is the molar absorptivity when the percent transmittance is 56.2%, the pathlength is 2.00 cm, and the concentration is 0.0000748 $M$?

60.  In each of the following, enough data are given to calculate the indicated parameter(s). Show your work.

(a)  Calculate absorbance given that the transmittance is 0.551.

(b)  Calculate the molar absorptivity given that the absorbance is 0.294, b is 1.00 cm, and the concentration is 0.0000351 $M$.

(c)  Calculate the transmittance given that the absorbance is 0.297.

(d)  Calculate transmittance given that the percent transmittance is 42.8%.

(e)  Calculate percent transmittance given that the absorptivity is 12,562 L mol$^{-1}$ cm$^{-1}$, the pathlength is 1.00 cm, and the concentration is 0.00000355 $M$.

61.  In each of the following, enough data are given to calculate the indicated parameter(s).

(a)  Calculate absorbance given that the transmittance is 0.651.

(b)  Calculate absorptivity given that the absorbance is 0.234, b is 1.00 cm, and the concentration is 0.0000391 $M$.

(c)  Calculate transmittance given that the absorbance is 0.197.

(d)  Calculate transmittance given that the percent transmittance is 62.8%.

(e)  Calculate percent transmittance given that the absorptivity is 13562 L mol$^{-1}$ cm$^{-1}$, b is 1.00 cm, and the concentration is 0.00000355 $M$.

62.  A standard 5 ppm iron sample gave a transmittance reading of 52.8%. What is the concentration of an unknown iron sample if its transmittance is 61.7%?

63.  A series of five standard copper solutions are prepared, and the absorbances are measured as indicated below. Plot the data and determine the concentration of the unknown.

| A     | C (ppm) |
|-------|---------|
| 0.104 | 1       |
| 0.198 | 2       |
| 0.310 | 3       |
| 0.402 | 4       |
| 0.500 | 5       |
| 0.334 | Unknown |

64. Match the items in the left-hand column to the items in the right-hand column by writing the appropriate letter in the blank. Each is used only once.

    __e__ T            (a) The intensity of light after having passed through a solution of an absorbing species.

    __f__ A            (b) A = abc

    __a__ I             (c) Molar absorptivity

    __g__ $I_o$           (d) Pathlength

    __h__ a            (e) $I/I_o$

    __d__ b            (f) $-\log T$

    __b__ Beer's law    (g) The intensity of light after having passed through a blank solution.

    __i__ %T           (h) Absorptivity

    __c__ ε            (i) T × 100

65. In an experiment in which the percent Mn in steel is determined, the Mn in the steel is oxidized to permanganate, giving the solution the light purple color measured in Figure 7.17, lower right.

    (a) Look at Figure 7.17 and tell what is the best choice for a wavelength for doing this analysis and why?

    (b) Look at Figure 7.17 and convert the percent transmittance at the wavelength you determined in (a) to absorbance.

    (c) Assuming a pathlength of 1 cm and a concentration of 0.0500 $M$, what is the molar absorptivity at this wavelength?

66. Compare the common sample containers for UV-VIS work, IR work, and atomic absorption work.

67. State the importance of the wavelength of maximum absorbance, $\lambda_{max}$.

68. **Report**

Find a real-world visible spectrophotometric analysis in a methods book (AOAC, USP, ASTM, etc.), in a journal, or on a website, and report on the details of the procedure according to the following scheme:

    (a) Title

    (b) General information, including the type of material examined, the name of the analyte, sampling procedures, and sample preparation procedures

    (c) Specifics, including whether a Beer's law plot is used, what color-developing reagent is used (if there is one), reactions involved to obtain the color, what wavelength is used, how the data are gathered, concentration levels for standards, how the standards are prepared, and potential problems

    (d) Data handling and reporting

    (e) References

# 8

# UV-VIS and IR Molecular Spectrometry

## 8.1 Review

Spectrochemical methods of analysis were introduced in Chapter 7. Ultraviolet–visible (UV-VIS) molecular spectrochemical methods utilize light in the ultraviolet and visible regions of the electromagnetic spectrum to analyze laboratory samples for molecular compounds and complex ions. Qualitative analysis (identification of unknowns and detection of impurities in knowns) is accomplished by comparing absorption or transmission spectra (molecular fingerprints) with known spectra. Quantitative analysis is accomplished with the use of Beer's law (Section 7.5). The fundamental instrument design is illustrated in Figure 7.12, and the sample containers are the cuvettes pictured in the left photograph in Figure 7.20.

UV-VIS spectra result from electronic transitions occurring in the analyte molecules and complex ions. Since vibrational levels are superimposed on the electronic levels, many electronic transitions are possible (refer to Figure 7.15 and accompanying discussion), resulting in continuous spectra rather than the line spectra that characterize spectra of atoms.

## 8.2 UV-VIS Instrumentation

The fundamental instrument design in Figure 7.12 is meant only as an illustration of the basic concept of the instrument design. As we will see in the sections to follow, the path from the light source, through the wavelength selector and sample, to the detector can involve significant other optical components in addition to the instrument components illustrated in Figure 7.12. We now proceed to describe the instruments in more detail.

### 8.2.1 Sources

First we discuss the possibilities for the source of the light used. Figure 7.12 shows an ordinary light bulb for this. In reality, special light sources are used in order to provide an optimum-quality light beam for the region of the spectrum utilized. An ideal light source is one that emits an intense continuous spectrum of light across an entire region of the spectrum, such as the visible region, while also exhibiting a long life. A light source used frequently for visible light absorption studies is the tungsten filament source. If an instrument is meant strictly for visible light studies, then this lamp is the only one present in the instrument. Such an instrument is often referred to as a **colorimeter**.

A light source used frequently for ultraviolet absorption studies is the deuterium lamp. If an instrument is meant strictly for ultraviolet work, then the deuterium lamp is the only light source present and the instrument is called a **UV spectrophotometer.** Often, both a tungsten filament lamp and a deuterium lamp are present and are individually selectable. Also, instead of having two independently selectable

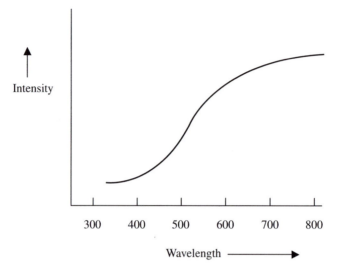

**FIGURE 8.1**  A graph showing approximately how the intensity of light from a tungsten filament source varies with wavelength.

sources, a light source that can be used for both ultraviolet and visible studies, the xenon arc lamp, may be present. In these latter two cases, the instrument is called a **UV-VIS spectrophotometer**.

### 8.2.1.1  Tungsten Filament Lamp

For the visible region, many instruments utilize a light bulb with a tungsten filament. Such a source is very bright and emits light over the entire visible region and into the near infrared region. However, the intensity of the light varies dramatically across this wavelength range (Figure 8.1). This creates a bit of a problem for the analyst; we will discuss this later.

### 8.2.1.2  Deuterium Lamp

Deuterium is the name given to the isotope of hydrogen that has one neutron in the atomic nucleus (as opposed to zero neutrons for hydrogen and two neutrons for tritium, the other hydrogen isotopes). The lamp contains deuterium at a low pressure. Electricity applied to electrodes in the lamp results in a continuous UV emission due to the presence of the deuterium. Its wavelength output ranges from 185 nm to about 375 nm, satisfactory for most UV analyses. Here again the intensity varies with wavelength.

### 8.2.1.3  Xenon Arc Lamp

This lamp contains xenon at a fairly high pressure and the light is formed via a discharge across a pair of electrodes. A continuous ultraviolet and visible emission is emitted due to the presence of the xenon. In some instruments, the electronic circuitry creates regular pulses of light that are very intense and therefore more useful. This results in a longer life for the lamp. The intensity varies with wavelength.

## 8.2.2  Wavelength Selection

In order to plot the absorption spectrum of a compound or complex ion, we must be able to carefully control the wavelengths from the broad spectrum of wavelengths emitted by the source so that we can measure the absorbance at each wavelength. Additionally, in order to perform quantitative analysis by Beer's law, we need to be able to carefully select the wavelength of maximum absorption, also from this broad spectrum of wavelengths, in order to plot the proper absorbance at each concentration. These facts dictate that we must be able to filter out the unwanted wavelengths and allow only the wavelength of interest to pass.

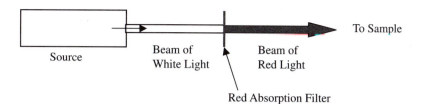

**FIGURE 8.2**   An illustration of how a red-colored glass filter isolates the red region of the visible spectrum.

As a point of clarification, however, we must recognize that, in reality, there is no such thing as a single wavelength. In the visible region, the spectrum of colors is continuous, meaning that there is no sharp delineation between green light and blue light, for example, or where one wavelength ends and the adjacent wavelength begins. The electromagnetic spectrum is a continuous wavelength band. Thus wavelength selection in a spectrochemical instrument actually consists of the selection of a narrow wavelength band from the larger band. The width of the band that is allowed to pass is called the **bandwidth**. The narrowness of the band that is allowed to pass varies from one design to another and is called the **resolution**. Thus a high resolution (narrow bandwidth) is ideal, although many applications do not require the best available resolution and money is often saved by purchasing instruments with a low resolution (wide bandwidth).

### 8.2.2.1   Absorption Filters

The most inexpensive way to isolate a wavelength band is with the use of absorption filters. For visible light, such a filter would consist of colored glass, the color of the glass indicating what region of the visible spectrum is passed. Thus, if the wavelength called for in a given method is in the red region of the visible spectrum, a red colored glass filter is chosen. See Figure 8.2. Absorption filters have the lowest resolution, having effective bandwidths varying from 30 to 250 nm. See Workplace Scene 8.1.

### 8.2.2.2   Monochromators

The word "monochromator" is derived from the Latin language, "mono" meaning "one" and "chromo" meaning "color." It is a device more sophisticated than an absorption filter that isolates the narrow band of wavelengths from visible and ultraviolet sources.

A monochromator is made up of three parts: an entrance slit, a dispersing element, and an exit slit. In addition to these, there is often a network of mirrors situated for the purpose of aligning or collimating a beam of light before and after it contacts the dispersing element. A slit is a small circular or rectangular hole cut in an otherwise opaque plate, such as a black metal plate. The size of the opening is often variable—a variable slit width. The entrance slit is where light enters the monochromator from the source. Its purpose is to create a unidirectional beam of light of appropriate intensity from the multidirectional light emanating from the source. Its slit width is usually variable so that the intensity of the beam can be adjusted—the wider the opening the more intense the beam.

After passing through the entrance slit, the beam encounters a **dispersing element**. The dispersing element disperses the light into its component wavelengths. For visible light, for example, this would mean that a beam of white light is dispersed into a spray of rainbow colors, the violet–blue wavelengths on one end to the red wavelengths on the other, with the green and yellow in between, like a rainbow. The narrow band of the spectrum is then selected by the **exit slit**. See Figure 8.3. As the dispersing element is rotated, the spray of colors moves across the exit slit such that a different narrow wavelength range emerges from the exit slit at each position of rotation. See Figure 8.4. The exit slit width can be variable too, but making it wider would result in a wider wavelength band (a wider **bandwidth**) passing through, which can be undesirable. The light emerging from this exit slit is therefore monochromatic and is passed on to the sample compartment to pass through the sample. The concept is the same for UV light.

# WORKPLACE SCENE 8.1

At the City of Lincoln, Nebraska, Wastewater Treatment Plant Laboratory, wastewater samples at various stages of treatment are analyzed for dissolved ammonia, nitrate, total Kjeldahl nitrogen, cyanide, and phosphates. All of these analyses are performed with the help of a **flow injection analyzer**, in which reagents and samples are drawn through small-diameter tubing, mixed to generate the characteristic color, and then analyzed spectrophotometrically while flowing. In this instrument, glass filters are used to isolate the region of the spectrum required and are interchangeable, meaning a different filter is used for each different analyte determined.

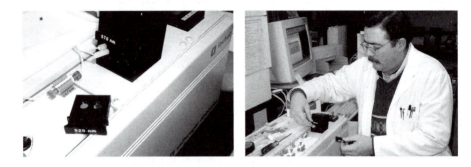

Left, interchangeable colored glass filters used in a flow injection analyzer for different analytes. Notice the value of the wavelength inscribed on the edge of each filter. Right, Steve Kruse of the City of Lincoln, Nebraska, Wastewater Treatment Plant Laboratory prepares to change the filter in the instrument as he prepares to test for a different analyte in wastewater samples.

The rotation of the dispersing element is accomplished by manually turning a knob on the face of the instrument or by internal programmed scanning controls. The position of the knob is coordinated with the wavelength emerging from the exit slit such that this wavelength is read from a scale of wavelengths on the face of the instrument, on a readout meter, or on a computer screen. For manual control, the operator thus simply dials in the desired wavelength. In more sophisticated instruments, the wavelength information is set with software on a computer screen. The bandwidth for slit/dispersing element/slit monochromators varies from 0.5 to 350 nm, depending on the quality of the dispersing element and the wavelength range in question. Some instruments improve (narrow) the bandwidth by using filters ahead of the dispersing element.

The dispersing element is either a diffraction grating or a prism. A **prism** is a three-dimensional triangularly shaped glass or quartz block. When the light beam strikes one of the three faces of the prism, the light emerging through another face is dispersed.

A diffraction grating is used more often than a prism. The dispersing element pictured in Figures 8.3 and 8.4 is a diffraction grating. A **diffraction grating** is like a highly polished mirror that has a large number of precisely parallel lines or grooves scribed onto its surface. Light striking this surface is reflected, diffracted, and dispersed into the component wavelengths as indicated in Figures 8.3 and 8.4.

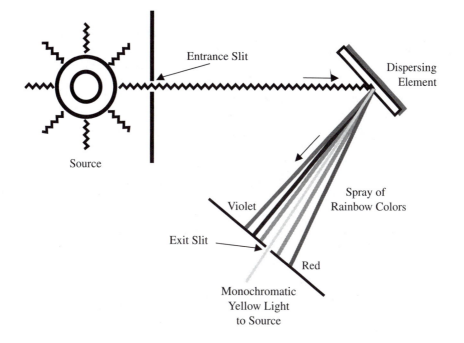

**FIGURE 8.3**  An illustration of a monochromator.

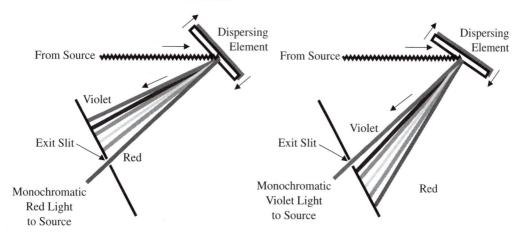

**FIGURE 8.4**  Rotation of the dispersing element directs different wavelengths to emerge from the exit slit.

## 8.2.3  Sample Compartment

Following the wavelength selection by the monochromator, the beam passes on to the sample compartment where the sample solution, held in the cuvette, is positioned in its path. The sample compartment is an enclosure with a lid that can be opened and closed in order to insert and remove the cuvette. When the lid is closed, the compartment should be relatively free of **stray light**, although this is not a requirement if a xenon arc lamp is used as the source. This is because of the high intensity of the xenon arc lamp. The cuvette is held snugly in a spring-loaded holder.

Some spectrophotometers are single-beam instruments, and some are double-beam instruments. In a double-beam instrument the light beam emerging from the monochromator is split into two beams at some point between the monochromator and the detector. The double-beam design provides certain advantages that we will discuss shortly.

### 8.2.3.1   Single-Beam Spectrophotometer

In a single-beam spectrophotometer, the monochromatic light beam created by the monochromator passes directly through the sample solution held in the cuvette and then proceeds to the detector. This is the most inexpensive design and is especially useful for routine absorbance measurements for which the wavelength of maximum absorbance (the wavelength to be used) is known in advance without having to scan a particular wavelength range to determine it.

In Section 6.6, we briefly discussed the need for a blank in instrumental analysis (all sample components except the analyte), saying that it is needed for a precalibration step — before calibrating the instrument with standards. For UV-VIS spectrophotometry, this precalibration step would consist of placing the blank in the cuvette in the path of the light. This would produce the most intense light beam possible at the detector and thus a readout of 100% T. If the readout is not 100% T, the electronics are tweaked to produce this readout, much like a pH meter's electronics is tweaked to produce a readout expected for a given buffer solution. The precalibration may also include a dark current control. In this case, with no cuvette in the cuvette holder, a shutter blocks all light from the detector and thus produces a 0% T reading. Having performed these precalibration steps at the wavelength to be used, the instrument is ready for making the measurements on the standards and samples.

If the wavelength of maximum absorbance is not known in advance, then it must be determined by scanning the wavelength range involved. With a single-beam instrument, this is a manual, tedious process because the precalibration step must be repeated each time the wavelength is changed. The reason for this is that the intensity of the light from the source (and therefore the intensity of the light at the detector after having passed through the blank) is not the same at all wavelengths. It is also true that the detector response is not constant at all wavelengths. (Refer again to Figure 8.1 and the accompanying discussion concerning source variations.) That being the case, the intensity of light that has passed through the blank will produce a percent transmittance readout different from 100% when the instrument is switched to the new wavelength, and so the electronics must again be tweaked to produce the 100% T reading assumed for the blank.

The tedious process (set wavelength, read blank, tweak to read 100%, read sample, change wavelength, read blank, tweak to read 100%, read sample, change wavelength, etc.) can be eliminated by one of two ways. One is to use a single-beam instrument with rapid scanning capability. Rapid scanning occurs by programming the instrument to steadily rotate the grating with a motor while simultaneously measuring the absorption data. In this case, the absorbance vs. wavelength information for the blank is quickly acquired when the blank is in the light path, and the results are stored and used later to adjust the results for the sample via computer for a similar rapid scan with the sample in the cuvette. While single-beam instruments with rapid scanning capability exist, the double-beam designs discussed below are more popular and offer an additional advantage.

For precise work, a second problem exists with a single-beam design, both with manual blank and sample switching and with single-beam rapid scanning. This is that the precalibration with the blank takes at least 10 sec, and perhaps as long as several minutes or more, before the sample is read. This may be a problem because there can be minor intensity fluctuations or drift in the light source or detector (even if the wavelength is not changed) that can undo the precalibration before reading a sample. The double-beam designs discussed below offer an advantage for this problem as well.

### 8.2.3.2   Beamsplitting and Chopping

Double-beam means that the light beam produced by the monochromator is split into two beams. This is usually done in one of two ways. One is to use a slotted mirror, called a **beamsplitter**, set at an angle in the path of the light (see Figure 8.5, drawing on the right). In this way, half the light intensity passes through the slots and half is reflected to another path (a second beam), creating a double beam in space. Another is to utilize a rotating circular partial mirror (often called a light **chopper**), also set at an angle (Figure 8.5, left). This creates a strobe effect, or a double beam in time, in which two beams rapidly alternate: one beam follows a path through the open half of the mirror at one moment in time and the other reflects to another path at another moment in time.

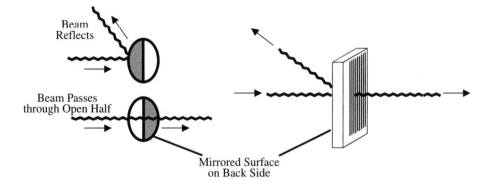

**FIGURE 8.5**    Left, a rotating light chopper. In one position, light passes through open half. After 180° rotation, light reflects from mirror half. Right, a beamsplitter consisting of a mirror with slots.

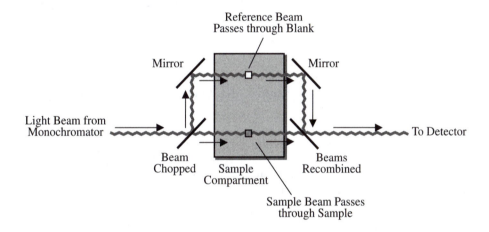

**FIGURE 8.6**    An illustration (top view) of the double-beam design utilizing two cuvette holders in the sample compartment.

### 8.2.3.3   Double-Beam Designs

Probably the most popular, and the most advantageous, double-beam design is one in which the light emerging from the monochromator is chopped into two beams that take parallel but separate paths through the sample compartment. In the sample compartment are two cuvette holders, one to hold the blank and one to simultaneously hold the sample being measured. One of the two beams passes through the blank and is called the reference beam. The other passes through the sample and is called the sample beam. The beams are recombined with a second chopper prior to reaching the detector. See Figure 8.6. The detector is programmed to the chopping frequency and can differentiate between the two beams. This design allows the instrument's electronics, or computer software, to self-adjust for the blank reading at each wavelength a split second before taking the sample reading. Thus both disadvantages of the single-beam systems (the tedium and the time factor) are done away with while also rapidly scanning the wavelength range of interest (by motor-driven dispersing element rotation) to obtain the molecular absorption spectrum.

Another, but less advantageous, double-beam design is one in which the beam emerging from the monochromator is split with a beamsplitter (Figure 8.5, right). However, the second beam is directed immediately to a second detector rather than through the blank. This design requires only one cuvette holder in the sample compartment. See Figure 8.7. Such a design adjusts for the different intensities at different wavelengths in a manner identical to that in the single-beam scanning instrument, i.e., by scanning the blank at a separate time via motor-driven dispersing element rotation, and then adjusting

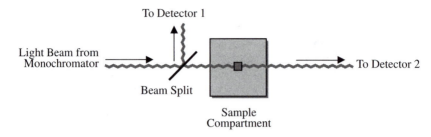

**FIGURE 8.7** The double-beam design in which the second beam passes directly to a detector without passing through the sample compartment.

the sample spectrum with the blank spectrum. Thus there is no advantage for the problem of different intensities at different wavelengths. However, monitoring the intensity of the second beam in the manner described does assist with the problem of source drift and fluctuations, or with changes that may occur because the operator changed the scan speed, because the instrument can immediately adjust.

### 8.2.4   Detectors

Photomultiplier tubes or photodiodes (light sensors) are used as detectors in UV-VIS spectrophotometers, while thermcouples (heat sensors) are used as detectors for infrared (IR) spectrometry. This is the reason UV-VIS instruments are called spectrophotometers while IR instrument are called spectrometers.

#### 8.2.4.1   Photomultiplier Tube

A photomultiplier tube is a light sensor combined with a signal amplifier. The light emerging from the sample compartment strikes the photosensitive surface and the resulting electrical signal is amplified.

   The photomultiplier tube consists of a **photocathode**, an anode, and a series of **dynodes** for multiplying the signal, hence its name. A dynode is an electrode that, when struck with electrons, emits other secondary electrons. The dynodes are situated between the photocathode and the anode. A high voltage is applied between the photocathode (an electrode that emits electrons when light strikes it) and the anode. When the light beam from the sample compartment strikes the photocathode, electrons are emitted and accelerated, because of the high voltage, to the first dynode, where more electrons are emitted. These electrons pass on to the second dynode, where even more electrons are emitted, etc. When the electrons finally reach the anode, the signal has been sufficiently multiplied as to be treated as any ordinary electrical signal able to be amplified by a conventional amplifier. This amplified signal is then sent on to the readout in one form or another. Figure 8.8 illustrates this process.

#### 8.2.4.2   Photodiodes

Photodiodes make use of the unique properties of semiconductors, such as silicon. Silicon can be doped with impurities to make it either electron rich (an **n-type semiconductor**) or electron poor (a **p-type semiconductor**). When an n-type semiconductor is in contact with a p-type semiconductor, electronic changes occur at the boundary, or junction. A photodiode is a p–n junction constructed with the top p layer so thin that it is transparent to light. Light shining through the p layer creates additional free electrons in the n layer that can diffuse to the p layer, thus creating an electrical current that depends on the intensity of the light. This small current is easily amplified and measured.

### 8.2.5   Diode Array Instruments

A **diode array** is a series of several hundred photodiodes arranged in a linear array. Single-beam spectrophotometers have been invented that utilize a diode array as the detector. In this case, the cuvette is positioned between the source and the dispersing element. Then, following the dispersion of the light, there is no exit slit. The spray of wavelengths created by the grating fall instead across the diode array,

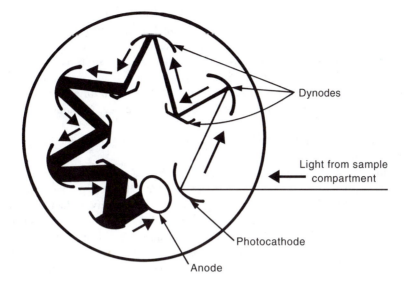

**FIGURE 8.8**    The photomultiplier tube.

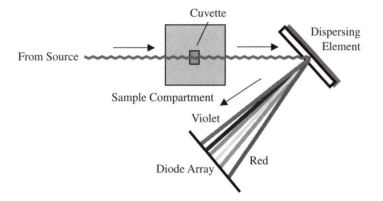

**FIGURE 8.9**    An illustration of a diode array spectrophotometer.

and the entire spectrum of light is measured at once. See Figure 8.9. The diode array spectrophotometer is a powerful instrument whenever a very rapid measurement of the absorption spectrum is needed, such as in the liquid chromatography instruments discussed in Chapter 13. The blank is read at a separate time like other single-beam instruments.

# 8.3   Cuvette Selection and Handling

Cuvettes used for UV-VIS spectrophotometry must be transparent to all wavelengths of light for which it is used. If visible light is used, the material must ideally be completely clear and colorless, which means that inexpensive materials, such as colorless plastic and ordinary colorless glass, are perfectly suitable. A case of 500 plastic 1-cm-square cuvettes may cost as little as $50. However, ordinary colorless glass and plastic are not transparent to light in the ultraviolet region. For ultraviolet spectrophotometry, the cuvettes must be made of quartz, which is more expensive. A matched set of two cuvettes to be used in a double-beam spectrophotometer may cost as much as $400.

   If two or more different cuvettes are used in an analysis, one should be sure that they are matched. Matched cuvettes are identical with respect to pathlength and reflective and refractive properties in the area where the light beam passes. If the pathlengths were different, or if the wall of one cuvette reflects more or less light than another cuvette, then the absorbance measurement could be different for that reason, and not because the solution concentration is different. Thus there would be an error. A general guideline with respect to matching cuvettes is that a solution that transmits 50% of the incident light should not have a percent transmittance reading differing of more than 1% in any cuvette. Also, such cuvettes must be placed in the instrument in exactly the same way each time, since the pathlength and reflective and refractive properties can change by rotating the cuvette. Obviously, cuvettes that have scratches or cleaning procedures that may cause scratching should be avoided. Cotton swabs can be used for scrubbing, rather than metal-handled brushes. Any liquid or fingerprints adhering to the outside wall must be removed with a soft, lint-free cloth or towel.

In addition to scratches, fingerprints, water spots, etc., the analyst should be aware that any foreign particulate matter suspended in the sample or standards would also be a problem. Particulates in the path of the light would reflect and scatter the light, lessening the intensity of the transmitted light, and would result in an error in the reading. Extracts of solids, such as soils, are especially susceptible to this difficulty. The problem is solved by making sure the samples are well filtered. Some spectrometric experiments, however, are designed to actually *measure* such particles. The analysis of samples for suspended particulate matter by spectrophotometry is an analysis for the **turbidity** of those samples.

## 8.4 Interferences, Deviations, Maintenance, and Troubleshooting

### 8.4.1 Interferences

Interferences are quite common in qualitative and quantitative analysis by UV-VIS spectrophotometry. An interference is a contaminating substance that gives an absorbance signal at the same wavelength or wavelength range selected for the analyte. For qualitative analysis this would show up as an incorrect absorption spectrum, thus possibly leading to erroneous conclusions if the contaminant was not known to be present. For quantitative analysis, this would result in a higher absorbance than one would measure otherwise. Absorbances are additive. This means that the total absorbance measured at a particular wavelength is the sum of absorbances of all absorbing species present. Thus, if an interference is present, the correct absorbance can be determined by subtracting the absorbance of the interference at the wavelength used, if it is known. The modern solution to these problems is to utilize separation procedures such as liquid–liquid extraction or liquid chromatography to separate the interfering substance from the analyte prior to the spectrophotometric measurement. Liquid–liquid extraction was discussed in Chapter 2 as a sample preparation technique and will be discussed in more detail in Chapter 11. Liquid chromatography will be discussed in Chapters 11 and 13.

### 8.4.2 Deviations

Deviations from Beer's law are in evidence when the Beer's law plot is not linear. This is probably most often observed at the higher concentrations of the analyte, as indicated in Figure 8.10. Such deviations can be either chemical or instrumental.

Instrumental deviations occur because it is not possible for an instrument to be accurate at extremely high or extremely low transmittance values—values that are approaching either 0 or 100% T. The normal working range is between 15 and 80%, corresponding to absorbance values between 0.10 and 0.82. It is recommended that standards be prepared to measure in this range and that unknown samples be diluted if necessary.

Deviations due to chemical interferences occur when a high or low concentration of the analyte causes chemical equilibrium shifts in the solution that directly or indirectly affect its absorbance. It may be necessary in these instances to work in a narrower concentration range than expected. This means that unknown samples may also need to be further diluted, as in the instrumental deviation case.

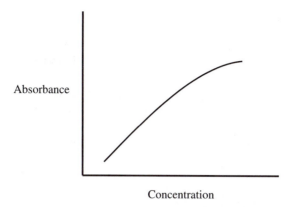

Absorbance

Concentration

**FIGURE 8.10** Deviations from Beer's law often manifest themselves by a nonlinear portion of the Beer's law plot at the higher concentrations.

## 8.4.3 Maintenance

The basic considerations for spectrophotometer maintenance are safety and cleanliness. Maintaining a safe and clean work environment ensures a longer life for the instruments and a lower likelihood for problems relating to contamination, broken or misaligned parts, or injury. Solutions spilled on sensitive electronic circuits can render them inoperative. Any spills should be cleaned up immediately according to local safety protocols.

Instrument operators may also want to conduct periodic wavelength calibrations. It is important to know that the wavelength displayed or dialed in really is the wavelength passing through the sample compartment. One way to check for proper wavelength calibration is to prepare a solution of an analyte for which the wavelength of maximum absorbance is well known. If the maximum absorbance does not occur at this wavelength, it may be that the wavelength control is out of calibration. In this case, the operator should contact the manufacturer's service organization.

## 8.4.4 Troubleshooting

Troubleshooting of spectrophotometers and procedures involving spectrophotometers can involve checking electrical components, checking the mechanics of a procedure, and checking for contamination. The failure of electrical components can involve something as simple as checking whether the instrument or its component is plugged in and switched on, whether a fuse has blown, whether a lamp switch is on, whether the light source has burned out, etc. But it may also involve much more complicated problems, such as the failure of an entire electronic module or circuit board due to a faulty component, such as an amplifier. In the latter case, the operator may need to contact the manufacturer's service organization.

Problems with the mechanics of a procedure can involve an improperly diluted sample (perhaps manifested by an absorbance reading that is greater than specified or expected), an obstruction in the sample or reference beam, an improperly aligned source or mirror, the incomplete programming of a scan, or improper or inappropriate software entry. In these cases, the operator will need to carefully examine his or her technique or procedure, or instrumental parameters, such as the optical path, perhaps with the help of the instrument troubleshooting guide, to solve the problem.

Contamination, as indicated previously, can manifest itself in the absorption spectrum. If a sample is contaminated with a chemical that has its own absorption spectrum in the range studied, the absorption pattern (spectrum) of the analyte will not match the expected pattern. If a contaminant is suspected, it may originate from any of the chemicals used in the procedure, including the solvent, the analyte used to prepare standards, or any other chemical added to the solutions tested. In that case, repeating the solution preparation using chemicals from fresh, unopened containers may help solve the problem.

## 8.5  Fluorometry

Fluorometry is an analytical technique that utilizes the ability of some substances to exhibit luminescence. **Luminescence** is a phenomenon in which a substance appears to glow when a light shines on it. In other words, light of a wavelength different from that of the irradiating light is released or emitted following the absorption of this light. The casual observer notices luminescence when the irradiating light is ultraviolet light and the emitted light is visible light, such as what can be seen, for example, when a black light shines on a poster or other material with fluorescent dyes in it. The phenomenon is explained based on light absorption theory and what can happen to a chemical species in order to revert back to the ground state once the absorption—the elevation to an excited state—has taken place. Luminescence can occur with molecules, complex ions, and atoms. The present discussion will focus on molecules and complex ions. Atomic fluorescence will be mentioned in Chapter 9.

All atoms, molecules, and complex ions seek to exist in their lowest possible energy state at all times. When a molecule is raised to an excited electronic energy state through the absorption of light, it is no longer in its lowest possible energy state and will subsequently lose the energy it gained to return to the lowest (ground) state. Most often, the energy is lost through mechanical means, such as through molecular collisions. However, there can be a direct jump back to the ground state with only intermediate stops at some lower vibrational states. Luminescence caused by such a direct jump is called **fluorescence**. Luminescence resulting from a jump back to the ground state after routing through other electronic states is called **phosphorescence**. Phosphorescence is often referred to as delayed fluorescence. In either case, the energy the molecule gained as a result of the absorption process is lost in the form of light. Since it is light of less energy due to the accompanying small energy losses in the form of vibrational loss, the wavelength of emission is longer than the wavelength of absorption and the material appears to glow. A simplified energy level diagram is shown in Figure 8.11. We will refer to fluorescence and phosphorescence collectively as fluorescence.

Figure 8.11 is a *simplified* energy level diagram because it shows only one wavelength of light being absorbed and only one wavelength being emitted. In reality, *many* wavelengths can be absorbed (giving rise to the molecular absorption spectrum as discussed in Chapter 7) and *many* wavelengths of light can be emitted (giving rise to a so-called **fluorescence spectrum**) due to the presence of many vibrational levels in the each of the electronic levels.

In the laboratory, solutions of analytes that fluoresce are tested by measuring the intensity of the light emitted. The instrument for measuring fluorescence intensity is called a **fluorometer**. Inexpensive instruments used for routine work utilize absorption filters similar to what was described previously for absorption spectrophotometers (see Figure 8.2 and accompanying discussion) and are called **filter fluorometers**. Two such filters are needed—one to isolate the wavelength from the source to be absorbed, the wavelength

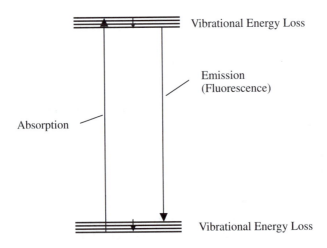

**FIGURE 8.11**  An energy level diagram depicting the phenomenon of fluorescence in a molecule or complex ion.

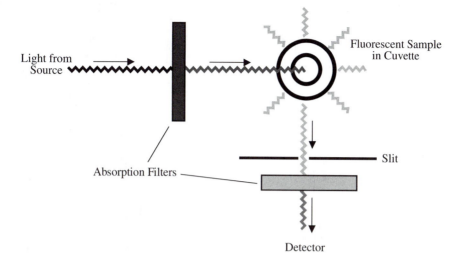

FIGURE 8.12   An illustration of a filter fluorometer.

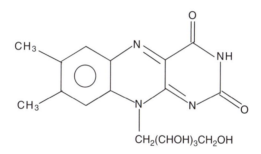

$$CH_2(CHOH)_3CH_2OH$$

FIGURE 8.13   Structure of riboflavin.

of maximum absorbance, and the other to isolate the wavelength of maximum fluorescence to be detected. In addition, the instrument components are arranged in a **right-angle configuration** so that the fluorescence intensity measurement is free of interference from transmitted light from the source. See Figure 8.12. Research grade instruments called **spectrophotofluorometers** utilize the slit/dispersing element/slit monochromators and are considerably more expensive. Such instruments are used for more precise work and can measure the absorption and fluorescence spectra.

When the concentration of a fluorescing analyte is small so as to result in a small absorbance value, the intensity of the resulting fluorescence is proportional to concentration and is therefore measured for quantitative analysis. Thus the usual procedure for quantitative analysis consists of the measurement of a series of standard solutions of the fluorescing analyte or other species proportional to the analyte. A graph of fluorescence intensity vs. concentration is expected to be linear in the concentration range studied.

The types of compounds that can be analyzed by fluorometry are rather limited. Benzene ring systems, such as the vitamins riboflavin (Figure 8.13) and thiamine, are especially highly fluorescent compounds and are analyzed in foods and pharmaceutical preparations by fluorometry. Metals can be analyzed by fluorometry if they are able to form complex ions by reaction with a ligand having a benzene ring system.

Fluorometry and absorption spectrophotometry are competing techniques in the sense that both analyze for molecular species and complex ions. Each offers its own advantages and disadvantages. As stated above, the number of chemical species that exhibit fluorescence is very limited. However, for those species that do fluoresce, the fluorescence is generally very intense. Thus we can say that while absorption spectrophotometry is much more universally applicable, fluorometry suffers less from interferences and

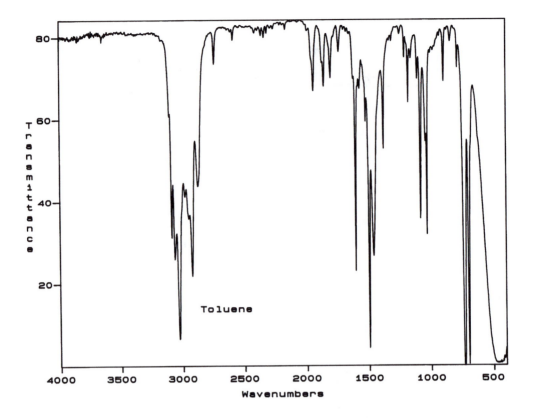

**FIGURE 8.14**   The IR spectrum of toluene. Note that the x-axis is the wavenumber. Also note the sharpness of the absorption bands.

is usually much more sensitive. Therefore when an analyte does exhibit the somewhat rare quality of fluorescence, fluorometry is likely to be chosen for the analysis. The analysis of foods and pharmaceutical preparations for vitamin content is an example, since vitamins such as riboflavin exhibit fluorescence and fluorometry would be relatively free of interference and would be very sensitive.

## 8.6   Introduction to IR Spectrometry

IR spectrometry differs from UV-VIS spectrophotometry in the following ways:

1. Absorption of IR light results in vibrational energy transitions rather than electronic transitions. The concept of vibrational energy transitions was introduced in Chapter 7.
2. While liquid solutions are often analyzed in IR work, pure liquids and undissolved solids, including polymer films, are also often analyzed, as are gases.
3. In IR work, the containers that hold liquids or liquid solutions in the path of the light are not called cuvettes. They are called liquid sampling cells.
4. The liquid sampling cells have extremely short pathlengths (often fractions of a millimeter), defined by the thickness of a thin polymer film spacer (see Figure 7.20).
5. Liquid sampling cells utilize large polished inorganic salt crystals as windows (cell walls) for the IR light. Water, in which these windows are soluble, must be scrupulously avoided, meaning that analyte solvents must be water-free organic liquids that do not dissolve the inorganic salt crystals. Glass and plastic have significant disadvantages and are not usually used.
6. IR spectra are usually transmission spectra rather than absorption spectra, and wavenumber, rather than wavelength, is plotted on the x-axis.

7. IR spectra are characterized by rather sharp absorption bands and each such band is characteristic of a particular covalent bond in the sample molecule. An example is shown in Figure 8.14. Thus, while IR spectra are molecular fingerprints, as are UV-VIS spectra, they have a greater worth to a qualitative analysis scheme due to the specificity of the absorption bands.

8. Modern IR spectrometers do not use light dispersion to acquire spectra. Rather, they utilize a device called an interferometer between the source and the sample. This design requires a signal processing circuit that performs a mathematical operation called a Fourier transformation to obtain the spectra.

## 8.7  IR Instrumentation

Let us begin with the instrumentation. Dispersive IR instruments, similar to the double-beam instruments described for UV-VIS spectrophotometry, have been used in the past but have become all but obsolete. While some laboratories may still use these instruments, we will not discuss them here.

The methodology that involves instruments that utilize the interferometer and Fourier transformation mentioned in Section 8.6 has come to be known as **Fourier transform infrared spectrometry** (FTIR).

In the FTIR instrument, the undispersed beam of IR light from the source first enters an interferometer. An **interferometer** is a device that creates a pattern of light resulting from the combined constructive and destructive interference of all component wavelengths. This combined interference is caused by first splitting the beam and then directing the two beams through paths of two different lengths, one of which is fixed and one of which is variable. The fixed length path utilizes a mirror that is fixed in position. The variable length path utilizes a mirror that is movable. When the two beams are recombined at the same beamsplitter (mirrored on both sides), they may be in phase (when the two paths are of the same length) or out of phase (when the two paths are not of the same length). See Figure 8.15.

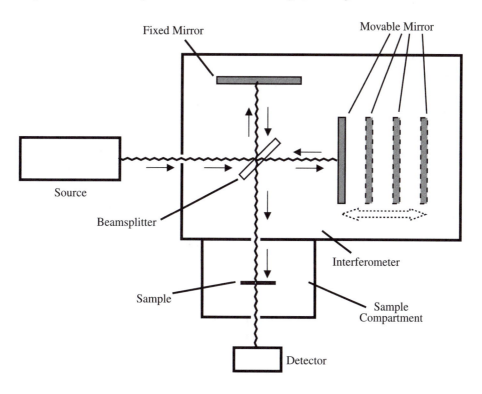

**FIGURE 8.15**  An illustration of an FTIR instrument showing the light source, the interferometer, the sample compartment, and detector.

When the two beams are in phase, all wavelengths combine constructively, resulting in a maximum intensity burst of radiant power. However, even a slight change in the variable length path due to a movement of the movable mirror results in a dramatic decrease in the intensity. A complete traversing of the movable mirror through its path gives a pattern to the light at the detector directly attributable to the combined constructive and destructive interference of all the wavelengths in the beam as a function of mirror position. This pattern is called an **interferogram**. The pattern is the same each time, provided there is nothing in the path of the light to absorb any given wavelength. When an absorbing sample is placed in the path of the light in the sample compartment, some wavelengths are absorbed sharply; this creates a different interferogram at the detector. Comparing the two interferograms using the Fourier transformation signal processing results in the absorption spectrum of the sample.

The advantages of FTIR over the dispersive technique are: 1) it is faster, making it possible to be incorporated into chromatography schemes, as will be seen in Chapters 12 and 13, and 2) the energy reaching the detector is much greater, thus increasing the sensitivity.

# 8.8   Sampling

**Sampling**, in the context of infrared spectrometry, refers to the method by which a sample is held in the path of the IR light. Since liquids (including pure liquids and liquid solutions), solids, and gases can be analyzed by infrared spectrometry, we have **liquid sampling**, **solid sampling**, and **gas sampling**. In any case, it is important to point out that glass and plastic are undesirable materials for the cells. The reason is that glass and plastic are molecular (covalent) materials and would absorb IR light and interfere with reading the sample. Inorganic salts, such as NaCl and KBr, are ionic materials and do not absorb IR light because ionic bonds cannot undergo vibrational energy transitions. Covalent bonds can vibrate while ionic bonds cannot.

## 8.8.1   Liquid Sampling

The liquid sampling cell previously mentioned (items 3 to 5 in Section 8.6 and Figure 7.20) is the primary means for liquid sampling. In other words, for pure liquids (often referred to as **neat liquids**) and liquid solutions, sandwiching a thin layer of liquid between two large NaCl or KBr crystals (the windows) is the classic procedure for mounting the sample in the path of the light. The windows are positioned such that the IR light beam from the interferometer passes directly through the assembly from one window, through the sample, and then through the other window to the detector, as shown in Figure 8.16.

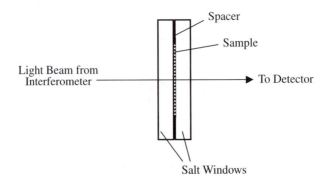

**FIGURE 8.16**    An illustration of a liquid sampling cell placed in the path of the light.

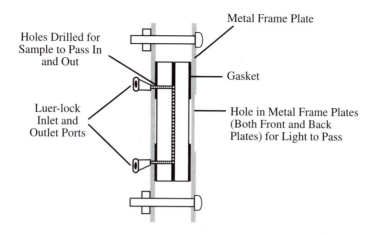

**FIGURE 8.17** An illustration of the sealed cell assembly.

Typical dimensions for such windows are about 2 cm wide × 3 cm long × 0.5 cm thick. Positioning or holding the windows (often called salt plates) in place is done using either a sealed cell, a demountable cell, or a combination sealed demountable cell. **Sealed cells** are intended to be permanent fixtures for the windows and should not be disassembled during their useful lifetime. They have a fixed pathlength and are very useful for quantitative analysis, since the pathlength is reproduced (same cell, same pathlength) from one standard to another and also for the unknown. The fixed pathlength is defined by the thin polymer film spacer mentioned in item 4 in Section 8.6 and is also illustrated in Figure 8.16. The spacer leaves a space in the center to be filled by a very small volume of sample, as shown in Figure 8.16. Sealed cells have inlet and outlet ports on a metal frame for introducing the sample via a Luer-lock syringe through one salt plate to the space between the salt plates. The window on the side of the ports has holes drilled in it to facilitate moving the sample from the syringe, through the inlet port to the space, and then out the outlet port. See Figure 8.17.

**Demountable cells** can be disassembled and may or may not use a spacer. Without a spacer, the liquid sample is simply applied to one salt plate, and the second plate is positioned over the first to smear the liquid out between the plates. Such a method cannot be useful for quantitative analysis because the pathlength is undefined and not reproducible. Also, the two salt plates may be positioned in the path of the light without a frame if there is a way to hold them there. If a spacer is used, it is positioned over one of the plates. The sample is applied to this plate in the cutout space in the spacer reserved for it, and the second plate is then positioned over the first with both spacer and sample in between. There are no inlet and outlet ports since the sample is not introduced that way. However, the plates, with sample, are placed in a frame, or holder, similar in appearance to the sealed cell, but without the ports. See Figure 8.18. Such a cell is also undesirable for quantitative analysis because of the difficulty in obtaining an identical pathlength each time the cell is reassembled.

A **sealed demountable cell** is a combination of the sealed cell and the demountable cell, meaning that it is like a sealed cell but is meant to be disassembled if desired. Disassembly would be for the purpose of changing the spacer and not for introducing the sample, since the sample is introduced through the inlet port with the syringe. A sealed demountable cell is useful for both qualitative and quantitative analysis since the pathlength is defined and is reproducible as long as the cell is not disassembled between standards and unknowns. A sealed demountable looks identical to a sealed cell (Figure 8.17).

All three cells utilize neoprene gaskets to cushion the fragile salt crystals from the metal frame. In the case of the sealed and sealed demountable cells, the top neoprene gasket and window have holes drilled in them to coincide with the inlet and outlet ports to facilitate filling the space, created by the spacer,

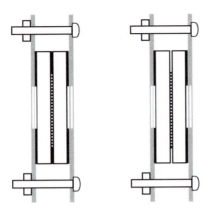

**FIGURE 8.18**   Left, a demountable cell with spacer. Right, a demountable cell without spacer. Note that there are no drilled holes for the sample and no inlet or outlet ports. Both metal frame plates (front and back) have holes for light to pass.

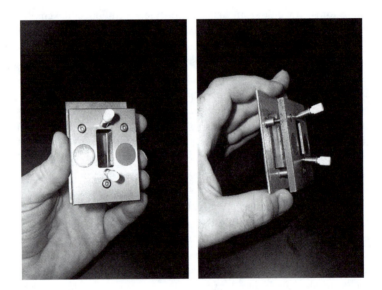

**FIGURE 8.19**   Photographs of a sealed (or sealed demountable) cell.

with the liquid sample using the syringe. The path of the liquid sample is shown in the figures. Photographs of a sealed cell, or an assembled sealed demountable cell, are shown in Figure 8.19.

Filling the cells with sample and eliminating the sample when finished can be troublesome. The sample inlet and outlet ports are each tapered to receive a syringe with a Luer (tapered) tip that locks in place. The useful procedure for filling is to raise the outlet end by resting it on a pencil or similar object (to eliminate air bubbles) and then to use a pressure–vacuum system with the use of two syringes, one in the outlet port and one in the inlet port, as shown in Figure 8.20. While pushing on the plunger of the syringe containing the liquid sample in the inlet port and pulling up on the plunger of the empty syringe in the outlet port, the cell can be filled without excessive pressure on the inlet side. This reduces the possibility of damaging the cell (by causing a leak in the seal between the salt plates and spacer, for example) when excessive pressure may be needed, especially when working with unusually viscous

**FIGURE 8.20**    Filling a liquid sampling cell.

samples and short pathlengths. If the sample can be loaded without excess pressure, then the outlet syringe can be eliminated. Tapered Teflon™ plugs are used to stopper the ports immediately after filling. The cell may be emptied and readied for the next sample by using two empty syringes and the same push–pull method. When refilling, an excess of liquid sample may be used to rinse the cell and eliminate the residue from the previous one. Alternatively, the cell may be rinsed with a dry volatile solvent and the solvent evaporated before introducing the next sample.

The analyst must be careful to protect the salt crystals from water during use and storage. Sodium chloride and potassium bromide are, of course, highly water soluble, and the crystals may be severely damaged with even the slightest contact with water. All samples introduced into the cell must be dry. This is important for another reason. Water contamination will be seen on the measured spectrum (as an alcohol; see Section 8.10) and cause erroneous conclusions. If the windows are damaged with traces of water, they will become fogged and will appear to become nontransparent. The windows may be repolished if this happens. Depending on the extent of the damage, various degrees of abrasive materials may be used, but the final polishing step must utilize a polishing pad and a very fine abrasive. Polishing kits are available for this. Finger cots should be used to protect the windows from finger moisture.

Liquids can be sampled as either the neat liquid (pure) or mixed with a solvent (solution). Neat liquids are tested when the purpose of the experiment is either identification or the determination of purity. Identification is possible because the spectrum is a fingerprint when no solvent or contaminant is present. Impurities are found when extraneous absorption bands or distortions in analyte absorption bands appear.

When a solution is tested, both analyte and solvent absorption bands will be present in the spectrum, and identification, if that is the purpose of the experiment, is hindered. Some solvents have rather simple IR spectra and are thus considered more desirable as solvents for qualitative analysis. Examples are carbon tetrachloride ($CCl_4$, only C–Cl bonds), choloroform ($CHCl_3$), and methylene chloride ($CH_2Cl_2$). The infrared spectra of carbon tetrachloride and methylene chloride are shown in Figure 8.21. There is a problem with toxicity with these solvents, however. For quantitative analysis, such absorption band interference is less of a problem because one needs only to have a single absorption band of the analyte isolated from the other bands. This one band can be the source of the data for the standard curve since the peak absorption increases with increasing concentration (see Section 8.11 and Experiment 25). See Workplace Scene 8.2.

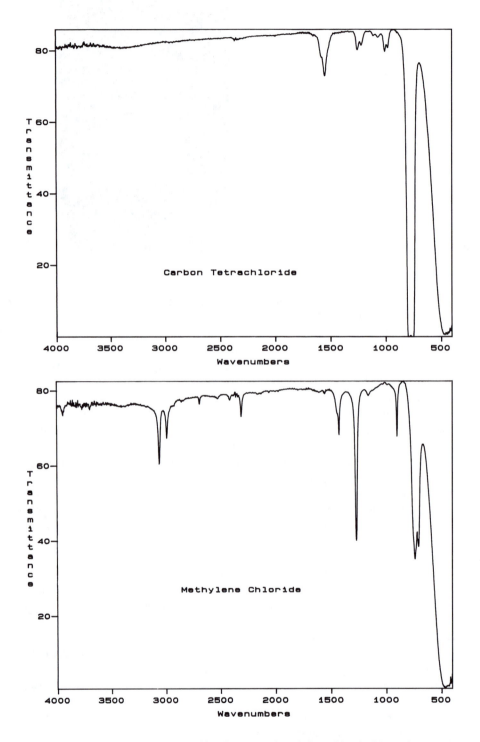

**FIGURE 8.21**    Infrared spectra of carbon tetrachloride (top) and methylene chloride (bottom).

# WORKPLACE SCENE 8.2

Simethicone is an antigas ingredient in many liquid and solid pharmaceutical preparations, and FTIR is used in quality assurance laboratories to determine whether its concentration is at the specified level. A sample of the product is dispersed in an HCl solution and the simethicone extracted from this solution with toluene. The toluene solutions are then run on the FTIR using a liquid sampling cell. For the quantitative analysis, a simethicone absorption band that is free from interference from the toluene absorption bands is used in a manner similar to that of the isopropyl alcohol band in Experiment 26.

Dave Segura prepares to use an FTIR instrument for simethicone analysis. Notice the liquid sampling cell in Dave's left hand.

## 8.9   Solid Sampling

Infrared spectrometry is one of the few analytical techniques that can measure both dissolved and undissolved solids. Thus there are some unique and interesting solid sampling methods.

### 8.9.1   Solution Prepared and Placed in a Liquid Sampling Cell

The most straightforward method for analyzing a solid material by infrared spectrometry is to dissolve it in a suitable solvent and then to measure this solution using a liquid sampling cell such as one of the several described in Section 8.8. Thus it becomes a liquid sampling problem, the experimental details of which have already been discussed (Section 8.8). It is the only method of solid sampling suitable for quantitative analysis because it is the only one that has a defined and reproduced pathlength.

### 8.9.2   Thin Film Formed by Solvent Evaporation

A simple method for solids is to prepare a solution of the solid, place several drops of the solution on the surface of a single salt plate, and then allow or force the solvent to evaporate, leaving a thin film of the solid on the plate. If this plate is fixed in the path of the light, the spectrum of the solid can be measured by passing the light through it as described for the liquid sampling devices. See Figure 8.22.

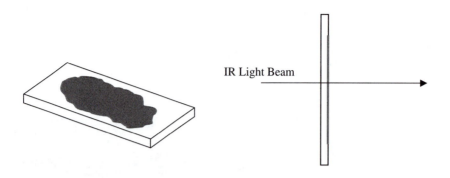

**FIGURE 8.22**  A solid may be prepared by dissolving it in an organic solvent, placing it on a salt crystal (left), and allowing the solvent to evaporate such that the film of the solid is cast on the surface of the salt plate, which can then be placed in the path of the light (right).

So-called disposable IR cards are also available for this. Such cards have polyethylene windows on which the sample solution may be applied and the solvent evaporated. Unless they are subtracted out, IR spectra from such cards would show the polyethylene absorption bands.

The solvent evaporation method is especially useful for polymers that can be cast onto the surface of the salt plate. Casting involves dissolving the polymer in an appropriate solvent at an elevated temperature and evaporating the solvent on the salt plate by heating the salt plate on a hot plate or under a heat lamp.

### 8.9.3  KBr Pellet

The KBr pellet technique is based on the fact that *dry*, finely powdered potassium bromide has the property of being able to be squeezed under very high pressure into a transparent disc — transparent to both infrared light and visible light. It is important for the KBr to be dry in order to obtain a good pellet and eliminate absorption bands due to water in the spectrum. If a small amount of the dry solid analyte (0.1 to 2.0%) is added to the KBr prior to pressing, then a wafer (pellet) can be formed from which a spectrum of the solid can be obtained. Such a wafer is simply placed in the path of the light in the instrument such that the IR beam passes from the interferometer, through the wafer, to the detector.

Pressing the KBr–sample mixture to make the transparent wafer is done in a special die made for this purpose. The die is placed in a laboratory press made for this purpose. One rather simple such die and press can be described as follows. The pellet die consisting of a threaded body and two bolts with polished faces may be used. One bolt is turned completely into the body of the die. A small amount of the powdered KBr–sample mixture, enough to cover the face of the bolt inside, is added and the other bolt turned down onto the sample, squeezing it into the pellet or wafer. See Figure 8.23. The two bolts are then carefully removed. If the pellet is well formed, it will remain in the die and appear transparent, or nearly so. The die is then placed in the instrument so that the light beam passes directly through the center of the die and through the pellet.

In addition, compression methods that utilize hydraulics and levers may be used. These sometimes have pressure gauges that allow the analyst to apply a certain optimal force in order to maximize the chances of making a quality pellet. To make a quality pellet, in addition to using the optimum pressure, it is important for the KBr and the sample to be dry, finely powdered, and well mixed. An agate mortar and pestle is recommended for the grinding and mixing of the KBr and sample prior to compression. See Workplace Scene 8.3.

### 8.9.4  Nujol Mull

The Nujol (mineral oil) mull is also often used for solids. In this method, a small amount of the finely divided solid analyte (one or two micron particles) is mixed together with an amount of mineral oil to form a mixture with a toothpaste-like consistency. This mixture is then placed (lightly squeezed) between

# WORKPLACE SCENE 8.3

A n important raw material for pharmaceutical formulations is polyethylene oxide because it is used in many and various products. Polyethylene oxide is a solid material purchased in bulk as a powder. Sampling technicians collect samples in glass bottles. In the laboratory, small portions are taken and mixed with potassium bromide for the purpose of producing a KBr pellet and obtaining the IR spectrum. The spectrum thus obtained is compared to a reference spectrum to confirm that the material is indeed polyethylene oxide.

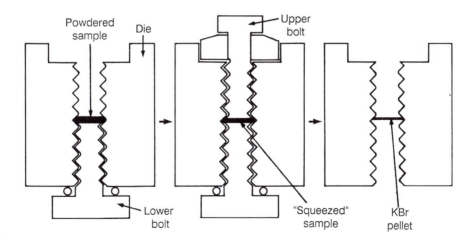

Left, Jessica Weinland fills a KBr pellet die with a KBr–polyethylene oxide sample. Right, Ms. Weinland uses a laboratory press to create the pellet. Notice the pellet die positioned in the press in the photo on the right.

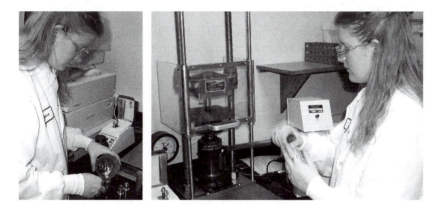

**FIGURE 8.23**    A simple method for making a KBr pellet.

two NaCl or KBr windows, which are similar to those used in the demountable cell discussed previously for liquids. If the particles of solid are not already the required size when received, they must be finely ground with an agate mortar and pestle and can be ground directly with mineral oil to create the mull to be spread on the window. Otherwise, a small amount (about 10 mg) of the solid is placed on one window along with one small drop of mineral oil. A gentle rubbing of the two windows together with a circular or back and forth motion creates the mull and distributes it evenly between the windows. The windows are placed in the demountable cell fixture and in the path of the light as described previously.

A problem with this method is the fact that mineral oil is a mixture of covalent substances (high-molecular-weight hydrocarbons) and its characteristic absorption spectrum will be found superimposed in the spectrum of the solid analyte, as with the solvents used for liquid solutions discussed previously. However, the spectrum is a simple one (Figure 8.24) and often does not cause a significant problem, especially if the solid is not a hydrocarbon.

## 8.9.5  Reflectance Methods

Rather than measure the light *transmitted through* a sample, as in all of the IR sampling methods discussed thus far, some methods measure the light *reflected from* a sample. There are various methods employed to measure this reflectance, depending on the properties of the sample.

### 8.9.5.1  Specular Reflectance

If the sample can be cast as a thin, smooth, mirror-like film on a flat surface, the IR spectrum can be measured by measuring an IR light beam made to reflect from this surface (see Figure 8.25). The sample is mounted in air and the light beam travels through air before and after striking the sample.

### 8.9.5.2  Internal Reflectance

In this method, the sample (such as a polymer film) is pressed against a transparent material having a high refractive index (called the **internal reflection element** (IRE)). The IR light beam passes through this material, rather than air, before and after reflecting from the sample, hence the reason for describing

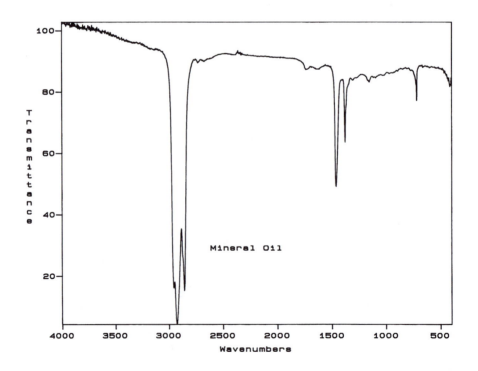

**FIGURE 8.24**  The IR spectrum of mineral oil.

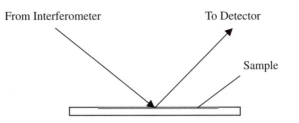

**FIGURE 8.25**    Specular reflectance.

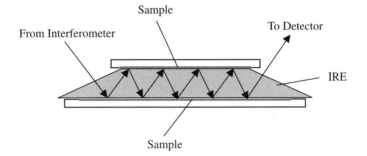

**FIGURE 8.26**    Multiple internal reflectance.

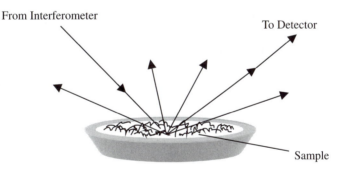

**FIGURE 8.27**    Diffuse reflectance.

the technique as internal. Devices that utilize multiple such reflections, with the sample pressed against both the upper and lower edges of the IRE, have been fabricated and the technique dubbed **multiple internal reflectance**. A large number of such reflections (due to a very long IRE and sample) results in the intensity of the light beam being attenuated, or systematically reduced, and thus multiple internal reflectance is sometimes called **attenuated internal reflectance**. See Figure 8.26.

### 8.9.5.3   Diffuse Reflectance

Powdered samples that cannot be cast as films can be measured by this technique. Quite simply, the IR light beam shines on a quantity of the powdered sample held in a cup. In this case, the light scatters, or reflects in all directions. Some of this light is captured by the detector. A disadvantage is that the light intensity is decreased substantially by the scattering, hence the term diffuse. See Figure 8.27.

## 8.9.6   Gas Sampling

Gases, such as automobile exhausts and polluted air, can be measured by IR spectrometry. A special cell is used for containing the gas. This cell is cylindrically shaped and has inlet and outlet stopcocks for introducing the sample. The side windows of the cylinder are nonabsorbing inorganic salt crystals.

## 8.10   Basic IR Spectra Interpretation

Once the spectrum is obtained, the question remains as to how to interpret it. We stated previously (Section 8.6) that IR spectra are characterized by rather sharp absorption bands (peaks) and each such peak is characteristic of a particular covalent bond in a molecule in the path of the light. Thus for qualitative analysis (identification and purity characterization), the analyst correlates the location (the wavenumber), the shape, and the relative intensity of a given peak observed in a spectrum with a particular type of bond.

The region of the electromagnetic spectrum involved here, of course, is the infrared region. This spans the wavelength region from about 2.5 $\mu$m to about 17 $\mu$m, or, in terms of wavenumber, from about 4000 cm$^{-1}$ to about 600 cm$^{-1}$. Infrared spectra are usually displayed as transmission spectra in which the x-axis is the wavenumber. Thus, we will be depicting the spectra with a 100% T baseline at the top of the chart (as the maximum y-axis value) and the peaks deflecting toward the 0% T level when an absorption occurs, as in the spectra shown previously (Figures 8.14, 8.21, and 8.25).

The region from 4000 cm$^{-1}$ to approximately 1500 cm$^{-1}$ is especially useful for correlating peak location with bonds. For this reason, we will refer to this region as the peak ID region. The region from 1500 to 600 cm$^{-1}$ is typically very busy (refer to Figures 8.14, 8.21, and 8.25 as examples) and is not as useful for such correlation. We will refer to this region as the fingerprint region because, while not generally useful for peak correlation, it remains very useful as the molecular fingerprint. This means that we can still use this region, as we can the peak ID region, for peak-for-peak matching with a known spectrum from a library of known spectra. See Figure 8.28. Thus, we look for characteristic bands in the 4000 to 1500 cm$^{-1}$ region to perhaps assign the unknown to a particular class of compounds, i.e., to narrow down the possible structures, and then look to the fingerprint region and the overall spectrum to make the final determination, matching peak-for-peak. In addition to the location (i.e., wavenumber) of the peaks, it can also be useful to examine their width and depth (strength). The descriptions broad and sharp are often used to describe the width of the peaks, and weak, medium, and strong are used to describe the depth or strength of the peaks. To understand these descriptions, refer to Figures 8.29 to 8.31. Figures 8.29 and 8.31 show broad, strong peaks centered around 3300 cm$^{-1}$. Figure 8.30 shows a sharp, strong

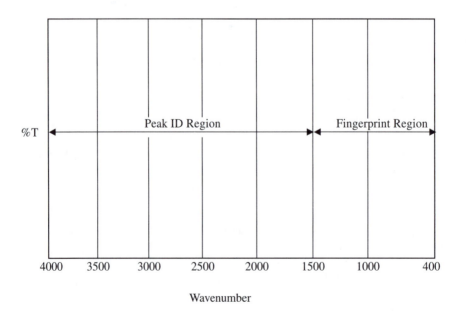

**FIGURE 8.28**   The Peak ID and fingerprint regions of the IR spectrum.

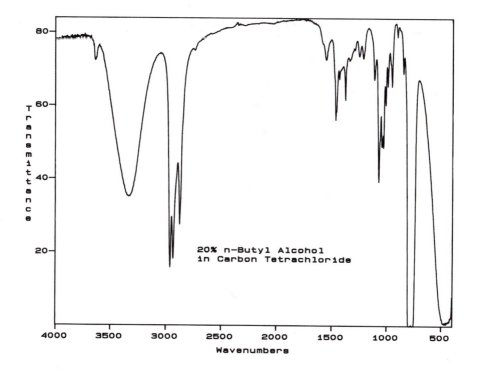

**FIGURE 8.29** The IR spectrum of an alcohol with no benzene rings.

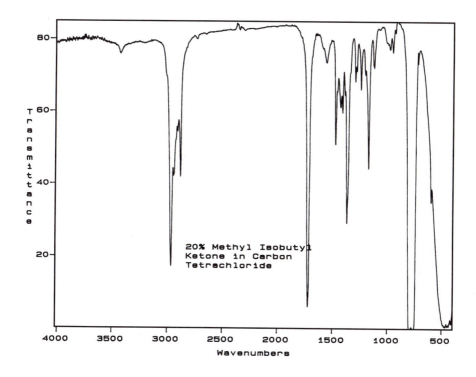

**FIGURE 8.30** The IR spectrum of a compound with a carbonyl group and no benzene ring.

**TABLE 8.1**   Some Easily Recognizable IR Absorption Patterns

| Bond | Description of Peak |
|------|---------------------|
| –C–H, where C is not part of a benzene ring | Sharp, strong peak on the low side of 3000 cm$^{-1}$, between about 2850 and 3000 cm$^{-1}$ |
| –C–H, where C is part of a benzene ring or double bond | Sharp, medium peak on the high side of 3000 cm$^{-1}$, between about 3000 and 3100 cm$^{-1}$ |
| –C–H, where C is part of a triple bond | Sharp, medium peak on the high side of 3000 cm$^{-1}$, between about 3250 and 3300 cm$^{-1}$ |
| –O–H, in alcohols, phenols, and water, for example | Broad, strong peak centered at about 3300 cm$^{-1}$ |
| –C=O, in aldehydes, ketones, etc. | Sharp, strong peak at about 1700 cm$^{-1}$ |
| –C–C–, in benzene rings | Two sharp, strong peaks near 1500 and 1600 cm$^{-1}$, a series of weak peaks (called overtones) between 1600 and 2000 cm$^{-1}$, the latter in the case of a mono-substituted benzene ring |

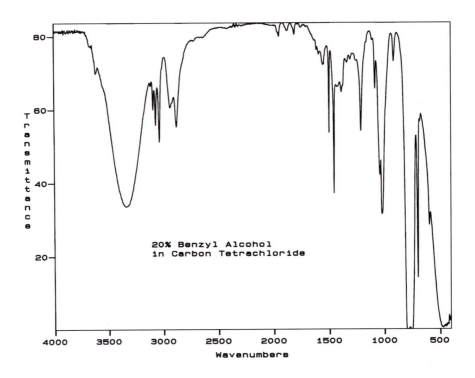

**FIGURE 8.31**   The IR spectrum of an alcohol with a benzene ring.

peak at 1685 cm$^{-1}$. Figure 8.31 shows a sharp, medium peak at 1220 cm$^{-1}$ and a series of weak peaks between 1700 and 2000 cm$^{-1}$.

   Table 8.1 correlates the location, width, and depth or strength of various IR absorption patterns for some common kinds of bonds. By comparing Figures 8.29 to 8.31 with Table 8.1, we can surmise that the spectrum in Figure 8.29 is that of an alcohol with no benzene rings. We can also surmise that the spectrum in Figure 8.30 is that of a compound containing a carbonyl group (such as an aldehyde) with no benzene ring. And we can surmise that the spectrum in Figure 8.31 is that of an alcohol with a benzene ring. So-called IR correlation charts, which are graphical displays showing the location (wavenumber ranges) of the IR absorption peaks of most bonds, can be helpful in the interpretation of spectra. These are found in the *CRC Handbook of Chemistry and Physics*, for example, and in some organic chemistry textbooks.

# MEASURING CELL PATHLENGTH
# AND POLYMER FILM THICKNESS

T he pathlength of IR liquid sampling cells and the thickness of polymer films can be measured in a rather convenient way using the interference fringe pattern observed on the baseline of an IR spectrum. An example baseline for an empty liquid sampling cell having a pathlength of 0.025 mm is shown in this figure. The constructive and destructive interference that occurs due to the IR light beam reflecting off the interior faces of the crystals causes the repetitive up-and-down wavy appearance. The pathlength (t) is calculated using the following formula:

$$t = \frac{5M}{\bar{v}_1 - \bar{v}_2}$$

where M is the number of minima in the pattern between wavenumbers $\bar{v}_1$ and $\bar{v}_2$. In this example, the pathlength is calculated to be 0.026 mm:

$$t = \frac{5 \times 10}{2900 - 980} = 0.026 \text{ mm}$$

Polymer film thicknesses can be determined similarly.

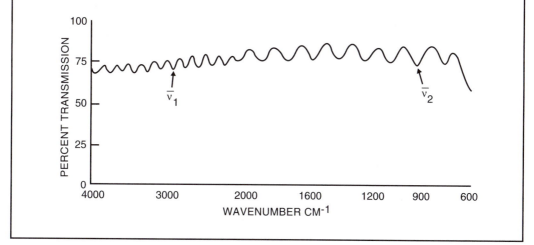

## 8.11  Quantitative Analysis

Quantitative analysis procedures using infrared spectrometry utilize Beer's law. Thus only sampling cells with a constant pathlength can be used. Once the percent transmittance or absorbance measurements are made, the data reduction procedures are identical with those outlined in Chapter 7 (preparation of standard curve, etc.).

Reading the percent transmittance from the recorded IR spectrum for quantitative analysis can be a challenge. In the first place, there can be no interference from a nearby peak due to the solvent or other component. One must choose a peak to read that is at least nearly, if not completely, isolated.

Secondly, the baseline for the peak must be well defined. The baseline for the entire spectrum is the portion of the trace where there are no peaks. It does not usually correspond to the 100% T line. Thus, two percent transmittance readings must be taken, one for the baseline (corresponding to where the baseline would be if the peak were absent—a blank reading) and one for the minimum percent transmittance, the tip of the peak. The two readings are then converted to absorbance, and the absorbance of the baseline is subtracted from the absorbance of the peak. The result is the absorbance of the sample.

It should be mentioned again that the pathlength of the sample cell used must be constant for all standards and the unknown. Using care when filling and cleaning cells is also important to avoid alterations in pathlength due to excessive pressure or leaking.

# Experiments

## Experiment 22: Spectrophotometric Analysis of a Prepared Sample for Toluene

Note: The slightest contamination with water or other solvent will result in gross error.

*Remember safety glasses.*

1. Prepare one 100-mL volumetric flask and five 25-mL volumetric flasks. These flasks should be clean and dry. They may be prepared by washing with soapy water and a brush, rinsing with water, removing the water with several rinses with acetone, and removing the acetone with several rinses with the alkane solvent to be used (e.g., hexane, heptane, or cyclohexane).

2. Prepare 100 mL of a stock standard solution of toluene in an alkane solvent by pipetting 0.1 mL of toluene into the flask and diluting to the mark with the alkane. Use the 100-mL flask prepared in step 1. Shake well. This solution has a concentration of 0.870 g/L.

3. Prepare four calibration standards with concentrations of 0.0435, 0.0870 , 0.174, and 0.261 g/L in the alkane solvent by diluting the stock solution prepared in step 1. Use the 25-mL volumetric flasks prepared in step 1.

4. Obtain an unknown from your instructor and dilute to the mark with the alkane solvent. Rinse the remaining 25-mL flask from step 1 with the control sample several times before filling the flask nearly to the mark with this sample.

5. Obtain an absorption spectrum of each of your standards, the unknown, and the control sample on a scanning UV-VIS spectrophotometer interfaced to a computer for data acquisition. Follow the instructions provided for your instrument and software.

6. Obtain the maximum absorbance value for each spectrum and prepare the standard curve (Beer's law plot) as in Experiment 18. Obtain the concentration of the unknown and the control sample.

7. Maintain the logbook for the instrument used in this experiment. Record the date, your name(s), the experiment name or number, the correlation coefficient, and the results for the control sample. Also plot the control sample results on a control chart for this experiment posted in the laboratory.

8. Dispose of all solutions as directed by your instructor.

## Experiment 23: Determination of Nitrate in Drinking Water by UV Spectrophotometry

*Remember safety glasses.*

1. Your instructor has prepared a 50 ppm stock solution of nitrogen in water by dissolving 0.361 g of $KNO_3$ in 1000 mL of solution.

2. Prepare calibration standards that are 0.5, 1.0, 3.0, 5.0, and 10.0 ppm N in 50-mL volumetric flasks by diluting the stock standard with distilled water. After diluting to the mark, pipet 1.0 mL of 1 $N$ HCl into each flask and shake well.

3. Filter the water samples if necessary. Then prepare them for measurement by adding 1 mL of 1 $N$ HCl to 50 mL of each. This can be done by filling (rinse first) 50-mL volumetric flasks with the samples to the 50-mL mark and then pipetting 1.0 mL of the HCl into these flasks. Shake well. Prepare the control sample and a blank (distilled water) in the same way as the samples.

4. Zero the UV spectrophotometer with the blank at 220 nm. Measure all solutions (calibration standards, unknowns, and control) as follows using quartz cuvettes. Measure the absorbances at

220 and 275 nm for each solution. Since dissolved organic matter may also absorb at 220 nm, it is necessary to correct for interference. Subtracting two times the absorbance at 275 from the absorbance at 220 is a sufficient correction procedure for this interference. Thus, subtract two times the absorbance at 275 from the absorbance at 220 for each.

5. If any samples are outside the range of the standards (high), dilute them by an amount that you feel will bring them into range and measure them again.

6. Create the standard curve as practiced in Experiment 18 by plotting the resulting absorbance vs. concentration and obtain the concentration of the samples and the control. If any water samples were diluted prior to the measurement, apply the dilution factor before reporting the results.

7. Maintain the logbook for the instrument used in this experiment. Record the date, your name(s), the experiment name or number, the R square value, and the results for the control sample. Also plot the control sample results on a control chart posted in the laboratory for this experiment.

## Experiment 24: Fluorometric Analysis of a Prepared Sample for Riboflavin

Note: Riboflavin solutions are light sensitive. Store them in the dark as you prepare them.
*Remember safety glasses.*

1. Your instructor has prepared a 50 ppm riboflavin stock solution in 5% (by volume) acetic acid. Prepare a series of calibration standard solutions that are 0.2, 0.4, 0.6, 0.8, and 1.0 ppm from the 50 ppm stock solution and again use 5% acetic acid for the dilution. Use 25-mL volumetric flasks for these standards and shake well.

2. Obtain the unknown from your instructor and again dilute with 5% acetic acid to the mark and shake well.

3. Obtain fluorescence measurements (F) on all of your standards, the unknown, and control. Use 5% acetic acid for a blank. Your instructor will demonstrate the use of the instrument, and explain how it was set up to measure riboflavin.

4. Create the standard curve, using the procedure practiced in Experiment 18, by plotting fluorescence intensity vs. concentration and determine the concentration of the unknown and the control sample.

5. Maintain the logbook for the instrument used in this experiment. Record the date, your name(s), the experiment name or number, the correlation coefficient, and the results for the control sample. Also plot the control sample results on a control chart for this experiment posted in the laboratory.

## Experiment 25: Qualitative Analysis by Infrared Spectrometry

### Introduction

In this experiment you will be given two unknown organic liquids to attempt to identify by infrared spectrometry. For one of the unknowns you will be given its molecular formula. The other must be identified by matching its infrared spectrum to a spectrum in a reference catalog of spectra (sometimes called a spectral library).

Safety note: Avoid contact of all organic compounds with your skin (wear gloves) and also avoid breathing the vapors as much as possible.
*Remember safety glasses.*

1. Obtain the two unknowns from your instructor. One will have an identifying number, and the other will have an identifying letter. For the one identified by a number, obtain its molecular formula from your instructor. Record all information.

2. Obtain the infrared spectra of the two liquids by whatever method suggested by your instructor (liquid sampling cell or disposable card).

3. Attempt to identify the unknown labeled with a letter by matching the spectrum with a spectrum from the IR spectral library. First, take a preliminary look at the spectrum to check for obvious signs of the various functional groups (refer to Table 8.1 or a correlation chart). This will help reduce the number of possibilities. Report your decision regarding its identity to your instructor and justify your decision.

4. Based on your molecular formula alone, you can make some decisions regarding the unknown labeled with a number. For example, there must be at least six carbons in order for it to have a benzene ring. If there is a benzene ring, chances are that it will have approximately the same number of hydrogens as carbons. If there is one oxygen, it is probably either an alcohol, ether, aldehyde, or ketone.

5. Look at the spectrum to further narrow down the possibilities. Record all possibilities (isomers) for the structure of this unknown based on the IR spectrum and the molecular formula alone.

6. As a final step, you may look at the catalog of spectra and identify it based on matching spectra. Was the identity of the unknown among the possibilities you determined just by the formula and spectrum alone? Report your decision regarding its identity to your instructor and justify your decision.

## Experiment 26: Quantitative Infrared Analysis of Isopropyl Alcohol in Toluene

Discussion: Isopropyl alcohol exhibits an infrared absorption peak at 817 cm$^{-1}$. This peak is well isolated from any other peak due to either the analyte or toluene, the solvent. Thus, this peak is appropriate for quantitative analysis of isopropyl alcohol dissolved in toluene. Your instructor may ask you to disassemble and reassemble the cell so as to have the appropriate spacer in place.

*Remember safety glasses.*

1. In *dry* 25-mL volumetric flasks, prepare four calibration standards of isopropyl alcohol in toluene solvent that are 20, 30, 40, and 50% in alcohol concentration. Obtain an unknown from your instructor and dilute to the mark with toluene. Shake well.

2. Obtain a liquid sampling cell from your instructor and fill it with the 20% solution. Place this cell in the FTIR instrument, and obtain the transmittance spectrum according to the instructions specific to your instrument. Record the percent transmittance value at the tip of the 817 cm$^{-1}$ peak and at the baseline. Repeat with the other three standards, the unknown, and the control sample.

3. Convert all percent transmittance readings to absorbance. Subtract the absorbance at the baseline from the absorbance at the tip of the 817 cm$^{-1}$ peak in each spectrum.

4. Create the standard curve, using the procedure practiced in Experiment 18, by plotting absorbance vs. concentration and determine the concentration of the unknown and control.

5. Maintain the logbook for the instrument used in this experiment. Record the date, your name(s), the experiment name or number, the correlation coefficient, and the results for the control sample. Also plot the control sample results on a control chart for this experiment posted in the laboratory.

## Experiment 27: Identifying Minor Components of Commercial Solvents[*]

### Introduction

Some commercial solvents, such as hexane, xylene, and ethyl alcohol, have minor components that can interfere with their effective use in a laboratory or industry. For example, denatured ethyl alcohol has purposely been contaminated with small amounts of poisonous organic liquids in order to deter someone from using it to make alcoholic drinks. Hexane and xylene are often labeled as the plural (hexanes and xylenes) to indicate that they are actually mixtures of isomers. In this experiment, the minor components of one or more of these solvents will be identified.

Note: It is necessary to have available the pure solvent as well as the mixtures in order to conduct this experiment as written.

*Remember safety glasses.*

1. Obtain a sample of the solvent and record its name. Look at some literature sources or a chemical catalog to try to get an idea of what the minor components of your solvent might be. Also obtain a sample of the pure solvent (no minor components).

---

[*]The idea for this lab activity came from Susan Marine of Miami University Middletown, Middletown, Ohio.

2. Obtain the spectra of your sample and also the pure solvent. Use computer software to subtract the spectrum of the pure solvent from that of the mixture.

3. Obtain the spectrum of a liquid that you suspect to be a minor component based on your research from step 1. Then identify the minor components by comparing spectra.

## Experiment 28: Measuring the Pathlength of IR Cells

1. Obtain several IR liquid sampling cells. Using a dry syringe, push and pull air back and forth through the cells to make sure they are free of solvents. Then obtain an IR spectrum of each of the empty cells.

2. Calculate the cell pathlength for each using the method described in the sidebox found in Section 8.11 on page 233.

# Questions and Problems

1. Draw a block diagram showing all the components of a basic spectrophotometer.
2. Define colorimeter, UV spectrophotometer, and UV-VIS spectrophotometer.
3. Briefly describe the light sources used for both visible and UV work.
4. Is the intensity of light emitted by the various light sources used for visible and ultraviolet work the same at all emitted wavelengths? How does intensity vary with wavelength for a tungsten filament lamp?
5. How is the wavelength called for in a procedure isolated from the many wavelengths present in the light from the light source in a spectrophotometer?
6. Define bandwidth and resolution.
7. What is an absorption filter? What is a monochromator?
8. What are the components of a monochromator? What is the function of each component?
9. Consider an experiment in which an analyst wants to change the wavelength used in a given colorimetry experiment from 392 nm (blue light) to 728 nm (red light). He or she turns a knob on the face of the instrument to do this. Tell exactly what is happening inside the instrument when this done.
10. What is a diffraction grating?
11. Why is the sample compartment a box isolated from the rest of the instrument?
12. What features of a single-beam spectrophotometer differentiate it from a double-beam spectrophotometer? In what situations is it used over a double-beam instrument?
13. What is meant by the precalibration of a single-beam spectrophotometer?
14. For what kind of experiment is the use of a single-beam spectrophotometer a slow, tedious process? Why is it slow and tedious?
15. Can a single-beam spectrophotometer be used for rapid scanning? Explain.
16. What problem exists if you want to use a single-beam instrument for precise work, even if it has a rapid scanning capability?
17. What is a light chopper? What is a beamsplitter?
18. What is a double-beam spectrophotometer?
19. What are two designs of a double-beam spectrophotometer and what are the advantages and disadvantages of each?
20. What are the advantages of a double-beam instrument over a single-beam instrument?
21. Why is a double-beam spectrophotometer preferred for rapid scanning?
22. Imagine an experiment in which the molecular absorption spectrum of a particular chemical species is needed. Which instrument is preferred—a single-beam or double-beam instrument? Why?
23. What is a photomultiplier tube? Describe what it does and how it works.
24. Define photocathode, dynode, and photodiode.
25. What is a diode array spectrophotometer? What advantages does it have?

26. Can cuvettes used for visible spectrophotometry be made of plastic? Explain.
27. What does it mean to say that the cuvettes used must be matched? Explain why the cuvettes need to be matched.
28. A certain pair of cuvettes is not matched. Name at least two things that may be different about them—things over which the analyst does not have any control.
29. What does it mean to say that a given substance interferes with a spectrochemical analysis?
30. What does it mean to say that there is a deviation from Beer's law?
31. What percent transmittance range and what absorbance range are considered to be the optimum working ranges for spectrochemical measurements?
32. Elaborate on the maintenance procedures required for spectrophotometers.
33. How can the wavelength calibration be checked on a spectrophotometer?
34. Discuss troubleshooting procedures for spectrophotometers for: 1) failure of electrical components, 2) unexpectedly high absorbance readings, and 3) an unexpected peak in an absorption spectrum.
35. Is the wavelength of fluorescence longer or shorter than the wavelength of absorption? Explain your answer with the help of an energy level diagram.
36. Is the energy of absorption more or less than the energy of fluorescence? Explain your answer with the help of an energy level diagram.
37. Why are there two wavelength selectors in a fluorometer?
38. What is meant by a right-angle configuration in a fluorometer and why is this instrument constructed this way?
39. A fluorometer differs in basic design from an absorption spectrophotometer in two major ways.
    (a) What are they?
    (b) Explain the need for these design differences.
40. Draw a diagram of a fluorescence instrument and point out the differences between it and the basic single-beam absorption spectrophotometer.
41. When performing a quantitative analysis procedure using fluorometry, what parameter is measured by the instrument and plotted vs. concentration?
42. Why is it that benzene ring systems such as riboflavin can be analyzed by fluorometry while uncomplexed metal ions cannot?
43. Fluorometry is more selective and more sensitive than absorption spectrometry. Tell what is meant by selectivity and sensitivity.
44. Under what circumstances would you want to use a fluorometric procedure rather than an absorption spectrometry procedure? Briefly explain.
45. What two advantages does fluorometry have over absorption spectrophotometry?
46. The fact that very few chemical species fluoresce works to both the advantage and disadvantage of fluorometry as a quantitative technique. Explain this.
47. How do absorption spectrophotometry and fluorometry compare in terms of:
    (a) Instrument design
    (b) Sensitivity
    (c) Applicability
48. Compare the energy transitions caused by infrared light absorption to those caused by UV-VIS light absorption.
49. Compare the cuvettes used for UV-VIS spectrophotometry to the liquid sampling cells used for IR spectrometry.
50. Compare typical IR spectra with typical UV-VIS spectra. What is similar and what is different?
51. Why is it that the infrared spectrum of an organic compound is more useful than the UV-VIS spectrum for qualitative analysis?
52. What do the letters FTIR stand for?
53. Fill in the blanks with either "double-beam dispersive" or "FTIR," whichever correctly completes the statement:
    (a) Uses an interferometer _____

    (b) Uses a moveable mirror _____

    (c) Utilizes the slit/dispersing element/slit arrangement _____

    (d) Measures the IR spectrum in a matter of seconds _____

    (e) Disperses IR light into a spray of wavelengths, a narrow band of which is selected _____

    (f) A Fourier transform is performed on the data in order to obtain the absorption data at all wavelengths _____

54. What is an interferometer and what is its function in an FTIR instrument?

55. Answer with either "double-beam dispersive" or "FTIR" in order to indicate which instrument design is described by each statement.

    (a) Utilizes a moveable mirror to create an interference pattern

    (b) Records the IR spectrum in a matter of seconds

    (c) Is an older design that is nearly obsolete

    (d) Utilizes an interferometer

56. Why are inorganic compounds useful as sample windows and matrix material for infrared analysis?

57. What is a neat liquid?

58. Name three methods for mounting a liquid sample in the path of the light in an infrared spectrometer.

59. What are the differences between a sealed cell, a demountable cell, and a sealed demountable cell for liquid sampling?

60. What defines the pathlength in a sealed demountable IR cell?

61. Describe the pressure–vacuum method of filling an IR cell equipped with inlet and outlet ports.

62. Name two separate problems that occur when a sample for infrared analysis is contaminated with water.

63. Why is a water-contaminated sample a problem for IR cells?

64. What problem is encountered with spectra interpretation when a solution of the analyte in a particular solvent is analyzed? Why does the use of carbon tetrachloride as the solvent minimize this problem?

65. Name at least five methods by which the IR spectrum of a solid can be obtained.

66. What is a KBr pellet?

67. Give two reasons why the potassium bromide used to make the KBr pellet must be dry.

68. Explain the use of a hydraulic press for IR laboratory work.

69. What is a Nujol mull?

70. What advantage do the KBr pellet method and the reflectance methods have over the solution and mineral oil mull methods for solids?

71. Why is it that the presence of mineral oil in the mineral oil mull usually does not cause a problem with spectral interpretation?

72. Name three reflectance methods for solid sampling and tell what the differences are.

73. Briefly describe the diffuse reflectance method for solids.

74. Can gases be measured by IR spectrometry? Explain.

75. For each of the four IR spectra shown in Figures 8.32 to 8.35, tell which of the six bonds in Table 8.1 are present and which are absent. When finished, suggest a total structure for each that fits your observations in each case.

76. Look at the three infrared spectra in Figures 8.36 to 8.38 and answer the following questions:

    (a) Are any of the spectra that of an alcohol? If so, which? What absorption pattern(s) at what wavelength(s) identifies an alcohol?

    (b) Are any of the spectra that of a compound containing a benzene ring? If so, which? What three absorption patterns at what wavelengths show that a compound has a benzene ring?

    (c) Are any of the spectra that of a compound containing only carbons and hydrogens? If so, which? Benzene rings contain only carbons and hydrogens. Might the spectrum or spectra you chose for your answer above indicate a benzene ring? (Tell what absorption patterns are present or not present that would support your answer.)

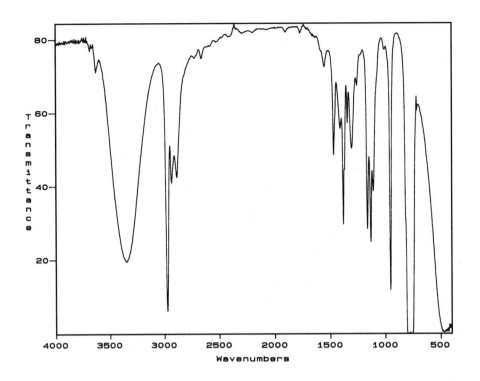

**FIGURE 8.32**    An infrared spectrum for problem 75.

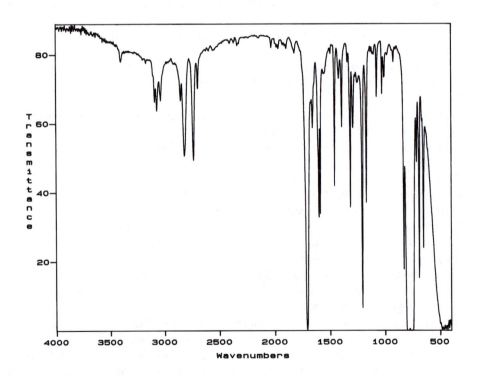

**FIGURE 8.33**    An infrared spectrum for problem 75.

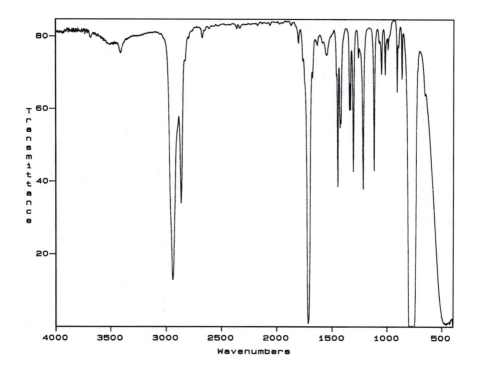

**FIGURE 8.34** An infrared spectrum for problem 75.

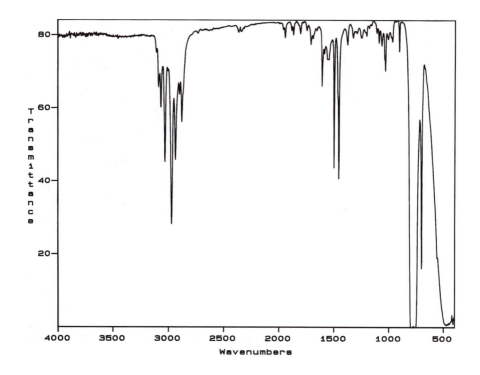

**FIGURE 8.35** An infrared spectrum for problem 75.

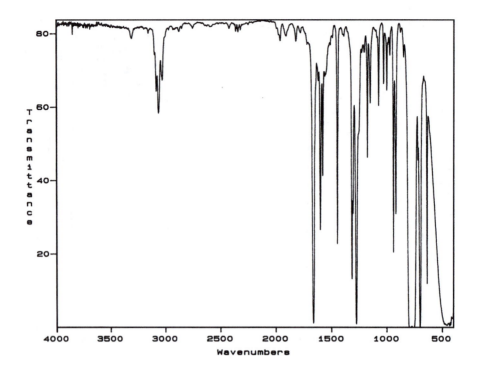

**FIGURE 8.36**   An infrared spectrum for problem 76.

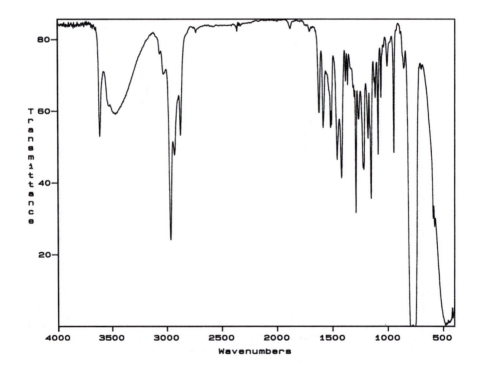

**FIGURE 8.37**   An infrared spectrum for problem 76.

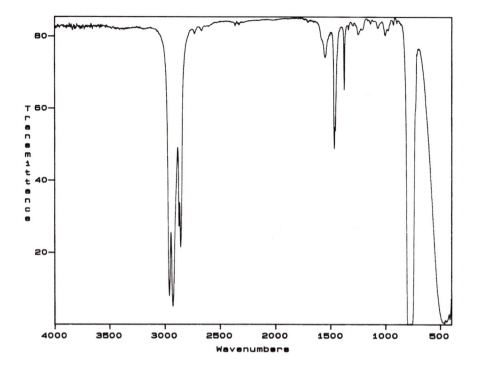

**FIGURE 8.38**   An infrared spectrum for problem 76.

(d) Do any of the three compounds have a carbonyl group (–C=O)? If so, which? What absorption pattern, or lack thereof, at what wavelength supports your answer?

(e) The three compounds whose spectra appear in these figures are n-heptane, 3-methylphenol, and benzophenone. Which is which?

77. Describe an experiment in which the *quantity* of an analyte is determined by infrared spectrometry.

78. **Report**

Find a real-world UV or IR spectrometric analysis in a methods book (AOAC, USP, ASTM, etc.), in a journal, or on a website and report on the details of the procedure according to the following scheme:

(a)  Title

(b) General information, including the type of material examined, the name of the analyte, sampling procedures, sample preparation procedures, and whether it is qualitative or quantitative

(c) Specifics, including whether a Beer's law plot is used, any chemical reactions needed in advance of the measurements, what wavelength is used (or how the absorption and transmittance spectra are used), how the data are gathered, concentration levels for standards, how the standards are prepared, and potential problems

(d) Data handling and reporting

(e) References

# *9*

# Atomic Spectroscopy

## 9.1 Review and Comparisons

**Atomic spectroscopy** refers to the absorption and emission of ultraviolet–visible (UV-VIS) light by atoms and monoatomic ions and is conceptually similar to the absorption and emission of UV-VIS light by molecules, as discussed in Chapter 8. Some differences have already been discussed in Chapter 7 (see Sections 7.4.3 and 7.4.4). In brief, absorption spectra for atoms are due to atoms in the gas phase absorbing UV-VIS light from a light source and are characterized by very narrow wavelength absorption bands called **spectral lines**. These very narrow bands result from energy transitions between electronic energy levels in which there are no vibrational levels superimposed (because covalent bonds are required for vibrational levels to exist and unbound gas phase atoms have no covalent bonds).

Similarly, emission spectra for atoms and monoatomic ions are due to gas phase atoms or monoatomic ions in excited states emitting UV-VIS light because they are returning to the ground state. Emission spectra are also characterized by very narrow wavelength bands (also called spectral lines) because they too involve electronic energy levels in which there are no vibrational levels superimposed. In fact, for a given kind of atom, the emission lines occur at the same wavelengths as the absorption lines because they involve the same energy levels. See Figure 9.1.

Besides these differences in electronic energy levels and spectra, atomic spectroscopy differs from UV-VIS molecular spectroscopy in the following ways:

1. Analysis for atoms means that atomic spectroscopy is limited to the elements. In fact, the key word for atomic spectroscopy is *metals*. The vast majority of methods involving atomic spectroscopy are methods for determining metals.
2. Sample preparation schemes for atomic spectroscopy usually place the metals in water solution. Since metals are present as ions in water solution, atomic spectroscopy methods must have a means for converting metal ions into free gas phase ground state atoms (a process called **atomization**) in order to measure them. Most of these methods involve a large amount of thermal energy.
3. Perhaps the most noticeable difference in instrumentation is the sample container used for atomic spectroscopy. This container is the source of the thermal energy needed for the conversion of ions in solution to atoms in the gas phase (and hence is called an **atomizer**) and in no way resembles a simple cuvette. Recall, for example, the brief discussion and photograph of the flame container in Section 7.5.
4. The need for and use of thermal energy as outlined above has resulted in the invention of a number of separate and distinctly different atomizer and instrument designs, albeit based on the same theory, under the heading of atomic spectroscopy.
5. **Spectral line sources** are used as light sources in atomic absorption instruments rather than the **continuum sources** used for UV-VIS molecular absorption instruments, and several atomic emission techniques require no light source at all apart from the thermal energy source.
6. Since the analytes for atomic spectroscopy are severely limited (elements only), compared to the large number of molecular and complex ion analytes for UV-VIS molecular absorption spectrometry,

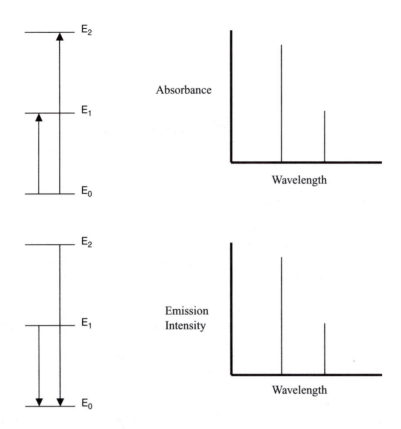

**FIGURE 9.1**    Top left, an energy level diagram depicting energy transitions for atoms from the ground state to two excited states. Top right, a line absorption spectrum with two spectral lines resulting from the transitions in the top left. Bottom left, an energy level diagram depicting energy transitions for atoms from two excited states to the ground state. Bottom right, a line emission spectrum with two spectral lines resulting from the transitions in the bottom left. Note what is plotted on the y-axis in the two spectra.

the wavelengths used for quantitation are well known and do not require the analyst to ever first measure the absorption or emission spectra. Hence, although atomic absorption and emission spectra are atomic fingerprints in the same context that molecular absorption and emission spectra are molecular fingerprints and can be used for identification purposes, they are not needed for quantitative work. The wavelengths needed are simply looked up in tables of spectral lines available in instrument manufacturer's literature or handbooks.

## 9.2    Brief Summary of Techniques and Instrument Designs

The most important of the techniques and instrument designs requiring thermal energy (item 4 above) are **flame atomic absorption** (flame AA), **graphite furnace atomic absorption** (graphite furnace AA), and **inductively coupled plasma atomic emission** (ICP). Each of these will be discussed in detail in this chapter. Less important techniques and instrument designs that require thermal energy are **flame emission** and **atomic fluorescence**. Two that do not require thermal energy or require only minimal thermal energy are the **cold vapor mercury system** and the **metal hydride generation** technique. One that requires electrical energy is **arc and spark emission**. Each of these techniques will be briefly discussed.

Flame AA utilizes a large flame as the atomizer. A photograph of this atomizer was first shown in Figure 7.20 and is reproduced here in Figure 9.2. The sample (a solution) is drawn into the flame by a vacuum mechanism that will be described. The atomization occurs immediately. The light beam for the

**FIGURE 9.2**   The flame atomizer.

**FIGURE 9.3**   A graphite furnace module. This module is approximately 6 in. across.

absorption measurements is directed through the width of the flame, right to left or left to right depending on the instrument orientation, and measured.

The atomizer for graphite furnace AA is, as the name implies, a furnace. It is actually a small graphite tube that can be quickly electrically heated to a very high temperature. A relatively small volume of the sample solution is placed in the tube either manually (with a micropipet) or drawn in with a vacuum mechanism. This furnace is then electrically brought to a very high temperature in order to atomize the sample inside. The light beam is directed through the tube, in which there is a cloud of atoms, and measured. Instruments that utilize a flame for the atomizer can also utilize the graphite furnace. It is a matter of replacing the flame module with the furnace module and lining it up with the light beam. Figure 9.3 is a photograph of a graphite furnace module.

ICP is an emission technique, which means that it does not use a light source. The light measured is the light emitted by the atoms and monoatomic ions in the atomizer. The ICP atomizer is an extremely hot plasma, which is a high-temperature ionized gas composed of electrons and positive ions confined by a magnetic field. The extremely high temperature means that the atoms and monoatomic ions undergo sufficient excitation (and de-excitation) such that relatively intense emission spectra result. The sample is drawn in with a vacuum mechanism that will be described. The intensity of an emission line is measured and related to concentration.

## 9.3 Flame Atomic Absorption

Flame AA is the oldest of the three techniques and has been widely used for many years. The instruments are relatively inexpensive and methods have been well tested and are well understood.

### 9.3.1 Flames and Flame Processes

When solutions of metal ions are introduced into a flame, several processes occur in extremely rapid succession. Refer to Figure 9.4 throughout the following discussion.

First, the solvent evaporates, leaving behind formula units of the formerly dissolved salt. Next, dissociation of the formula units of salt into atoms occurs—the metal ions atomize, or are transformed into atoms. Then, if the atoms are easily raised to excited states by the thermal energy of the flame, a **resonance** process occurs in which the atoms resonate back and forth between the ground state and the excited states.

Actually, only a small percentage of the atoms (less than 0.1%, depending on the temperature of the flame) are found in the excited state at any particular moment. As these atoms drop back to the ground state (a natural process), the emission spectrum is emitted. For those atoms that are easily excited under these conditions, the emitted wavelengths are in the visible region of the spectrum, and the entire flame takes on a color characteristic of the element that is in the flame. It is a characteristic color because each element has its own characteristic line spectrum, the atomic fingerprint, resulting from the particular energy transitions that element has. Easily excited elements include sodium (yellow), calcium (orange), lithium (red), potassium (violet), and strontium (red). It is possible to quantitate these elements using the flame emission technique mentioned in Section 9.2; this will be discussed later.

The unexcited atoms in the flame (the 99.9%) are available to be excited by a light beam. Thus, as shown in Figure 9.4, a light source is used and a light beam is directed through the flame. The experiment is a Beer's law experiment, the width of the flame being the pathlength.

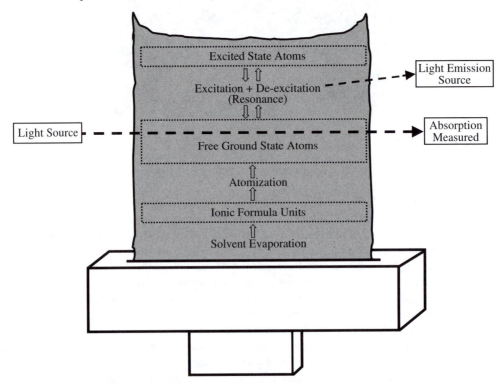

**FIGURE 9.4** An illustration of the processes occurring in a flame into which a solution of a metal ion has been introduced. See text for discussion.

# THE AUTO-SAMPLER

A common accessory seen on many laboratory instruments, including flame AA, graphite furnace AA, and ICP instruments, in busy analytical laboratories is the so-called **auto-sampler**, such as the one pictured here. Although the design can vary, the typical auto-sampler is a circular carousel having round slots to hold vials that contain the solutions to be analyzed—standards, samples, and controls. In operation, the carousel slowly rotates and a robotic sampling tube dips into the vials one at a time and draws sample solutions into the instrument to which it is attached. This automatic action allows the operator to perform other tasks while the measurements are being taken. The auto-sampler pictured here is attached to a graphite furnace AA. The sampling tube and robotic arm can be seen in the upper left in the photograph. See also Workplace Scenes 12.1, 12.2, and 16.2.

We implied in the discussion above that the flame temperature is important. This is true for both the atomization and excitation processes.

All flames require both a fuel and an oxidant in order to exist. Bunsen burners and Meker (Fisher) burners utilize natural gas for the fuel and air for the oxidant. The maximum temperature of an air–natural gas flame is 1800 K. However, in order to sufficiently atomize most metal ions and excite a sufficient number of atoms of the more easily excited elements for quantitation purposes, and therefore achieve a desirable sensitivity for quantitative analysis by atomic spectroscopy, a hotter flame is desirable. Most flames used for flame AA are air–acetylene flames—acetylene the fuel, air the oxidant. A maximum temperature of 2300 K is achieved in such a flame. Ideally, pure oxygen with acetylene would produce the highest temperature (3100 K), but such a flame suffers from the disadvantage of a high burning velocity, which decreases the completeness of the atomization and therefore actually lowers the sensitivity. Nitrous oxide ($N_2O$) used as the oxidant, however, produces a higher flame temperature (2900 K) while burning at a low rate. Thus, $N_2O$–acetylene flames are fairly popular. The choice is made based on which flame temperature–burning velocity combination works best with a given element. Since all elements have been studied extensively, the recommendations for any given element are available. Table 9.1 lists most metals and the recommended flame for each. Air–acetylene flames are the most commonly used.

## 9.3.2 Spectral Line Sources

As mentioned in item 5 of Section 9.1, the light sources used in atomic absorption instruments are sources that emit spectral lines. Specifically, the spectral lines used are the lines in the line spectrum of the analyte being measured. These lines are preferred because they represent the precise wavelengths that are needed for the absorption in the flame, since the flame contains this analyte. Spectral line sources emit these wavelengths because they themselves contain the analyte to be measured, and when the lamp is on, these internal atoms are raised to the excited state and emit their line spectrum when they return

**TABLE 9.1**   Listing Showing Which Oxidant Is Recommended for the Various Metals and Nonmetals Analyzed by Flame AA

| Air |
|---|
| Antimony, arsenic, bismuth, cadmium, calcium, cesium, chromium, cobalt, copper, gold, indium, iridium, iron, lead, lithium, magnesium, manganese, mercury, nickel, palladium, platinum, potassium, rhodium, rubidium, ruthenium, selenium, silver, sodium, tellurium, thallium, zinc |

| Nitrous Oxide |
|---|
| Aluminum, barium, beryllium, boron, dysprosium, erbium, europium, gadolinium, gallium, germanium, hafnium, holmium, lanthanum, molybdenum, neodymium, niobium, phosphorus, praseodymium, rhenium, samarium, scandium, silicon, strontium, tantalum, terbium, thulium, tin, titanium, tungsten, uranium, vanadium, ytterbium, yttrium, zirconium |

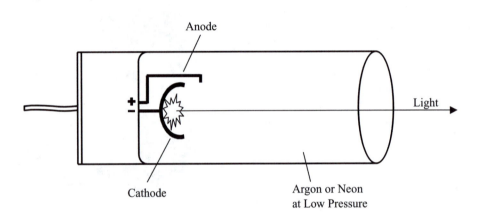

**FIGURE 9.5**   The hollow cathode lamp.

to the ground state. It is this emitted light, the same light that will be absorbed in the flame, that is directed at the flame.

### 9.3.2.1   Hollow Cathode Lamp

The most widely used spectral line source for atomic absorption spectroscopy is the hollow cathode lamp. An illustration of this lamp is shown in Figure 9.5. The internal atoms mentioned above are contained in a **cathode**, a negative electrode. This cathode is a hollowed cup, pictured with a C shape in the figure. The internal excitation and emission process occurs inside this cup when the lamp is on and the anode (positive electrode) and cathode are connected to a high voltage. The light is emitted as shown.

The lamp itself is a sealed glass tube and is filled with an inert gas, such as neon or argon, at a low pressure. This inert gas plays a role in the emission process, as shown in Figure 9.6. Argon (Ar) is the inert gas shown in this figure. When the lamp is on, the argon atoms undergo ionization, as shown. The positively charged argon ions ($Ar^+$ in the figure) then crash into the interior of the cathode because of the strong attraction of their positive charge for the negative electrode. The violent contact of the argon ions with the interior surface of the cathode causes **sputtering** (transfer of surface atoms in the solid phase to the gas phase due to the collisions). Additional collisions of the argon ions with the gas phase metal atoms (M in the figure) cause the metal atoms to be raised to the excited state ($M^*$ in the figure); when they drop back to the ground state, the light is emitted.

The hollow cathode lamp must contain the element being measured. A typical atomic absorption laboratory has a number of different lamps in stock that can be interchanged in the instrument, depending on what metal is being determined. Some lamps are multielement, which means that several different specified kinds of atoms are present in the lamp and are all raised to the excited state when the lamp is on. The light emitted by such a lamp consists of the line spectra of all the kinds of atoms present. One may think that the lines of the elements other than the analyte might interfere with the measurement of

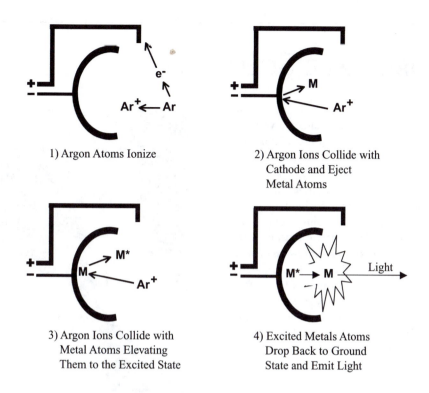

1) Argon Atoms Ionize

2) Argon Ions Collide with
Cathode and Eject
Metal Atoms

3) Argon Ions Collide with
Metal Atoms Elevating
Them to the Excited State

4) Excited Metals Atoms
Drop Back to Ground
State and Emit Light

**FIGURE 9.6** The process occurring in a hollow cathode lamp providing the light used in an atomic absorption instrument. See text for discussion and definitions of symbols.

the analyte. This is not usually a problem because a monochromator is used following the flame to isolate the spectral line of the analyte. See Workplace Scene 9.1.

### 9.3.2.2 Electrodeless Discharge Lamp

A light source known as the electrodeless discharge lamp (EDL) is sometimes used. In this lamp, there is no anode or cathode. Rather, a small, sealed quartz tube containing the metal or metal salt and some argon at low pressure is wrapped with a coil for the purpose of creating a radio frequency (RF) field. The tube is thus inductively coupled to an RF field, and the coupled energy ionizes the argon. The generated electrons collide with the metal atoms, raising them to the excited state. The characteristic line spectrum of the metal is thus generated and is directed at the flame just as with the hollow cathode tube. EDLs are available for 17 different elements. Their advantage lies in the fact that they are capable of producing a much more intense spectrum and thus are useful for those elements whose hollow cathode lamps can produce only a weak spectrum.

## 9.3.3 Premix Burner

The burner used for flame AA is a premix burner. It is called that because all the components of the flame (fuel, oxidant, and sample solution) are premixed, as they take a common path to the flame. The fuel and oxidant originate from pressurized sources, such as compressed gas cylinders, and their flow to the burner is controlled at an optimum rate by flow control mechanisms that are part of the overall instrument unit.

The sample solution is aspirated (drawn by vacuum) from its original container through a small tube and converted to an aerosol, or fine mist, prior to the mixing. These steps (aspiration and conversion to an aerosol) are accomplished with the use of a **nebulizer** at the head of the mixing chamber. The nebulizer is a small (3 cm long, 1 cm in diameter) adjustable device resembling the nozzle one places on the end of a garden hose to create a water spray. There are two inlets to the nebulizer. One inlet is a small plastic

# WORKPLACE SCENE 9.1

Fertilizers and animal feeds used in the agriculture industry contain a variety of minerals that are the source of important chemical elements. At the Nebraska State Agriculture Laboratory, flame atomic absorption spectroscopy is used to analyze samples for many of these elements, including calcium, magnesium, manganese, sodium, copper, zinc, and phosphorus. Each element must be within a standard range of concentrations based on the guaranteed analysis label for the product in question. If the concentration is found to be outside this range, the supplier is notified that the product is in violation of state law and the product must be either relabeled or sent back to the manufacturer.

Cindy Wiebeck of the Nebraska State Agriculture Laboratory aspirates calcium standards into the atomic absorption flame in order to calibrate the instrument for the analysis of fertilizer for calcium. In the background, note the hollow cathode lamps held in a rotating turret in the instrument.

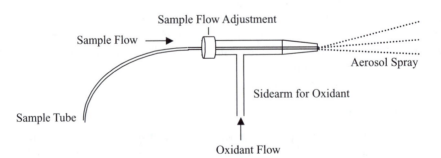

**FIGURE 9.7**    An illustration of a nebulizer.

tube (the "garden hose") protruding from the back, and the other is a side arm entering at a right angle. With the oxidant flowing, the small plastic tube is dipped into the sample solution by the operator in order to initiate the aspiration. The oxidant, due to its rapid flow through the sidearm, creates suction in the sample line (called the Venturi effect) and draws the sample solution from its container and into the mixing chamber. See Figure 9.7.

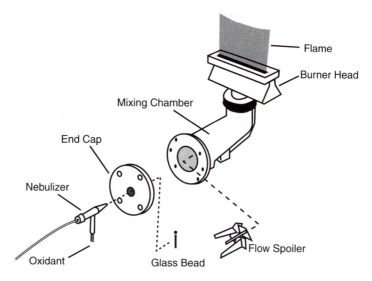

**FIGURE 9.8** An illustration of the complete premix system.

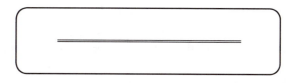

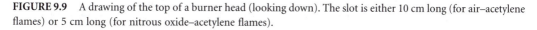

**FIGURE 9.9** A drawing of the top of a burner head (looking down). The slot is either 10 cm long (for air–acetylene flames) or 5 cm long (for nitrous oxide–acetylene flames).

The aerosol spray, as it emerges from the nebulizer, contains variable-sized solution droplets. The larger droplets are undesirable because solvent evaporation from such droplets in the flame is inefficient. Hence, they should be removed. In addition, there must be some sort of mechanism for the thorough mixing of the aerosol spray, the oxidant, and the fuel in the mixing chamber. For these reasons, an impact device is used in the mixing chamber. This impact device is typically a flow spoiler, or series of baffles, placed in the path of the flow. It may also include a glass bead positioned near the tip of the nozzle. See Figure 9.8 for an illustration of the complete system: nebulizer, glass bead, flow spoiler, and burner head. With this design, the larger droplets in the nebulizer spray fall to the bottom of the mixing chamber. A drain hole (shown in the figure) allows the accumulated solution to drain out to a waste receptacle. Approximately 90% of aspirated aqueous solution is actually eliminated through the drain and never reaches the flame.

The burner head itself is interesting. The flame must be wide (for a long pathlength), but does not have to be deep. In other words, the burner has a long slot cut in it (5 or 10 cm long) to shape the flame in a wide but shallow contour. Typically, the flame is 10 cm wide for air–acetylene and 5 cm wide for $N_2O$–acetylene. See Figure 9.9.

## 9.3.4 Optical Path

The optical path for flame AA is arranged in this order: light source, flame (sample container), monochromator, and detector. Compared to UV-VIS molecular spectrometry, the sample container and monochromator are switched. The reason for this is that the flame is, of necessity, positioned in an open area of the instrument surrounded by room light. Hence, the light from the room can leak to the detector and therefore must be eliminated. In addition, flame emissions must be eliminated. Placing the monochromator between the flame and the detector accomplishes both. However, flame emissions that are the

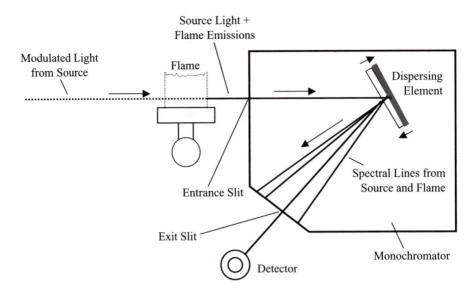

**FIGURE 9.10**    An illustration of the single-beam flame atomic absorption optical path. The modulated light from the source is created either by a chopper or through electronic pulsing.

same wavelength as the source wavelength are not blocked by the monochromator but also must be eliminated. These emissions exist because the analyte atoms are in the flame and are emitting the same wavelength as the wavelength from the source that the analyst wishes to measure. Modulating the light from the source either through electronic pulsing or by placing a light chopper (no mirror) in the optical path ahead of the flame solves this dilemma. The detector sees alternating signals of: 1) source light combined with flame emissions, and 2) flame emissions only. Knowing the frequency of the modulation, the detector is able to eliminate the flame emissions by subtraction.

The optical path for flame AA can be either single-beam or double-beam. However, as we explained in Section 9.1, item 6, the absorption spectra are seldom measured. Thus, the disadvantages of a single-beam instrument are not as severe. In a given experiment, the wavelength is seldom changed and so there is no need to recalibrate with the blank in the tedious manner described for UV-VIS molecular absorption instruments in the previous chapter. However, the problems of source drift and fluctuations still exist, although in modern instruments, these are minimized by improved electronics. Thus the single-beam design is often used. The optical path for a single-beam instrument, including chopper, is shown in Figure 9.10.

The double-beam design uses a beamsplitter (or chopper) to divert the light from the source around the flame. The two beams are then joined again before entering the monochromator. The second beam does not pass through a blank. If it did, it would require a second flame matched to the first. As stated previously, atomic absorption analyses seldom require wavelength scans, so there would be no advantage to such a design. The primary function of a double-beam design is to eliminate problems due to source drift and noise. A side advantage is that source warm-up time is eliminated since changes in intensity during warm-up are immediately compensated, and thus very rapid changeover of lamps in automated instruments is possible. Both the sample beam and the reference beam may be chopped so that the detector can differentiate and eliminate emissions from the flame. See Figure 9.11.

## 9.3.5    Practical Matters and Applications

### 9.3.5.1    Slits and Spectral Lines

There is more than one spectral line in the line spectrum of an element and therefore more than one line to choose from when setting the monochromator. For each element, there is one line that gives the optimum absorptivity for that element, and this line is therefore the most sensitive and useful. This line

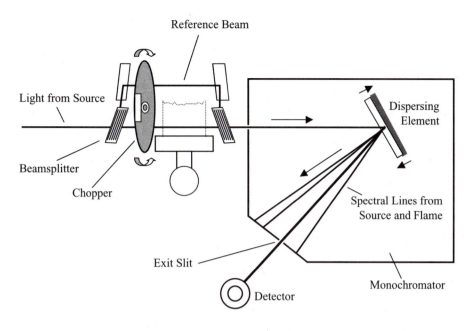

**FIGURE 9.11** An illustration of a double-beam optical path. See text for explanation.

is called the **primary line** and is the line most often chosen—the monochromator control is set at that wavelength. Other lines are called **secondary lines**. A secondary line may be chosen if, for some reason, the primary line is inappropriate, such as if there is another element in the sample with a line too close to the primary line of the analyte. Primary and secondary lines are listed in manufacturers' literature. The more automated instruments automatically set the monochromator control to the primary line as part of the instrument setup process during power-up.

The potential problem of one line being too close to another line is solved by a slit control. The wider a slit, the greater the bandpass and the more apt an interfering line would be allowed to enter and exit the monochromator and be measured along with the primary line. Atomic absorption monochromators are typically operated with bandpass values ranging from 0.2 to 2.0 nm. The required bandpass for a given element is selected by choosing the appropriate slit width setting. The slit width settings, like the wavelength settings, are listed in manufacturers' literature. For a given element, the operator sets the slit value using a slit control on the instrument. The more automated instruments automatically set the slit to the optimum value as part of the instrument setup during the power-up process. The slit value represents the bandpass for both the entrance and exits slits. If there is an interfering line at the optimum setting, either the slit width is narrowed or a secondary line is chosen. Both of these choices result in less desirable sensitivity.

### 9.3.5.2 Hollow Cathode Lamp Current

The electrical power or current the hollow cathode lamp receives is adjustable. There is an optimum setting (the operating current) representing the most intense light without significantly shortening the life of the lamp. The operator sets the value of the current when setting up the instrument. In the more automated instruments, the lamp current is set automatically during the power-up process. Setting the lamp current above the designated operating current shortens the life of the lamp, an item of significant expense. Setting the lamp current below the operating current may produce satisfactory results and extends the life of the lamp. It is recommended that a logbook be maintained for each lamp in the laboratory so that intensity, sensitivity, and hours used records are kept.

### 9.3.5.3 Lamp Alignment

The hollow cathode lamp must be aligned properly to provide the optimum intensity throughout the optical path. This is done during setup by monitoring the light energy reaching the detector while moving the lamp horizontally and vertically. Each lamp is slightly different in shape and internal alignment, and thus this external alignment procedure must be performed each time the lamp is changed. Highly automated instruments can perform this function automatically.

### 9.3.5.4   Aspiration Rate

There is a sample flow adjustment nut on the nebulizer (see Figure 9.7). Turning this nut in or out changes the flow rate of the sample solution through the nebulizer and can even result in oxidant being drawn back and bubbled through the sample solution if it is not adjusted properly. The setting is optimized by monitoring the absorbance of an analyte standard during the adjustment. A maximum absorbance would indicate an optimum flow rate.

### 9.3.5.5   Burner Head Position

The burner head can be adjusted vertically, horizontally (toward and away from the operator), and rotationally. Initially, it should be adjusted vertically so that the light beam passes approximately 1 cm above the center slot and horizontally so that the light beam passes directly through the center of the flame, end to end. All three positions can be optimized by monitoring the absorbance of an analyte standard while making the adjustments. A maximum absorbance would indicate the optimum position.

### 9.3.5.6   Fuel and Oxidant Sources and Flow Rates

The sources of acetylene, nitrous oxide, and sometimes air are usually steel cylinders of the compressed gases purchased from specialty gas or welders' gas suppliers. Thus, several compressed gas cylinders are usually found next to atomic absorption instrumentation and the analyst becomes involved in replacing empty cylinders with full ones periodically. Safety issues relating to storage, transportation, and use of these cylinders will be addressed in Section 9.3.7. The acetylene required for atomic absorption is a purer grade of acetylene than that which welders use.

There is an optimum fuel and oxidant flow rate to the flame, or, more precisely, an optimum fuel–oxidant flow rate ratio. If the flame is oxidant-rich, it is too cool. If it is fuel-rich, it is too hot. Again, monitoring the absorbance of an analyte standard while varying the flow rates helps to find the optimum ratio. Instrument manufacturers' literature will also provide assistance. Safety issues relating to the proper flow rate of these gases will be addressed in Section 9.3.7.

### 9.3.5.7   Standard Curve

Quantitative analysis in flame atomic absorption spectroscopy utilizes Beer's law. The standard curve is a Beer's law plot, a plot of absorbance vs. concentration. The usual procedure, as with other quantitative instrumental methods, is to prepare a series of standard solutions over a concentration range suitable for the samples being analyzed, i.e., such that the expected sample concentrations are within the range established by the standards. The standards and the samples are then aspirated into the flame and the absorbances read from the instrument. The Beer's law plot will reveal the useful linear range and the concentrations of the sample solutions. In addition, information on useful linear ranges is often available for individual elements and instrument conditions from manufacturers' and other literature.

## 9.3.6   Interferences

Interferences can be a problem in the application of flame AA. Interferences can be caused by chemical sources (chemical components present in the sample matrix that affect the chemistry of the analyte in the flame) or spectral sources (substances present in the flame other than the analyte that absorb the same wavelength as the analyte).

### 9.3.6.1   Chemical Interferences

Chemical interferences are the result of problems with the sample matrix. For example, viscosity and surface tension affect the aspiration rate and the nebulized droplet size, which, in turn, affect the measured absorbance. The most useful solution to the problem is **matrix matching**, matching the matrix (all components except the analyte) of standards (and blank) as these solutions are prepared with that of the sample (e.g., same solvent, same components at the same concentrations, etc.) so that the same matrix effect is seen in all solutions measured and therefore becomes inconsequential. Of course, the complete qualitative and quantitative composition of the sample matrix must be known in order to match it by measuring out individual components for the preparation of the standards and blank. Unfortunately, the analyst usually does not have such sweeping information about the sample.

An alternative is the **standard additions method**. In this method, a certain volume of the sample solution itself is present in the same proportion in all standard solutions. It is equivalent to adding standard amounts of analyte to the sample solution, hence the term standard additions. This solves the interference problem because the sample matrix is always present at the same component concentrations as in the sample—the matrix is matched. In addition, the sample solution components need not be identified.

The best way to accomplish this is to prepare standards in the usual way—add increasing volumes of a standard analyte solution to a series of volumetric flasks (include zero added)—but also add a volume of the sample solution to each before diluting to the mark with solvent. Thus you would have a series of standards in which the concentration of analyte *added* would be known, the smallest concentration added being zero. Exactly how much sample solution is used and what concentration added values would be prepared would be dictated by what concentration levels, with additions, would produce a linear standard curve. In any case, a diluted sample matrix is present in each standard and the matrices are matched. A disadvantage is that it is impossible to prepare a blank with a matched matrix. Thus, a pure solvent blank, or other approximation, must be used.

The standard curve takes on a slightly different look in the standard additions method. It is a plot of absorbance vs. concentration *added* (to the sample), rather than just concentration. The y-axis in such a plot is not at its true position. It is offset to the right by the concentration in the solution with zero-added concentration, which is the sample solution. The concentration in this solution is the concentration sought. To show it on the graph, the standard curve is extrapolated to intersect with the x-axis. This intersection point is the true position of the y-axis, and the concentration of the unknown is represented by the length of the x-axis between the two y-axes. See Figure 9.12. The precise concentration in the zero-added solution is found using the equation of the straight line.

Extrapolation means that the curve is extended lower than the lowest standard into a range in which its linearity has not been tested. This is another disadvantage of this method.

One notable chemical interference occurs when atomization is hindered due to an unusually strong ionic bond between the ions in the ionic formula unit. A well-known example occurs in the analysis of a sample for calcium. The presence of sulfate or phosphate in the sample matrix along with the calcium suppresses the reading for calcium because of limited atomization due to the strong ionic bond between calcium and the sulfate and phosphate ions. This results in a low reading for the calcium in the sample in which this interference exists. The usual solution to this problem is to add a substance to the sample that would chemically free the element being analyzed, calcium in our example, from the interference.

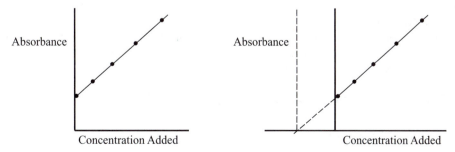

**FIGURE 9.12** Illustration of the manipulation of the standard curve in the standard additions method. Left, the standard curve before extrapolation. Note that the line does not pass through the origin. Right, after extrapolation. The vertical dashed line represents the true y-axis.

With our calcium example, the substance that accomplishes this is lanthanum. Lanthanum sulfates and phosphates are more stable than the corresponding calcium salts, and thus the calcium is free to atomize when lanthanum is present. Thus, analyses for calcium usually include addition of a lanthanum salt to the sample and standards.

Other chemical interferences may be overcome by changing oxidants (air or nitrous oxide are the choices) or by changing the fuel–oxidant flow ratio, giving a lower or higher flame temperature.

### 9.3.6.2  Spectral Interferences

Spectral interferences are due to substances in the flame that absorb the same wavelength as the analyte, causing the absorbance measurement to be high. The interfering substance is rarely an element, however, because it is rare for another element to have a spectral line at exactly the same wavelength, or near the same wavelength, as the primary line of the analyte. However, if such an interference is suspected, the analyst can tune the monochromator to a secondary line of the analyte to solve the problem.

Absorption due to the presence of light-absorbing molecules in the flame and light dimming due to the presence of small particles in the flame are much more common spectral interferences. Such phenomena are referred to as **background absorption**. To solve the problem, **background correction** techniques have been employed. These techniques are designed to isolate the background absorption and subtract it out. A common technique is to use a deuterium arc lamp, a light source that we previously mentioned for its use in UV-VIS molecular spectrophotometers, to direct a continuum light beam through the optical path simultaneously with the light from the spectral line source (the hollow cathode lamp or electrodeless discharge lamp). Both light beams are absorbed by the sample in the flame, but the continuum source absorption is measured separately (with the use of chopping frequencies) and subtracted from the total in a manner similar to that of flame emissions.

### 9.3.7  Safety and Maintenance

There are a number of important safety considerations regarding the use of atomic absorption equipment. These center around the use of highly flammable acetylene, as well as the use of a large flame, and the possible contamination of laboratory air by combustion products.

All precautions relating to compressed gas cylinders must be enforced—the cylinders must be secured to an immovable object, such as a wall; they must have approved pressure regulators in place; they must be transported on approved carts; etc. Tubing and connectors must be free of gas leaks. There must be an independently vented fume hood in place over the flame to take care of toxic combustion products. Volatile flammable organic solvents and their vapors, such as ether and acetone, must not be present in the lab when the flame is lit.

Precautions should be taken to avoid flashbacks. Flashbacks result from improperly mixed fuel and air, such as when the flow regulators on the instrument are improperly set or when air is drawn back through the drain line of the premix burner. Manuals supplied with the instruments when they are purchased give more detailed information on the subject of safety.

Finally, periodic cleaning of the burner head and nebulizer is needed to ensure minimal noise level due to impurities in the flame. Scraping the slot in the burner head with a sharp knife or razor blade to remove carbon deposits and removing the burner head for the purpose of cleaning it in an ultrasonic cleaner bath are two commonplace maintenance chores. The nebulizer should be dismantled, inspected, and cleaned periodically to remove impurities that may be collected there.

## 9.4  Graphite Furnace Atomic Absorption

### 9.4.1  General Description

Another very important atomic absorption technique that we briefly mentioned in Section 9.2 is graphite furnace atomic absorption. As stated in the earlier section, the atomizer for graphite furnace AA is a small hollow graphite cylinder (see Figure 9.13) that can be quickly electrically heated to a very high

**FIGURE 9.13** The small graphite cylinder that is the graphite furnace. The small hole is for introducing samples to the interior of the cylinder.

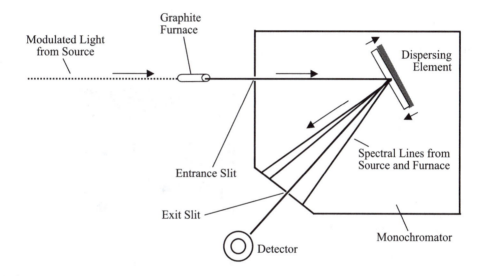

**FIGURE 9.14** An illustration of the light path for a single-beam atomic absorption spectrophotometer with a graphite furnace as the atomizer.

temperature. A small volume of the sample solution is placed inside the cylinder. This furnace is electrically brought to a very high temperature in order to atomize the sample inside. The cylinder becomes filled with an atomic cloud, the light beam is directed through it and absorbed, and the absorption is measured.

Instruments that utilize a flame for the atomizer can also utilize the graphite furnace. It is a matter of replacing the flame module with the furnace module and lining it up with the light beam. The same light source (usually a hollow cathode lamp) and the same optical path are used. See Figure 9.14 and compare with Figure 9.10.

The furnace module is much more than a small graphite cylinder, however (refer back to Figure 9.3). The furnace must be protected from thermal oxidation (conversion of graphite carbon to oxides of carbon in the presence of air at high temperatures). Thus, an inert gas (argon) must be circulated inside and outside the cylinder to protect it from contact with air. A source of argon and purge controls must be provided. Flowing water must be circulated through an enclosed housing for the purpose of rapidly cooling the furnace between runs. A source of cold water and the appropriate plumbing must be provided. In addition, a relatively high-voltage power source is used and electrical contacts to the tube must be provided.

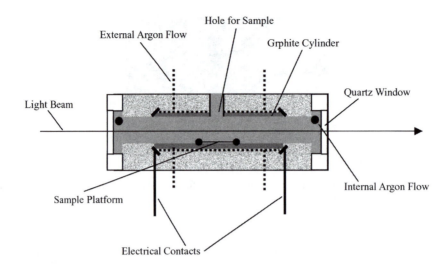

FIGURE 9.15    The graphite furnace assembly. Channels for water cooling are not shown.

FIGURE 9.16    A photograph looking inside the graphite cylinder. The sample platform is visible.

The graphite cylinder must therefore be encased in a rather complex maze of argon flow channels (for flow both internal and external to the furnace), channels for water flow, and electrical contacts. Despite the apparent clutter, the unit must still be open on each end for the light beam to pass through. Quartz windows cap each end of the assembly, and a hole is included for introducing the sample to an internal platform through the hole in the cylinder. See Figure 9.15 for an illustration. The internal platform is shown in Figure 9.16.

The temperature of the furnace is programmed to take into account several effects of heating to optimize measurement. First, since the evaporation of the solvent occurs at a much lower temperature than the atomization, the temperature is initially ramped to the solvent boiling point for a short period. After this, the temperature is ramped still further to a temperature at which organic matter is charred, vaporized, and swept away. It is important to note that the flowing argon continuously sweeps through the interior of the furnace such that solvent molecules and the "smoke" from the charring are swept from the furnace.

Finally, the temperature is stepped to the atomization temperature (approximately 2500 K). It is at this temperature that the atomic cloud is present in the furnace and the absorbance is measured. This absorbance is monitored over the period of time for which the atomic cloud is in residence and is usually recorded as an absorbance vs. time peak. The flow of argon continues to sweep through the interior of the furnace, and therefore the absorbance signal is a transient one. With some units, it is possible to momentarily stop the argon flow to increase the residence time of the atomic cloud. See Figure 9.17 for an illustration of a temperature vs. time program and the absorbance vs. time signal that results. A final step (not illustrated in the figure) to a clean-out temperature prepares the furnace for the next sample.

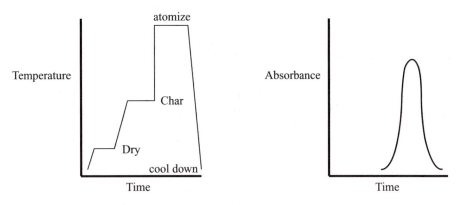

**FIGURE 9.17** An illustration of a temperature program for a graphite furnace experiment (left), and the absorbance signal that results (right). The absorbance signal corresponds to the third temperature plateau. See text for a more detailed explanation.

### 9.4.2 Advantages and Disadvantages

The flame atomizer continuously sweeps a given analyte concentration across the light path such that the number of ground state atoms present in the light path at a given moment is comparatively small. This means that this atomizer is not highly efficient, does not produce a result that is as sensitive as may be desired, and, at a minimum, requires several milliliters of sample solution to obtain a reliable absorbance reading. The graphite furnace, on the other hand, confines the atomic vapor produced from a small sample volume (microliters) to a very small space in the path of the light, resulting in a much larger concentration of ground state atoms in this space than in that of the flame. This results in a much more sensitive measurement, meaning that much smaller concentrations in the sample solution can be detected and measured. The detection limit is also much improved over flame AA (see Section 9.7).

On the other hand, absorbance readings from the graphite furnace are less reproducible due to a less-controlled pathlength and physical presence in the furnace. In addition, interferences (background absorptions) are more common and more severe. So-called **matrix modifiers** are often used to control interferences. These are chemicals added to the sample to assist in the separation of the analyte from the matrix prior to atomization such that the interference is not present at the time of atomization. Background absorptions can also be eliminated with the use of **background correction** techniques, such as the use of a continuum source, as we discussed for flame AA. However, a more common background correction procedure for graphite furnace AA is the **Zeeman background correction** technique. In this technique, a powerful magnetic field is used to shift the energy levels of atoms and molecules and therefore to shift the wavelengths that are absorbed. Pulsing the magnetic field on and off allows subtraction of the background absorption. See Workplace Scene 9.2.

## 9.5 Inductively Coupled Plasma

Besides flame AA and graphite furnace AA, there is a third atomic spectroscopic technique that enjoys widespread use. It is called **inductively coupled plasma spectroscopy**. Unlike flame AA and graphite furnace AA, the ICP technique measures the emissions from an atomization/ionization/excitation source rather than the absorption of a light beam passing through an atomizer.

Recalling Figure 9.4, we know that thermal energy sources, such as a flame, atomize metal ions. But we also know that that these atoms experience resonance between the excited state and ground state such that the emissions that occur when the atoms drop from the excited state back to the ground state can be measured. While there are several techniques that measure such emissions, including flame emissions

# WORKPLACE SCENE 9.2

At SACHEM Inc. in Clayburne, Texas, ICP coupled to mass spectrometry (see Section 10.5.7) has been an important tool for determining elements present at trace levels in their ultrapure products. However, graphite furnace AA has better precision and detection limits for some elements and there is no interference from argon isotopes like there is with ICP–mass spectrometry (ICP-MS) (see Section 10.5.7). There can be an interference because ICP-MS uses argon gas and the argon isotopic ion with mass number 40 ($^{40}Ar^+$) interferes with the measurement of other isotopic ions, such as the calcium isotopic ion, also with mass number 40 ($^{40}Ca^+$), during the mass spectrometric aspect of the analysis. Graphite furnace AA, on the other hand, provides excellent sensitivity for calcium without the argon problem. Thus, graphite furnace AA and ICP-MS complement each other very well in this laboratory.

Robert Krause of SACHEM Inc. analyzes trace elements in ultrapure chemical products at the sub-parts per billion level using graphite furnace AA spectrophotometry.

(see Section 9.6), ICP is the most important because the excitation source is much hotter, resulting not only in atomizaton, but also in ionization and many more emissions, and more intense emissions, than from the flame.

ICP does not use a flame. Rather, as briefly mentioned in Section 9.2, this atomization/ionization/excitation source is a high-temperature plasma. A plasma, in this context, is a gaseous mixture of atoms, cations, and electrons that is directed though an induced magnetic field, causing coupling of the ions in the mixture with the magnetic field, hence the name inductively coupled plasma. The interaction with the magnetic field causes atoms and ions to undergo extreme resistive heating. The ICP torch resembles a flame as the plasma emerges from the magnetic field. See Figure 9.18. The temperature of the ICP torch is in excess of 6000 K.

Refer to Figure 9.19 as you read the following more detailed description. A quartz tube serves as the conduit for the plasma through the magnetic field. The quartz tube is actually three concentric tubes in one. The plasma (and sample) path is the center tube. The other tubes contain flowing auxiliary and coolant gases. The plasma itself consists of a flowing stream of argon gas that has been partially ionized with the use of an in-line electrical discharge device (not shown in the figure) prior to reaching the

**FIGURE 9.18**   A photograph of the ICP torch. From http://icp-oes.com. Reproduced with permission of Jobin-Yvon Horiba, France.

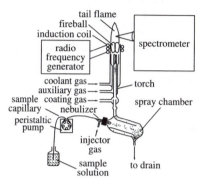

**FIGURE 9.19**   An illustration of the ICP system. See text for description. From http://icp-oes.com. The Spectroscopy Net, 2002. Reproduced with permission of Jobin-Yvon Horiba, France.

magnetic field. The magnetic field is induced with the use of a radio frequency generator and an induction coil by passing an alternating radio frequency current through the coil. This coil consists of a three- or four-turn heavy copper wire. This coil is also visible in Figure 9.18.

The sample is drawn from its container with a peristaltic pump to a nebulizer. There are several designs for the nebulizer, but it performs the same function as the flame AA nebulizer, converting the sample solution into fine droplets (with the larger droplets flowing to a drain) that flow with the argon to the torch. The emissions are measured by the spectrometer at a particular zone in the torch, often called the viewing zone or analytical zone, as shown in Figure 9.19.

Unlike a flame, in which only a very limited number of metals emit light because of the low temperature, virtually all metals present in a sample emit their line spectrum from the ICP torch. Not only does this make for a very broad application for ICP, but it also means that a given sample may undergo very rapid and simultaneous multielement analysis. With this in mind, it is interesting to consider the options for the optical path for the ICP instrument.

One is the sequential optical path in which a monochromator is used and the emissions are measured one wavelength after the other sequentially by rotating the dispersing element while the individual lines emerge from the exit slit. In this case, one phototube measures all emission lines. See Figure 9.20. The advantages are: 1) all wavelengths of all elements can be utilized, and 2) spectral interferences can be avoided by careful wavelength selection. The disadvantage is that it is slower than the alternative (see below).

The other option is comparable to the diode array design used with UV-VIS molecular absorption instruments previously described in Chapter 8. This design is known as the simultaneous direct reading polychromator design, in which there are a number of exit slits and phototubes for measuring a number of lines at once. See Figure 9.21. The wavelengths to be measured are set by the manufacturer. The advantages of this design are: 1) it is fast, since all the desired wavelengths are measured simultaneously, and 2) it is good for samples with known matrices because it can be preset to the optimum wavelengths. The disadvantages are: 1) the wavelength choices are changeable, but expensive, and 2) spectral interferences can be a problem.

To repeat a statement in our opening paragraph of this section, ICP measures the emission of light from the atomization/ionization/excitation source (ICP torch) rather than absorption of light by atoms

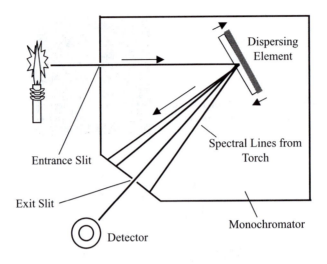

**FIGURE 9.20**  An illustration of the optical path for an ICP instrument that utilizes a monochromator for the sequential measurement of spectral lines.

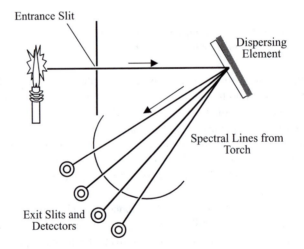

**FIGURE 9.21**  An illustration of the optical path of an ICP instrument that utilizes the simultaneous direct reading polychromator for the measurement of spectral lines.

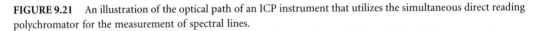

in this source, as was the case with flame AA and graphite furnace AA. For this reason, it should be stressed that quantitative analysis is not based on Beer's law. Rather, the intensity of the emissions is proportional to concentration—the higher the concentration, the more intense the emission. Thus for quantitative analysis, a series of standard solutions that bracket the expected concentration in the unknown are prepared and the intensity for each is measured. The standard curve is then a plot of intensity vs. concentration, as shown in Figure 9.22.

The advantage of ICP is that the emissions are of such intensity that it is usually more sensitive than flame AA (but less sensitive than graphite furnace AA). In addition, the concentration range over which the emission intensity is linear is broader. These two advantages, coupled with the possibility of simultaneous multielement analysis offered by the direct reader polychromator design, make ICP a very powerful technique. The only real disadvantage is that the instruments are more expensive. See Workplace Scene 9.3.

# WORKPLACE SCENE 9.3

The determination of trace metallic elements is important to material processing, recovery, environmental sampling, and remediation programs at Los Alamos National Laboratory. Samples are received in the liquid or solid state and are prepared and analyzed according to approved Environmental Protection Agency (EPA) methods or procedures. This analysis procedure involves the standardization and calibration of the inductively coupled plasma atomic emission (ICP-AES) system with known concentrations of analytes of interest. The calibration and standardization data obtained are then used to determine the concentration of trace metallic elements (analytes) in the samples. During the treatment, preparation, segregation, and analysis of samples using the ICP-AES system, significant quantities of residue and chemical waste are generated. Thus the proper disposal of waste and residue according to waste disposal is also an important part of this effort.

MaryAnn Abeyta, a technician with the Los Alamos National Laboratory, analyzes samples using an IRI-SICP-AES spectrometer.

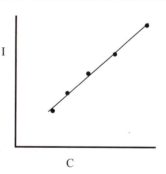

**FIGURE 9.22** The standard curve for ICP quantitative analysis: intensity vs. concentration.

ICP has been coupled with another analytical technique known as **mass spectrometry**. This technique, as well as the technique coupling it with ICP, will be described in Chapter 10.

## 9.6 Miscellaneous Atomic Techniques

### 9.6.1 Flame Photometry

Flame photometry is the name given to the technique that measures the intensity of the light emitted by analyte atoms in a flame. It is the oldest of all the atomic techniques. It is not highly applicable because of the low temperature of the flame. Only a handful of elements can be measured with this technique, including sodium, potassium, lithium, calcium, strontium, and barium. The technique was formerly used by hospital and clinical laboratories for measuring these elements in body fluids, but has been supplanted by the use of ion-selective electrodes (see Chapter 14). Instruments designed for flame atomic absorption can be used as flame photometers. It is a mode of operation in which the hollow cathode is not used and the detector measures the flame emissions. Advantages are simple, inexpensive instruments and acceptable sensitivities for the metals listed. Disadvantages include less sensitivity than other techniques. The standard curve for quantitative analysis is intensity vs. concentration, as shown in Figure 9.22 for ICP.

### 9.6.2   Cold Vapor Mercury

Mercury is the only metal that is a liquid at ordinary temperatures. It is therefore also the only metal that has a significant vapor pressure at ordinary temperatures. For this reason, it is possible to obtain mercury atoms in the gas phase for measurement by atomic absorption without the use of thermal energy. It is a matter of chemically converting mercury ions in the sample into elemental mercury, getting it in the gas phase, and channeling it into the path of the light of an atomic absorption instrument.

Reducing agents typically used for the conversion of mercury ions in solution to elemental mercury include stannous chloride ($SnCl_2$) and sodium borohydride ($NaBH_4$). Elemental mercury, the product of the reduction, is converted to the gas phase by simply bubbling air through the solution. The mercury–air mixture is channeled to a quartz absorption cell, a long-path-length tube positioned in the path of the light from a mercury hollow cathode lamp, and the absorption is measured. The gas phase mixture is ultimately channeled to waste. The sensitivity is far greater than what can be achieved with flame AA.

### 9.6.3   Hydride Generation

A useful technique for the otherwise difficult elements of arsenic, bismuth, germanium, lead, antimony, selenium, tin, and tellurium is the hydride generation technique. This technique is similar to the cold vapor mercury technique in that a reducing agent, in this case sodium borohydride, is used to convert ions of these metals to a more useful state—in this case, to hydrides, such as $AsH_3$. Hydrides of the metals listed above are volatile and can be channeled to an absorption cell using moving air, as in the cold vapor mercury technique. Once in the absorption cell, however, the gaseous mixture must be heated in order to achieve atomization. Heating can be accomplished with a low-temperature flame or with an electrical heating system. As with cold vapor mercury, the vapor in the cell is ultimately channeled to waste. Absorbance is measured and Beer's law applies.

Advantages include a very acceptable method for the elements listed when other techniques for these elements fail. Sensitivity for these elements is also very good.

### 9.6.4   Spark Emission

A technique that utilizes a solid sample for light emission is spark emission spectroscopy. In this technique, a high voltage is used to excite a solid sample held in an electrode cup in such a way that when a spark is created with a nearby electrode, atomization, excitation, and emission occur and the emitted light is measured. Detection of what lines are emitted allows for qualitative analysis of the solid material. Detection of the intensity of the lines allows for quantitative analysis.

### 9.6.5   Atomic Fluorescence

Molecular fluorescence, described in Chapter 8, has a counterpart in atomic spectroscopy called atomic fluorescence. In order for atoms to emit light of a different wavelength than that which was absorbed, they must detour to an intermediate excited state before returning to the ground state. This can occur when the light absorbed elevates the atoms to a higher excited state than the first excited state, so that the route back to the ground state includes the first excited state and thus a fluorescence—light of less energy being emitted. The light must be measured at right angles, as with the molecular instruments. This technique is not popularly used.

## 9.7 Summary of Atomic Techniques

We have spoken frequently in this chapter about sensitivity and detection limit in reference to advantages and disadvantages of the various techniques. Sensitivity and detection limit have specific definitions in atomic absorption. **Sensitivity** is defined as the concentration of an element that will produce an absorption of 1% (absorptivity percent transmittance of 99%). It is the smallest concentration that can be determined with a reasonable degree of precision. **Detection limit** is the concentration that gives a readout level that is double the electrical noise level inherent in the baseline. It is a qualitative parameter in the sense that it is the minimum concentration that can be detected, but not precisely determined, like a blip that is barely seen compared to the electrical noise on the baseline. It would tell the analyst that the element is present, but not necessarily at a precisely determinable concentration level. A comparison of detection limits for several elements for the more popular techniques is given in Table 9.2.

The description and application of the techniques discussed in this chapter are given in Table 9.3.

**TABLE 9.2**   Detection Limits for Selected Elements ($\mu$g/L)

| Element | Flame AA | Graphite Furnace AA | ICP | Cold Vapor Hg | Hydride |
|---------|----------|---------------------|-----|---------------|---------|
| Arsenic | 150 | 0.05 | 2 | — | 0.03 |
| Bismuth | 30 | 0.05 | 1 | — | 0.03 |
| Calcium | 1.5 | 0.01 | 0.05 | — | — |
| Copper | 1.5 | 0.014 | 0.4 | — | — |
| Iron | 5 | 0.06 | 0.1 | — | — |
| Mercury | 300 | 0.6 | 1 | 0.009 | — |
| Potassium | 3 | 0.005 | 1 | — | — |
| Zinc | 1.5 | 0.02 | 0.2 | — | — |

*Source*: From Perkin Elmer Instruments literature. With permission.

**TABLE 9.3**   Summary of the Techniques Described in This Chapter

| Technique | Principle | Comments |
|-----------|-----------|----------|
| Flame AA | Light absorbed by atoms in a flame is measured | A well-established technique that remains very popular for a wide variety of samples and analytes |
| Graphite furnace AA | Light absorbed by atoms in a small graphite cylinder furnace is measured | A popular and very sensitive technique that is applicable to small volumes of sample solution; precision not as good as other techniques |
| ICP | Light emitted by atoms and monoatomic ions in an inductively coupled plasma is measured | A popular technique useful over a broad concentration range; multielement analysis is possible; instruments are costly |
| Hydride generation | Light absorbed by atoms derived from hydrides that are generated by chemical reaction are measured | An excellent technique for a limited number of analytes that are difficult to measure otherwise |
| Cold vapor mercury | Light absorbed by atoms of mercury generated by chemical reaction at room temperature is measured | An excellent technique for mercury analysis |
| Flame photometry | Light emitted by atoms in a flame is measured | A well-established technique for a limited number of elements; not used much due to emergence of other techniques |
| Spark emission | Light emitted by atoms generated from a powdered solid sample in a spark source is measured | Useful for qualitative analysis of solid materials |
| Atomic fluorescence | Light emitted by atoms excited by the absorption of light in a flame is measured | Not a popular technique, but can offer sensitivity advantages |

# Experiments

## Experiment 29: Quantitative Flame Atomic Absorption Analysis of a Prepared Sample

Note: Your instructor will select the metal to be the analyte in this experiment.
*Remember safety glasses.*

1. Prepare calibration standards of the selected metal, with concentrations suggested by your instructor, from a 1000 ppm stock solution. Dilute each to the mark with distilled water and shake well.
2. Obtain an unknown from your instructor, and dilute to the mark with distilled water and shake well.
3. Select the hollow cathode lamp to be used, and install it in the instrument. Your instructor may wish to demonstrate the lamp installation first. Also, check the mixing chamber drain system to be sure it is ready.
4. A number of instrument variables need to be set prior to making measurements. These include slit, wavelength, lamp current, lamp alignment, amplifier gain, aspiration rate, burner head position, acetylene pressure, air pressure, acetylene flow rate, and air flow rate. Some instruments are rather automated in the setup process, while others are not. Your instructor will provide detailed instructions for the particular instrument you are using. Be sure to turn on the fume hood above the flame.
5. The blank for this experiment is distilled water. Follow the instructions provided for your instrument for zeroing with the blank and reading the standards, unknown, and control.
6. Follow the instructions provided for instrument shutdown.
7. Create the standard curve, using the procedure practiced in Experiment 18, by plotting absorbance vs. concentration, and determine the concentration of the unknown and control.
8. Maintain the logbook for the instrument used in this experiment. Record the date, your name(s), the experiment name or number, the correlation coefficient, and the results for the control sample. Also plot the control sample results on a control chart for this experiment posted in the laboratory.

## Experiment 30: Verifying Optimum Instrument Parameters for Flame AA

### Introduction

All atomic absorption analyses require that a number of parameters and instrument settings be optimized prior to taking readings. These include slit setting, wavelength setting, lamp current, lamp alignment, gain setting, aspiration rate, burner head position, fuel flow rate, and oxidant flow rate. In this experiment, all of these will be tentatively set for optimum use according to the manufacturer's and instructor's suggestions. Then each will be varied slightly to observe the effect on the measured absorbance. *Safety glasses are required at all times* while operating the instrument and the flame must *never* be left unattended.

### Procedure

1. Prepare a standard solution of a metal by diluting a stock standard with water to give a concentration that will give a reasonable absorbance when conditions are optimal, such as 5 ppm copper or zinc. Your instructor may have a suggestion for which metal and what concentration.
2. Set up the atomic absorption instrument according to your instructor's and the manufacturer's suggestions so that all settings are presumably optimum for the metal chosen (see steps 3 and 4 of Experiment 29). This includes all parameters mentioned in the introduction above. After aspirating the distilled water blank to zero the readout, obtain an absorbance reading for the standard solution prepared in step 1. The reading should be between 0.5 and 0.8. If it is not in this range, either adjust the concentration or expand the absorbance readout scale.
3. Now vary the parameters below one at a time as directed. After each variation, zero the readout and measure the absorbance. Return to the optimum setting before going on the next parameter. **Slit setting.** Vary the slit setting as many times as possible. Record the absorbance values in a table along with the corresponding slit settings.

**Wavelength setting.** Carefully change the wavelength setting by, say, 0.1 nm each time to several values both less than and greater than the optimum and measure the corresponding absorbances. Record the absorbance values in a table along with the wavelength settings if possible.

**Lamp current.** Vary the hollow cathode lamp current 1 mA at a time in both directions from the optimum and record the absorbance reading each time. Be careful not to exceed the maximum current rating indicated on the label of the lamp. Record the absorbance values in a table along with the corresponding values of the lamp current.

**Lamp alignment.** Vary the position of the lamp alignment screws, both the vertical and the horizontal, individually, half a turn in each direction from the optimum and record the absorbance with each half turn. After recording the data for one, either the vertical or the horizontal, reset to the optimum then vary the other. This will result in two sets of data, one for the vertical and one for the horizontal with different absorbance values recorded for each half turn away from the optimum in both directions.

**Gain setting.** If your instrument has separate readout meters for the light energy intensity and absorbance, it will be easy to perform this portion of the experiment. Simply turn the gain setting to vary the energy readout by five units at a time. Then aspirate the sample and record the absorbance readout each time. If your instrument does not have separate meters, you will need to switch back and forth from energy display to absorbance display in order to get both readings. In either case, you should record a set of absorbance readings that correspond to a set of energy readings resulting from varying the gain setting.

**Aspiration rate.** The aspiration rate is varied with an adjustment on the nebulizer. Your instructor will demonstrate how to do this. It is possible to change this setting to the extent that air will actually be blown out of the sample tube rather than sucking the solution in. Check this out ahead of time without the flame lit so that the adjustment will not be made to that extreme when the flame is lit. It is possible to measure the aspiration rate by using distilled water in a 10-mL graduated cylinder and a stopwatch. Fill a 10-mL graduated cylinder to the 10-mL line for each measurement. Then aspirate for 20 sec and determine how many milliliters are consumed. In this way, you can calculate a milliliters per minute aspiration rate for each absorbance reading. Note: If during other experiments a slower-than-expected aspiration rate is observed, this may signal an obstruction in the sample tube or nebulizer orifice and indicate that a cleaning step is in order.

**Burner head position.** The burner head can be adjusted up and down and in and out. The optimum position is such that the light beam passes squarely through the flame about 1 cm above the burner head. This can be checked out before lighting the flame using an opaque white card, or similar material, placed perpendicular to the light beam. For this experiment, vary the position of the burner head in both directions from the optimum, both up and down and in and out, by a quarter turn of the control knob each time. You should then have two sets of data, one for the up-and-down variable and one for the in-and-out variable, showing absorbance readings varying with the number of quarter turns both clockwise and counterclockwise from the optimum.

**Fuel flow rate.** Your instructor will show you how to vary the fuel flow rate to the burner head and also how to read the flow rate on the flow meter. Measure absorbance values at ten different flow rates, selecting a flow rate range that will maintain a flame while bracketing what the manufacturer's literature or your instructor may suggest as the optimum.

**Oxidant flow rate.** Repeat the above, but for the oxidant flow rate rather than the fuel flow rate. Use extreme caution here, since a flashback may result from too much or too little oxidant flow.

4. Follow the posted instructions for instrument shutdown.
5. Make nine graphs, one for each of the varied parameters, plotting absorbance vs. the parameter setting or reading and comment on what was discovered in each case. Also comment on what would happen in each case if the analyte metal were changed to some other metal. Would the optimum settings found be different or the same? Explain.

## Experiment 31: The Analysis of Soil Samples for Iron Using Atomic Absorption

Note: Soil samples may be dried and crushed ahead of time. If not, you may report the results on an as received basis.

It is presumed that you have previously performed Experiment 29 or another introductory experiment on the atomic absorption equipment. If you have not done this, carefully read through the Experiment 29 procedure so that you are aware of certain precautions and instrument setup requirements.

*Remember safety glasses.*

1. Prepare 500 mL of extracting solution by diluting 25 mL of 1 $N$ HCl and 12.5 mL of 1 $N$ H$_2$SO$_4$ to 500 mL with distilled water.
2. Weigh 5.00 g of each soil sample to be tested into *dry* 50-mL Erlenmeyer flasks. Pipet 20 mL of extracting solution into each flask. Stopper and shake on a shaker for at least 20 min.
3. Filter all samples simultaneously into small, *dry* beakers using *dry* Whatman #42 filter paper, transferring only the supernatant. Do not rinse the soil samples in the filters, since that would dilute the sample extracts by an inexact amount.
4. While waiting for the samples to be shaken and filtered, prepare standards that are 0.5, 1.0, 3.0, and 5.0 ppm iron. Use a 1000 ppm iron solution and prepare one 100 ppm intermediate stock solution. Use 50-mL volumetric flasks, and use the extracting solution as the diluent. A control may also be provided. Dilute it to the mark with the extracting solution.
5. Set up the atomic absorption instrument and obtain absorbance readings for all solutions using the extracting solution for the blank. Follow the instructions provided for instrument shutdown.
6. Create the standard curve, using the procedure practiced in Experiment 18, by plotting absorbance vs. concentration and determine the concentration of the unknown and control. Calculate the parts per million (ppm) Fe in the soil by the following calculation:

$$\text{ppm Fe (extractable)} = \text{ppm (in solution)} \times 4$$

7. Maintain the logbook for the instrument used in this experiment. Record the date, your name(s), the experiment name or number, the correlation coefficient, and the results for the control sample. Also plot the control sample results on a control chart for this experiment posted in the laboratory.

## Experiment 32: The Analysis of Snack Chips for Sodium by Atomic Absorption

Note: It is presumed that you have previously performed Experiment 29 or another introductory experiment on the atomic absorption equipment. If you have not done this, carefully read through the Experiment 29 procedure so that you are aware of certain precautions and instrument setup requirements.

*Remember safety glasses.*

1. Prepare an HCl solution by diluting 100 mL of concentrated HCl with 44 mL of water.
2. Prepare sodium standards (by diluting 1000 ppm Na) that are 1, 3, 5, and 7 ppm. A control may also be provided. These should be approximately 0.5% HCl by volume.
3. Prepare samples (maximum of two) by grinding and weighing 5 g of each into 500-mL Erlenmeyer flasks. Add 50 mL of the HCl solution prepared in step 1 to each. Bring each to a boil on a hot plate, and then simmer for 5 min. Cool and transfer the supernatents for each to separate 100-mL volumetric flasks. Dilute to the mark with distilled water and shake. Filter each through Whatman #1 filter paper into *dry* 250-mL Erlenmeyer flasks. Pipet 1 mL of each of these extracts into other clean 100-mL volumetric flasks, and dilute to the mark with water. Save the original extracts in case more dilutions are needed.
4. Set up the atomic absorption instrument and obtain absorbance readings for all solutions using the extracting solution for the blank. Follow the instructions provided for instrument shutdown.

5. Create the standard curve, using the procedure practiced in Experiment 18, by plotting absorbance vs. concentration and determine the concentration of the unknown and control. Calculate the milligrams of Na in 5 g of chips by the following calculation:

$$\text{ppm found} \times 0.1 \times 100$$

Also calculate the milligrams of sodium in one bag of chips.

6. Maintain the logbook for the instrument used in this experiment. Record the date, your name(s), the experiment name or number, the correlation coefficient, and the results for the control sample. Also plot the control sample results on a control chart for this experiment posted in the laboratory.

## Experiment 33: The Atomic Absorption Analysis of Water Samples for Iron Using the Standard Additions Method

Note: It is presumed that you have previously performed Experiment 29 or another introductory experiment on the atomic absorption equipment. If you have not done this, carefully read through the Experiment 29 procedure so that you are aware of certain precautions and instrument setup requirements.
*Remember safety glasses.*

1. Prepare 100 mL of a 100 ppm stock iron solution from the available 1000 ppm.
2. Prepare a set of five standards in 25-mL flasks such that the iron concentrations added are 0, 1, 3, 5, and 7 ppm. A control may also be provided. Do not dilute to the mark until step 3.
3. Pipet 20.00 mL of the sample into each flask from step 2 and then dilute each to the mark with distilled water and shake.
4. Prepare the atomic absorption instrument for iron analysis, and measure all standards, the control, and the water sample. Use iron-free hard water for a blank.
5. Follow the instructions provided for instrument shutdown.
6. Create the standard curve, using the procedure practiced in Experiment 18, by plotting absorbance vs. concentration. Obtain the concentration of the control in the usual manner. Obtain the concentration of the sample by extrapolating to zero absorbance.
7. Maintain the logbook for the instrument used in this experiment. Record the date, your name(s), the experiment name or number, the correlation coefficient, and the results for the control sample. Also plot the control sample results on a control chart for this experiment posted in the laboratory.

## Experiment 34: The Determination of Sodium in Soda Pop

Note: It is presumed that you have previously performed Experiment 29 or another introductory experiment on the atomic absorption equipment. If you have not done this, carefully read through the Experiment 29 procedure so that you are aware of certain precautions and instrument setup requirements.
*Remember safety glasses.*

1. Prepare 100 mL of a 100 ppm sodium solution from the available 1000 ppm. Obtain soda pop samples and degas approximately 5 to 10 mL of each.
2. From the 100 ppm Na stock, prepare standards of 1, 3, 5, and 7 ppm in 25-mL volumetric flasks. A control sample may also be provided. Dilute to the mark with distilled $H_2O$.
3. Pipet 1 mL of each soda pop sample into separate 25-mL volumetric flasks and dilute to the mark with distilled water. Shake.
4. Obtain absorbance values for all standards, samples, and the control using an atomic absorption instrument.
5. Follow the instructions provided for instrument shutdown.
6. Create the standard curve, using the procedure practiced in Experiment 18, by plotting absorbance vs. concentration and determine the concentration of the unknown and control. Multiply the concentrations by 25 to get the parts per million Na in the soda pop. Also calculate the milligrams of sodium in one 12-oz can of the soda pop.

7. Maintain the logbook for the instrument used in this experiment. Record the date, your name(s), the experiment name or number, the correlation coefficient, and the results for the control sample. Also plot the control sample results on a control chart for this experiment posted in the laboratory.

# Questions and Problems

1. Absorption spectra for atoms are characterized by spectral lines. Explain this statement.
2. (a) With help of an energy level diagram, tell what is the source of the lines in the line spectra of metals.
   (b) We speak of line spectra often as both emission spectra and absorption spectra. Name two common sources of atomic emission spectra.
3. Explain why emission lines occur at the same wavelengths as absorption lines.
4. What are three major differences between UV-VIS molecular spectrophotometry and atomic absorption spectrophotometry?
5. What is an atomizer? Identify at least four atomizers used in atomic spectroscopy.
6. Name two techniques that are atomic absorption techniques and two that are atomic emission techniques.
7. Atomization occurs in a flame as one of several processes after a solution of a metal ion is aspirated. What other processes occur and in what order?
8. (a) The percentage of atoms in a typical air–acetylene flame that are in the excited state at any point in time is less than 0.01%. What does this have to do with the fact that flame AA is a useful technique?
   (b) Define resonance as it pertains to atomic spectroscopy.
9. What temperature can be achieved by each of the following flames?
   (a) Air–natural gas
   (b) Air–acetylene
   (c) $N_2O$–acetylene
   (d) Oxygen–acetylene
10. Which of the flames listed in question 9 are most commonly used. Why?
11. Even though an oxygen–acetylene flame can produce the hottest temperature, what disadvantage does it possess that limits its usefulness in practice?
12. There is no monochromator placed between the light source and the flame in an atomic absorption experiment. Why is this?
13. The hollow cathode lamp must contain the metal to be analyzed in the cathode. Explain.
14. A typical atomic absorption laboratory has many hollow cathode lamps available and interchanges them in the instrument frequently. Explain why this is in terms of the processes occurring in the lamp and in the flame.
15. What is an EDL?
16. What is a nebulizer? Describe its use in conjunction with a premix burner.
17. Describe the design and use of the premix burner in flame AA.
18. Why must a premix burner have a drain line attached? What safety hazard exists because of this drain line and how do we deal with it?
19. What purpose does the light chopper serve in an atomic absorption instrument?
20. How does a single-beam atomic absorption instrument differ from a double-beam instrument? What advantages does one offer over the other?
21. How does a double-beam atomic instrument differ from a double-beam molecular instrument?
22. Since the sample "cuvette" is the flame located in an open area of the atomic absorption instrument, rather than a glass container held in the light tight box in the case of molecular instruments, how is room light prevented from reaching the detector and causing an interference?
23. Define what is meant by the primary and secondary lines in a line spectrum.

24. Name five instrument controls that must be optimized when setting up an atomic absorption spectrophotometer for measurement. Explain why each must be optimized.
25. Does Beer's law apply in the case of flame AA? Explain.
26. What is the difference between a chemical interference and a spectral interference?
27. What is meant by matrix matching? Why is this important in a flame AA experiment?
28. What is the standard additions method and why does it help with the problem of chemical interferences?
29. What are two disadvantages of the standard additions method?
30. Why is lanthanum used in an analysis for calcium?
31. Define background absorption and background correction. How does the use of a continuum light source help with background correction?
32. What are three safety issues with regard to flame AA and how are they dealt with?
33. Give a brief description of the graphite furnace method of atomization.
34. Describe exactly how the atomic vapor is produced in a graphite furnace.
35. What is the difference in the optical paths of graphite furnace AA and flame AA?
36. The graphite furnace assembly utilizes argon gas, cold water, and a source of high voltage. Explain.
37. Describe the temperature program applied to a graphite furnace and explain each of the processes involved.
38. Why must the graphite furnace be protected from air?
39. Why is the absorbance signal developed in the case of the graphite furnace AA technique said to be transient?
40. What are the advantages and disadvantages of the graphite furnace atomizer?
41. What is meant by Zeeman background correction?
42. What do the letters ICP stand for? Is the ICP technique more closely related to atomic absorption or flame photometry? Explain.
43. Describe the ICP analysis method in detail.
44. What are the advantages of the ICP technique?
45. What are the advantages of the following atomic techniques over the standard flame atomic absorption?
    (a) Graphite furnace AA
    (b) ICP
46. Why is the cold vapor mercury technique good only for mercury?
47. Describe the hydride generation technique. Why is it useful?
48. Compare atomic absorption (both flame and graphite furnace), ICP, flame photometry, cold vapor mercury, hydride generation, atomic fluorescence, and spark emission in terms of:
    (a) the process measured
    (b) instrumental components and design
    (c) data obtained
49. Match each statement with one or more of the following: flame AA, flame photometry, atomic fluorescence, ICP, graphite furnace AA, and spark emission.
    (a) The technique that uses a partially ionized stream of argon gas called a plasma
    (b) The technique in which a light source is used to excite atoms, the emission from which is then measured
    (c) The three techniques that require light to be directed at the atomized ions
    (d) The technique that utilizes a solid sample
    (e) The four techniques that measure some form of light emission by excited atoms present in an atomizer
    (f) The absorption technique in which the atoms are in the path of the light for a relatively short time
    (g) The technique in which the emitting source may reach a temperature of 6000 K

50. Answer the following questions true or false:
    (a) In flame atomic absorption, the flame serves solely as an atomizer (in addition to being the sample "container").
    (b) Line spectra are emitted by atoms in a flame.
    (c) In ICP, one needs a light source to excite the atoms in a flame.
    (d) In ICP and flame photometry, the standard curve is a plot of emission intensity vs. concentration.
    (e) In flame atomic absorption, the monochromator is placed between the light source and the flame.
    (f) The population of atoms excited by the flame is what is measured in atomic absorption.
    (g) In flame AA, the pathlength is the width of the flame.
    (h) The hollow cathode lamp has a tungsten filament.
    (i) Atoms are raised to the excited state within a hollow cathode lamp.
    (j) Detection limits are generally better with graphite furnace AA than with any other atomic technique.
    (k) A premix burner has a drain line attached.
    (l) In a premix burner, the fuel, oxidant, and sample solution meet at the base of the flame.
    (m) The graphite furnace is an example of a nonflame atomizer.
    (n) A flashback results from an improperly mixed sample solution.
    (o) The fuel used in a flame AA unit is typically natural gas.
    (p) Lanthanum is frequently used to prevent chemical interferences when analyzing for calcium.
    (q) In the method of standard additions, standard quantities of the analyte are added to the blank in increasing amounts.
    (r) The method of standard additions utilizes Beer's law.
    (s) The atomic fluorescence technique requires a light source.
    (t) The spark emission technique requires a dispersing element, but not a monochromator.
51. Differentiate between detection limit and sensitivity. Compare flame AA, graphite furnace AA, ICP, cold vapor mercury, and hydride generation in regard to applicability and detection limit.
52. **Report**
    Find a real-world flame AA, graphite furnace AA, or ICP analysis in a methods book (AOAC, USP, ASTM, etc.), in a journal, or on a website and report on the details of the procedure according to the following scheme:
    (a) Title
    (b) General information, including which atomic spectroscopic technique you chose, the type of material examined, the name of the analyte, sampling procedures, and sample preparation procedures
    (c) Specifics, including whether it is the series of standard solution procedure or standard additions; what wavelength is used; if flame AA, what fuel–oxidant combination is used; if graphite furnace AA, what furnace program is used; if ICP, what specific instrument design is suggested; slit settings; how the data are gathered; concentration levels for standards; how the standards are prepared; and potential problems
    (d) Data handling and reporting
    (e) References

# 10

# Other Spectroscopic Methods

## 10.1  Introduction to X-Ray Methods

X-ray methods include x-ray diffraction, x-ray absorption, and x-ray fluorescence. X-ray diffraction is a technique for determining ultrasmall spacings in materials, such as the spacings between the atoms or ions in a crystal structure, or the thickness of a thin electroplated material. An example of the former is in soil laboratories in which the minerals in various soils need to be characterized. X-ray absorption is limited in application, but has been used to determine heavy elements in a matrix of lighter elements, such as determining lead in gasoline. X-ray fluorescence is much more popular and is used to determine elements in a wide variety of solid materials.

X-rays have extremely short wavelengths and high energy. A source of a wide band of x-ray wavelengths is the **x-ray tube**, in which x-rays are generated by bombarding a metal anode with high-energy electrons. These electrons are of sufficient energy as to cause inner-shell electrons (n = 1, or K shell) to be ejected from the atoms, which means the atoms are ionized. Subsequently, higher-energy electrons drop into the vacancies thus created and a sort of domino effect follows, with electrons from higher levels dropping into lower-level vacancies created when electrons in those levels dropped to lower levels. The x-rays are generated in this process because the loss in energy that occurs when the electrons drop into the vacancies corresponds to the energy of x-rays.

Some applications utilize specific x-ray wavelengths generated by fluorescence (see also Section 10.3). Such x-rays are generated by irradiating a given elemental material, such as copper, with the x-rays emitted by the x-ray tube described above. These x-rays are of higher quality because they do not include background radiation. The higher-energy x-rays from the x-ray tube are of sufficient energy to accomplish the ejection of electrons from the inner shells of the copper atoms, which, in turn, results in the domino effect again. However, now the wavelengths emitted constitute a high-quality x-ray line spectrum of copper and very specific x-ray wavelengths can be isolated.

In either the x-ray tube or the fluorescence source, the inner-shell electrons involved are usually those in the K shell (n = 1), the L shell (n = 2), the M shell (n = 3), and the N shell (n = 4). The absorption–emission domino effect occurs specifically as follows: 1) the complete ejection of an inner-shell electron (e.g., K or L) from an atom (ionization) occurs, creating a vacancy in this inner shell, 2) an electron from an outer shell (e.g., L, M, or N) drops to the vacancy in the inner shell, thus losing energy—the energy of a low-energy x-ray, and 3) electrons from outer shells (e.g., M or N) drop to the vacancies in the lower outer shells (e.g., L and M), also losing energy—also the energy of low-energy x-rays. See Figure 10.1.

The x-ray emissions are categorized as K, L, M, etc., emissions and as alpha ($\alpha$) and beta ($\beta$) emissions. It is a K emission if the electron drops from any higher level to the K shell. It is an L emission if it drops from any higher level to the L shell, etc. The $\alpha$ emissions are those that involve electrons that drop just one principle level, such as from the L shell to the K shell ($K_\alpha$ emissions) or from the M shell to the L shell ($L_\alpha$ emissions). The $\beta$ emissions are those in which electrons drop two levels, such as from the

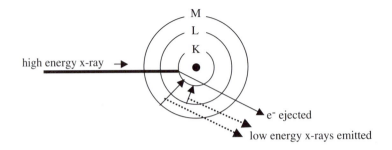

**FIGURE 10.1**    In x-ray fluorescence, high-energy x-rays are absorbed and low-energy x-rays are emitted.

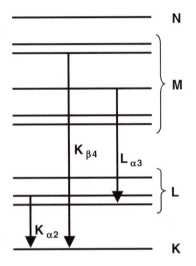

**FIGURE 10.2**    An energy level diagram showing examples of the different kinds of emissions and terminology.

M shell to the K shell ($K_\beta$ emissions) or from the N shell to the L shell ($L_\beta$ emissions). Emissions are further characterized by the subscript numbers 1, 2, 3, etc., to indicate the sublevel in the higher level (where the electron originated). Thus we have, for example, $K_{\alpha1}$, $K_{\alpha2}$, etc. See Figure 10.2. Beta emissions are higher-energy emissions (shorter wavelength) than alpha emissions because the energy jump is greater.

## 10.2   X-Ray Diffraction Spectroscopy

When a beam of light is directed at a structure that has very small regularly spaced lines or points in it, the light scatters, forming many beams of light having a particular regular pattern that appears to be the original light beam expanded in space. This scattering has come to be known as **diffraction**. The effect occurs when the wavelength of the light and the spacings between the lines or points are equivalent and constructive and destructive interference takes place due to the reflection from the different planes in the structure. The spacings between ions or atoms in crystal structures are approximately the same diameters as the wavelengths of x-rays. These spacings then serve to diffract x-rays such that a particular pattern of diffraction results from shining x-rays on a given atomic or ionic structure. A particular structure can be expected to give a characteristic, reproducible diffraction pattern. See Figure 10.3. The study of crystal structure by x-ray diffraction is an important spectroscopic technique in analytical chemistry.

In 1913, William and Lawrence Bragg, who were father and son, determined how the spacings of layers in crystal structures give different x-ray diffraction patterns. Today, it is possible to observe the diffraction

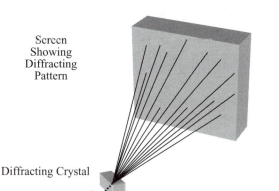

Screen
Showing
Diffracting
Pattern

Diffracting Crystal

X-ray Source

**FIGURE 10.3**    An illustration of the diffraction of an x-ray by a crystal.

pattern derived from a particular crystal and determine the structure of the crystal. The Braggs determined that the following factors are involved:

1. The distance between similar atomic planes in a crystalline structure. This distance is called the d-spacing.
2. The angle of diffraction, also called the theta angle. A diffractometer measures an angle that is double the theta angle. We refer to it as 2-theta.
3. The wavelength of the incident light.

These factors are combined in what has come to be known as **Bragg's law**.

$$n\lambda = 2d \sin\theta \qquad (10.1)$$

In this equation, n is an integer (1, 2, 3, ... ), the order of diffraction; $\lambda$ (Greek letter, lambda) is the wavelength in angstroms (Å);* d is the interplanar spacing in angstroms; and $\theta$ (Greek letter, theta) is the diffraction angle in degrees. An illustration of these parameters is given in Figure 10.4.

In order to determine the interplanar spacing, which is critical to characterizing a crystal structure, the wavelength and the diffraction angle must be known (so that d can be calculated from Equation (10.1)). The diffraction angle can be obtained from the instrument used (a **diffractometer**). In a diffractometer, in order to measure the angles, either the stage on which the sample is held is rotated or the source or detector is moved through an arc. The latter is illustrated in Figure 10.5.

The data obtained are usually displayed in a plot of signal intensity vs. 2-theta. A hypothetical example is given in Figure 10.6. Intensity peaks appear at certain 2-theta values, indicating constructive interference at that particular value. The pattern is specific to a particular interplanar spacing. Hence, crystal structures and film thicknesses can be determined from the pattern.

The wavelength used is dependent on the element used in the x-ray source. A common element for the source is copper, and the wavelength isolated from the copper emission is 1.539 Å. See Workplace Scene 10.1.

---

*1 angstrom (Å) = $10^{-10}$ m.

# WORKPLACE SCENE 10.1

Molex, Inc., of Lincoln, Nebraska, manufactures modular telephone jack connectors and other electronic connectors. These connectors are gold plated to optimize their electrical conductivity and to make them last longer. The electroplating process begins with a brass substrate. Two layers of metal are electroplated onto this substrate, a layer of nickel and a layer of gold. The reason for the nickel layer is that electroplated gold adheres better to nickel than to brass. Different products require different layer thicknesses, and thickness data must be made available to customers. For these reasons, and for internal quality assurance purposes, the laboratory at the Molex plant uses x-ray diffraction spectroscopy to measure the thicknesses of the two layers. The diffractometer is computer controlled, and a magnified view of the exact spot measured is shown on the screen as the measurement is made. A typical gold thickness is 20 $\mu$in., and a typical nickel thickness is 100 $\mu$in.

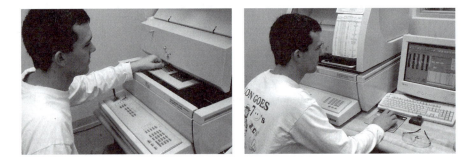

Troy Heany of the Molex, Inc., positions an electroplated specimen in the x-ray diffractometer (left) and then observes the computer image of the location on the specimen where the thickness is measured (right).

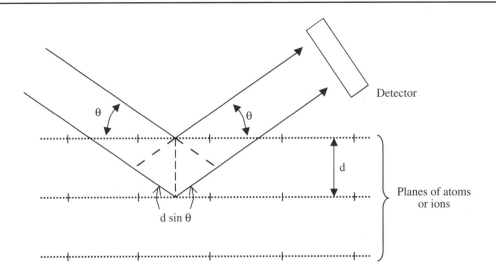

**FIGURE 10.4** An illustration of d, $\theta$, and d sin $\theta$ in Bragg's law. The distance traveled by the x-ray reflected from the second plane is greater than that reflected from the first plane by 2d sin $\theta$ in order for constructive interference to occur and a light intensity to be observed at the detector.

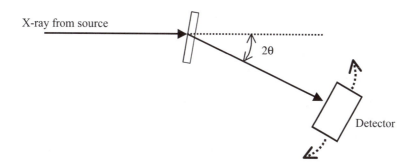

**FIGURE 10.5**    An illustration of a diffractometer in which the detector is moved through an arc in order to measure the angle. For convenience reasons, 2-theta is measured as illustrated.

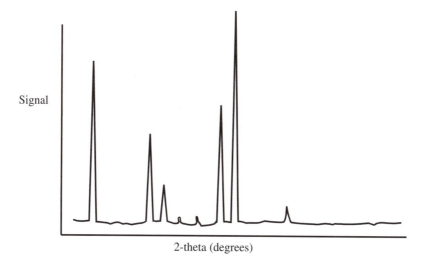

**FIGURE 10.6**    A hypothetical diffractogram, or plot of signal intensity vs. 2-theta, that results from an x-ray diffraction scan. The peaks represent theta angles at which constructive interference has occurred, resulting in a significant x-ray intensity.

## Example 10.1

The wavelength used in a particular x-ray diffraction analysis is 1.539 Å. If an intense peak occurs at a 2-theta value of 39.38°, what is the d-spacing in this structure?

### *Solution 10.1*

Since the 2-theta value is given, we must divide by 2 to obtain theta.

$$\theta = \frac{39.38}{2} = 19.69°$$

Then, using Bragg's law, assuming n = 1, we have the following:

$$n\lambda = 2d \sin\theta$$

$$1.539 = 2d \sin(19.69)$$

$$d = \frac{1.539}{2\sin(19.69)} = 2.318 \text{ Å}$$

# 10.3 X-Ray Fluorescence Spectroscopy

## 10.3.1 Introduction

The fluorescence process used in some x-ray sources, as described in Section 10.1, can also be used as an analytical tool. One can direct either high-energy electron beams or x-rays at an unknown sample and perform qualitative and quantitative analysis by making measurements on the lower-energy x-ray emissions that occur. Let us first briefly review what we have discussed to this point concerning the concept of fluorescence.

In the past two chapters, we have considered fluorescence in two different contexts: molecular fluorescence and atomic fluorescence. In molecular fluorescence, ultraviolet light is absorbed by molecules or complex ions and outermost electrons are elevated to a higher level, an excited state. These electrons drop back to the ground electronic level, but because of vibrational energy losses in both electronic states, the energy lost (emitted) is less than the energy that was absorbed (refer to Figure 8.11). In atomic fluorescence, ultraviolet light is absorbed by atoms, rather than molecules and complex ions, but again outermost electrons are elevated to a higher energy level, an excited state. However, in this case, the elevation is to an excited state higher in energy than one or more other excited states. The energy loss that subsequently occurs involves stops at the intermediate levels (before the final drop to the ground state) such that, once again, lower-energy light is emitted (Section 9.6.5). Thus fluorescence, whether molecular or atomic, is the emission of light of a lower energy than the light that was absorbed due to outermost electrons losing less energy in returning to lower levels than they had gained. The x-ray emission process described in Section 10.1 is also fluorescence because the emissions are of lower energy than the absorbed energy. The emissions represent energy drops from the higher inner electron shells to lower ones following the absorption of the very high energy used to eject the K shell electron that starts the process.

**X-ray fluorescence** is a type of atomic spectroscopy since the energy transitions occur in atoms. However, it is distinguished from other atomic techniques in that it is nondestructive. Samples are not dissolved. They are analyzed as solids or liquids. If the sample is a solid material in the first place, it only needs to be polished well, or pressed into a pellet with a smooth surface. If it is a liquid or a solution, it is often cast on the surface of a solid substrate. If it is a gas, it is drawn through a filter that captures the solid particulates and the filter is then tested. In any case, the solid or liquid material is positioned in the fluorescence spectrometer in such a way that the x-rays impinge on a sample surface and the emissions are measured. The fluorescence occurs on the surface, and emissions originating from this surface are measured.

There are two different instrument designs. These are the **energy-dispersive system** and the **wavelength-dispersive system**. In the energy-dispersive system, x-rays from the source impinge on the surface of the sample, the fluorescence occurs, and the intensity is measured by a detector. In the wavelength-dispersive system, x-rays impinge on the surface of the sample, the fluorescence occurs, and these emissions are further dispersed by a diffraction crystal (known as the analyzer crystal) of known interplanar spacing such that intensities at the 2-theta angles of diffraction are measured (see Section 10.2). Illustrations of these two designs are shown in Figure 10.7.

Each of the two designs offers advantages. The energy-dispersive system measures all elements simultaneously, meaning that the analysis is fast. However, the resolution is not optimum since the emissions do not undergo the dispersion by the analyzer crystal. Also, the energy-dispersive analysis is more sensitive. The advantage of the wavelength-dispersive system is therefore higher resolution, but the analysis is slower and less sensitive.

## 10.3.2 Applications

Elemental qualitative analysis is a popular application of x-ray fluorescence spectroscopy. The values of the wavelengths reaching the detector are indicative of what elements are present in the sample. This is so because the inner-shell transitions giving rise to the wavelengths are specific to the element. Qualitative analysis

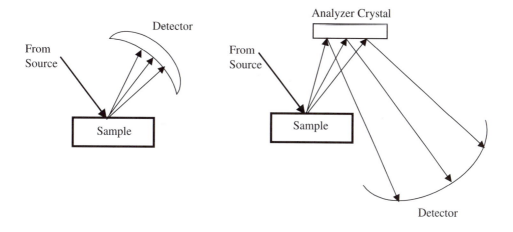

**FIGURE 10.7**  The two instrument designs for x-ray fluorescence spectroscopy. Left, the energy-dispersive system. Right, the wavelength-dispersive system.

applications include detection of impurity metals in alloys, electroplated materials, and other products from industrial manufacturing processes. In addition, the detection and characterization of minerals in geological and soil samples by x-ray fluorescence is commonplace. Also important are the analyses of corrosion products for the purpose of identifying elements that may pinpoint the cause of the corrosion (and failure) of construction materials, and the analysis of rare or archeological objects for which a nondestructive method is important.

Quantitative analysis is possible by measuring the intensity of the x-ray emissions. Thus, if quantitative analysis is important in all the above cited qualitative applications, for example, then this intensity is measured and compared to standards.

### 10.3.3  Safety Issues Concerning X-Rays

The need for safety consciousness around x-ray equipment is well known. In health and dental care facilities, rooms in which x-ray equipment is used are shielded so that x-ray technicians need only to step out of the room behind a shielded wall in order to be protected. Patients are shielded with lead aprons.

Analytical laboratories housing x-ray instrumentation must follow strict safety rules. X-ray equipment must be constructed so as to minimize radiation leaks. Operators of the equipment must be properly trained and the training documented and updated. The person designated as the safety officer must be aware of the presence of the equipment and must follow state and federal guidelines relating to registration of the equipment, proper labeling of equipment, room and hallway postings, training of employees, and communication of health risks, such as to pregnant employees. The equipment must be tested periodically to check the operation of safety devices and to check for radiation leaks.

## 10.4 Nuclear Magnetic Resonance Spectroscopy

### 10.4.1  Introduction

Up to now, we have studied many instrumental analysis techniques that are based on the phenomenon of light absorption. We have discussed atomic and molecular techniques utilizing light in the ultraviolet and visible regions of the spectrum and involving electronic energy transitions—the elevation of electrons to higher energy states. We have also discussed techniques utilizing light in the infrared region involving molecular vibrational and rotational energy transitions. In the current chapter, we have looked at techniques utilizing the high energy of x-rays capable of exciting inner-shell electrons in atoms. We now introduce the concepts of **nuclear magnetic resonance** spectrometry (NMR). This technique utilizes

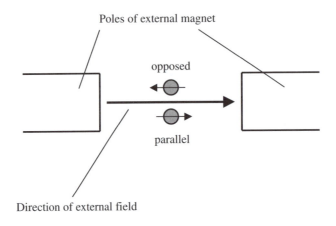

**FIGURE 10.8**    An illustration of the magnetic fields of spinning–precessing $^1$H nuclei aligned parallel to or opposed to an external magnetic field. The opposed alignment is shown above the parallel alignment in order to represent it as a higher energy state.

light in the very low energy radio wave region of the spectrum and involves nuclear spin energy transitions that occur in a magnetic field.

The technique is based on the theory and experimental evidence that atomic nuclei with an odd number of either protons or neutrons behave as small magnets by virtue of the fact that they are spinning and precessing like a top or gyroscope. Since a nucleus is positively charged, the spinning–precession motion creates a small magnetic field around it. The nucleus most often studied and measured is the nucleus of the hydrogen atom, symbolized $^1$H. If we were to position a compound with $^1$H nuclei in its structure (true of the vast majority of organic compounds) in an external magnetic field, i.e., between the poles of a magnet, the magnetic fields of the spinning–precessing $^1$H nuclei, representing smaller magnetic fields, will align themselves to the external field. They can align parallel to or opposed to the external field. The opposed alignment represents a slightly higher energy state, and so there are fewer nuclei in this state at a given point in time. See Figure 10.8. The two states are in resonance, meaning that they continually flip from one state to the other and back.

The energy difference between the two different alignments is on the order of approximately $6.6 \times 10^{-26}$ J. The frequency corresponding to this energy is $1 \times 10^8$ sec$^{-1}$, or 100 MHz, which is a radio wave frequency. Thus, with more nuclei in the lower energy state at any given moment, radio frequency (RF) light, or light in the radio frequency region of the electromagnetic spectrum, can be absorbed by the molecules in a magnetic field so as to cause the nuclear spin energy transition and more resonance, hence the name nuclear magnetic resonance. See Figure 10.9. While this phenomenon applies to nuclei of certain other elements, NMR has found its most useful application in the measurement of the $^1$H nucleus. For this reason, it is sometimes also referred to as proton magnetic resonance (PMR). The application lies mostly in the determination of the structure of organic compounds, and it is a powerful qualitative analysis tool for such compounds, as we shall see.

## 10.4.2   Instrumentation

A unique situation exists with NMR because the transition from one spin state to the other depends on both the energy of the RF and the strength of the external magnetic field. The greater the field strength, the larger the energy spread between the two energy levels, as shown in Figure 10.10, and therefore the greater the RF required. Thus, one could say that the precise RF that will be absorbed depends on the strength of the magnetic field used. But, since the strength of the magnetic field is variable, one can also say that the strength of the magnetic field required depends on the RF to be used. In the traditional instrument design, the RF is constant and the external magnetic field is varied in order to generate the

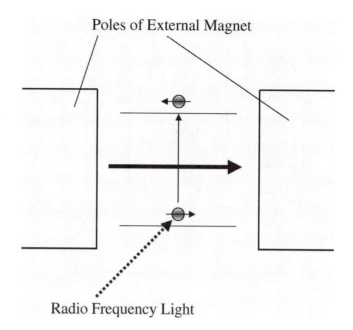

Poles of External Magnet

Radio Frequency Light

**FIGURE 10.9** Those $^1$H nuclei with a magnetic field parallel to the external field can absorb radio frequency light and be promoted to the higher energy state in which their magnetic field is opposed to the external field.

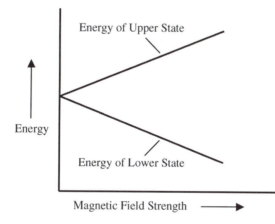

**FIGURE 10.10** The energy difference between the two spin states depends on the magnetic field strength.

NMR spectrum. Fourier transform nuclear magnetic resonance (FTNMR) instruments, which are similar in principle to Fourier transform infrared spectrometry (FTIR) instruments, are popular today. We will briefly describe these instruments later in this section.

The available NMR instruments are classified according to the RF employed. Instruments can utilize different RFs. Examples are the 60-MHz and the 100-MHz instruments. Different frequencies require different field strengths. For example, a 60-MHz instrument requires a field strength of 14,092 G (the gauss (G) is a unit of field strength), while a 100-MHz instrument requires 23,486 G. In the traditional instruments, the field strength is varied over a very narrow range. To accomplish this variability, a pair of sweep coils are used as described below.

The basic components of the traditional NMR instrument therefore include a large-capacity superconducting magnet, a sample holder, an RF transmitter, an RF receiver/detector, a sweep generator with

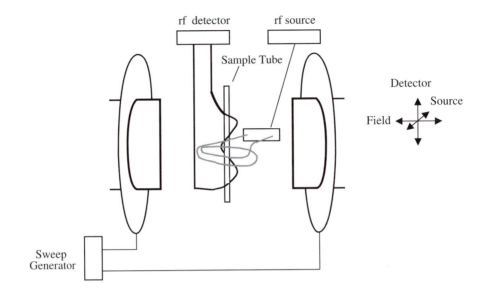

**FIGURE 10.11**   An illustration of the traditional NMR spectrometer as described in the text. The magnetic field, RF generator, and RF detector are mutually perpendicular, as indicated on the right.

the sweep coils, and a data system. The external magnetic field, the RF transmitter, and the RF detector are all mutually perpendicular. The sweep coils consist of a pair of metal wire coils encircling the poles of the magnet. A small current flowing through the coils from the sweep generator permits the precise scanning of the magnetic field over a very narrow range. The RF transmitter emits the RF, the RF receiver/detector receives the frequency and detects the absorption, and the data system acquires the data and displays the NMR spectrum. The sample is held in a 5-mm-outside-diameter (O.D.) glass tube containing less than half a milliliter of liquid, which in turn is held in a fixture called the sample probe. This tube rotates rapidly in the sample probe as the spectrum is being measured. This rotation ensures that all nuclei in the sample are affected equally by the applied magnetic field. An illustration of this design is shown in Figure 10.11.

As stated above, the modern NMR instrument uses Fourier transform technology similar to that of the FTIR instrument described in Chapter 8. It is therefore accurately described as Fourier transform nuclear magnetic resonance. In FTNMR instruments, the design is identical to the older instruments, except that very brief repeating pulses of RF energy are applied to the sample while holding the magnetic field constant (no sweep generator or coils). As with FTIR, the resulting detector signal contains the absorption information for all transitions occurring in the sample, which in this case are all the nuclear energy transitions of all $^1H$ nuclei. The Fourier transformation, performed by computer, then sorts out and plots the absorption as a function of frequency, giving the same result, the NMR spectrum, as the traditional instruments, but in much less time. In addition, it is more sensitive and gives better resolution.

## 10.4.3   The NMR Spectrum

Absorption of various RFs occurs at a constant field strength. Absorption of a particular RF occurs at various field strengths. The traditional NMR spectrum is a plot of the latter—absorption vs. field strength.

Intuitively, one may expect all hydrogen nuclei to absorb a particular RF at a particular field strength, reflecting an assumption that all hydrogen nuclei are identical. If this were true, the proton NMR spectrum of all compounds would be identical—hardly useful for any analytical characterization. Fortunately, this is not the case.

### 10.4.3.1 Chemical Shifts

Electrons, as well as nuclei (other than $^1$H) with an odd number of protons or neutrons, are, of course, present around and near the hydrogen nuclei in a molecule, and these are generating individual magnetic fields too. These very small fields change the value of the applied field slightly, giving an effective applied field somewhat smaller than expected and therefore a slightly shifted absorption pattern for the $^1$H nuclei, the so-called chemical shift. If all hydrogen nuclei in a molecule were shielded by electrons and neighboring nuclei equally, we would still have the situation of just one major absorption peak. Due to the different environments surrounding the different hydrogens in an organic molecule, however, we find major absorption peaks at field strengths representing each of these environments.

In other words, all hydrogens with a given environment give one characteristic major absorption peak at a specific field value. Methyl alcohol, $CH_3OH$, for example, would give two major peaks, one representing the methyl hydrogens (all three hydrogens have the same environment) and one representing the hydroxyl hydrogen (a different environment from the other three and therefore not equivalent), while cyclohexane, which has 12 equivalent hydrogens, gives just one major peak. The importance of this information is that: 1) from the number of different major peaks, we can tell how many different kinds of hydrogen there are, and 2) from the amount of shielding shown, we can determine the structure near each hydrogen.

### Example 10.2

How many major peaks can be expected in the NMR spectrum of: (a) ethyl alcohol, (b) n-propyl alcohol, (c) isopropyl alcohol, (d) acetone, and (e) benzene?

### Solution 10.2

(a) In ethyl alcohol, $CH_3CH_2OH$, the three $CH_3-$ hydrogens are equivalent and different from the others, the two $-CH_2-$ hydrogens are equivalent and different from the others, and the hydroxyl hydrogen is different from the others. Therefore, there would be three major peaks.

(b) In n-propyl alcohol, $CH_3CH_2CH_2OH$, there are four different kinds of hydrogen and therefore four major peaks.

(c) In isopropyl alcohol, $CH_3CHCH_3$, there are only three different kinds of hydrogen (the six hydrogens
$\quad\quad\quad\quad\quad\quad \mid$
$\quad\quad\quad\quad\quad\quad OH$
on the two methyl groups are equivalent), so only three major peaks.

(d) In acetone, $CH_3CCH_3$, there is only one kind of hydrogen, so only one major peak.
$\quad\quad\quad\quad\quad \mid\mid$
$\quad\quad\quad\quad\quad O$

(e) In benzene, there are six equivalent hydrogens, so only one major peak.

Since the shifts can be extremely slight (on the order of parts per million (ppm) of the external field strength), the actual field strength of the peak is difficult to measure precisely. For this reason, a reference compound, typically tetramethylsilane (TMS), is added to the sample. All the hydrogens in the TMS structure are equivalent (Figure 10.12). It is then convenient to modify the x-axis to show the difference between the single TMS peak and the compound's peaks. The x-axis thus typically represents a difference factor, symbolized by the lowercase Greek delta ($\delta$), in which the single TMS peak is 0.0 ppm field strength

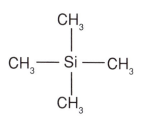

**FIGURE 10.12** The structure of tetramethylsilane (TMS). All the hydrogens are equivalent, giving one NMR peak, which is used as a reference peak.

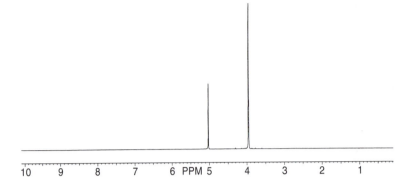

**FIGURE 10.13**   The spectrum of methyl alcohol.

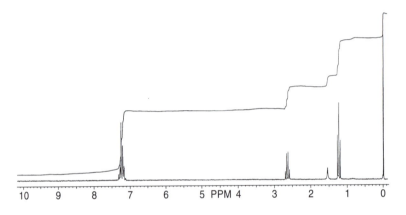

**FIGURE 10.14**   The NMR spectrum of ethylbenzene with integration trace.

difference and other peaks then are a certain parts per million away from the TMS peak. Figure 10.13 shows such an NMR spectrum for methyl alcohol.

### 10.4.3.2  Peak Splitting and Integration

Additional qualitative information is possible with NMR spectra. First, a given major absorption peak, while apparently arising from one particular kind of hydrogen, may appear to be split into two or more peaks, giving what are termed doublets, triplets, etc., a phenomenon called **spin–spin splitting**. The splitting is due to the effect that $^1$H nuclei on adjacent carbons have on the nucleus in question. Splitting occurs because the magnetic fields that these different hydrogens generate become magnetically coupled to each other. In general, the number of peaks resulting from the splitting are predicted according to a rule called the **n + 1 rule**, which states that if there are n equivalent hydrogens on the adjacent carbon, then the number of peaks that appear will be n + 1. So, for example, if there are no hydrogens on the adjacent carbon, then 0 + 1, or only one, peak appears. If there is one hydrogen on the adjacent carbon, then 1 + 1, or two, peaks will appear as a result of the splitting. If two equivalent hydrogens, then three peaks, etc. For example, Figure 10.14 is the NMR spectrum of ethylbenzene. The triplet at a d range of approximately 1.2 to 1.3 is due to the three hydrogens on the –CH$_3$ half of the ethyl group. It is a triplet because the adjacent carbon has two equivalent hydrogens and n + 1 = 3. The quartet at a d range of approximately 2.6 to 2.7 is due to the –CH$_2$ half of the ethyl group. It is a quartet because the adjacent carbon has three equivalent hydrogens and n + 1 = 4. An important point to remember is that if all the hydrogens on both carbons are equivalent, then there will be no splitting.

Second, the area under a peak is indicative of the number of hydrogens it represents. This, too, is important qualitative information, and so NMR spectrometer data systems are designed to determine the areas under the peaks to determine the number of hydrogens of each type. An integration trace is often displayed and can then be used to determine the number of hydrogens of each type. Figure 10.14 (ethylbenzene) shows integration trace. Notice the rise in the trace above the $-CH_3$ half compared to the rise in the trace above the $-CH_2$ half of the ethyl group. They are in an apparent ratio of 3:2.

## 10.4.4  Solvents and Solution Concentration

The 5-mm tube held in the sample probe is indicative of the sample size used in an NMR instrument—typically some fraction of a milliliter. Typical analyte quantities dissolved in this volume range from 10 to 50 mg.

A popular solvent for the sample is deuterated chloroform, $CDCl_3$. The D is the symbol for deuterium, the isotope of hydrogen that has one proton and one neutron. This solvent has no $^1H$ nuclei and does not absorb RFs. It will therefore not interfere in the analysis.

## 10.4.5  Analytical Uses

The primary use of NMR is in the determination of the structure of unknown organic compounds. It is often used in conjunction with other spectrometric techniques (FTIR and mass spectrometry, for example) in this determination. NMR spectra are molecular fingerprints, however, and by comparison with data files of known spectra, the structure of an unknown can be determined independent of other data.

# 10.5  Mass Spectrometry

## 10.5.1  Introduction

The final spectroscopic technique we will discuss is mass spectrometry. Unlike all the others, this technique does not use light at all. Very briefly, the instrument known as a mass spectrometer utilizes a high-energy electron beam to cause total destruction and fragmentation of the molecules of a gaseous sample. This fragmentation results in small, positively charged pieces or fragments of the molecules that are detected individually on the basis of their mass-to-charge ratios. The details of exactly how these positively charged fragments are separated and detected differ according to the specific design of the mass analyzer portion of the instrument. The different designs are discussed below. In any case, the information acquired and displayed by the data system (the so-called mass spectrum) allows the analyst to reconstruct the original molecule and thereby identify it.

## 10.5.2  Instrument Design

There are three different instrument designs that we will discuss: the magnetic sector mass spectrometer, the quadrupole mass spectrometer, and the time-of-flight mass spectrometer. All three designs consist of a system for sample introduction, an electron beam to cause the fragmentation, and the mass analyzer for separating and detecting the charged fragments. In each case, the entire path of the fragments, including the inlet system, must be evacuated to $10^{-4}$ to $10^{-8}$ torr. This requirement means that a strong and sophisticated vacuum system must also be part of the setup. The reason for the vacuum is to avoid collisions of both the electron beam and the sample fragments with foreign particles that would alter the results.

## 10.5.3  The Magnetic Sector Mass Spectrometer

A diagram of a magnetic sector mass spectrometer is shown in Figure 10.15. In this instrument, the magnet is a powerful, variable-field electromagnet, the poles of which are shaped to cause a bending of the path of the fragments through a specific angle, such as 90°. It is possible to vary the field strength of the magnet in such a way as to scan the magnetic field and to focus the ion fragments of variable mass-to-charge ratio

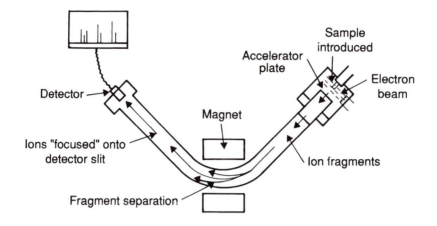

**FIGURE 10.15**   An illustration of a magnetic sector mass spectrometer.

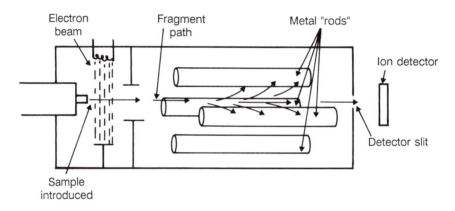

**FIGURE 10.16**   An illustration of a quadrupole mass spectrometer.

onto the detector slit one at a time. In this way, specific fragments created at the electron beam can be separated from other fragments and detected individually.

## 10.5.4   The Quadrupole Mass Spectrometer

A diagram of the quadrupole mass spectrometer is shown in Figure 10.16. Here, four short, parallel metal rods (poles) with a diameter of about half a centimeter each are utilized. These rods are aligned parallel to and surrounding the fragment path as shown. Two nonadjacent rods, such as those in the vertical plane, are connected to the positive poles of variable dc and ac power sources, while the other two are connected to the negative poles. Thus, a variable electric field is created, and as the fragments enter the field and begin to pass down the center area, they deflect from their path. Varying the field creates the ability to focus the fragments one at a time onto the detector slit, as in the magnetic sector instruments. The quadrupole instrument is newer and more popular since it is much more compact and provides a faster scanning capability.

## 10.5.5   The Time-of-Flight Mass Spectrometer

The time-of-flight mass spectrometer is different from the others in that it does not use a magnetic field. Rather, it separates the charged fragments on the basis of how fast they move through a drift tube in which a strong electric field is present due to a high-voltage metal grid positioned in the path. The lighter fragments travel through the tube faster than the heavier fragments, and the different masses are detected

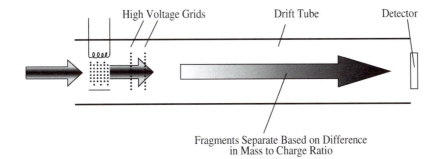

**FIGURE 10.17** An illustration of a time-of-flight mass spectrometer.

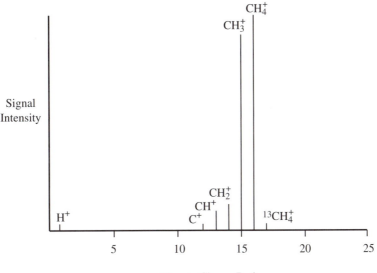

**FIGURE 10.18** The mass spectrum of methane.

at different times, depending on their time of flight through the tube. See Figure 10.17. The signal is thus measured vs. time, but can be translated into the traditional mass spectrum.

## 10.5.6 Mass Spectra

The data collected by the data system is displayed as a **mass spectrum**, which is a plot of signal intensity vs. mass-to-charge ratio. At specific mass-to-charge ratios, sharp peaks are displayed, each representing a particular fragment having that ratio. The key is that a given molecule always disintegrates into the same fragments and therefore always displays the same mass spectrum, a molecular fingerprint. This occurs because certain bonds within a molecule are weaker than others, and the mass spectral pattern is due to the fact that it is these same bonds that always break—and thus the same fragments always form—when subjected to the bombardment of the electron beam. Also noteworthy is the fact that different naturally occurring isotopes of a given element are always present in the various fragments and are thus always apparent in the mass spectrum.

A simple example is the mass spectrum of methane, $CH_4$ (Figure 10.18). A methane molecule breaks into the following fragments (in order of relative abundance): $CH_4^+$, $CH_3^+$, $CH_2^+$, $CH^+$, $H^+$, $C^+$, and $^{13}CH_4^+$. In the figure, each peak is identified as being due to one of these fragments. Notice that one of the peaks

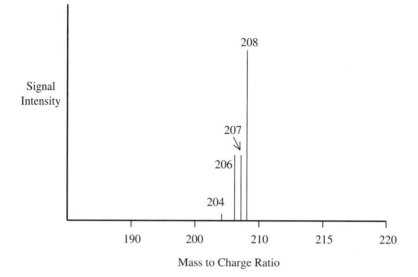

**FIGURE 10.19**   The mass spectrum of elemental lead.

is due to a fragment with the $^{13}$C isotope (the fragment identified as $^{13}$CH$_4^+$). A $^{13}$C atom is one of about every one hundred carbon atoms in nature.

Because the fragmentation pattern produced by a mass spectrometer can be used as a fingerprint of molecule, the mass spectrum reveals, for example, whether the correct compound has been synthesized and whether contaminants are present. One can see that it is a molecular fingerprint, just as absorption spectra are molecular fingerprints, and that it is a powerful tool for identification purposes.

Besides the obviously significant applicability to molecular compound identification, mass spectrometry also finds application in elemental analysis, such as to determine what isotopes of an element might be present in a sample or, combined with inductively coupled plasma (Section 10.5.7), what the concentration is of an element in a sample. The isotopes of radioactive elements may be detected in a variety of samples, including biological matrices. Different isotopes are detected because the differing counts of neutrons that characterize atoms of different isotopes produce ions of different mass-to-charge ratios in the mass spectrometer. For example, the mass spectrum of elemental lead is presented in Figure 10.19. Isotopes of lead with atomic weights of 204, 206, 207, and 208 are known and have the relative abundance indicated by the signal heights in Figure 10.19. See Workplace Scene 10.2.

## 10.5.7   Mass Spectrometry Combined with Inductively Coupled Plasma

The inductively coupled plasma (ICP) technique described in Chapter 9 (Section 9.5) has been coupled with mass spectrometry. The ICP source provides the mass spectrometer with a source of charged monatomic ions such that the electron beam is not needed. The "exhaust" of the ICP source is fed into the mass spectrometer for analysis by the mass analyzer portion of the mass spectrometer (Figure 10.20).

The intensity of the signal due to the analyte ion is proportional to the concentration of the analyte in the sample. In most cases there will be more than one analyte ion for a given element (due to the presence of more than one naturally occurring isotope) and therefore more than one signal for that element. It may happen that more than one element in a sample will produce a signal with a particular mass–charge ratio. There would therefore be a spectral interference if that signal were to be measured. Nearly all elements, however, have at least one isotope mass that no other element has. Thus, a signal with this spectral interference is ignored in favor of one that is due to just one element.

# WORKPLACE SCENE 10.2

Personnel working in some programs at the Los Alamos National Laboratory (LANL) may handle radioactive materials that, under certain circumstances, could be taken into the body. Employees are monitored for such intakes through a series of routine and special bioassay measurements. One such measurement involves a thermal ionization mass spectrometer. In this technique, the metals in a sample are electroplated onto a rhenium filament. This filament is inserted into the ion source of the mass spectrometer and a current is passed through it. The ions of the plutonium isotopes are thus formed and then accelerated through the magnetic field. The number of ions of each isotope are counted and the amount of Pu-239 in the original sample calculated by comparison to a standard.

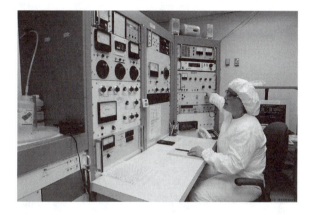

Kim Israel, a technician at LANL, Chemistry Division Bioassay Program, runs a thermal ionization mass spectrometry (TIMS) instrument.

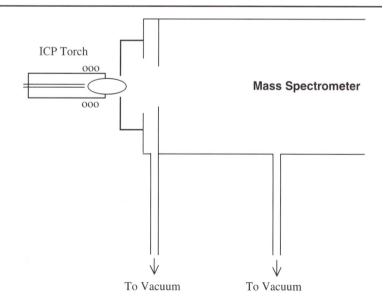

**FIGURE 10.20** An illustration of a mass spectrometer coupled to an inductively coupled plasma instrument.

# WORKPLACE SCENE 10.3

The analysis of tetramethylammonium hydroxide (TMAH) solutions manufactured by SACHEM Inc. of Cleburne, Texas, includes the determination of trace elements. These elements cause less-than-optimum performance of integrated circuit boards manufactured by SACHEM's customers that use these solutions in their processes. Alkali and alkaline earth metals (e.g., Li, Na, K, Mg, Ca, and Ba) can reduce the oxide breakdown voltage of the devices. In addition, transition and heavy metal elements (e.g., Ti, Cr, Mn, Fe, Co, Ni, Cu, Zn, Ag, Au, and Pb) can produce higher dark current. Doping elements (e.g., B, Al, Si, P, As, and Sn) can alter the operating characteristics of the devices. In SACHEM's quality control laboratory, ICP coupled to mass spectrometry is used to simultaneously analyze multiple trace elements in one sample in just 1 to 4 min. This ICP-MS instrument is a state-of-the-art instrument that can provide high throughput and low detection limits at the parts per thousand level. Trace elemental determination at the parts per thousand level must be performed in a clean room so that trace elemental contamination from airborne particles can be minimized.

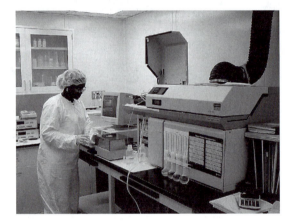

Katrice Harris of SACHEM Inc. measures trace elements in ultrapure TMAH solutions at the parts per thousand level by using ICP-MS in a clean room.

The detection limits for elements analyzed by ICP–mass spectrometry (ICP-MS) are significantly lower in most cases than the detection limits for other atomic techniques. See Table 10.1. See Workplace Scene 10.3.

## 10.5.8  Mass Spectrometry Combined with Instrumental Chromatography

Chromatography instruments (Chapters 11 to 13) are instruments that separate and quantitate samples of complex mixtures with surprising ease and accuracy in a very short time. The separation occurs in a column, which is a tube that contains a stationary material phase, either a liquid or solid, that selectively impedes the progress of the mixture components as they move through it. This forward movement occurs because a second phase, a moving gas or liquid phase called the mobile phase, is passing through the column and also influences the rate of movement of the mixture components. Mixture components then emerge from the end of the column one at a time and are electronically detected and measured one at a time. A mass spectrometer can be used as the detector system.

**TABLE 10.1** Detection Limits for ICP-MS for the Elements Listed in Table 9.2[a]

| Element | ICP-MS Detection Limit ($\mu$g/mL) | Best Detection Limit from Table 9.2 |
|---|---|---|
| Arsenic | 0.0006 | 0.03 (hydride) |
| Bismuth | 0.0006 | 0.03 (hydride) |
| Calcium | 0.0002 | 0.01 (GFAA) |
| Copper | 0.0002 | 0.014 (GFAA) |
| Iron | 0.0003 | 0.06 (GFAA) |
| Mercury | 0.016 | 0.009 (cold vapor mercury) |
| Potassium | 0.0002 | 0.005 (GFAA) |
| Zinc | 0.0003 | 0.02 (GFAA) |

*Note:* GFAA = graphite furnace atomic absorption.

[a]Compared to the best detection limits for these elements found in Table 9.2 for other atomic techniques.

*Source:* From Perkin-Elmer Instruments literature. With permission.

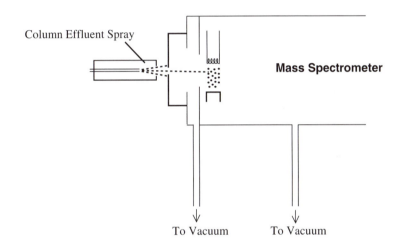

**FIGURE 10.21** An illustration of a mass spectrometer coupled to a chromatography instrument.

When the mobile phase is a gas, the technique is called gas chromatography–mass spectrometry (GC-MS). In this case, the mixture is in the gas phase as it moves through the column, and the individual separated mixture components emerge in the gas phase and mix with the gaseous mobile phase, which is usually helium. When the mobile phase is a liquid, or a mixture of liquids, the technique is called liquid chromatography–mass spectrometry (LC-MS). In this case, the mixture is in the liquid phase as it moves through the column and individual separated mixture components emerge dissolved in this liquid phase.

In either case, there is a need to eliminate significant amounts of gas or liquid as the sample passes into the high-vacuum environment of the mass spectrometer. Thus a special interface is needed to accomplish this. But, indeed, reliable interfaces have been invented as evidenced by the extensive use of these techniques in modern analytical laboratories. See Figure 10.21. Also, see additional discussions in Chapters 12 and 13 (Sections 12.6.7 and 13.6.6), as well as Workplace Scene 13.5.

Mass spectrometry offers a special advantage over other detection schemes for instrumental chromatography because the mass spectra of the mixture components are acquired, providing a very powerful qualitative analysis tool for these components.

# Questions and Problems

1. Name three analytical methods involving x-rays and give some information about the application of each.

2. Name two sources of x-rays, and explain how the x-rays are generated in each case.

3. Define $K_\alpha$ emission, $L_\beta$ emission, and $K_{\alpha 2}$ emission.

4. What is diffraction? How does constructive and destructive interference result in a diffraction pattern?

5. What is Bragg's law? Define each of the parameters in Bragg's law.

6. The wavelength used in a particular x-ray diffraction analysis is 1.539 Å. If an intense peak occurs at a 2-theta value of 44.28°, what is the d-spacing in this structure?

7. The wavelength used in a particular x-ray diffraction analysis is 1.539 Å. If a standard crystal with a d-spacing of 0.844 Å is used, at what angle can a light intensity be expected to be observed?

8. Why is it that the x-ray emission depicted in Figure 10.1 is considered fluorescence?

9. Is x-ray fluorescence molecular, atomic, or neither? Explain.

10. What are the advantages and disadvantages to using an analyzer crystal in an x-ray fluorescence instrument, as depicted in Figure 10.7?

11. What are some safety rules to be followed in laboratories in which x-ray equipment is used?

12. What does NMR stand for?

13. Why is a large magnet needed as part of an NMR instrumentation?

14. What wavelength region of the electromagnetic spectrum is needed for an NMR experiment? Why?

15. What two energy states of spinning–precessing nuclei present in a magnetic field are involved in an NMR experiment? Which state represents the higher energy?

16. Why is NMR sometimes referred to as PMR?

17. What does FTNMR stand for?

18. What is meant by the terms hertz and megahertz?

19. What is a typical value of the RF (in megahertz) used in an NMR experiment? What is the magnitude of the magnetic field (in gauss) needed for an instrument utilizing this RF?

20. What is a gauss?

21. Name seven components of the traditional NMR instrument and briefly describe the function of each.

22. How does a modern FTNMR experiment differ from an experiment using a traditional instrument?

23. What parameters are represented on the y- and x-axes of an NMR spectrum?

24. What is the chemical shift and what causes it?

25. How is it that peaks representing different hydrogens in an organic molecule appear at different locations in an NMR spectrum?

26. How many major peaks can be expected in the NMR spectrum of: (a) dimethyl ether, (b) diethyl ether, (c) methyl ethyl ketone, (d) acetone, and (e) t-butyl alcohol?

27. What is it about the structure of methyl alcohol that would result in two major peaks in its NMR spectrum (Figure 10.13)?

28. Explain the use of tetramethylsilane (TMS) in an NMR experiment.

29. Look at the NMR spectrum in Figure 10.22. How many different kinds of hydrogen are represented? Explain your answer.

30. What does the integrator trace in Figure 10.22 tell you about the number of the different kinds of hydrogen present in the structure?

31. If you were told that the spectrum in Figure 10.22 was either ethyl alcohol or diethyl ether, what evidence would you cite to conclude that it is ethyl alcohol and not diethyl ether?

32. If you were told that the spectrum in Figure 10.22 resulted from either ethyl alcohol or methyl ethyl ether, what evidence would you cite to conclude that it is ethyl alcohol and not methyl ethyl ether?

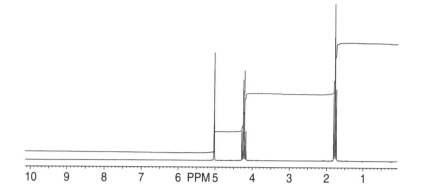

**FIGURE 10.22** The NMR spectrum for problems 29 to 32.

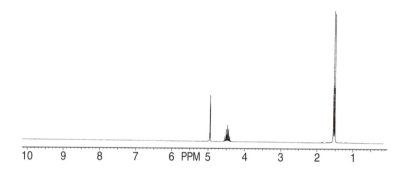

**FIGURE 10.23** The NMR spectrum for problems 34 and 35.

33. In Figure 10.22, the major peak at 5 ppm is a singlet, the major peak at 4.2 ppm is a quartet, and the major peak at 1.8 is a triplet. How does the n + 1 rule confirm that the compound may be ethyl alcohol?

34. In Figure 10.23, the major peak at 5 ppm is a singlet, the major peak at 4.5 ppm is a septet (split into seven peaks), and the major peak at 1.5 is a doublet. The spectrum is that of either n-propyl alcohol, isopropyl alcohol, or acetone. Decide which and justify your decision.

35. After answering problem 34, describe what the integrator trace would look like on this spectrum.

36. What causes NMR peaks to be split into doublets, triplets, etc.? How does the presence of split peaks assist with qualitative analysis?

37. What is the purpose of the high-energy electron beam utilized in a mass spectrometer?

38. Without specific references to mass analyzer design, explain how a mass spectrometer can determine that the different molecular fragments created have different charges and masses (or mass-to-charge ratios)? Why is this fact important in its use in qualitative analysis?

39. Why must a mass spectrometer utilize a sophisticated high-vacuum system?

40. Describe the differences between a magnetic sector mass spectrometer, a quadrupole mass spectrometer, and a time-of-flight mass spectrometer.

41. What is plotted on the y- and x-axes of a mass spectrum?

42. Briefly discuss the value of the mass spectrum as a tool for qualitative analysis.

43. Figure 10.24 shows the mass spectrum of acetone. Based on the mass-to-charge ratio of each line, write a formula for as many of the fragments as you can, as was done for methane in Figure 10.18.

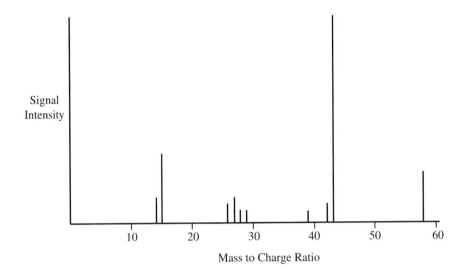

**FIGURE 10.24**    The mass spectrum of acetone.

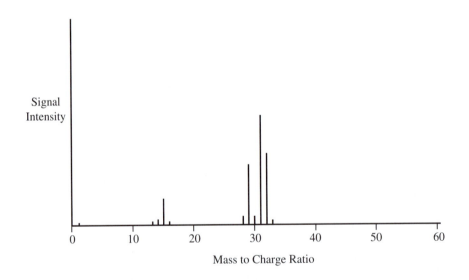

**FIGURE 10.25**    The mass spectrum of an unknown for problem 44.

44. The mass spectra given in Figures 10.25 to 10.28 are those of benzene, methanol, ethanol, and methyl ethyl ketone. Tell which spectrum is of which compound, and justify your answers.
45. How can the number of isotopes and their relative abundance be determined by mass spectroscopy?
46. Why is no electron beam needed in an ICP-MS instrument?
47. What advantage does ICP-MS usually have over graphite furnace atomic absorption, for example?
48. What do GC-MS and LC-MS refer to?
49. What role does mass spectrometry play in instrumental chromatography? What advantage does mass spectrometry offer instrumental chromatography over other detection systems?

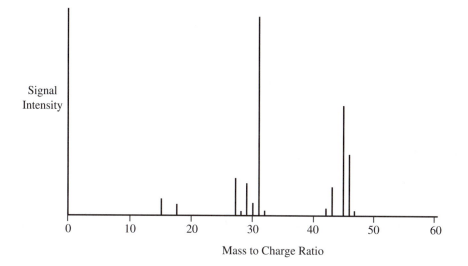

**FIGURE 10.26**   The mass spectrum of an unknown for problem 44.

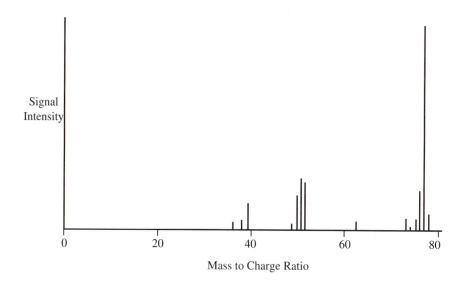

**FIGURE 10.27**   The mass spectrum of an unknown for problem 44.

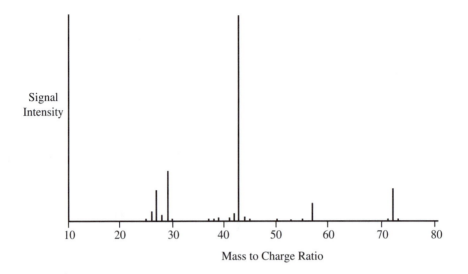

**FIGURE 10.28**   The mass spectrum of an unknown for problem 44.

# 11

# Analytical Separations

## 11.1 Introduction

Modern-day chemical analysis can involve very complicated material samples—complicated in the sense that there can be many substances present in the sample, creating a myriad of problems with interferences when the lab worker attempts the analysis. These interferences can manifest themselves in a number of ways. The kind of interference that is most familiar is one in which substances other than the analyte generate an instrumental readout similar to the analyte, such that the interference adds to the readout of the analyte, creating an error. However, an interference can also suppress the readout for the analyte (e.g., by reacting with the analyte). An interference present in a chemical to be used as a standard (such as a primary standard) would cause an error, unless its presence and concentration were known (determinant error, or bias). Analytical chemists must deal with these problems, and chemical procedures designed to effect separations or purification are now commonplace.

This chapter and Chapters 12 and 13 describe modern analytical separation science. First, purification procedures known as recrystallization and distillation will be described. Then the separation techniques of extraction and chromatography are discussed. This is followed by, in Chapters 12 and 13, instrumental chromatography techniques that can resolve very complicated samples and quantitate usually in one easy step.

## 11.2 Recrystallization

**Recrystallization** is a purification technique for a solid, usually organic. The separation is based on the solid's solubility in a liquid solvent. The solid must be sparingly soluble in this solvent at room temperature, but highly soluble at higher temperatures. Both soluble and insoluble impurities are considered to be present, and the procedure removes both if the concentrations are not too large.

The key to the procedure is to use a *minimum* amount of solvent, such that the solid will just dissolve at the elevated temperature (usually the boiling point, if the solid is stable at that temperature). While maintaining this elevated temperature, any impurity that has not dissolved can be filtered out. The insoluble impurities are thus removed.

Soluble impurities, however, are still present in the hot filtrate. These are removed by cooling the filtrate. Cooling the filtrate causes the solid being purified to crystallize because its solubility is exceeded at the lower temperature. If the minimum amount of solvent is used, the maximum possible amount of the solid being purified will crystallize and be recovered via a second filtering step. If the solubilities of the soluble impurities have not been exceeded during the cooling step, they will stay dissolved and be separated during the second filtering. It may be necessary to perform the recrystallization several times in order to get the desired purity.

# WORKPLACE SCENE 11.1

Because of the quantity of pure water needed for the laboratory work, many analytical laboratories use a deionization system rather than a distillation system for purifying water. Such a system utilizes cartridges packed with an ion exchange resin to exchange ions in the raw water with hydrogen and hydroxide ions, which in turn react to form more water, thus removing all ions from the water. For most routine analyses, general reagent preparation, and glassware washing, a general water deionization system is satisfactory. However, where parts per billion analyses are run, the water used to make standard solutions, etc., must be even more pure. In these cases, an ultrapure deionization system, consisting of additional ion exchange cartridges, is used to further enhance the purity of the water.

Eric Lee of the City of Lincoln, Nebraska, Water Treatment Plant Laboratory prepares to replace a spent deionization cartridge with a new one.

## 11.3   Distillation

**Distillation** is a method of purification of liquids contaminated with either dissolved solids or miscible liquids. The method consists of boiling and evaporating the mixture followed by recondensation of the vapors in a condenser, which is a tube usually cooled by isolated cold tap water. The theory is that the vapors (and thus recondensed liquids) will be purer than the original liquid. The separation is based on the fact that the contaminants have different boiling points and vapor pressures than the liquid to be purified. Thus, when the liquid is boiled and evaporated, the vapors (and recondensed liquids) created have a composition different from the original liquid. The substances with lower boiling points and higher vapor pressures are therefore separated from substances that have high boiling points and low vapor pressures.

Distillation of water to remove hardness minerals is an example and probably the most common application in an analytical laboratory. Of all the applications of distillation, it is one of the easiest to perform. While water is known to have a relatively high boiling point and low vapor pressure, the dissolved minerals are ionic solids that generally have extremely high boiling points (indeed, extremely high *melting* points) and extremely low vapor pressures. Thus, a simple distilling apparatus and a single distillation,

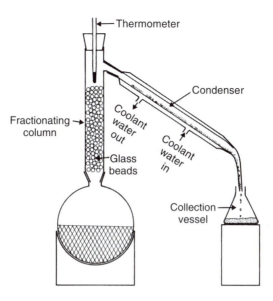

**FIGURE 11.1** An illustration of a distillation apparatus fitted with a fractionating column.

or, at most, two (doubly distilled) or three (triply distilled) distillations, will produce very pure water.[*] See Workplace Scene 11.1.

Organic liquids that are contaminated with other organic liquids usually constitute a much more difficult situation. Such liquids probably have such similar boiling points and vapor pressures that a distillation of a mixture of two or more would result in all being present in the distillate (the condensed vapors)—an unsuccessful separation. However, the liquid that has the highest vapor pressure or lowest boiling point, while not being completely separated, would be present in the initial distillate at a higher concentration level than before the distillation. It follows that if this distillate were then to be redistilled, perhaps over and over again, further enrichment of this component would take place such that an acceptable separation would eventually take place. However, the time involved in such a procedure would be prohibitive. A procedure known as fractional distillation solves the problem.

**Fractional distillation** involves repeated evaporation–condensation steps before the distillate is actually collected. These repeated steps occur in a fractionating column (tube) above the original heated container—a column that contains a high surface area of inert material for condensing the vapors. As the vapors condense on this material, the material itself heats up and the condensate reevaporates. The reevaporated liquid then moves further up the column and contacts more cold inert material, and the process occurs again—and again and again as the liquid makes its way up the column. If a fractionating column were used that is long enough and contains a sufficient quantity of the high-surface-area material, any purification based on differences in boiling point and vapor pressure can be effected. An illustration of a distillation apparatus fitted with a fractionating column is shown in Figure 11.1. The high-surface-area packing material in a fractionating column typically consists of glass beads, glass helices, or glass wool.

Each time a single evaporation–condensation step occurs in a fractionating column, the condensate has passed through what has been called a theoretical plate. A **theoretical plate** is thus that segment of fractionating column in which one evaporation–condensation step occurs. The name is derived from the image that the condensate is captured on small plates inside the fractionating column from which it is again boiled and evaporated. A fractionating column used for a given liquid mixture is then identified

---

[*]Water is often "deionized" using an ion exchange (Section 11.6) cartridge to remove hardness minerals. Such **deionization** of water is often done in conjunction with distillation, such that the water is both deionized and distilled prior to use. Also, if the water is contaminated with organics, or other low boiling substances, a charcoal filter cartridge is often used as well.

as having a certain number of theoretical plates, and given liquid mixtures are known to require a certain number of theoretical plates in order to achieve a given purity. The **height equivalent to a theoretical plate** (HETP) is the length of fractionating column corresponding to one theoretical plate. If the number of theoretical plates required is known, then the analyst can select a height of column that would contain the proper number of plates according to the manufacturer's specifications, or according to his own measurements of a homemade column. Height selection is not entirely experimental, however. The use of liquid–vapor composition diagrams to predict the theoretical plates required can help. These diagrams are based on boiling point and vapor pressure differences in a pair of liquids. Further discussion of the use of these diagrams is beyond the scope of this book.

The concept of theoretical plates is applicable to instrumental chromatography as well. This will be discussed in Section 11.8.4.

# 11.4   Liquid–Liquid Extraction

## 11.4.1   Introduction

One popular method of separating an analyte species from a complicated liquid sample is the technique known as **liquid–liquid extraction** or **solvent extraction**, first mentioned in Chapter 2. In this method, the sample containing the analyte is a liquid solution, typically a water solution, that also contains other solutes. The need for the separation usually arises from the fact that the other solutes, or perhaps the original solvent, interfere in some way with the analysis technique chosen. An example is a water sample that is being analyzed for a pesticide residue. The water may not be a desirable solvent and there may be other solutes that may interfere. It is a selective dissolution method—a method in which the analyte is removed from the original solvent and subsequently dissolved in a different solvent (extracted) while most of the remainder of the sample remains unextracted, i.e., remains behind in the original solution.

The technique obviously involves two liquid phases—one the original solution and the other the extracting solvent. The important criteria for a successful separation of the analyte are: 1) that these two liquids be immiscible, and 2) that the analyte be more soluble in the extracting solvent than the original solvent. See Workplace Scene 11.2.

## 11.4.2   The Separatory Funnel

As mentioned in Chapter 2, the extraction usually takes place in a specialized piece of glassware known as a separatory funnel. The **separatory funnel** is manufactured especially for solvent extraction. It has a teardrop shape with a stopper at the top and a stopcock at the bottom (see Figure 11.2 and also refer to Figure 2.2). The sample and solvent are placed together in the funnel, and the funnel is tightly stoppered and, while holding the stopper in with the index finger, shaken vigorously for a moment.

Following this, the funnel may need to be vented, since one of the liquids is likely to be a volatile organic solvent, such as methylene chloride. Venting is accomplished by opening the stopcock while inverted. This shaking–venting step is then repeated several times such that the two liquids have plenty of opportunity for the intimate contact required for the analyte to pass into the extracting solvent to the maximum possible extent. See Figure 11.3 for shaking–venting illustrations.

Following this procedure, the funnel is positioned in a padded ring in a ring stand and left undisturbed for a period of time to allow the two immiscible layers to once again separate. The purpose of the specific design of the separatory funnel is mostly to provide for easy separation of the two immiscible liquid layers after the extraction takes place. All one needs to do is remove the stopper, open the stopcock, allow the bottom layer to drain, and then close the stopcock when the interface between the two layers disappears from sight in the stopcock. The denser of the two liquids is the bottom layer and will be drained through the stopcock first. The entire process may need to be repeated several times, since the

# WORKPLACE SCENE 11.2

The analysis of human plasma for acetaminophen, the active ingredient in some pain relievers, involves a unique extraction procedure. Small-volume samples (approximately 200 $\mu$L) of heparinized plasma, which is plasma that is treated with heparin, a natural anticoagulant found in biological tissue, are first placed in centrifuge tubes and treated with 1 $N$ HCl to adjust the pH. Ethyl acetate is then added to extract the acetaminophen from the samples. The tubes are vortexed, and after allowed to separate, the ethyl acetate layer containing the analyte is decanted. The resulting solutions are evaporated to dryness and then reconstituted with an 18% methanol solution, which is the final sample preparation step before HPLC analysis. The procedure is a challenge because the initial sample size is so small.

Randy Karl of MDS Pharma Services in Lincoln, Nebraska, examines a centrifuge tube containing a heparinized plasma sample prior to performing an extraction procedure using ethyl acetate.

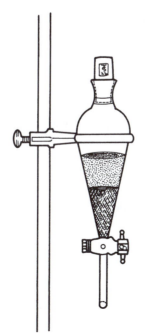

**FIGURE 11.2** A drawing of a separatory funnel, containing two immiscible liquids, held in an iron ring clamped to a ring stand.

**FIGURE 11.3**    Illustrations of the shaking and venting procedures with the separatory funnel.

extraction is likely not to be quantitative. This means that another quantity of fresh extracting solvent may need to be introduced into the separatory funnel with the sample and the shaking procedure repeated. Even so, the experiment may never be completely quantitative. See the next section for the theory of extraction and a more in-depth discussion of this problem.

### 11.4.3   Theory

The process of a solute dissolved in one solvent being pulled out, or "extracted" into a new solvent actually involves an equilibrium process. At the time of initial contact, the solute will move from the original solvent to the extracting solvent at a particular rate, but, after a time, it will begin to move back to the original solvent at a particular rate. When the two rates are equal, we have equilibrium. We can thus write the following:

$$A_{orig} \rightleftharpoons A_{ext} \tag{11.1}$$

in which A refers to analyte and orig and ext refer to original solvent and extracting solvent, respectively. If the analyte is more soluble in the extracting solvent than in the original solvent, then, at equilibrium, a greater percentage will be found in the extracting solvent and less in the original solvent. If the analyte is more soluble in the original solvent, then the greater percentage of analyte will be found in the original solvent. Thus, the amount that gets extracted depends on the relative distribution between the two layers, which, in turn, depends on the solubilities in the two layers. A **distribution coefficient** analogous to an equilibrium constant (also called the **partition coefficient**) can be defined as follows:

$$K = \frac{[A]_{ext}}{[A]_{orig}} \tag{11.2}$$

Often, the value of K is approximately equal to the ratio of the solubilities of A in the two solvents. If the value of K is very large, the transfer of solute to the extracting solvent is considered to be quantitative. A value around 1.0 would indicate equal distribution and a small value would indicate very little transfer.

Uses of the distribution coefficient include: 1) the calculation of the amount of a solute that is extracted in a single extraction step, 2) the determination of the weight of the solute in the original solute (important if you are quantitating the solute in this solvent), 3) the calculation of the optimum volumes of both the extracting solvent and the original solution to be used, 4) the number of extractions needed to obtain a particular quantity or concentration in the extracting solvent, and 5) the percent extracted. The following expansion of Equation (11.2) is useful for these:

$$K = \frac{W_{ext} / V_{ext}}{W_{orig} / V_{orig}} \tag{11.3}$$

## Example 11.1

The distribution coefficient for a given extraction experiment is 98.0. If the concentration in the extracting solvent is 0.0127 $M$, what is the concentration in the original solvent?

*Solution 11.1*

$$K = \frac{[A]_{ext}}{[A]_{orig}} = 98.0$$

$$98.0 \times [A]_{orig} = [A]_{ext}$$

$$[A]_{orig} = \frac{[A]_{ext}}{98.0}$$

$$[A]_{orig} = \frac{0.0127\ M}{98.0} = 0.000130\ M$$

## Example 11.2

What weight of analyte is found in 50.0 mL of an extracting solvent when the distribution coefficient is 231, and the weight of analyte found in 75.0 mL of the original solvent after extraction is 0.00723 g?

*Solution 11.2*

$$K = \frac{W_{ext}/V_{ext}}{W_{orig}/V_{orig}} = \frac{\dfrac{W_{ext}}{50.0\ mL}}{\dfrac{0.00723\ g}{75.0\ mL}} = \frac{\dfrac{W_{ext}}{50.0\ mL}}{0.0000964\ g/mL} = 231$$

$$\frac{W_{ext}}{50.0\ mL} = 231 \times 0.0000964\ g/mL$$

$$W_{ext} = 231 \times 0.0000964\ g/mL \times 50.0\ mL = 1.11\ g$$

More complicated chemical systems may require a more universally applicable quantity called the distribution ratio to describe the system. These involve situations in which the analyte species may be found in different chemical states and different equilibrium species, some of which may be extracted while others are not. An example is an equilibrium system involving a weak acid. In such a system, there may be one (or several) protonated species and one unprotonated species. The **distribution ratio**, D, then takes into account all analyte species present:

$$D = \frac{C_A(ext)}{C_A(orig)} \tag{11.4}$$

in which $C_A(ext)$ and $C_A(orig)$ represent the total concentration of all analyte species present in the two phases regardless of chemical state. Further treatment of this situation is beyond the scope of this text.

## 11.4.4  Percent Extracted

The fraction extracted, also known as the **extraction efficiency**, is a measure of the success of an extraction. It is defined as the weight of the analyte found in the extracting solvent after extraction divided by the total weight in the original solvent before extraction. **Percent extracted** is then defined as

$$\% \text{ extracted} = \frac{W_{ext}\ (\text{after extraction})}{W_{orig}\ (\text{before extraction})} \times 100 \tag{11.5}$$

It may be possible to evaluate the percent extracted by a separate experiment. A solution of the analyte in the original solvent may be prepared such that $W_{orig}$ (before extraction) is known. Following this, an extraction is performed on this solution using a particular volume of extracting solvent ($V_{ext}$). This volume of extract is then analyzed quantitatively for the analyte by some appropriate analysis technique. Knowing the concentration of the analyte and the volume of extract converted to liters ($L_{ext}$), one can calculate the percent extracted:

$$\% \text{ extracted} = \frac{[A]_{ext} \text{ (after extraction)} \times L_{ext}}{W_{orig} \text{ (before extraction)}} \times 100 \qquad (11.6)$$

Experiment 35 in this chapter is such an experiment.

### Example 11.3

If 0.239 g of a compound is dissolved in water and this solution is extracted with 50.0 mL of an organic solvent such that the concentration of the compound in the extract is 0.340 g/L, what is the percent extracted?

### *Solution 11.3*
Substituting into Equation (11.6) we have the following:

$$\% \text{ extracted} = \frac{0.340 \text{ g/L} \times 0.0500 \text{ L}}{0.239 \text{ g}} \times 100 = 7.11\%$$

## 11.4.5  Countercurrent Distribution

Just one extraction performed on a solution of a complicated sample will likely not result in total or at least sufficient separation of the analyte from other interfering solutes. Not only will these other species also be extracted to a certain degree with the analyte, but some of the analyte species will likely be left behind in the original solvent as well. Thus, the analyst will need to perform additional extractions on both the extracting solvent, to remove the other solutes that were extracted with the analyte, and the original solution, to remove additional analyte that was not extracted the first time. One can see that dozens of such extractions may be required to achieve the desired separation. Eventually, however, there would be a separation. The process is called **countercurrent distribution**.

In countercurrent distribution, the extracting solvent, after first being in contact with the original solution, is moved to another separatory funnel in which there is fresh original solvent, while fresh extracting solvent is brought into the original funnel and the extractions performed. Then the extracting solvent from the second funnel is moved to a third funnel containing fresh original solvent; the extracting solvent from the first is moved to the second funnel, and fresh extracting solvent is introduced into the first funnel. The process continues in this manner until the desired separation occurs. The concept is illustrated in Figure 11.4. The top half in each segment (a, b, etc.) represents extracting solvent while the lower half represents original solvent. A mixture of ▪ and · is being separated. In each segment, fresh extracting solvent is introduced on the upper left, while fresh original solvent is introduced on the lower right. This illustration shows a complete separation in segments g and h. Some chromatography methods, which are very popular separation methods, are based on this concept and will be discussed in Sections 11.6 to 11.8.

## 11.4.6  Evaporators

Following an extraction procedure, the analyte concentration may not be high enough to be detected with the method chosen, e.g., gas chromatography (Chapter 12). In that case, it is necessary to evaporate some solvent to concentrate the extracted analyte in a small volume of solution, thus dramatically increasing its concentration. It is not uncommon for several hundred milliliters of solution to be reduced to just a few milliliters in this process. The modern procedure is to use a commercial evaporator that gently heats the solution while blowing a stream of nitrogen gas toward the surface. If the solvent is

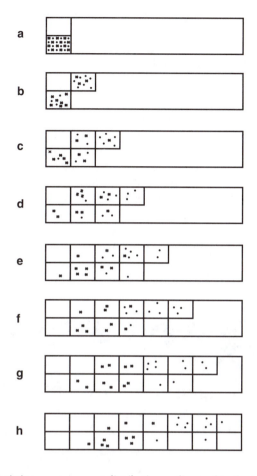

**FIGURE 11.4** A drawing depicting countercurrent distribution as discussed in the text.

volatile, the required evaporation takes place in a rather short period of time and the solution is transferred to a small vial for the analysis. See Workplace Scene 11.3.

## 11.5 Solid–Liquid Extraction

There are instances in which the analyte needs to be extracted from a solid material sample rather than a liquid (see also Section 2.5.3). As in the above discussion for liquid samples, such an experiment is performed either because it is not possible or necessary to dissolve the entire sample or because it is undesirable to do so because of interferences that may also be present. In these cases, the weighed solid sample, preferably finely divided, is brought into contact with the extracting liquid in an appropriate container (not a separatory funnel) and usually shaken or stirred for a period of time, sometimes at an elevated temperature, such that the analyte species is removed from the sample and dissolved in the liquid. The time required for this shaking is determined by the rate of the dissolving. A separatory funnel is not used since two liquid phases are not present, but rather a liquid and a solid phase. A simple beaker, flask, or test tube usually suffices. See Workplace Scene 11.4 for a unique alternative.

Following the extraction, the undissolved solid material is then filtered out and the filtrate analyzed. Examples of this would be soil samples to be analyzed for metals, such as potassium or iron, and polymer film or insulation samples to be analyzed for formaldehyde residue. The extracting liquid may or may not be aqueous. Soil samples being analyzed for metals, for example, utilize aqueous solutions of appropriate

# WORKPLACE SCENE 11.3

When analyzing water from the well field at the City of Lincoln, Nebraska, Water Treatment Plant Laboratory for herbicides, the herbicides are extracted from the water by passing 1 L of the water through a solid phase extractor disk, where the herbicides are adsorbed (see also Section 2.6.1). The adsorbed chemicals are then washed from this disk with a mixture of ethyl acetate and dichloromethane. Because this organic mix always retains some water, the solution is then passed through an anhydrous sodium sulfate column for drying.

The concentration of the herbicide in the resulting solutions is too low to be measured by the standard method, and so the concentrator described in Section 11.4.6 is used to remove most of the organic solvents. The solution is placed in a concentrator sample tube. The concentrator uses heat and a flow of nitrogen gas to evaporate the solvent from the tube such that all the herbicide is then found in approximately 1 mL of solution, thus dramatically increasing the concentration of the herbicide.

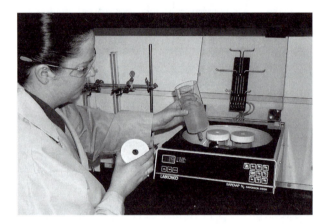

Allison Trentman of the City of Lincoln, Nebraska, Water Treatment Plant Laboratory is placing a concentrator sample tube containing an extracted, dried sample into the unit that uses heat and a stream of nitrogen gas to evaporate the solvent down to just 1 mL.

inorganic compounds, sometimes acids, while soil samples, or the polymer films or insulation samples referred to above, being analyzed for organic compounds utilize organic solvents for the extraction. As with the liquid–liquid examples, the extract is then analyzed by whatever analytical technique is appropriate—atomic absorption for metals and spectrophotometry or gas or liquid chromatography for organics. Methods for increasing the concentration of the analyte in the extract may also be required, especially for organics.

It may be desirable to try to keep the sample exposed to *fresh* extracting solvent as much as possible during the extraction in order to maximize the transfer to the liquid phase. This may be accomplished by pouring off the filtrate and reintroducing fresh solvent periodically during the extraction and then combining the solvent extracts at the end. There is a special technique and apparatus, however, that have been developed, called the **Soxhlet extraction**, which accomplish this automatically (previously described

# WORKPLACE SCENE 11.4

A mechanical vacuum extractor is sometimes used in soil chemistry laboratories to extract chemical components from soil. This extractor is a device that has 24 slots for mounting individual plastic syringes in inverted positions around a carousel. A downward force actually pulls the syringe plungers out and draws an extracting solution downward through the soil samples held in extraction vessels at the top of the device. These extraction vessels are actually other syringe barrels, but in a normal position (not inverted). The resulting extracts are used for a variety of soil chemistry analyses using a variety of instruments, including atomic absorption, inductively coupled plasma, and chromatography.

Patty Jones checks the operation of an automatic mechanical vacuum extractor in the chemistry section of the Natural Resources Conservation Service (NRCS) Soil Survey Laboratory in Lincoln, Nebraska.

in Section 2.5.3). The Soxhlet apparatus is shown in Figure 11.5. The extracting solvent is placed in the flask at the bottom, while the weighed solid sample is placed in the solvent-permeable thimble in the compartment directly above the flask. A condenser is situated directly above the thimble. The thimble compartment is a sort of cup that fills with solvent when the solvent in the flask is boiled, evaporated, and condensed on the condenser. The sample is thus exposed to freshly distilled solvent as the cup fills. When the cup is full, the glass tube next to the cup is also full, and when it (the tube) begins to overflow, the entire contents of the cup are siphoned back to the lower chamber and the process repeated. The advantages of such an apparatus are: 1) fresh solvent is continuously in contact with the sample (without having to introduce more solvent, which would dilute the extract), and 2) the experiment takes place unattended and can conveniently occur overnight if desired.

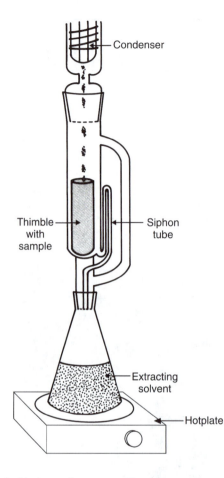

Condenser

Thimble with sample

Siphon tube

Extracting solvent

Hotplate

**FIGURE 11.5**    A drawing of a Soxhlet extraction apparatus.

## 11.6  Chromatography

A myriad of techniques used to separate complex samples come under the general heading of chromatography. The nature of chromatography allows much more versatility, speed, and applicability than any of the other techniques discussed in this chapter, particularly when the modern instrumental techniques of gas chromatography (GC) and high-performance liquid chromatography (HPLC) are considered. These latter techniques are covered in detail in Chapters 12 and 13. In this chapter, we introduce the general concepts of chromatography and give a perspective on its scope. Since there are many different classifications, this will include an organizational scheme covering the different types and configurations that exist.

**Chromatography** is the separation of the components of a mixture based on the different degrees to which they interact with two separate material phases. The nature of the two phases and the kind of interaction can be varied, giving rise to the different types of chromatography that will be described in the next section. One of the two phases is a moving phase (the mobile phase), while the other does not move (the stationary phase). The mixture to be separated is usually introduced into the mobile phase, which then is made to move or percolate through the stationary phase either by gravity or some other force. The components of the mixture are attracted to and slowed by the stationary phase to varying degrees, and as a result, they move along with the mobile phase at varying rates, and are thus separated. Figure 11.6 illustrates this concept.

The mobile phase can be either a gas or a liquid, while the stationary phase can be either a liquid or solid. One classification scheme is based on the nature of the two phases. All techniques which utilize a

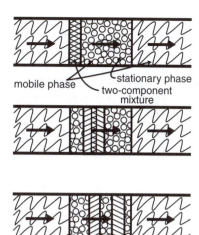

mobile phase — stationary phase
two-component mixture

**FIGURE 11.6** Mixture components separate as they move through the stationary phase with the mobile phase.

gas for the mobile phase come under the heading of gas chromatography. All techniques that utilize a liquid mobile phase come under the heading of liquid chromatography (LC). Additionally, we have gas–liquid chromatography (GLC), gas–solid chromatography (GSC), liquid–liquid chromatography (LLC), and liquid–solid chromatography (LSC), if we wish to stipulate the nature of the stationary phase as well as the mobile phase. It is more useful, however, to classify the techniques according to the nature of the interaction of the mixture components with the two phases. These classifications we refer to in this text as types of chromatography.

## 11.7 Types of Chromatography

### 11.7.1 Partition Chromatography

In Section 11.4 we stated that some chromatography methods are based on the concept of countercurrent solvent extraction. You will recall that this is the technique in which a large number of extractions are performed, with fresh extracting solvent being brought into contact with previously extracted samples, while fresh sample solvent is brought into contact with solvent extract from previous extractions (see Figure 11.4 and accompanying discussion). The extracting solvent can be thought of as continuously moving across the sample solvent while the latter remains stationary. The mixture components, which were initially found in the first segment of sample solvent, then distribute back and forth between the two phases as the extracting liquid moves and are found individually separated at different points along the way according to their individual solubilities in the two solvents. See Figure 11.7.

The extracting solvent in this scenario is the chromatographic mobile phase, while the sample solvent is the stationary phase. Liquid–liquid **partition chromatography** is based on this idea. The mobile phase is a liquid that moves through a liquid stationary phase as the mixture components partition or distribute themselves between the two phases and become separated. The separation mechanism is thus one of the dissolving of the mixture components to different degrees in the two phases according to their individual solubilities in each.

It may be difficult to imagine a liquid mobile phase used with a liquid stationary phase. What experimental setup allows one liquid to move through another liquid (immiscible in the first) and how can one expect partitioning of the mixture components to occur? The stationary phase actually consists of a thin liquid film chemically bonded to the surface of finely divided solid particles, as shown in Figure 11.8. It is often referred to as **bonded phase chromatography** (BPC). Such a stationary phase cannot be removed from the solid substrate by heat, reaction, or dissolving in the mobile phase.

Since the separation depends on the relative solubilities of the components in the two phases, the polarities of the components and of the stationary and mobile phases are important to consider. If the stationary phase is somewhat polar, it will retain polar components more than it will nonpolar components, and thus

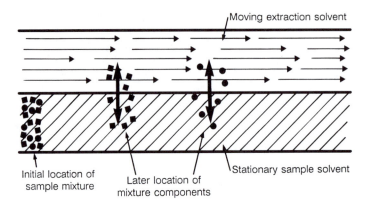

**FIGURE 11.7**   An illustration of how two mixture components separate, by the movement of an extracting solvent across a stationary sample solvent, due to the differing solubilities of the components in the two phases.

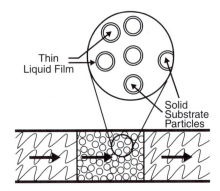

**FIGURE 11.8**   An illustration of partition chromatography. A thin liquid film chemically bonded to the surface of finely divided solid particles is the stationary phase.

the nonpolar components will move more quickly through the stationary phase than the polar components. The reverse would be true if the stationary phase were nonpolar. Of course, the polarity of a liquid mobile phase plays a role too.

The mobile phase for partition chromatography can also be a gas (GLC). In this case, however, the mixture components' solubility in the mobile phase is not an issue—rather, their relative vapor pressures are important. This idea will be expanded in Chapter 12.

In summary, partition chromatography is a type of chromatography in which the stationary phase is a liquid chemically bonded to the surface of a solid substrate, while the mobile phase is either a liquid or gas. The mixture components dissolve in and out of the mobile and stationary phases as the mobile phase moves through the stationary phase, and separation occurs as a result. Examples of mobile and stationary phases will be discussed in Chapters 12 and 13.

## 11.7.2   Adsorption Chromatography

Another chromatography type is **adsorption chromatography**. As the name implies, the separation mechanism is one of adsorption. The stationary phase consists of finely divided solid particles packed inside a tube, but with no stationary liquid substance present to function as the stationary phase, as is the case with partition chromatography. Instead, the solid itself is the stationary phase, and the mixture components, rather than dissolve in a liquid stationary phase, adsorb or stick to the surface of the solid. Different mixture components adsorb to different degrees of strength, which also depends on the mobile phase, and thus again they become separated as the mobile phase moves. The nature of the adsorption involves the interaction of polar molecules, or molecules with polar groups, with a very polar solid stationary phase. Thus, hydrogen bonding, or similar molecule–molecule interactions, is involved.

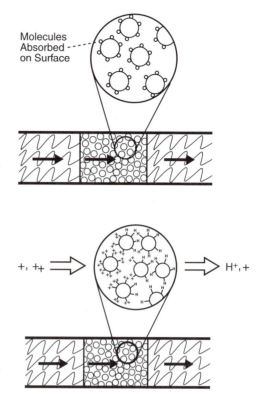

**FIGURE 11.9** An illustration of adsorption chromatography.

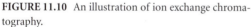

**FIGURE 11.10** An illustration of ion exchange chromatography.

This very polar solid stationary phase is typically silica gel or alumina. The polar mixture components can be organic acids, alcohols, etc. The mobile phase can be either a liquid or a gas. This type of chromatography is depicted in Figure 11.9.

### 11.7.3 Ion Exchange Chromatography

A third chromatography type is **ion exchange chromatography** (IEC). As the name implies, it is a method for separating mixtures of ions, both inorganic and organic. The stationary phase consists of very small polymer resin beads that have many ionic bonding sites on their surfaces. These sites selectively exchange ions with certain mobile phase compositions as the mobile phase moves. Ions that bond to the charged site on the resin beads are thus separated from ions that do not bond. Repeated changing of the mobile phase can create conditions that will further selectively dislodge and exchange bound ions, which then are also separated. This stationary phase material can be either an **anion exchange resin**, which possesses positively charged sites to exchange negative ions, or a **cation exchange resin**, which possesses negatively charged sites to exchange positive ions. The mobile phase can only be a liquid. Further discussion of this type can be found in Chapter 13. Figure 11.10 depicts ion exchange chromatography. See Workplace Scene 11.5.

A special application of ion exchange resins is in the deionizaton of water. Wide columns packed with a mixture of an anion exchange resin that exchanges dissolved anions for hydroxide ions, and a cation exchange resin that exchanges dissolved cations for hydrogen ions are used because water that is passed through such a column becomes free of ions (deionized) since the hydrogen and hydroxide ions combine to form more water.

### 11.7.4 Size Exclusion Chromatography

**Size exclusion chromatography** (SEC), also called **gel permeation chromatography** (GPC) or **gel filtration chromatography** (GFC), is a technique for separating dissolved species on the basis of their size.

# WORKPLACE SCENE 11.5

At Los Alamos National Laboratory in New Mexico the Analytical Chemistry Group (C-AAC) supports the Pu-238 Heat Source Project that fabricates heat sources for use in the space industry. These heat sources have been used on NASA's deep-space probes and on instruments exploring the surface of Mars. The chemical and isotopic purity of the heat sources are critically controlled to ensure dependable service. The Radiochemistry Task Area performs analyses of the heat source material for four radioisotopes: americium-241, plutonium-238, neptunium-237, and uranium-235.

The analysis for Np involves a two-step process. The first step involves an ion exchange column to separate the Np from Pu and U. A neptunium-239 tracer is added before the ion exchange column to track Np recovery. First, the Pu sample is evaporated with hydrobromic acid to produce the Pu(III) and Np(IV) species. The Pu sample is then treated with bromine to cause partial oxidation of Pu(III) to Pu(IV) and of Np(IV) to Np(V) before loading on the ion exchange column. After elution with dilute hydrobromic acid, the eluate is treated with stannous chloride to tie up any remaining bromine. An organic solvent is added to the eluate to remove any remaining Pu by liquid–liquid extraction. After plating and flaming a fraction of the organic phase (to burn off remaining solvent) on a glass slide, the sample is radio-counted using a gross gamma detector, a gross alpha detector, and an alpha spectrometer. The Np concentration can now be calculated.

All of the work is performed in a negative-pressure glove box or open-front hood in a radiological control area. Proper personal protective equipment (PPE) is used, including a lab coat, booties, safety glasses, disposable sleeves, and gloves.

Lisa Townsend, a technician in the Radiochemistry section of the Actinide Analytical Chemistry Group, analyzes bulk components and impurities in plutonium-238 materials used to fabricate heat sources used in space exploration. She utilizes a combination of ion exchange and solvent extraction techniques and determines component concentrations using alpha and gamma radio-counting instrumentation.

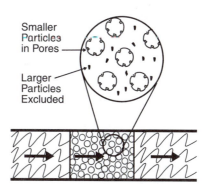

**FIGURE 11.11** An illustration of size exclusion chromatography.

The stationary phase consists of porous polymer resin particles. The components to be separated can enter the pores of these particles and be slowed from progressing through this stationary phase as a result. Thus, the separation depends on the sizes of the pores relative to the sizes of the molecules to be separated. Small particles are slowed to a greater extent than larger particles, some of which may not enter the pores at all, and thus the separation occurs. The mobile phase for this type can also only be a liquid, and it too is discussed further in Chapter 13. The separation mechanism is depicted in Figure 11.11.

## 11.8    Chromatography Configurations

Chromatography techniques can be further classified according to configuration—how the stationary phase is contained, how the mobile phase is configured with respect to the stationary phase in terms of physical state (gas or liquid) and positioning, and how and in what direction the mobile phase travels in terms of gravity, capillary action, or other forces.

Configurations can be broadly classified into two categories: the planar methods and the column methods. The planar methods utilize a thin sheet of stationary phase material and the mobile phase moves across this sheet, either upward (**ascending chromatography**), downward (**descending chromatography**), or horizontally (**radial chromatography**). Column methods utilize a cylindrical tube to contain the stationary phase, and the mobile phase moves through this tube either by gravity, with the use of a high-pressure pump, or by gas pressure. Additionally, with the exception of paper chromatography, those that utilize a liquid for the mobile phase are capable of utilizing all of the types reviewed above. Paper chromatography utilizing unmodified cellulose sheets is strictly partition chromatography (see the next section). If the mobile phase is a gas (**gas chromatography**), the type is limited to adsorption and partition methods. Table 11.1 summarizes the different configurations. Let us consider each individually.

### 11.8.1    Paper and Thin-Layer Chromatography

Paper chromatography and thin-layer chromatography (TLC) constitute the planar methods mentioned above. Paper chromatography makes use of a sheet of paper having the consistency of filter paper (cellulose) for the stationary phase. Since such paper is hydrophilic, the stationary phase is actually a thin film of water unintentionally adsorbed on the surface of the paper. Thus, paper chromatography represents a form of partition chromatography only. The mobile phase is always a liquid.

With thin-layer chromatography, the stationary phase is a thin layer of material spread across a plastic sheet or glass or metal plate. Such plates or sheets may be either purchased commercially already prepared or prepared in the laboratory. The thin-layer material can be any of the stationary phases described earlier, and thus TLC can be any of the four types, including adsorption, partition, ion exchange, and size exclusion. Perhaps the most common stationary phase for TLC, however, is silica gel, a highly polar stationary phase for adsorption chromatography, as mentioned earlier. Also common is pure cellulose, the same material for paper chromatography, and here also we would have partition chromatography. The mobile phase for TLC is always a liquid.

The most common method of configuring a paper or thin-layer experiment is the ascending configuration shown in Figure 11.12. The mixture to be separated is first spotted (applied as a small spot)

**TABLE 11.1**    Summary of the Different Chromatography Configurations

| Geometry | Configuration | Migration Direction | Applicable Types |
|----------|---------------|---------------------|------------------|
| Planar | Paper | Ascending, descending, radial | Partition |
| Planar | Thin layer | Ascending, descending, radial | Adsorption, partition, ion exchange, size exclusion |
| Column | Open column | Descending | Adsorption, partition, ion exchange, size exclusion |
| Column | GC | N/A | Adsorption, partition |
| Column | HPLC | N/A | Adsorption, partition, ion exchange, size exclusion |

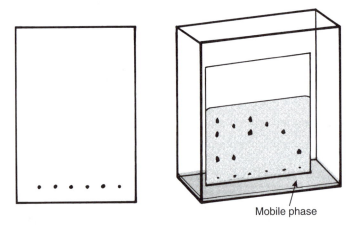

Mobile phase

**FIGURE 11.12**    The paper or thin-layer chromatography configuration. Left, the drawing shows the paper or TLC plate with spots applied. Right, the drawing shows the chromatogram in the developing chamber nearing complete development.

within 1 in. of one edge of a 10-in.-square paper sheet or TLC plate. A typical experiment may be an attempt to separate several spots representing different samples and standards on the same sheet or plate. Thus, as many as eight or more spots may be applied on one sheet or plate. So that all spots are aligned parallel to the bottom edge, a light pencil mark can be drawn prior to spotting. The size of the spots must be such that the mobile phase will carry the mixture components without streaking. This means that they must be rather small—they must be applied with a very small diameter capillary tube or micropipet. An injection syringe with a 25-$\mu$L maximum capacity is usually satisfactory.

Following spotting, the sheet or plate is placed spotted edge down in a developing chamber that has the liquid mobile phase in the bottom to a depth lower than the bottom edge of the spots. The spots must not contact the mobile phase. The mobile phase proceeds upward by capillary action (or downward by both capillary action and gravity if in the descending mode) and sweeps the spots along with it. At this point, chromatography is in progress, and the mixture components will move with the mobile phase at different rates through the stationary phase, and if the mixture components are colored, evidence of the beginning of a separation is visible on the sheet or plate. The end result, if the separation is successful, is a series of spots along a path immediately above the original spot locations, each representing one of the components of the mixture spotted there. See Figure 11.13.

If the mixture components are not colored, any of a number of techniques designed to make the spots visible may be employed. These include iodine staining, in which iodine vapor is allowed to contact the plate. Iodine will absorb on most spots, rendering them visible. Alternatively, a fluorescent substance may be added to the stationary phase prior to the separation (available with commercially prepared plates), such that the spots, viewed under an ultraviolet light, will be visible because they do not fluoresce while the stationary phase surrounding the spots does.

The visual examination of the chromatogram can reveal the identities of the components, especially if standards were spotted on the same paper or plate. Retardation factors (so-called $R_f$ factors) can also

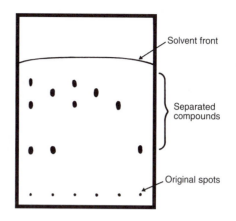

**FIGURE 11.13**   A developed paper or thin-layer chromatogram.

be calculated and used for qualitative analysis. These factors are based on the distance the mobile phase has traveled on the paper (measured from the original spot of the mixture) relative to the distances the components have traveled, each measured from either the center or leading edge of the original spot to the center or leading edge of the migrated spot:

$$R_f = \frac{\text{distance mixture component has traveled}}{\text{distance mobile phase has traveled}} \quad\quad (11.7)$$

These factors, which are fractions less than or equal to 1, are compared to those of standards to reveal the identities of the components.

Quantitative analysis is also possible. The spot representing the component of interest can be cut (in the case of paper chromatography) or scraped from the surface (TLC), dissolved, and quantitated by some other technique, such as spectrophotometry. Alternatively, modern scanning densitometers, which utilize the measurement of the absorbance or reflectance of ultraviolet or visible light at the spot location, may be used to measure quantity.

Using the TLC concept to prepare pure substances for use in other experiments, such as standards preparation or synthesis experiments, is possible. This is called preparatory TLC and involves a thicker layer of stationary phase so that larger quantities of the mixture can be spotted and a larger quantity of pure component obtained.

Additional details of planar chromatography—methods of descending and radial development, how to prepare TLC plates, tips on how to apply the sample, what to do if the spots are not visible—and the details of preparatory TLC, etc., are beyond our scope.

## 11.8.2   Classical Open-Column Chromatography

Another configuration for chromatography consists of a vertically positioned glass tube in which the stationary phase is placed. It is typical for this tube to be open at the top (hence the name **open-column chromatography**), to have an inner diameter on the order of a centimeter or more, and to have a stopcock at the bottom, making it similar to a buret in appearance. With this configuration, the mixture to be separated is placed at the top of the column and allowed to pass onto the stationary phase by opening the stopcock. The mobile phase is then added and continuously fed into the top of the column and flushed through by gravity flow. The mixture components separate on the stationary phase as they travel downward and, unlike the planar methods, are then collected as they elute from the column. In the classical experiment, a fraction collector is used to collect the eluting solution. A typical fraction collector consists of a rotating carousel of test tubes positioned under the column such that fractions of eluate are collected over a period of time, e.g., overnight or a period of days, in individual test tubes. See Figure 11.14. This makes qualitative or quantitative analysis possible through the examination of these fractions by some other technique, such as ultraviolet–visible (UV-VIS) spectrophotometry.

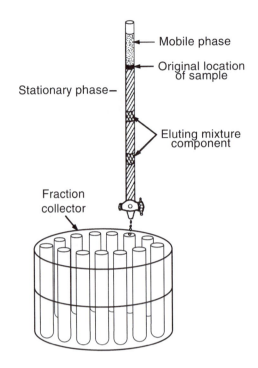

Mobile phase

Original location
of sample

Stationary phase—

Eluting mixture
component

Fraction
collector

**FIGURE 11.14**   The classic open-column chromatography configuration with fraction collector.

The length of the column is determined according to the degree to which the mixture components separate on the stationary phase chosen. Difficult separations would require more contact with the stationary phase and thus may require longer columns. Again, all four types (adsorption, partition, ion exchange, and size exclusion) can be used with this technique.

It is well known that classical open-column chromatography has been largely displaced by the modern instrumental techniques of liquid chromatography. However, open columns are still used where extensive sample cleanup in preparation for the instrumental method is necessary. One can imagine that dirty samples originating, for example, from animal feed extractions or soil extractions, etc., may have large concentrations of undesirable components present. Since only very small samples (on the order of 1 to 20 $\mu$L) are needed for the instrumental method, the time required for obtaining a clean sample by this method, assuming the components of interest are not retained, is the time it takes for an initial amount of mobile phase to pass through from top to bottom. Compared to the overnight time frame, such a cleanup time is quite minimal and does not diminish the speed of the instrumental methods. See Workplace Scene 11.6.

## 11.8.3   Instrumental Chromatography

The concept of the finely divided stationary phase packed inside a column to allow the collection of the individual components as they elute, as discussed in the last section, presents a useful, more practical alternative. One can imagine such a column along with a continuous mobile phase flow system (that does not use gravity for the flow), a device for introducing the mixture to the flowing mobile phase, and an electronic sensor at the end of the column—all of this incorporated into a single unit (instrument) used for repeated, routine laboratory applications. There are two such chromatography configurations that are in common use today: **gas chromatography** and **high-performance liquid chromatography**. These techniques essentially can incorporate all types of column chromatography discussed thus far (HPLC), as well as those types in which the mobile phase can be a gas (GC). Both add a degree of efficiency and speed to the chromatography concept. HPLC, for example, is such a high-performance

# Workplace Scene 11.6

The urine of people who are heavy smokers contains mutagenic chemicals, chemicals that cause mutations in biological cells. Bioanalytical laboratories can analyze urine samples for these chemicals, but the samples must be cleaned up first prior to extraction with methylene chloride. The procedure for this cleanup utilizes an open column chromatography. Columns several inches tall and about an inch wide are prepared by packing them with an adsorbing resin that has been treated with methyl alcohol. The urine samples are passed through these columns as part of the sample preparation scheme.

Stephanie Marquardt, a technician in the bioanalytical laboratory at the MDS Pharma Services laboratory in Lincoln, Nebraska, prepares open columns for the preparation of urine samples for analysis for mutagens.

technique for liquid mobile phase systems that a procedure that might normally take hours or days with open columns can usually be accomplished in a few minutes. The full details of these instrumental techniques are discussed in Chapters 12 and 13.

## 11.8.4 The Instrumental Chromatogram

Common to both GC and HPLC is the fact that they utilize an electronic sensor for detecting mixture components as they elute from the column. These sensors (more commonly called detectors) generate electronic signals to produce a chromatogram. A **chromatogram** is the graphical representation of the separation, a plot of the electronic signal vs. time. The chromatogram is traced on a computer screen or other recording device as the experiment proceeds. For the following discussion, refer to Figure 11.15 throughout.

Initially, in the time immediately after the sample is introduced to the flowing mobile phase and before any mixture component elutes (time 1 in Figure 11.15), the electronic sensor sees only the flowing mobile

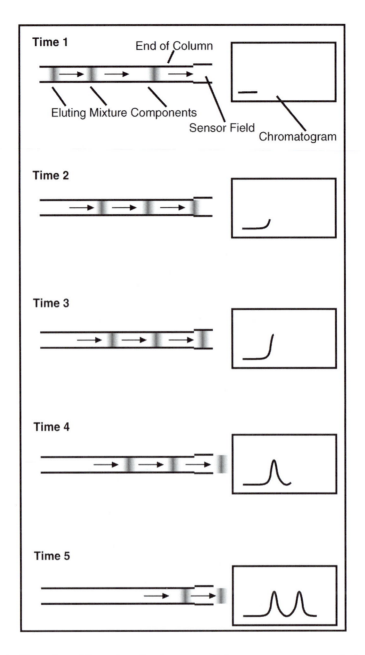

**FIGURE 11.15**  An illustration of the tracing of an instrumental chromatogram as separated mixture components elute from the column and pass through an electronic sensor.

phase. The mobile phase is used as a blank such that the signal generated by the sensor is set to zero. Thus, when only mobile phase is eluting from the column, a baseline, at the zero electronic signal level, is traced on the chromatogram.

When a mixture component emerges and passes into the sensor's detection field (time 2), an electronic signal different from zero is generated and the chromatogram trace deflects from zero. When the concentration of the mixture component is a maximum in the sensor's detection field (time 3), the deflection reaches its maximum. Then, when the mixture component, continuing to move with the mobile phase, has cleared the sensor's detection field (time 4), the sensor sees only mobile phase once again and the electronic signal generated is back to zero. When the second mixture component emerges, the process

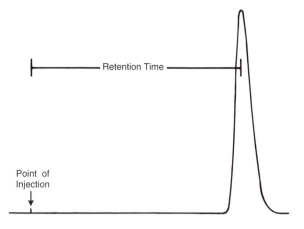

**FIGURE 11.16**   Retention time is the time from when the sample is injected into the flowing mobile to the time the apex of the peak appears on the chromatogram.

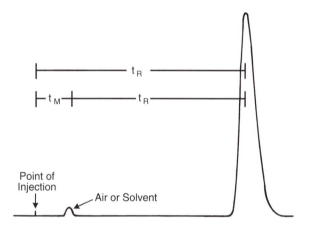

**FIGURE 11.17**   A chromatogram showing the definitions of $t_R$, $t'_R$, and $t_M$.

occurs again (time 5), and when the third component emerges, again, etc. The result is a peak observed on the chromatogram trace for each mixture component to elute.

The terminology associated with the chromatogram is the same for both gas chromatography and high-performance liquid chromatography. For example, the time that passes from the time the sample is first introduced into the flowing mobile phase until the apex of the peak is seen on the chromatogram is called the **retention time** of that component, or the time that that mixture is retained by the column. The retention time is useful for a qualitative analysis of the mixture, as we shall see in the next chapter. It is given the symbol $t_R$. See Figure 11.16. Typically, retention times vary from a small fraction of a minute to about 20 min, although much longer retention times are possible.

Another parameter often measured is the **adjusted retention time**, $t'_R$. This is the difference between the retention time of a given component and the retention time of an unretained substance, $t_M$, which is often air for GC and the sample solvent for HPLC. Thus, the adjusted retention time is a measure of the exact time a mixture component spends in the stationary phase. Figure 11.17 shows how this measurement is made. The most important use of this retention time information is in peak identification, or qualitative analysis. This subject will be discussed in more detail in Chapter 12.

Other parameters sometimes obtained from the chromatogram, which are mostly measures of the degree of separation and column efficiency, are resolution (R), the number of theoretical plates (N), and

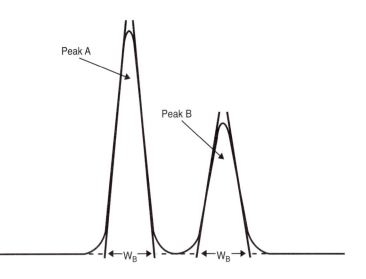

**FIGURE 11.18**   The measurement of width at the base, used in resolution and theoretical plate calculations.

the height equivalent to a theoretical plate (HETP or H). These require the measurement of the width of a peak at the peak base. This measurement is made by first drawing the tangents to the sides of the peaks and extending these to below the baseline, as shown for the two peaks in Figure 11.18. The width at peak base, $W_B$, is then the distance between the intersections of the tangents with the baseline, as shown. **Resolution** is defined as the difference in the retention times of two closely spaced peaks divided by the average widths of these peaks:

$$R = \frac{[t_R(B) - t_R(A)]}{\dfrac{[W_B(A) + W_B(B)]}{2}} \tag{11.8}$$

In this equation, $t_R(B)$ is the retention time for peak B, $t_R(A)$ is the retention time for peak A, $W_B(A)$ is the width at the base of peak A, and $W_B(B)$ is the width at the base of peak B.

### Example 11.4

Two chromatography peaks show retention times of 1.3 and 2.5 min. The width at the base of the first peak is 0.25 min, and the width at the base of the second peak is 0.29 min. What is the resolution?

### *Solution 11.4*

Utilizing Equation (11.8), we have the following:

$$R = \frac{(2.5 - 1.3)}{(0.25 + 0.29)/2} = 4.4$$

R values of 1.5 or more would indicate complete separation. Peaks that are not well resolved would inhibit satisfactory qualitative and quantitative analysis. An example of a chromatogram showing unsatisfactory resolution of two peaks is shown in Figure 11.19.

The **number of theoretical plates**, N, is also of interest. The concept of theoretical plates was discussed briefly in Section 11.3 for distillation. For distillation, one theoretical plate was defined as one evaporation–condensation step for the distilling liquid as it passes up a fractionating column. In chromatography, one theoretical plate is one extraction step along the path from injector to detector. You will recall in Section 11.4 that we spoke of countercurrent distribution (Figure 11.4) and mentioned that chromatography methods are based on this concept. Chromatography is analogous to a series of many extractions, but with one

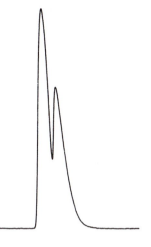

**FIGURE 11.19**   Unresolved peaks.

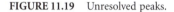

solvent (the mobile phase) constantly moving through the other solvent (the stationary phase), rather than being passed along through a series of separatory funnels, as in countercurrent distribution. The equilibration that would occur in the fictional separatory funnel is one theoretical plate in chromatography. The number of theoretical plates can be calculated from the chromatogram using one of the peaks as follows:

$$N = 16 \left( \frac{t_R}{W_B} \right)^2 \tag{11.9}$$

The **height equivalent to a theoretical plate**, H, is that length of column that represents one theoretical plate, or one equilibration step. Obviously, the smaller value of this parameter, the more efficient the column. The more theoretical plates packed into a length of column, the better the resolution. It is calculated by dividing the column length by N:

$$H = \frac{\text{length}}{N} \tag{11.10}$$

## Example 11.5

The retention time of a certain component on a particular 2.0-m gas chromatography column is 3.1 min. The width at the base for the peak is 0.39 min. How many theoretical plates are in this column and what is the height equivalent to a theoretical plate?

*Solution 11.5*

First, for the number of theoretical plates, we use Equation (11.9):

$$N = 16 \left( \frac{3.1}{0.39} \right)^2 = 1.0 \times 10^3 \text{ plates}$$

Next, for the height equivalent to a theoretical plate, we use Equation (11.10):

$$H = \frac{2.0}{1.0 \times 10^3} = 0.0020 \text{ m, or } 2.0 \text{ mm}$$

A parameter defined as the **relative retention**, or **selectivity**, is often reported for a given instrumental chromatography system as a number that ought to be able to be reproduced from instrument to instrument and laboratory to laboratory regardless of slight differences that might exist in the systems (column lengths, temperature, etc.). This parameter compares the retention of one component (1) with another

(2) and is given the symbol alpha ($\alpha$). It is defined as follows:

$$\alpha = \frac{t'_R(1)}{t'_R(2)} \qquad (11.11)$$

A selectivity equal to 1 would mean that the 2 retention times are equal, which means no separation at all.

**Example 11.6**

The retention time for a mixture component A is 5.4 min, and the retention time for a mixture component B is 3.3 min. The retention time for the solvent is 1.1 min. What is the selectivity for component A relative to B?

*Solution 11.6*

$$\alpha = \frac{t'_R(A)}{t'_R(B)} = \frac{[t_R(A) - t_M]}{[t_R(B) - t_M]} = \frac{(5.4 - 1.1)}{(3.3 - 1.1)} = 2.0$$

Finally, a parameter known as the **capacity factor** may be determined. The capacity factor, symbolized k′ (k-prime) is the adjusted retention time divided by the retention time of an unretained substance, $t_M$, such as air in GC or the sample solvent in HPLC.

$$k' = t'_R / t_M \qquad (11.12)$$

The capacity factor is a measure of the retention of a component per column volume, since the retention time is referred to the time for the unretained component. The greater the capacity factor, the longer that component is retained and the better the chances for good resolution. An optimum range for k′ values is between 2 and 6.

**Example 11.7**

What is the capacity factor if the retention time for the component of interest is 3.2 min and the retention time of the sample solvent is 0.70 min?

*Solution 11.7*

$$k' = \frac{t'_R}{t_M} = \frac{(t_R - t_M)}{t_M} = \frac{(3.2 - 0.70)}{0.70} = 3.6$$

## 11.8.5 Quantitative Analysis with GC and HPLC

The physical size of a peak, or area under the peak, traced on the chromatogram is directly proportional to the amount of that particular component passing through the detector.[*] This, in turn, is proportional to the concentration of that mixture component in the sample solution. It is also proportional to the amount of solution injected, since this too dictates how much passes through the detector. The more material being detected, the larger the peak. Thus, for quantitative analysis, it is important that we have an accurate method for determining the areas of the peaks.

The most popular method of measuring peak area is by integration. Integration is a method in which the series of digital values acquired by the data system as the peak is being traced are summed. The sum is thus a number generated and presented by the data system and is taken to be the peak area. See Figure 11.20. We will discuss in Chapters 12 and 13 exactly how this area is converted to the quantity of analyte in GC and HPLC and the issues involved. See Workplace Scene 11.7.

---

[*]A given amount of one component will not produce a peak the same size as an equal amount of another component, however. This will be discussed further in Chapter 12, Section 12.8.2.

# WORKPLACE SCENE 11.7

The modern analytical laboratory employing instrumental chromatography uses a computer data collection system and associated software to acquire the data and display the chromatogram on the monitor. Parameters important for qualitative and quantitative analysis, including retention times and peak areas, are also measured and displayed. The software can also analyze the data to determine resolution, capacity factor, theoretical plates, and selectivity.

B.J. Kronschnabel of the City of Lincoln, Nebraska, Water Treatment Plant Laboratory examines a chromatogram displayed on a computer monitor.

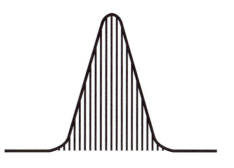

**FIGURE 11.20**  An illustration of how the area of an instrumental chromatography peak is determined by integration. The series of digital values acquired by the data system, represented by the vertical lines, are summed.

## 11.9  Electrophoresis

### 11.9.1  Introduction

Another separation technique utilizes an electric field. An electric field is an electrically charged region of space, such as between a pair of electrodes connected to a power supply. The technique utilizes the varied rates and direction with which different organic ions (or large molecules with charged sites) migrate while under the influence of the electric field. This technique is called **electrophoresis**. **Zone electrophoresis** refers to the common case in which a medium such as cellulose or gel is used to contain the solution. A schematic diagram of the electrophoresis apparatus resembles an electrochemical apparatus in many

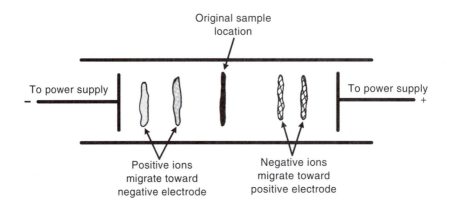

**FIGURE 11.21**    An illustration of the concept of electrophoresis.

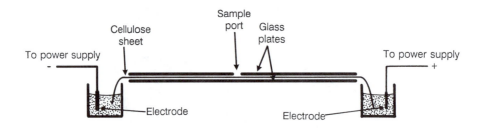

**FIGURE 11.22**    A paper electrophoresis apparatus.

respects. A power supply is needed for connection to a pair of electrodes to create the electric field. The medium and sample to be separated are positioned between the electrodes. The basic concept of the technique and apparatus is illustrated in Figure 11.21.

Electrophoresis is for separating ions, since only ions will migrate under the influence of an electric field, negative ions to the positive electrode and positive ions to the negative electrode. Scientists have found electrophoresis especially useful in biochemistry experiments in which charged amino acid molecules and other biomolecules need to be separated. Thus, application to protein and nucleic acid analysis has been popular (see Chapter 16).

The principles of separation are: 1) ions of opposite charge will migrate in different directions and become separated on that basis, and 2) ions of like charge, while migrating in the same direction, become separated due to different migration rates. Factors influencing migration rate are charge values (e.g., $-1$ as opposed to $-2$) and different mobilities. The mobility of an ion is dependent on the size and shape of the ion as well as the nature of the medium through which it must migrate. The biomolecules referred to above can vary considerably in size and shape; thus electrophoresis is a powerful technique for separating them. As for the medium used, there are some options, including the use of an electrolyte-soaked cellulose sheet (**paper electrophoresis**), a thin gel slab (**gel electrophoresis**), or a capillary tube (**capillary electrophoresis**). The nature of the electrolyte solution used and its pH are also variable.

## 11.9.2   Paper Electrophoresis

Figure 11.22 represents a paper electrophoresis apparatus. The soaked cellulose sheet is sandwiched between two horizontal glass plates with the ends dipped into vessels containing more electrolyte solution. The electrodes are also dipped into these vessels, as shown. The sample is spotted in the center of the sheet, and the oppositely charged ions then have room to migrate in opposite directions on the sheet. Qualitative analysis is performed much as with paper chromatography, by comparing the distances the

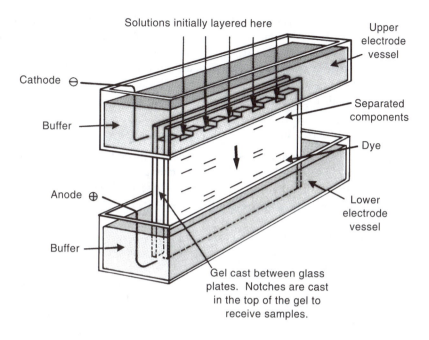

Solutions initially layered here

Upper electrode vessel

Cathode ⊖

Buffer

Separated components

Dye

Anode ⊕

Lower electrode vessel

Buffer

Gel cast between glass plates. Notches are cast in the top of the gel to receive samples.

**FIGURE 11.23**   A gel electrophoresis apparatus.

individual components have migrated to those for standards spotted on the same sheet. (It may be necessary to render the spots visible prior to the analysis, as with paper chromatography.)

Problems associated with paper electrophoresis include the siphoning of electrolyte solution from one vessel, through the paper, to the other vessel when the levels of solution in the two vessels are different, causing the spots to possibly migrate in the wrong direction. The solution to this problem is to ensure that the levels in the two vessels are the same. Another problem stems from the fact that oxidation–reduction processes are occurring at the surfaces of the electrodes. This may introduce undesirable contaminants to the electrolyte solution. These contaminants may in turn migrate onto the sheet. The solution to this problem is to isolate the electrodes while still allowing electrical contact, such as with the use of baffles, to keep the contaminants from diffusing from the vessels.

## 11.9.3   Gel Electrophoresis

A typical gel electrophoresis apparatus is shown in Figure 11.23. The thin gel slab referred to above is contained between two glass plates. The slab is held in a vertical position and has notches at the top where the samples to be separated are spotted or streaked. In the configuration shown in the figure, only downward movement takes place, and thus only one type of ion, cation or anion, can be separated, since there is only one direction to go from the notch.

A tracking dye can be added to the sample so that the analyst can know when the experiment is completed (the leading edge of the sample solvent is visible via the tracking dye). Also, the slab can be removed from the glass plates and a staining dye can be applied that binds to the components, rendering them visible. Components with different mobilities through the gel show up as different bands or streaks on the gel. Qualitative analysis is performed as with paper electrophoresis—standards are applied alongside the samples and the components identified by their positions relative to the standards. Different types of gels can be used.

Finally, **isoelectric focusing** has been a useful extension of basic gel electrophoresis in protein analysis. In this technique, a series of ampholytes is placed on the slab via electrophoresis. An **ampholyte** is a substance whose molecule contains both acidic and basic functional groups. Solutions of different ampholytes have different pH values. Different ampholyte molecules differ in size and therefore will have varying mobilities in the electrophoresis experiment. Thus, these molecules migrate into the slab, take up different positions along the height of the slab, and create a pH gradient through the height of the slab. Amino acid molecules have different mobilities in different pH environments and also have their charges neutralized at particular pH values, rendering them immobile at some position in the gel. Thus mixture components migrate to a particular location in the gel. The pH at which the sample component is neutralized is called the isoelectric point, and this technique is called isoelectric focusing, since samples are separated according to their components' isoelectric points.

### 11.9.4   Capillary Electrophoresis

As the name implies, **capillary electrophoresis** is electrophoresis that is made to occur inside a piece (50 to 100 cm) of small-diameter capillary tubing, similar to the tubing used for capillary GC columns. The tubing contains the electrolyte medium, and the ends of the tube are dipped into solvent reservoirs, as is the paper in paper electrophoresis. Electrodes in these reservoirs create the potential difference across the capillary tube. An electronic detector, such as those described for HPLC (Chapter 13), is on-line and allows detection and quantitative analysis of mixture components.

The very small volume of sample (5 to 50 nL) is injected on one end of the tube, typically the end dipped into the reservoir containing the positive electrode. When the experiment begins, the positive ions migrate quickly through the capillary tube toward the negative electrode in the opposite reservoir and separate on the basis of their mobilities, as in the other electrophoresis techniques. Ultimately, they pass individually through the detector and generate peaks on a recorder trace in a way very similar to that of HPLC. Applications of this technique are, like those described above, very important in biochemistry. However, capillary electrophoresis offers some important advantages that point to a very bright future. These include the use of smaller quantities of samples, minimization of electrical heating during the separation, potential for automation, less solvent to dispose of, and the completion of qualitative and quantitative analysis in a much shorter period of time.

# Experiments

## Experiment 35: Extraction of Iodine with Heptane

Important note: Do not use an analytical balance to weigh iodine because its vapors are corrosive. Be cautious with iodine crystals, being careful not to spill any crystals on the floor or bench top. Wear gloves.
  *Remember safety glasses.*

1. Prepare 100.0 mL of a 0.0075 $M$ solution of iodine ($I_2$) in heptane solvent. Use this solution for the stock solution in step 4.
2. Your instructor may elect to perform this step for you ahead of time. Prepare 25.0 mL of a water solution of iodine that also has potassium iodide added. The potassium iodide is added to increase the solubility so that the iodine dissolves completely in the water. This solution should be 0.025 $M$ in iodine and 0.15 $M$ in KI.
3. From this solution, prepare 25.0 mL of a 0.00025 $M$ solution of iodine in water. Pour this entire solution into a 125-mL separatory funnel. Stopper the funnel.
4. Prepare a series of standard solutions of iodine in heptane, 25.0 mL each, by diluting the stock solution from step 1. The concentrations should be 0.00015, 0.00030, 0.00045, 0.00060, and 0.00075 $M$.
5. Extract the solution in the separatory funnel in step 3 with 10 mL of heptane. After extraction, let stand at least 5 min in the separatory funnel undisturbed.
6. Measure the molecular absorption spectrum in the visible region (400 to 700 nm) of each of the standards. Also measure the molecular absorption spectrum of the heptane layer in the separatory funnel (the extract). You can fill the cuvette conveniently by using a dropper to draw the solution (top layer) out of the separatory funnel. Obtain the maximum absorbance for each and create the standard curve. Determine the concentration of iodine in the extract.
7. Calculate the percent extracted and the distribution coefficient.

## Experiment 36: Solid–Liquid Extraction: Chlorophyll from Spinach Leaves

Chlorophyll is a common chemical found in most green leaves. It is important in photosynthesis.
  *Remember safety glasses.*

### Part A: Determination of the Most Effective Extraction Solvent[*]

The purpose of this experiment is to test four organic solvents of varying polarity to determine which solvent is more effective in extracting chlorophyll from spinach leaves. The solvents are ethyl acetate, dichloromethane, acetone, and ethanol.

### Procedure

1. Obtain four spinach leaves of equal size. Cut or break each into small pieces and place in separate 50-mL beakers. Place 20 mL of ethyl acetate in one beaker, 20 mL of dichloromethane in another, etc., labeling each beaker with the name of the solvent used. Stir each mixture thoroughly with a glass stirring rod (do not cross-contaminate) and then let stand at room temperature for 15 min.
2. After the 15 min, filter the solutions using quantitative filter paper into separate spectrophotometry cuvettes labeled with small labels near the top, indicating which solvent is in the cuvette. While waiting for the filtering, prepare blanks for the spectrophotometer by filling four additional cuvettes with the pure solvents. Again, make sure the cuvettes are appropriately labeled so that you know which cuvette contains which solvent.
3. Using the individual solvents as blanks, acquire a visible absorption spectrum of each sample. Superimpose the four spectra, and determine which solvent is the most effective solvent for extracting chlorophyll from spinach leaves.
4. You should see a peak in the absorbance at about 660 nm. Obtain the precise maximum absorbance at the peak for each solution, and prepare a bar graph (use spreadsheet software) of absorbance vs. solvent.

### Part B: Quantitative Determination of Total Chlorophyll in Spinach Leaves

It has been suggested[**] that 80% acetone in water is a good extraction solvent for chlorophyll in plant leaves. The absorptivity of total chlorophyll in this solvent is 34.5 mL mg$^{-1}$ cm$^{-1}$. Design an experiment, and write a procedure (including final calculation) in which the total chlorophyll in spinach leaves is determined and expressed in milligrams per gram. The procedure should include a quantitative extraction procedure (grinding a weighed, cut-up leaf together with a small volume of the solvent with a mortar and pestle) and a quantitative transfer from the extraction vessel used to a volumetric flask. Ask your instructor to review your procedure before beginning the work.

## Experiment 37: Solid–Liquid Extraction: Determination of Nitrite in Hot Dogs[***]

Manufacturers place nitrite in hot dogs to prevent spoilage and keep the hot dog reddish in color. But when nitrite gets into the blood, it destroys the blood's ability to carry oxygen by converting hemoglobin to methemoglobin. The maximum concentration of nitrite allowed by law in hot dogs is 150 to 160 ppm $NO_2^-$.

  Nitrite is determined by diazo-coupling with sulfanilamide and NEDA (see below for full name), forming a reddish color that can be determined with a visible spectrophotometer.
  *Remember safety glasses.*

---

[*]The idea for this experiment came from Prof. David Baker of Delta College, University Center, Michigan.
[**]http://www.grinnel.edu/courses/bio/qubitmanual/Labs/CO2/Carbon_Dynamics-Part_2.htm (8-25-01)
[***]The idea for this experiment came from Mike Carlson at the Veterinary Diagnostic Center at the University of Nebraska, Lincoln.

**Reagent Preparation**

**Sulfanilamide solution.** Dissolve 10 g of sulfanilamide in 350 mL of water to which has been added 25 mL of concentrated phosphoric acid. Dilute to 500 mL with water. Store in refrigerator. Solution is stable for at least 1 month.

**N-(1-naphthyl) ethylenediamine dihydrochloride (NEDA).** Prepare 100 mL of a 0.5% aqueous solution of NEDA. The solution is stable for at least a month if refrigerated and protected from light. It is a poison—use gloves.

**Nitrite stock solution.** Prepare a 10 mg/L $NO_2^-$ stock solution of potassium nitrite.

**Procedure**

1. Weigh the selected hot dog and cut into three equal pieces. Select one piece and record its weight. Thoroughly grind the piece you just weighed using a mortar and pestle or homogenizer.
2. Add 30 mL of water to hot dog and continue grinding.
3. Pour supernatant into a test tube and centrifuge.
4. Prepare the following nitrite standards in water from the 10 mg/L stock: 0.1, 0.2, 0.3, 0.4, and 0.5 mg/L.
5. Pipet 1 mL of the sample supernatant into a 25-mL beaker. Also pipet 1 mL of each standard and 1 mL of a control sample into separate 25-mL beakers.
6. Add 5.0 mL of sulfanilamide solution, mix, and allow to stand for at least 3 min.
7. Add 0.5 mL of NEDA, mix, and allow to stand for 20 min.
8. Measure the absorbance of standards, control, and samples at 540 nm.
9. Create the standard curve, using the procedure practiced in Experiment 18, by plotting absorbance vs. concentration and determine the concentration of the unknown and control. Calculate the concentration of nitrite in one hot dog.
10. Maintain the logbook for the instrument used in this experiment. Record the date, your name(s), the experiment name or number, the R square value, and the results for the control sample. Also plot the control sample results on a control chart for this experiment posted in the laboratory.

## Experiment 38: The Thin-Layer Chromatography Analysis of Cough Syrups for Dyes

Note: Many cough medicine formulations exhibit the color they have because a food colorant has been added. Such colorants are standard Food, Drug, and Cosmetic (FD&C) dyes, typically blue #1; red #3, #33, and #40; or yellow #5, #6, and #10. In this experiment, identification of such dyes is accomplished by TLC.

*Remember safety glasses.*

1. Weigh 10 mg of each dye standard into 50-mL volumetric flasks, dilute to the mark with distilled water, and shake. If the dye is not water soluble (Lake dyes are insoluble dyes), dissolve it directly in the n-butanol in step 3.
2. Pipet 5 mL of each standard and 5 mL of each sample into separate 50-mL Erlenmeyer flasks.
3. Add five drops of diluted HCl (prepared by diluting 23.6 mL of concentrated HCl to 100 mL in the distilled water) and 5 mL of n-butanol to each flask and shake on a shaker for 30 min.
4. Pour each into test tubes and allow the layers to separate. Spot the butanol (top) layer onto a TLC plate (consult instructor).
5. Place the development solvent (1-propanol:ethyl acetate:concentrated $NH_4OH$, 1:1:1) in the developing tank, and develop the chromatography plate so that the solvent migrates for 12 cm.
6. Allow the plate to dry. Observe and measure $R_f$ values, if possible, and identify the dyes in the cough syrups.

### Experiment 39: The Thin-Layer Chromatography Analysis of Jelly Beans for Food Coloring

*Remember safety glasses.*

1. Add about 25 mL of warm water to three to six jelly beans of each color. Allow to stand until they look white (5 to 10 min). If allowed to stand longer, too much sugar will be extracted with the colors.
2. Remove the candy from the water and filter the extractant through Whatman #1 or equivalent grade filter paper.
3. Transfer the liquid filtrate into a 125-mL separatory funnel. Add 25 mL of butanol and about 1 mL of 10% phosphoric acid. Shake for about 45 sec and then allow the layers to separate.
4. Transfer the organic (top) layer to a small shaker using a transfer pipet. Evaporate the solvent in a steam bath.
5. Take up each dye residue in one or two drops of ammonia solution (prepared by diluting 1 mL of concentrated ammonium hydroxide to 100 mL). Using open-ended capillary tubes, spot each color onto both a cellulose TLC plate and a silica gel TLC plate at about 3 cm from the bottom edge.
6. Spot solutions of commercial food colorings on the same plates. Allow all the spots to dry completely and spot again any that are not dark enough in intensity.
7. Place the plates into a developing chamber and develop the plates using the developing solvent: butanol:ethanol:2% ammonia (3:1:2). Run for 1.5 h or until the solvent has progressed about 15 cm.
8. After drying the plates, identify the food dyes by matching the spot colors and positions with those of the commercial colorants. Your instructor may request that you measure the $R_f$ factors for this identification.

## Questions and Problems

1. Why is a study of modern separation science important in analytical chemistry?
2. Define recrystallization, distillation, fractional distillation, extraction, liquid–liquid extraction, solvent extraction, countercurrent distribution, liquid–solid extraction, and chromatography.
3. In recrystallization, how is it that both soluble and insoluble impurities are removed from the solid being purified?
4. In recrystallization, why is it good to use the minimum amount of solvent at the elevated temperature that will completely dissolve the substance being purified.
5. What does vapor pressure have to do with purification by distillation?
6. Water has a very high vapor pressure compared to the hardness minerals that may be dissolved in it. Why is that a good thing when wanting to purify tap water by distillation?
7. Why does one simple distillation remove most of the dissolved hardness minerals from tap water, while many distillations (or a fractionating column) are required to separate a mixture of two liquids?
8. What special problems exist when trying to separate two organic liquids by distillation?
9. What is a theoretical plate and the height equivalent to a theoretical plate (HETP) as they pertain to distillation?
10. Describe what happens in a liquid–liquid extraction experiment.
11. What are the two important criteria for a successful separation by solvent extraction?
12. Describe the glassware article called the separatory funnel and tell for what purpose and how it is used.
13. Iodine dissolved in water gives the water a brown color. Iodine dissolved in hexane gives the hexane a pink color. Hexane and water are immiscible. How can these facts be used to demonstrate a liquid–liquid extraction?
14. If the distribution coefficient, K, for a given solvent extraction is 169:

(a) What is the molar concentration of the analyte found in the extracting solvent if the concentration in the original solvent *after* the extraction is 0.027 *M*?

(b) What is the molar concentration of the analyte found in the extracting solvent if the concentration in the orginal solvent *before* the extraction was 0.045 *M*?

15. How many moles of analyte are extracted if 50 mL of extracting solvent is brought into contact with 50 mL of original solvent, the concentration of analyte in the original solvent is 0.060 *M*, and the distribution coefficient is 238?

16. In an extraction experiment, it is found that 0.0376 g of an analyte are extracted into 50 mL of solvent from 150 mL of a water sample. If there was orginally 0.192 g of analyte in this volume of the water sample, what is the distribution coefficient? What percent of analyte was extracted?

17. The distribution coefficient for a given extraction experiment is 527. If 0.037 g of analyte are found in 75 mL of the extracting solvent after extraction and 100 mL of original solvent was used, how many grams of analyte were in this volume of original solvent before extraction?

18. How does the distribution coefficient differ from the distribution ratio?

19. What is the purpose of an evaporator? Describe a modern way to evaporate excess solvent.

20. What is solid–liquid extraction? Without going into great detail, what are two different ways to accomplish a solid–liquid extraction?

21. Is a separatory funnel appropriate for a solid–liquid extraction? Explain.

22. Give some examples of the types of samples and analytes to which solid–liquid extraction would be applicable.

23. Give a general definition of chromatography that would apply to all types and configurations. (To say that it is a separation technique is important but not sufficient.)

24. Name the four types of chromatography described in this chapter and give the details of the separation mechanism of each.

25. How does partition chromatography differ from adsorption chromatography?

26. Find, in a reference book, a description of the Craig countercurrent distribution apparatus and discuss its design as it relates to the description of countercurrent extraction presented in this chapter.

27. Consider a mixture of compound A, a somewhat nonpolar liquid, and compound B, a somewhat polar liquid. Tell which liquid, A or B, would emerge from a chromatography column first under the following conditions and why:

(a) A polar liquid mobile phase and a nonpolar liquid stationary phase

(b) A nonpolar liquid mobile phase and a polar liquid stationary phase

28. We have studied four chromatography types. One of these is partition chromatography. Answer the following questions concerning partition chromatography yes or no:

(a) Can the mobile phase be a solid?

(b) Can the mobile phase be a liquid?

(c) Can the mobile phase be a gas?

(d) Can the stationary phase be a solid?

(e) Can the stationary phase be a liquid?

(f) Can the stationary phase be a gas?

29. (a) List all chromatography types that utilize a liquid stationary phase. If there are none, say so.

(b) List all chromatography types that can utilize a gaseous stationary phase. If there are none, say so.

(c) List all chromatography types that can be used with open-column chromatography. If there are none, say so.

(d) List all chromatography configurations that can utilize partition chromatography.

(e) List all chromatography configurations that can utilize size exclusion chromatography.

30. Tell what each of the following refer to: GC, LC, GSC, LSC, GLC, LLC, BPC, IEC, SEC, GPC, and GFC.

31. One type of chromatography separates small molecules from large ones. Name this type and tell how such a separation occurs.

32. Differentiate between the use of a cation exchange resin and an anion exchange resin in terms of whether the charged sites are positive or negative and whether cations or anions are exchanged.

33. (a) Name four chromatography configurations.
    (b) Choose one of your answers to (a) and tell how the stationary phase is configured relative to the mobile phase and what force is used to move the mobile phase through the stationary phase.

34. Compare high-performance liquid chromatography with open-column chromatography in terms of
    (a) How the stationary phase is contained
    (b) The force that moves the mobile phase through the stationary phase
    (c) How quantitative analysis takes place once the separation is completed

35. Compare thin-layer chromatography to gas chromatography in terms of
    (a) What force moves the mobile phase through the stationary phase
    (b) Whether it is described as planar or column
    (c) How qualitative analysis is performed once the separation is completed

36. What does the abbreviation LSC stand for? Give two examples of chromatography types that can be abbreviated as LSC.

37. Look at Figure 11.15 and explain how it is that a peak is traced on a chromatogram as separated mixture components elute from an instrumental chromatography column.

38. Match each item in the left column to the single item in the right column that most closely associates with it.
    (a) Paper chromatography
    (b) Ion exchange chromatography
    (c) Electrophoresis
    (d) Gas chromatography
    (e) Adsorption chromatography
    (f) Gel permeation chromatography
    (g) TLC
    (h) Open-column chromatography

    (i) Thin layer chromatography
    (j) Size exclusion chromatography
    (k) Stationary phase is water
    (l) Technically not chromatography
    (m) Column effluent is collected and analyzed
    (n) Uses a stationary phase that trades ions with the mobile phase
    (o) An instrumental method
    (p) Involves a mechanism in which the mixture components selectively stick to a solid surface

39. Fill in the blanks with a term from the following list:

    | | |
    |---|---|
    | Stationary phase | Mobile phase |
    | Adsorption | Partition |
    | Paper | Thin layer |
    | HPLC | Ion exchange |
    | Size exclusion | Electrophoresis |
    | Detector | |

    (a) In gas chromatography, the material packed within the column is usually a powdered solid that has a thin liquid film adsorbed on the surface. This thin liquid film is called the ____. This type of gas chromatography falls into the general classification of ___ chromatography.
    (b) In one type of chromatography, the components of the mixture are separated on the basis of the relative sizes of the molecules. This is called ____ chromatography.
    (c) The technique in which separation of charged species is effected by the use of an electric field is called ___.
    (d) The fact that gravity flow of a liquid through a packed column is time-consuming led to the development of ____.

**TABLE 11.2**   Table for Problem 40

| Configuration | Type | Stationary Phase | To Separate Mixtures of |
|---|---|---|---|
| Paper | Partition | | |
| HPLC | | | Different size molecules |
| | Ion exchange | | |
| | Partition | | Gases and volatile liquids |
| | Adsorption | Layer of adsorbent spread on glass plate | |

    (e)  GSC is a type of ___ chromatography.

    (f)  A type of chromatography in which a layer of adsorbent is spread on a glass or plastic plate is called _____ chromatography.

40. Fill in the blanks in Table 11.2.

41. Answer the following questions either true or false.

    (a)  The stationary phase percolates through a bed of finely divided solid particles in adsorption chromatography.

    (b)  The mobile phase can be either a liquid or a gas.

    (c)  The mobile phase is a moving phase.

    (d)  Partition chromatography can only be used when the mobile phase is a liquid.

    (e)  Adsorption includes LSC.

    (f)  In partition chromatography, the mobile phase partitions or distributes itself between the sample solution and the stationary phase.

    (g)  If the stationary phase is a polar liquid substance, nonpolar components will elute first.

    (h)  In GLC, the separation mechanism is partitioning.

    (i)  Size exclusion chromatography separates components on the basis of their charge.

    (j)  Gel permeation chromatography is another name for size exclusion chromatography.

    (k)  Ion exchange chromatography is a technique for separating inorganic ions in a solution.

    (l)  Paper chromatography is a type of LLC.

    (m)  Thin-layer chromatography and open-column chromatography are two completely different configurations of GSC.

    (n)  It is useful to measure $R_f$ values in open-column chromatography.

    (o)  $R_f$ values are used for quantitative analysis.

    (p)  TLC refers to thin-layer chromatography.

    (q)  HPLC refers to high-performance liquid chromatography.

42. Match each term to one of the statements:

| | |
|---|---|
| Partition chromatography | Adsorption chromatography |
| Ion exchange chromatography | Size exclusion chromatogaphy |
| Paper chromatography | Thin-layer chromatography |
| Open-column chromatography | Gas chromatography |
| High-performance liquid chromatography | Electrophoresis |

    (a)  A chromatography configuration in which the stationary phase is spread across a glass or plastic plate.

    (b)  A chromatography type in which the stationary phase is a liquid.

    (c)  A chromatography type designed to separate dissolved ions.

    (d)  A chromatography configuration that utilizes a high-pressure pump to move the mobile phase through the stationary phase.

    (e)  One of two chromatography types that have application in GC.

    (f)  A chromatography configuration that utilizes a fraction collector.

    (g)  The only chromatography type described by the letters GLC.

(h) One of two chromatography configurations in which the mobile phase moves by capillary action opposing gravity.

43. What is the difference between retention time and adjusted retention time?

44. Draw an example of an instrumental chromatogram showing one peak, label the x- and y-axes, and show clearly how the retention time and adjusted retention time are measured.

45. Consider the separation of components A and B on a GC column. The retention time for component A is 1.4 min and for component B is 2.1 min. The width at the base for the peak due to A is 0.38 min and for the peak due to B is 0.53 min. Calculate the resolution, R, and tell whether the two peaks are considered to be resolved and why.

46. Calculate the number of theoretical plates indicated for the data for component A in problem 45 and also calculate the height equivalent to a theoretical plate given that the column length is 72 in. Comment on the significance of your answers.

47. List three ways to increase the number of theoretical plates of a GC experiment.

48. Calculate the capacity factor when the retention time of a mixture component is 4.51 min and the retention time of the sample solvent is 0.95 min. Is it within the optimum range? Explain.

49. Calculate the selectivity for propyl paraben compared to methyl paraben given the following data:

$$t_R \text{ (methyl paraben)} = 3.23 \text{ min}$$

$$t_R \text{ (propyl paraben)} = 5.16 \text{ min}$$

$$t_M = 0.87 \text{ min}$$

Is this a good separation? Explain.

50. Calculate the capacity factor for the two mixture components in problem 49.

51. Draw an example of an instrumental chromatography peak and show in your drawing and describe in words the specific method by which peak area is measured by integration.

52. Describe the following: electrophoresis, zone electrophoresis, paper electrophoresis, gel electrophoresis, isoelectric focusing, and capillary electrophoresis.

53. Describe in detail the general mechanism of the separation of ions by electrophoresis.

54. What is capillary electrophoresis and what advantages does it have over other conventional electrophoresis techniques.

55. **Report**

Find a real-world noninstrumental chromatographic analysis in a methods book (AOAC, USP, ASTM, etc.) or journal and report on the details of the procedure according to the following scheme:

(a) Title

(b) General information, including type of material examined, the name of the analyte, sampling procedures, and sample preparation procedures

(c) Specifics, including chromatography configuration (paper, thin layer, open column), type of chromatography (include the specific stationary phase used), mobile phase characteristics, how the data are gathered, concentration levels for standards, how standards are prepared, and potential problems

(d) Data handling and reporting

(e) References

# 12

# Gas Chromatography

## 12.1 Introduction

One instrumental chromatography configuration discussed briefly in Chapter 11 is one in which the mobile phase is a gas and the stationary phase is either a liquid or a solid. This configuration is called **gas chromatography**, known simply as GC. It is also known as either GLC (**gas–liquid chromatography**) or GSC (**gas–solid chromatography**) in order to stipulate whether the stationary phase is a liquid or a solid, respectively. Most gas chromatography procedures utilize a liquid stationary phase, or GLC, and thus the chromatography type (see Chapter 11 for the distinction between type and configuration) is partition chromatography most of the time. The only other possible type that is applicable here is adsorption chromatography. Thus, GSC refers to this latter type.

For the novice having just read Chapter 11, it may be difficult to visualize a chromatography procedure that utilizes a gas for a mobile phase. The mobile phase gas, often called the **carrier gas**, is typically purified helium or nitrogen. It flows from a compressed gas cylinder via the regulated pressure of the cylinder and a flow controller through the column containing the stationary phase, where the separation takes place. The sample or standard is injected into the flowing carrier gas ahead of the column, and following the column, separated mixture components are electronically detected and peaks appear on the chromatogram, as discussed in Chapter 11. The analytical strategy for GC is presented in Figure 12.1.

The fact that the mobile phase is a gas creates additional unique features, one of which has to do with the mechanism of the separation. When we discussed partition chromatography in Chapter 11, we were, for the most part, assuming a liquid mobile phase. The partitioning mechanism in that case involved only the relative solubilities of the mixture components in the two phases. With GLC, the partitioning mechanism does involve the solubilities of the mixture components in one phase, the liquid stationary phase. However, it also involves the relative vapor pressures of the components since we must have partitioning between one phase that is a liquid and another phase that is a gas. Further, since the mobile phase is a gas, the mixture components must also be gases, or at least liquids with relatively high vapor pressures (elevated temperatures are used). This is true in order for the mixture components to be carried through the column as gases using a gaseous mobile phase.

Let us briefly review the concept of vapor pressure. Simply defined, **vapor pressure** is the pressure exerted by the molecules of a gas in equilibrium with its liquid present in the same sealed container. It is a measure of the tendency of the molecules of a liquid substance to escape the liquid phase and become gaseous. Compounds that have a low molecular weight and are nonpolar have a high vapor pressure. Compounds that have a high molecular weight and are polar have a low vapor pressure. Thus, example liquid substances with high vapor pressures are diethyl ether, acetone, and chloroform. Example liquid substances with low vapor pressures are water, high-molecular-weight alcohols, and aromatic halides. In gas chromatography, substances with high vapor pressures will be strongly influenced by the moving gaseous mobile phase and will emerge from the column quickly (short retention time) if their solubilities

**TABLE 12.1**   A Summary of Retention Concepts for GC

| Component's Vapor Pressure | Component's Solubility in Stationary Phase | Retention Time |
|---|---|---|
| High | Low | Short |
| High | High | Intermediate |
| Low | Low | Intermediate |
| Low | High | Long |

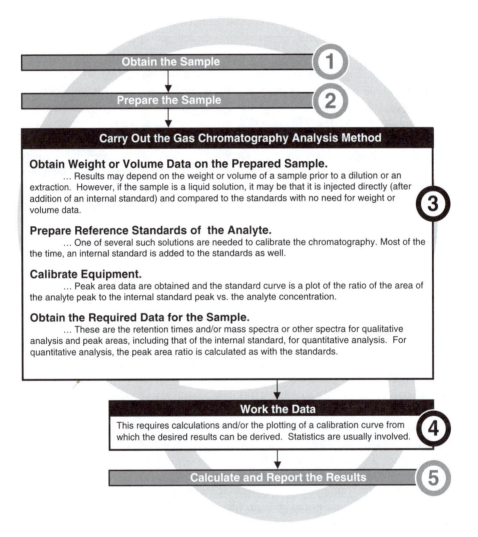

**FIGURE 12.1**   The analytical strategy for gas chromatography.

in the stationary phase are low. If their vapor pressures are high, but their solubilities in the stationary phase are also high, then they will emerge more slowly (intermediate retention time). If their vapor pressure is low, but they have a high solubility in the stationary phase, the time required for emergence from the column will be long (long retention time). Figure 12.2 and Table 12.1 summarize the vapor pressure concepts discussed here.

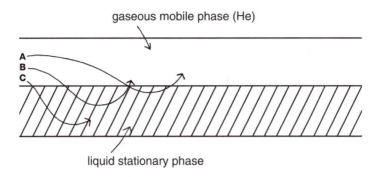

gaseous mobile phase (He)

liquid stationary phase

**FIGURE 12.2**    An illustration of the vapor pressure and solubility effects discussed in the text. Components A, B, and C have vapor pressures decreasing from A to B to C and have solubilities in the stationary phase increasing from A to B to C.

## 12.2    Instrument Design

Vapor pressures vary substantially with temperature. Because of this, it is useful to be able to use an elevated column temperature and to be able to carefully control the temperature of the column. For this reason, the column is always placed in a thermostated oven inside the instrument. Additionally, a sample introduction system is needed at the head of the column that will allow either samples of gases to be introduced into the flowing carrier gas (gas sampling valves) or samples of volatile liquids to be introduced so that they are immediately vaporized and carried on to the column in the gas phase. This latter system utilizes an injection configuration that includes a high-temperature **injection port** equipped with a rubber septum. The sample to be separated is drawn into a microliter syringe with a sharp beveled tip. This sharp tip is then used to pierce the septum so that the sample is introduced into the flowing carrier gas, flash vaporized due to the high temperature, and then carried onto the column.

Finally, a detection system is required at the opposite end of the column that will detect when a substance other than the carrier gas elutes. This detector can consist of any one of a number of different designs, but the purpose is to generate the electronic signal responsible for the chromatogram displayed on the data system screen and from which the qualitative and quantitative information is obtained.

The entire system is drawn in Figure 12.3. By way of summary, the helium or nitrogen flows from the compressed gas cylinder (pressure regulated by a pressure regulator on the gas cylinder) through the injection port, where the sample is introduced, onto the column, and subsequently through the detector. The electronic signal generated by eluting components is fed into the data system, where the peaks appear on the screen as described in Chapter 11 (see Section 11.8.4 and Figure 11.15). A sample **chromatogram** for a four-component mixture is shown in Figure 12.4.

Finally, it is important to recognize that there are several heated zones within the instrument. We have already mentioned the column oven. Temperatures here are typically in the 100 to 150°C range, but much higher temperatures are sometimes used. We have also mentioned the fact that the injection port area is heated (to cause vaporization of volatile liquid samples). The temperature in this zone depends on the volatility of the components but is typically in the 200 to 250°C range. Lastly, the detector must also be heated, mostly to prevent the condensation of the vapors as they pass through. GC detectors are designed to detect gases and not liquids. The temperature here is also usually in the 200 to 250°C range.

## 12.3    Sample Injection

The injection port is designed to introduce samples quickly and efficiently. Most GC work involves the separation of volatile liquid mixtures. In this case, the injection port must be heated to a rather high temperature in order to flash vaporize small amounts of such samples so that the entire amount is immediately carried to the head of the column by the flowing helium. The most familiar design consists

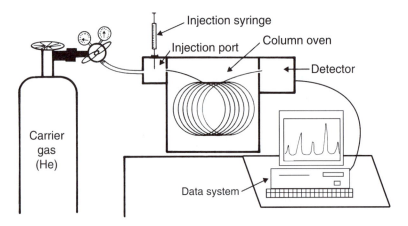

**FIGURE 12.3**    A drawing of a gas chromatograph.

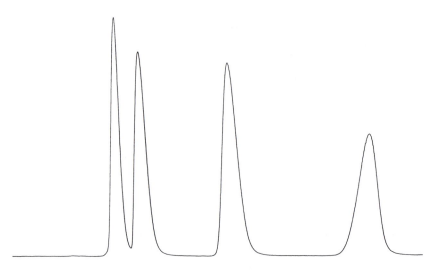

**FIGURE 12.4**    A sample gas chromatogram of a four-component mixture.

of a small glass-lined or metal tube equipped with a rubber septum on one end to accommodate injection with a syringe. As the helium "blows" through the tube, a small volume of injected liquid (typically on the order of 0.1 to 3 $\mu$L) is flash vaporized and immediately carried onto the column. A variety of sizes of syringes and some additional features (such as the use of a septum purge gas) are available to make the injection easy and accurate. The rubber septum, after repeated sample introduction, can be replaced easily. Sample introduction systems for gases (gas sampling valves) and solids are also available.

A volume of liquid as small as 0.1 to 3 $\mu$L (appearing as an extremely small drop) may seem to be extraordinarily small. When vaporized, however, such a volume is much larger and will occupy an appropriate volume in the column. Also, the detection system, as we will see later, will detect very small concentrations even in such a small volume. As a matter of fact, too large a volume is a concern to the operator, since columns, especially capillary columns, can become overloaded even with volumes that are very small. Overloading means that the entire vaporized sample will not fit onto the column all at once and will be introduced over a period of time. A good separation would be in jeopardy in such an instance. To guard against overloading of capillary columns, **split injectors** have been invented. In these injectors, only a fraction of the liquid from the syringe actually is passed to the column. The remaining

**TABLE 12.2**  Typical Injection Volumes for
Various Column Diameters

| Column Diameters | Maximum Injection Volumes ($\mu$L) |
|---|---|
| 1/4 in. (packed column) | 100 |
| 1/8 in. (packed column) | 20 |
| Capillary (open tubular) | 0.1 |

portion is split from the sample and vented to the air. A popular split ratio is 1% to the column and 99% to the split vent and septum purge. Most split injectors are also capable of being **splitless**, meaning that the split valve is closed and all the injected volume goes onto the column. These are called **split–splitless** injectors. There are basically four kinds of injectors: 1) **direct injection**, in which the injected sample is vaporized in the hot injector and 100% of it is carried onto the column, 2) **split injection**, in which the injected sample is vaporized in the hot injector and then split, 3) **splitless injection**, in which nearly all of the sample is vaporized in the hot injector and carried onto the column because the split valve is closed, and 4) **on-column injection**, in which the tip of the syringe is actually inside the column when the injection is made.

As implied above, the appropriate range of sample injection volume depends on column diameter. As we will see in the next section, column diameters vary from capillary size (0.2 to 0.3 mm) to 1/8 and 1/4 in. Table 12.2 gives the typical injection volumes suggested for these column diameters. The capillary columns are those in which the overloading problem mentioned above is most relevant. Injectors preceding the 1/8 in. or larger columns are not split.

The accuracy of the injection volume measurement can be very important for quantitation, since the amount of analyte measured by the detector depends on the concentration of the analyte in the sample as well as the amount injected. In Section 12.8, a technique known as the **internal standard** technique will be discussed. Use of this technique negates the need for superior accuracy with the injection volume, as we will see. However, the internal standard is not always used. Very careful measurement of the volume with the syringe in that case is paramount for accurate quantitation. Of course, if a procedure calls only for identification (Section 12.7), then accuracy of injection volume is less important. See Workplace Scene 12.1 for an example of a purge-and-trap procedure for injecting a GC sample.

## 12.4 Columns

### 12.4.1 Instrument Logistics

GC columns, unlike any other type of chromatography column, are typically very long. Lengths varying from 2 up to 300 ft or more are possible. Additionally, as mentioned previously, it is important for the column to be kept at an elevated temperature during the run in order to prevent condensation of the sample components. Indeed, maintaining an elevated temperature is very important for other reasons, as we shall see in the next section. The obvious logistical problem is how to contain a column of such length and be able to simultaneously control its temperature.

Such a long column is wound into a coil and fits nicely into a small oven, perhaps 1 to 3 ft$^3$ in size. This oven probably constitutes about half of the total size of the instrument. See Figure 12.3. Connections are made through the oven wall to the injection port and the detector. The temperatures of column ovens typically vary from 50 to 150°C, with higher temperatures possible in procedures that require them. A more thorough discussion of this subject is found in Section 12.5.

GC instruments are designed so that columns can be replaced easily by disconnecting a pair of brass fittings inside the oven. This not only facilitates changing to a different stationary phase altogether, but

# WORKPLACE SCENE 12.1

We mentioned in Chapter 2 (Section 2.6.1) that a **purge-and-trap** procedure sometimes precedes an analysis by gas chromatography. An example of this procedure is found in the City of Lincoln, Nebraska, Water Treatment Plant Laboratory. Water treatment includes chlorination. When water is chlorinated, chlorine reacts with organic matter to form trihalomethanes (THMs), such as chloroform, bromoform, bromodichloromethane, and chlorodibromomethane. THMs in water are regulated by the Safe Drinking Water Act, and so the laboratory must analyze the treated water to determine their concentration.

In the purge-and-trap procedure, vials filled to the brim with the water samples are loaded into an auto-sampler, and then when the unit is operating, samples are drawn, one by one, into a tube where helium sparging occurs. Because the THMs are volatile, the helium sparging draws them out of the samples. The helium–THM gaseous mixture then flows through a trap in which the THMs are adsorbed and concentrated. This is followed by a desorption step in which the desorbed THMs are guided to the GC column. A Hall detector is used.

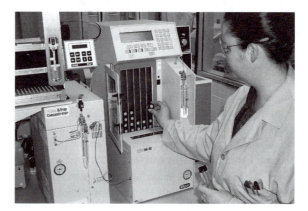

Allison Trentman of the City of Lincoln, Nebraska, Water Treatment Plant Laboratory loads vials of water samples into the auto-sampler in preparation for analysis for trihalomethanes. The helium sparging tube is the vertical tube on the front of the unit on the left.

also allows the operator to replace a given column with a longer one containing the same stationary phase. The idea here is to allow more contact with the stationary phase, which in turn is bound to improve the separation. For example, a 12-ft column has more resolving power than a 6-ft column simply because the mixture components have more contact with the stationary phase.

## 12.4.2  Packed, Open-Tubular, and Preparative Columns

It was indicated earlier that column lengths of up to 300 ft are not unusual. It should be mentioned here that the longer a column with the stationary phase tightly packed (a so-called **packed column**), the greater the gas pressure required to sustain the flow of carrier gas. A 20-ft length is approximately the

**FIGURE 12.5**   A comparison of GC packed and capillary columns.

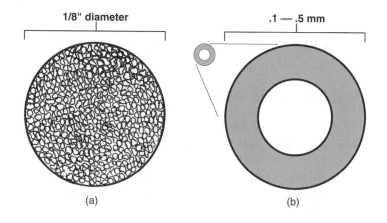

**FIGURE 12.6**   Cross sections of (a) a packed column and (b) an open-tubular capillary column.

upper limit for the length of a packed column. A more popular alternative is the **open-tubular capillary column**. Instead of tightly packed solid substrate particles holding the liquid stationary phase inside the column (see Section 11.7), the stationary phase is made to adsorb on the inside wall of a small-diameter capillary tube (0.1 to 0.53 mm) so that the tube remains open to gas flow in the center. A design such as this offers very little resistance to gas flow and can be made hundreds of feet long without having to utilize a large pressure. Such columns are extremely popular. Columns with a diameter of 0.53 mm are known as **megabore** columns. See Figures 12.5 and 12.6.

In addition to the **analytical columns** (columns used mainly for analytical work), so-called **preparative columns** may also be encountered. Preparative columns are used when the purpose of the experiment is to prepare a pure sample of a particular substance (from a mixture containing the substance) by GC for use in other laboratory work. The procedure for this involves the individual condensation of the mixture components of interest in a cold trap as they pass from the detector and as their peak is being traced on the recorder. While analytical columns can be suitable for this, the amount of pure substance generated is typically very small, since what is being collected is only a fraction of the extremely small volume injected. Thus, columns with very large diameters (on the order of inches) and capable of very large injection volumes (on the order of milliliters) are manufactured for the preparative work. Also, the detector used must not destroy the sample, like the flame ionization detector (Section 12.6) does, for example. Thus, the thermal conductivity detector (Section 12.6) is used most often with preparative gas chromatography.

## 12.4.3 The Nature and Selection of the Stationary Phase

The liquid stationary phase in a GLC packed column is adsorbed on the surface of a solid substrate (also called the support). This material must be inert and finely divided (powdered). The typical diameter of a substrate particle is 125 to 250 $\mu$, creating a 60- to 100-mesh material. These particles are of two general types: diatomaceous earth and Teflon™. **Diatomaceous earth**, the decayed silica skeletons of algae, is most commonly referred to by the manufacturer's (Johns Manville's) trade name, **Chromosorb**™. Various types of Chromosorb, which have had different pretreatment procedures applied, are available, such as Chromosorb P, Chromosorb W, and Chromosorb 101-104. The nature of the stationary phase as well as the nature of the substrate material are both usually specified in a chromatography literature procedure, and columns are tagged to indicate each of these as well.

Since the interaction of the mixture components with the liquid stationary phase plays the key role in the separation process, the nature of the stationary phase is obviously important. Several hundred different liquids useful as stationary phases are known. This means that the analyst has an awesome choice when it comes to selecting a stationary phase for a given separation. It is true, however, that relatively few such liquids are in actual common use. Their composition is frequently not obvious to the analyst because a variety of common abbreviations have come to be popular for the names of some of them. Table 12.3 lists a number of common stationary phases, their abbreviated names, a description of their structures, and the classes of compounds (in terms of polarity) for which each is most useful.

The selection of a stationary phase depends largely on trial and error or experience, with consideration given to the polar nature of the mixture, as noted in Table 12.3 or a similar table. The usual procedure is to select a stationary phase, based on such literature information, and attempt the separation under the various conditions of column temperature, length, carrier gas flow rate, etc., to determine the optimum capability for separating the mixture in question. If this optimum resolution is not satisfactory (see Section 12.6), then an alternate selection is apparently required.

More experienced chromatographers may refer to the McReynolds constants for a given stationary phase as a measure of its resolving power. A complete discussion of this subject, however, is beyond the scope of this text.

**TABLE 12.3**   Some Stationary Phases for GLC

| Abbreviated or Non-Descriptive Name | Structure, Descriptive Name, or Other Description | Useful for Mixtures of Compounds That are |
|---|---|---|
| FFAP | A teflon-based material | Highly polar |
| Casterwax | $CH_3(CH_2)_5$—CH—$CH_2$—CH=CH—$(CH_2)_{17}$—C(=O)OH, with OH | Highly polar |
| Carbowax (variety of molecular weights) | HO—[$CH_2$—$CH_2$—O]$_n$—H | Polar |
| XE-60 also XF-1150, SF-1125 | $Si(CH_3)_3$—O—[Si—O —— $Si(CH_3)_3$]$_n$ with $CH_3$ and $CH_2$—$CH_2$—$CH_2$—C≡N | Polar |
| OV-17 | Methyl, phenyl, silicone (a silicone oil) | Somewhat polar |
| OV-101 | Liquid methyl silicone | Nonpolar |
| OV-1 | Methyl silicone | Nonpolar |
| SE-30 | [—Si($CH_3$)—O—Si($CH_3$)—O—]$_n$ | Nonpolar |
| Apiezon (Various types) | A grease | Nonpolar |
| Squalane | High-molecular-weight hydrocarbon ($C_{30}$) | Nonpolar |

## 12.5    Other Variable Parameters

### 12.5.1    Column Temperature

Both the vapor pressure of a substance and the solubility of a substance in another substance change with temperature. Figure 12.7 shows, for example, how the vapor pressure can change with temperature, and Figure 12.8 shows how the solubility can change with temperature. It should not be surprising then that the precise control of the temperature of a GLC column is very important, since, as we have indicated, the separation depends on both vapor pressure and solubility. Both **isothermal** (constant) and **programmed** (continuously changing) temperature experiments are possible. For simple separations, the isothermal mode may well be sufficient — there may be sufficient differences in the mixture components' vapor pressures and solubilities to effect a good separation at the chosen temperature. However, for more complicated mixtures, a complete separation is less likely in the isothermal mode.

For example, consider gasoline, which has a good number of highly volatile components as well as a significant number of less volatile components. It is possible that at a temperature of, say, 100°C, some of the less volatile components will be resolved, but the more volatile ones will pass through unresolved and have very short retention times. A lower temperature of, say, 40°C, may cause complete resolution of these more volatile components, but would result in unwanted long retention times for the less volatile components and perhaps also result in poorly shaped peaks for these. If we could increase the temperature from 40 to 100°C or higher in the middle of the run, however, we could have the best of both worlds — complete resolution and reasonable retention times for all peaks. Thus, temperature-programmable ovens have been developed and are now commonplace on virtually all modern GC units. Temperature programming can consist of simple programs, such as that suggested above—a single linear increase from a low temperature to a higher temperature—and more complex programs. For example, a chromatography researcher may find that several temperature increases, and perhaps even a decrease, must be used in some instances to effect an acceptable separation. Most modern GC units are capable of at least a slow temperature decrease in the middle of the run since they are equipped with venting fans that bring ambient air into the oven to cool it. A simple temperature program is shown in Figure 12.9. See Workplace Scene 12.2.

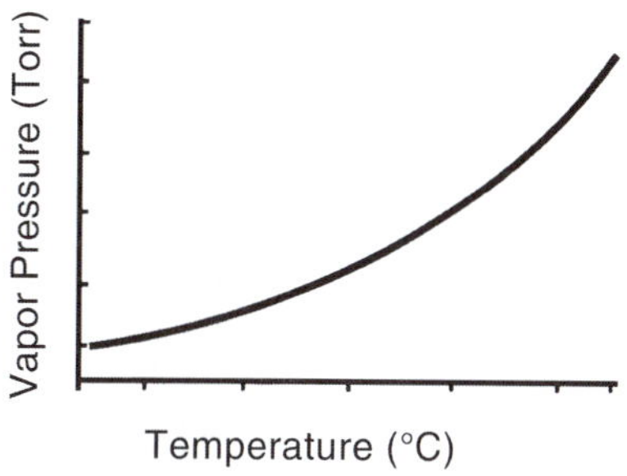

**FIGURE 12.7**    A graph showing a typical example of how the vapor pressure can change with temperature.

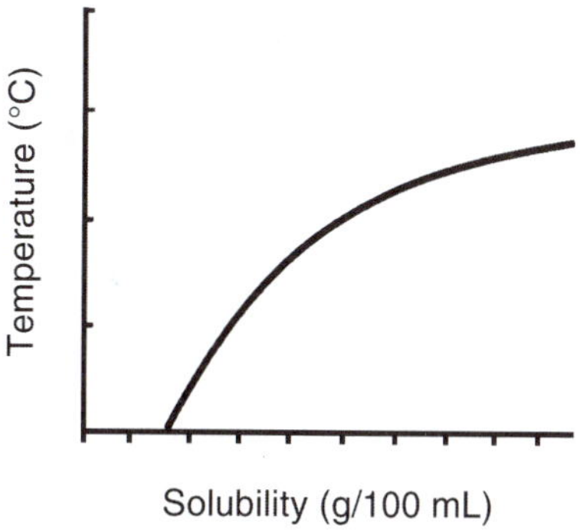

**FIGURE 12.8**    A graph showing how the solubility of a substance (in a hypothetical solvent) can change with temperature.

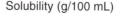

# WORKPLACE SCENE 12.2

Technicians that work at the state-of-the-art gas chromatography laboratory at the Midwest Research Institute in Kansas City, Missouri, have more than 20 temperature-programmable gas chromatographs with auto-sampler capabilities available to help them handle the many interesting projects with which they are involved. The setup includes a gas manifold system to allow easy access to the wide range of high-purity gases that are required for carrier gases and detectors. Among the interesting projects that utilize this facility is one that tests the levels of the toxic compound hexachlorobenzene in animal tissue. The analysts determine the milligrams of hexachlorobenzene contained in a kilogram of tissue using this equipment. The work also involves sample preparation, development and validation of methods, data compilation, calculations, and report preparation.

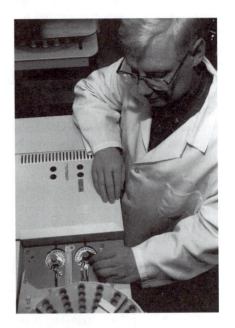

Richard Mathias makes adjustments on the auto-sampler attached to a gas chromatograph at the Midwest Research Institute.

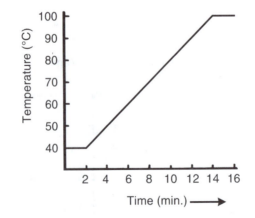

**FIGURE 12.9**    A simple temperature from 40 to 100°C at 5°C/min.

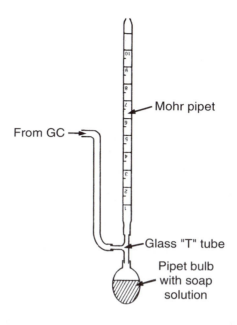

From GC →

Mohr pipet

Glass "T" tube

Pipet bulb
with soap
solution

**FIGURE 12.10**   A homemade soap bubble flow meter constructed from an old Mohr pipet, a piece of glass tubing, and a pipet bulb.

## 12.5.2   Carrier Gas Flow Rate

The rate of flow of the carrier gas affects resolution. A simple analogy here will make the point. Wet laundry hung out on a clothesline to dry will dry faster if it is a windy day. The components of the mixture will "blow" through the column more quickly (regardless of the degree of interaction with the stationary phase) if the carrier gas flow rate is increased. Thus, a minimum flow rate is needed for maximum resolution. It is known, however, that at extremely slow flow rates resolution is dramatically reduced due to factors such as packing irregularities, particle size, column diameter, etc.

It is obvious that the flow rate must be precisely controlled. The pressure from the compressed gas cylinder of carrier gas, while sufficient to force the gas through a packed column, does not provide the needed flow control. Thus a flow controller valve is built into the system. The flow rate of the carrier gas, as well as other gases used by some detectors, must be able to be carefully measured so that one can know what these flow rates are and be able to optimize them. Flow meters are commercially available. However, a simple soap bubble flow meter is often used and can be constructed easily from an old measuring pipet, a piece of glass tubing, and a pipet bulb. See Figure 12.10. With this apparatus, a stopwatch is used to measure the time it takes a soap bubble squeezed from the bulb to move between two graduation lines, such as the 0- and 10-mL lines. The commercial version uses an electronic sensor to measure the flow rate based on the bubble movement. See Workplace Scene 12.3.

## 12.6   Detectors

Detectors in gas chromatography are designed to generate an electronic signal when a gas other than the carrier gas elutes from the column. There have been a number of detectors invented to accomplish this. Not only do these detectors vary in design, but they also vary in their ability to measure small concentrations and in selectivity. Selectivity refers to the type of compound for which a signal can be generated. The flame ionization detector, for example, can measure very small concentrations but does not detect everything, i.e., it is selective for only a certain class of compounds. The thermal conductivity detector, on the other hand, detects virtually everything, i.e., it is a "universal" detector, but does not detect very

# WORKPLACE SCENE 12.3

At the City of Lincoln, Nebraska, Water Treatment Plant Laboratory, the concentrations of various agricultural chemicals found in the raw water, especially herbicides (such as atrazine) in the water, are important to determine. The method of choice is a gas chromatography method with a nitrogen–phosphorus detector (NPD).

A GC with an NPD detector utilizes four different gases: hydrogen, air, nitrogen (the makeup gas for the detector), and helium. The flow rates of each of these gases are important to individually measure and control, as is the flow rate of the mixture of gases at the outlet of the detector. Standard flow rates are hydrogen, 3 mL/min; air, 55 mL/min; nitrogen, 15 mL/min; and helium, 1 mL/min (when using a microbore column). The outlet of the detector should show a summation of all flow rates, 74 mL/min. An electronic digital bubble flow meter can be used to measure these flow rates. In this flow meter, a soap bubble is manually squeezed into the path of the flowing gas and carried through an electronic sensor field where the rate of flow is measured and displayed.

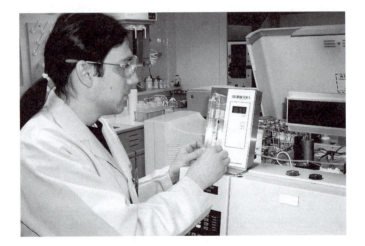

B.J. Kronschnabel of the City of Lincoln, Nebraska, Water Treatment Plant Laboratory measures the flow rate of the gases at the outlet of the NPD detector using a electronic digital bubble flow meter. Notice the open door at the top of the GC in the background where the detectors and flow channels are visible.

small concentrations. What follows is a brief description of the designs of the detectors that are in common use, along with some indication of their sensitivity and selectivity.

## 12.6.1  Thermal Conductivity

The thermal conductivity detector (TCD) operates on the principle that gases eluting from the column have thermal conductivities different from that of the carrier gas, which is usually helium. Present in the flow channel at the end of the column is a hot filament, hot because it has an electrical current passing through it. This filament is cooled to an equilibrium temperature by the flowing helium, but it is cooled differently by the mixture components as they elute, since their thermal conductivities are different from

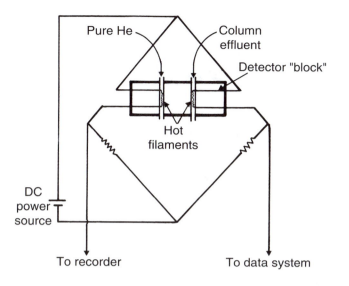

**FIGURE 12.11**   The thermal conductivity detector.

helium. This change in the cooling process causes the filament's electrical resistance to change and thus causes the current flowing through it and the voltage drop across it to change each time a mixture component elutes. A peak is then recorded on the chromatogram.

The actual design includes a second filament, within the same detector block. This filament is present in a different flow channel, however, one through which only pure helium flows. Both filaments are part of a Wheatstone Bridge circuit as shown in Figure 12.11, which allows a comparison between the two resistances and a voltage output to the data system, as shown. Such a design is intended to minimize effects of flow rate, pressure, and line voltage variations.

A flow-modulated design of the TCD has become popular. In this design, a single filament is used and the column effluent is alternated with the pure helium through the flow channel where the filament is located. This eliminates the need to use two matched filaments.

The thermal conductivity detector is universal (detects everything) and nondestructive (can be used with preparative GC), but it does not detect very small concentrations, compared to other detectors.

## 12.6.2   Flame Ionization Detector

Another very important GC detector design is the flame ionization detector (FID). In this detector, the column effluent is swept into a hydrogen flame, where the flammable components are burned. In the burning process, a very small fraction of the molecules become fragmented and the resulting positively charged ions are drawn to a collector (negatively charged) electrode, a metal cylinder above and encircling the flame, while electrons flow to the positively charged burner head. The negatively charged collector and the positively charged burner head are part of an electrical circuit in which the current changes when this process occurs, and the change is amplified and seen as a peak on the chromatogram. Figure 12.12 shows a drawing of this detector. The design includes the hydrogen flame burner nozzle, the collector electrode, an inlet for air to surround the flame, and an ignitor coil for igniting the hydrogen as it emerges from the nozzle.

Apparent near the bench on which the GC unit sits are pressure-regulated compressed gas cylinders of hydrogen and air (in addition to the carrier gas, helium or nitrogen). Metal tubing, typically 1/8-in. diameter, connect the cylinders to the detector. A needle valve is used for flow control. These valves are located in the instrument for easy access and control by the operator.

The FID detects very small concentrations, but is not universal; it also destroys (burns) the sample. It detects only organic substances that burn and fragment in a hydrogen flame. These facts preclude its use

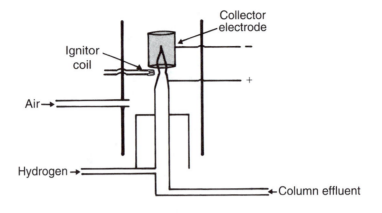

**FIGURE 12.12**   The flame ionization detector.

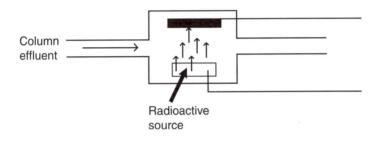

**FIGURE 12.13**   A diagram of an electron capture detector.

for preparative GC or for inorganic substances that do not burn, such as water, carbon dioxide, etc. Still, it is a very popular detector, given its sensitivity to small concentrations and given the fact that most analytical work involves flammable organic substances.

### 12.6.3  Electron Capture Detector

A third type of detector, required for some environmental and biomedical applications, is the electron capture detector (ECD). This detector is especially useful for large halogenated hydrocarbon molecules since it is the only one that has an acceptable sensitivity for such molecules. Thus, it finds special utility in the analysis of halogenated pesticide residues found in environmental and biomedical samples.

The electron capture detector is another type of ionization detector. Specifically, it utilizes the beta emissions of a radioactive source, often nickel-63, to cause the ionization of the carrier gas molecules, thus generating electrons that constitute an electrical current. As an electrophilic component, such as a pesticide, from the separated mixture enters this detector, the electrons from the carrier gas ionization are captured, creating an alteration in the current flow in an external circuit. This alteration is the source of the electrical signal that is amplified and sent on to the recorder. A diagram of this detector is shown in Figure 12.13. The carrier gas for this detector is either pure nitrogen or a mixture of argon and methane.

An additional consideration regarding pesticides warrants mentioning here. Most of these compounds decompose on contact with hot metal surfaces. This problem has been adequately solved for most pesticides by constructing the entire path of the sample out of glass or glass-lined materials. Thus, glass or glass-lined injection ports and all-glass columns are used.

In terms of advantages and disadvantages, the ECD detects extremely small concentrations, but only for a very select group of compounds—halogenated hydrocarbons. Other gases will not give a peak. It does not destroy the sample and thus may be used for preparative work.

## 12.6.4  The Nitrogen–Phosphorus Detector

While the ECD is useful for chlorinated hydrocarbon pesticides, the nitrogen–phosphorus detector (NPD), also known as the thermionic detector, is useful for phosphorus- and nitrogen-containing pesticides, organophosphates, and carbamates. The design represents a slight alteration of the design of the FID. In the NPD, we basically have an FID with a bead of alkali metal salt positioned just above the flame. The hydrogen and air flow rates are lower than those in the ordinary FID; this minimizes the fragmentation of other organic compounds. These changes result in a somewhat mysterious increase in both the selectivity and the ability to measure small concentrations of the pesticides.

## 12.6.5  Flame Photometric Detector

A detector that is specific for organic compounds containing sulfur or phosphorus is the flame photo-metric detector (FPD). A flame photometer (Section 9.6.1) is an instrument in which a sample solution is aspirated into a flame and the resulting emissions from the flame are measured with a photomultiplier tube. The FPD is a flame photometer positioned to accept the effluent from the column rather than an aspirated sample. The flame in this case is a hydrogen flame, as in the FID. The basic operating principle is that the sulfur or phosphorus compounds burn in the hydrogen flame and produce light-emitting species. A wavelength selector, typically a glass filter, makes this detector specific for these compounds. The signal for the recorder is the signal proportional to light intensity that is produced by the phototube.

The advantages are that it is very selective and detects small concentrations. Disadvantages include the problems associated with the need to carefully control the flame conditions so that the correct species are produced (S=S for the sulfur compounds and HPO for the phosphorus compounds). Such conditions include the gas flow rates and the flame temperature. Due to the rather hot flame, there can be problems with temperature-sensitive electronics.

Also, stray light creates an electronically noisy signal. It is a destructive detector.

## 12.6.6  Electrolytic Conductivity (Hall) Detector

The Hall detector converts the eluting gaseous components into ions in liquid solution and then measures the electrolytic conductivity of the solution in a conductivity cell. The solvent is continuously flowing through the cell, and thus the conducting solution is in the cell for only a moment while the conductivity is measured and the peak recorded before it is swept away with fresh solvent. The conversion to ions is done by chemically oxidizing or reducing the components with a reaction gas in a small reaction chamber made of nickel positioned between the column and the cell. The nature of the reaction gas depends on what class of compounds is being determined. Organic halides, the most common application, use hydrogen gas at 850°C or higher as the reaction gas. The strong HX acids are produced, which give highly conductive liquid solutions.

The Hall detector has an excellent ability to detect small concentrations and excellent selectivity, giving a peak for only those components that produce ions in the reaction chamber. It is a destructive detector.

## 12.6.7  GC-MS and GC-IR

We discussed the fundamentals of mass spectrometry in Chapter 10 and infrared spectrometry in Chapter 8. The quadrupole mass spectrometer and the Fourier transform infrared spectrometer have been adapted to and used with GC equipment as detectors with great success. **Gas chromatography–mass spectrometry** (GC-MS) and **gas chromatography–infrared spectrometry** (GC-IR) are very powerful tools for qualitative analysis in GC because not only do they give retention time information, but, due to their inherent speed, they are also able to measure and record the mass spectrum or infrared (IR) spectrum of the individual sample components as they elute from the GC column. It is like taking a photograph of each component as it elutes. See Figure 12.14. Coupled with the computer banks of mass and IR spectra, a component's identity is an easy chore for such a detector. It seems the only real

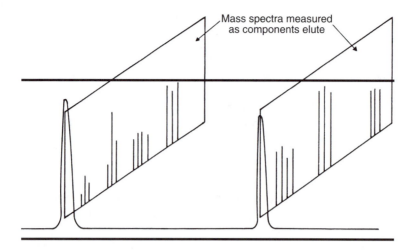

**FIGURE 12.14** The mass spectrum or IR spectrum are obtained with GC-MS or GC-IR as the mixture components elute—like taking a photograph of the eluting compounds.

disadvantage is that they are more expensive. The only other slight disadvantage is the fact that large amounts of computer memory space are required to hold the amount of spectral information required for a good qualitative analysis.

Both the GC-MS and GC-IR instruments obviously require that the column effluent be fed into the spectrometer detection path. For the IR instrument, this means that the IR cell, often referred to as a light pipe, be situated just outside the interferometer (Chapter 8) in the path of the light, of course, but it must also have a connection to the GC column and an exit tube where the sample may possibly be collected. The infrared detector is nondestructive. With the mass spectrometer detector, we have the problem of the low pressure of the mass spectrometry unit coupled with the ambient pressure of the GC column outlet. A special method is used to eliminate carrier gas while retaining sufficient amounts of the mixture components so that they are measurable with the mass spectrometer.

### 12.6.8  Photoionization

The photoionization detector (PID), as the name implies, involves the ionization of eluting mixture components by light, specifically, ultraviolet (UV) light. The UV source emits a wavelength characteristic of the gas (either helium or argon) inside. This light passes into an ionization chamber through a metal fluoride window and into the path of the column effluent there. This is where the mixture components absorb the light and ionize. The resulting ions are detected through the use of a pair of electrodes in the ionization chamber, the current from which constitutes the signal to the recorder. The specific lamp and window are chosen according to the ionization energy needed for the compounds in the sample.

Since different lamps and windows are available, this detection method can often be selective for only some of the components present in the sample. Its ability to detect small concentrations is especially good for aromatic hydrocarbons and inorganics. It is a nondestructive detector.

## 12.7  Qualitative Analysis

As mentioned in Section 11.8.4, the parameters that are most important for a qualitative analysis using most GC detectors are retention time, $t_R$; adjusted retention time, $t'_R$; and selectivity, $\alpha$. Their definitions were graphically presented in Figures 11.16 and 11.17. Under a given set of conditions (the nature of the stationary phase, the column temperature, the carrier flow rate, the column length and diameter, and the instrument dead volume), the retention time is a particular value for each component. It changes only when one or more of the above parameters changes. Thus, when the temperature changes, retention time will change, when the carrier gas flow rate changes, the retention time will change, etc.

Repeated injections into a given system under a given set of conditions should always yield a particular retention time for a given component. When one of the parameters changes, the retention time for that component will be slightly different. This is true, for example, when an analyst in another laboratory is trying to duplicate a given qualitative analysis and sets up with a different **dead volume**. The dead volume is the volume of the space between the injection port and the column packing and the space between the column packing and the detector. It also occurs in this scenario if the stationary phase has a slightly different composition than the original. The adjusted retention time will correct for changes in the dead volume, but it will not correct for any other change. Selectivity, however, does adjust for other changes. Thus the selectivity is an important parameter for qualitative analysis if the work involves other setups with other instruments and columns that do not exactly match the original.

The usual qualitative analysis procedure, then, is to establish the conditions for the experiment, perhaps by trial and error in one's own laboratory or by matching conditions outlined in a given procedure, that would resolve all compounds that may potentially be in the unknown. The idea is to match the retention time data, either ordinary retention time or the selectivity, whichever is appropriate, for standards (pure samples) with that for the unknown. The analyst can then proceed to match the retention time data for the unknown to those of the pure samples to determine which substances are present in the unknown (Experiment 40).

One important caution, however, is that there may be more than one component with the same retention time (no separation), and thus further experimentation may be required. For example, when working with a complex mixture whose components are perhaps not all known, it may be necessary to change the experimental conditions to determine whether a given peak is due to one component (known) or more (e.g., one known and one unknown). Changing the stationary phase may prove useful. Such a change would produce a chromatogram with completely different retention times and probably a different order of elution. Thus two components that were co-eluted before may now be separated, evidence for which would be a different peak size for the known component.

## 12.8 Quantitative Analysis

### 12.8.1 Quantitation Methods

Several different approaches exist as to what peaks are measured and how the mixture component of interest is actually quantitated. We now discuss three of the more popular methods.

### 12.8.2 The Response Factor Method

Consider a four-component mixture to be analyzed by GC. The chromatogram may look something like that shown in Figure 12.15. One might think it logical that in order to quantitate the mixture for, say, component B, all one would need to do is to measure the sizes of all four peaks and divide the size of the peak representing B by the total of all four:

$$B = \frac{\text{area}_B}{\text{area}_A + \text{area}_B + \text{area}_C + \text{area}_D} \qquad (12.1)$$

The problems with this approach are: 1) without comparing the peaks to a standard or a set of standards, it is not known whether the result is a weight, volume, or mole percent, and 2) the instrument detector does not respond to all components equally. For example, not all components will have the same thermal conductivity, and thus the thermal conductivity detector will not give equal sized peaks for equal concentrations of any two components. Thus, the sum of all four peaks would be a meaningless quantity, and the size of peak B by itself would not represent the correct fraction of the total.

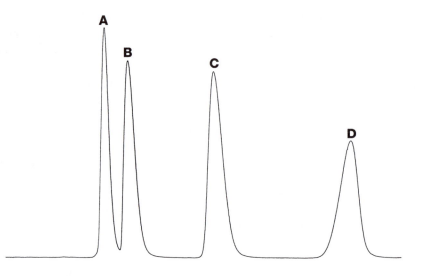

**FIGURE 12.15**   A chromatogram of a four-component mixture, A, B, C, and D.

It is possible, however, to measure a so-called response factor for the analyte, which is the area generated by a unit quantity injected, such as a microliter or microgram. The procedure is to inject a known quantity of the analyte, measured by the position of the plunger in the syringe (microliters) or by weighing the syringe before and after injection (micrograms). The peak size that results is measured and divided by this quantity:

$$\text{response factor} = \frac{\text{peak size}}{\text{quantity of pure sample injected}} \tag{12.2}$$

The quantity of analyte in an unknown sample is then determined by measuring the peak size of the analyte resulting from an injection of a known quantity of an unknown sample and dividing by the analyte's response factor:

$$\text{quantity of analyte} = \frac{\text{peak size}}{\text{response factor}} \tag{12.3}$$

The percent of the analyte can then be calculated as follows:

$$\% \text{ of analyte} = \frac{\text{quantity of analyte (from Equation (12.3))}}{\text{total quantity injected}} \tag{12.4}$$

In this method, only the peak of the analyte needs to be measured in the four-component mixture in order to quantitate this component.

### 12.8.3   Internal Standard Method

Since the peak size is directly proportional to concentration, one may think that one could prepare a series of standard solutions and obtain peak sizes to be used for a standard curve of peak size vs. concentration, a method similar to Beer's law in spectrophotometry, for example. But since peak size also varies with amount injected, there can be considerable error due to the difficulty in injecting consistent volumes, as discussed above and in Section 12.3. A method that does away with this problem is the internal standard method. In this method, all standards and samples are spiked with a constant known amount of a substance to act as what is called an internal standard. The purpose of the internal standard is to serve as a reference for the peak size measurements, so that slight variations in injection technique and volume injected are compensated for by the fact that the internal standard peak and the analyte peak are both affected by the slight variations.

The procedure is to measure the peak sizes of both the internal standard peak and the analyte peak and then to divide the analyte peak area by the internal standard peak area. The area ratio thus determined is then plotted vs. concentration of the analyte. The result is a method in which the volume injected is not as important and, in fact, can vary substantially from one injection to the next because this ratio does not change as the volume injected changes, since both peaks are affected equally by the changes.

Can just any substance serve as an internal standard? There are certain characteristics that the internal standard should have. They are listed below:

1. Its peak, like the analyte's, must be completely resolved from all other peaks.
2. Its retention time should be close to that of the analyte.
3. It should be structurally similar to the analyte.

### 12.8.4   Standard Additions Method

Increasing standard amounts of analyte are added to the sample and the resulting peak areas, which should show an increase with concentration added, are measured. This method is not as useful in GC as it would be in atomic absorption (see Chapter 9), since the sample matrix is not an issue in GC as it is in atomic absorption, due to the fact that matrix components become separated. However, standard additions may be useful for convenience's sake, particularly when the sample to be analyzed already contains a component capable of serving as an internal standard. Thus, standard additions could be used in conjunction with the internal standard method (see Experiment 45), and the internal standard would not have to be independently added to the sample and to the series of standards — it is already present, a convenient circumstance. Area ratio would then be plotted vs. concentration added and the unknown concentration determined by extrapolation to zero area ratio. Please refer to Chapter 9 for other details of the method of standard additions.

## 12.9   Troubleshooting

Problems that arise during a GC experiment usually manifest themselves on the chromatogram. Examples of such manifestations are peak shapes being distorted, peak sizes diminishing for reasons other than quantity of analyte, the baseline drifting, the retention times changing for no apparent reason, etc. These kinds of problems can usually be traced to injection problems, problems with the column, or problems with the detector. There can, of course, be problems associated with the electronics of the instrument. However, we will not be concerned with those here because of the large number of different instrument designs that have been manufactured over the years. The operator can usually find assistance for these in the troubleshooting section of the manuals that accompany the instrument.[*]

In the following paragraphs, we will address some of the most common problems encountered, pinpoint possible causes, and suggest methods of solving the problems.

### 12.9.1   Diminished Peak Size

We could also refer to this as reduced sensitivity. The peaks are smaller than expected based on previous observations when equal or greater quantities of a particular sample were injected. Such an observation usually means a problem with injection (less injected than assumed) or a problem with the detector such that a smaller electronic signal is sent to the recorder. One should check for a leaky or plugged syringe, a worn septum, a leak in the pre- and postcolumn connections, or a contaminated detector. Of course,

---

[*]See also the "GC Troubleshooting" column published regularly in the monthly journal LCGC, The Magazine of Separation Science.

# WORKPLACE SCENE 12.4

For various reasons, organic additives or inhibitors may be added to high-purity solvents during the manufacturing process. However, the quantity of these additives or inhibitors found in the final product must meet certain specifications. In the quality control laboratory of SACHEM Inc., Cleburne, Texas, gas chromatography is used to determine the concentration of these impurities, which is typically at the parts per million level. At SACHEM, GC is considered to be a high-precision instrument that requires well-trained analysts to operate and maintain. For example, frequent changes of the injector septum are necessary to ensure that there are no leaks. In addition, this laboratory provides comprehensive educational opportunities to analysts, including internal and external training classes and seminars.

Reagan Boles of SACHEM Inc. measures organic impurities in ultrapure solvent products at the parts per million level using gas chromatography.

detector attenuation, recorder sensitivity settings, electrical connections, and other associated hardware problems are potential causes. See Workplace Scene 12.4.

## 12.9.2  Unsymmetrical Peak Shapes

Peak fronting or peak tailing (Figure 12.16) are typical examples of unsymmetrical peaks. These could be indicators of poor injections, meaning too large an injection volume for the diameter of the column in use, too slow with the syringe manipulation during injection, or not fully penetrating the septum. It may indicate a decomposition of thermolabile components in contact with the hot system components such as the metal walls of the injection port and column. It may also mean contamination of the injection port or column.

## 12.9.3  Altered Retention Times

This is usually caused by changes in the carrier gas flow rate or column temperature. Flow rate changes can be caused by leaks in the system upstream from the column inlet, such as in the injection port (e.g., the septum); by low pressure in the system due to an empty or nearly empty carrier supply; or by faulty hardware,

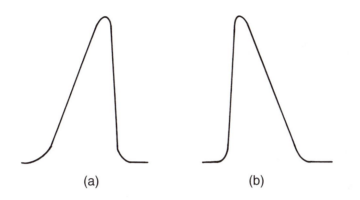

(a)                                    (b)

**FIGURE 12.16**   Peaks exhibiting (a) fronting and (b) tailing.

such as the flow control valve or pressure regulator. Temperature changes can be caused by a faulty temperature controller, an improperly set temperature program, a too short cool-down period prior to the next injection in a temperature-programmed experiment, etc. Altered retention times could also be caused by overloading the column or by diminished effectiveness (decomposition?) of the stationary phase.

## 12.9.4   Baseline Drift

This occurs when a new column has not been sufficiently conditioned, when the detector temperature has not reached its equilibrium value, or when the detector is contaminated or otherwise faulty. New columns need to be conditioned, usually with an overnight bakeout at the highest recommended temperature for that column. Detector signals may very well change when the detector temperature changes. One should be sure that sufficient time has been given for the detector temperature to level off. The nature of detector problems depends of course on the detector. TCD filaments may become oxidized due to an air leak, ionization detectors may be leaking, there may be a crack in the FID burner nozzle, etc. Baseline drift may also occur in cases when a temperature program covers a large range.

## 12.9.5   Baseline Perturbations

If the perturbations are in the form of spikes of an irregular nature, the problem is likely to be detector contamination. Such spikes are especially observed when dust particles have settled into the FID flame orifice. Of course, the problem may also be due to interference from electrical pulses from some other source nearby. Regular spikes can be due to condensation in the flow lines causing the carrier, or hydrogen (FID), to pulse, or they can be due to a bubble flow meter attached to the outlet of the TCD, as well as the electrical pulses referred to above. Baseline perturbations can also be caused by pulses in the carrier flow due to a faulty flow valve or pressure regulator.

## 12.9.6   Appearance of Unexpected Peaks

Unexpected peaks can arise from components from a previous injection that moved slowly through the column, contamination from either the reagents used to prepare the sample or the standards, or a contaminated septum, carrier, or column. Solutions to these problems include a rapid bakeout via temperature programming after the analyte peaks have eluted, use of pure reagents, and replacement or cleaning of septa, carrier, or column.

# Experiments

## Experiment 40: A Qualitative Gas Chromatographic Analysis of a Prepared Sample

### Introduction

This experiment is designed to be your first experience with a GC. The following conditions are suggested. Your instructor may choose other conditions.

Column: packed, 3% FFAP, 2 m in length

Carrier gas flow rate: 20 mL per min

Temperature program: 1.5 min at 80°C; increase to 120°C at the rate of 50° per minute; hold at 120°C for 7 min

Detector: FID set to appropriate range and attenuation; a TCD may be used if desired

Injector temperature: 250°C

Detector temperature: 200°C

Attenuation setting: probably either 128 or 256, depending on which gives the best peak size without going off the scale, as described in step 6 below

Under these conditions, a sample mixture consisting of benzene, toluene, ethylbenzene, chlorobenzene, bromobenzene, cyclohexane, and acetone should be well resolved.

*Remember safety glasses and gloves.*

### Procedure

1. Examine the instrument to which you are assigned. Locate the source of the carrier gas and trace the line to the instrument. If an FID is to be used, also locate the source of the hydrogen and air, and trace the lines of each to the instrument. Locate the injection port. Note any gauges and controls on the front of the instrument, and try to identify their functions. Open the column oven and locate the column. Note the proximity of the inlet end of the column to the injection port. Note the outlet end of the column and locate the detector.

2. Locate the data system for your instrument. Turn on the computer and start the data acquisition software. Your instructor will provide instructions on the use of the software. Also turn on the printer and ensure that there is plenty of paper in the paper tray.

3. Open the valve on the pressure regulator of the carrier gas bottle and ensure that there is flow through the system. Turn on the instrument. Set the temperatures of the column, injector, and detector, and also set up the temperature program indicated in the introduction above. Allow time for all components to come to the set temperatures.

4. If an FID is used, open the valves on the pressure regulators of the hydrogen and air bottles. After a minute or so, light the flame.

5. Obtain the unknown mixture contained in a vial from your instructor. It contains the mixture of organic liquids to be separated.

6. The purpose of the experiment is to identify the liquids in your unknown by retention time data. Your unknown may contain any number of the liquids listed in the introduction. Obtain chromatograms and determine the retention times of the pure liquids. Your instructor will demonstrate how to inject the sample (1 $\mu$L) and use the instrument and data system to determine the retention times of each. The pure liquids should also be contained in small vials. The temperature program may be interrupted and the temperature reset in any given run once the peak has been traced on the monitor. If any peak is so large that it goes off the scale (goes all the way to the top of the screen and displays a flat top), you should readjust your attenuation to a less sensitive setting to give a smaller peak.

7. Obtain a chromatogram of your unknown. The peaks will be smaller since the concentrations are not 100%, as they were for the pure liquids. You may have to adjust the attenuation to get large, well-defined peaks. Measure the retention times for each peak, and compare these to your data for the pure liquids. Those liquids that have retention times matching a peak in your unknown are contained in your unknown.

8. Report what organic liquids are in your unknowns and evidence to support this conclusion.

9. Maintain the logbook for the instrument used in this experiment. Record the date, your name(s), and the experiment name or number.

## Experiment 41: The Quantitative Gas Chromatographic Analysis of a Prepared Sample for Toluene by the Internal Standard Method

### Introduction

The conditions for this experiment are the same as those given in the introduction to Experiment 40. Perform steps 1 to 4 of Experiment 40 to familiarize yourself with the instrument and prepare it for the experiment. All flasks and pipets should be free of water and other solvents.

*Remember safety glasses and gloves.*

### Procedure

1. Prepare a series of standard solutions of toluene in cyclohexane that are 0.5, l, 2, and 3% toluene by volume. Use 25-mL volumetric flasks. Add exactly 0.50 mL of ethylbenzene to each flask before diluting to the mark with the cyclohexane. Shake well. The ethylbenzene is the internal standard.

2. The unknown is contained in a 25-mL volumetric flask and has been prepared by your instructor to have a concentration in the range of the standards when diluted to the mark with cyclohexane. Obtain the unknown, add the internal standard, and dilute with cyclohexane to the mark.

3. Inject 1 $\mu$L and obtain a good chromatogram of each solution. Since the conditions are the same as in Experiment 40, the retention times will be the same as those you discovered in that experiment. You will have to adjust the attenuation setting to get the proper-sized peaks for toluene and ethylbenzene. The cyclohexane peak will be much larger than the others, since it is the solvent and is present at a much larger concentration. This peak will not enter into the determination of the toluene concentration and can be allowed to go off the scale and ignored.

4. Determine the areas of the toluene and ethylbenzene peaks on each chromatogram. Your instructor will demonstrate how to obtain the peak areas with the software used. Calculate the ratio of the area of the toluene peak to the area of the ethylbenzene peak.

5. Plot the standard curve using the spreadsheet procedure used in Experiment 18. The y-axis is the area ratio and the x-axis is the toluene concentration. Obtain the correlation coefficient and the concentrations of the unknown and control, if there is one.

6. Maintain the logbook for the instrument used in this experiment. Record the date, your name(s), the experiment name or number, the correlation coefficient, and the results for the control sample. Also plot the control sample results on a control chart for this experiment posted in the laboratory.

## Experiment 42: The Determination of Ethanol in Wine by Gas Chromatography and the Internal Standard Method

### Introduction

The conditions for this experiment are the same as those given in the introduction to Experiment 40, except that the temperature is not programmed. Perform steps 1 to 4 of Experiment 40 (except for the temperature program) to familiarize yourself with the instrument and prepare it for the experiment. All flasks and pipets should be free of water and other solvents.

*Remember safety glasses and gloves.*

**Procedure**

1. Prepare four to six standard solutions of ethanol in water such that the alcohol content of the wine (as indicated on the wine label) is in the middle. For example, if the wine is 15% ethanol (volume percent assumed), standards of 5, 10, 20, and 25% are appropriate. Use 25-mL volumetric flasks, and pipet the ethanol accurately. Dilute to the mark with water and then add 1.00 mL of acetone (the internal standard) above the mark. Shake well.

2. Prepare each sample by filling a 25-mL flask with the wine and adding 1.00 mL of acetone above the mark.

3. There are three solution components, ethanol, acetone, and water. If a flame ionization detector is used, there will be only two peaks, since water will not give a peak. Set up the instrument for isothermal operation at 100°C. The two peaks should be nicely resolved and each run should take only a few minutes.

4. Obtain chromatograms for all standards and all wine samples. Obtain the areas of the ethanol and acetone peaks and calculate the area ratio: ethanol peak area to acetone peak area.

5. Plot the standard curve using the spreadsheet procedure used in Experiment 18. The y-axis is the area ratio, and the x-axis is the ethanol concentration. Obtain the correlation coefficient and the concentrations of the unknown and control, if there is one.

6. Maintain the logbook for the instrument used in this experiment. Record the date, your name(s), the experiment name or number, the correlation coefficient, and the results for the control sample. Also plot the control sample results on a control chart for this experiment posted in the laboratory.

## Experiment 43: Designing an Experiment for Determining Ethanol in Cough Medicine or Other Pharmaceutical Preparation

Design an experiment based on your experience in Experiment 42 to determine the alcohol content in cough medicine or other pharmaceutical preparation. Ask your instructor to approve your approach before beginning the work.

*Remember safety glasses.*

## Experiment 44: A Study of the Effect of the Changing of GC Instrument Parameters on Resolution

Note: A mixture of two organic liquids is used for this study. The specific liquids will be selected by your instructor. They should be available in a vial equipped with a rubber septum. Your instructor has also selected the initial column packing and length. The two liquids should be in a ratio of approximately 1:1 by volume.

*Remember safety glasses.*

1. Set the carrier gas flow rate for your instrument to 20 mL/min. The method of measuring this flow rate will be demonstrated by your instructor. Set the column temperature to 70°C. Inject 1.0 $\mu$L of the mixture and observe the resolution. Now change the column temperature to 80°C, wait 5 min for the oven temperature to become stable, and inject 1.0 $\mu$L again, observing the resolution. Repeat at 90, 100, 110, and 120°C, observing the effect on resolution.

2. Set the column oven temperature to 100°C and wait for the temperature to become stable. Set the carrier gas flow rate to 10 mL/min. Inject 1.0 $\mu$L of the mixture and observe the resolution. Now increase the carrier gas flow rate by 5 mL/min to observe the resolution at 15, 20, 25, 30, and 35 mL/min.

3. Set the carrier gas flow rate to 20 mL/min and obtain a series of chromatograms, each at a different volume injected: 0.5, 1.0, 1.5, 2.0, and 2.5 $\mu$L. The attenuation should be set so that the peaks from the larger injection volumes will not be off-scale.

4. If a longer column with the same stationary phase is available, change columns. Set the temperature and carrier gas flow rate to some combination of values that gave poor resolution in one of the previous steps. Inject 1.0 $\mu$L and observe the effect of a longer column on resolution.

5. Change column to one with some other stationary phase, perhaps one suggested by your instructor. Set the column temperature to 100°C and the carrier gas flow rate to 20 mL/min. Inject 1.0 $\mu$L of the mixture. Assuming good resolution, observe the order of elution. Is it different from that observed with the former stationary phase? If so, explain how that could be. If not, compare the resolution here with that of a previous injection in which all the parameters were equal. Comment on the difference.

6. As discussed in Chapter 11, a numerical value for resolution (R) can be calculated as follows:

$$R = \frac{2(t_2 - t_1)}{(w_1 + w_2)}$$

in which $t_2$ is the retention time for the component with the longest retention time (component two), $t_1$ is the retention time for the other component, and $w_1$ and $w_2$ are the respective widths at the base bisected by the tangents to the sides (see discussion in Section 11.8.4). As directed by your instructor, calculate resolution for some or all of the above data and construct graphs of resolution vs. volume injected, resolution vs. flow rate, and resolution vs. temperature. Comment on the results.

## Experiment 45: The Gas Chromatographic Determination of a Gasoline Component by Method of Standard Additions and an Internal Standard

Note: Refer to the text to refresh your memory concerning the method of standard additions and the internal standard method. Use a good fume hood when preparing the standards. All flasks and pipettes should be water-free.

*Remember safety glasses and gloves.*

1. Consult your instructor for proper separation conditions. Obtain a chromatogram of the gasoline sample and locate the peak to be identified and quantitated. Identify this peak by matching retention times with some pure samples, or consult your instructor.
2. Prepare three standard solutions in 25-mL volumetric flasks, using the gasoline to be tested as the diluent:
   a. 0.25 mL of the chosen component (1% addition)
   b. 0.50 mL of the chosen component (2% addition)
   c. 0.75 mL of the chosen component (3% addition)
3. Obtain chromatograms of the three standards and also the pure gasoline (0% added). Give the column plenty of time to clear between injections—temperature programming may be useful.
4. Select another peak that is well resolved from the others and use it as an internal standard. Obtain the peak areas and area ratios (analyte to internal standard) and plot peak area ratio vs. concentration added.
5. Extrapolate the line to zero peak area ratio and obtain the concentration in pure gasoline.

# Questions and Problems

1. What do the abbreviations GC, GLC, and GSC refer to?
2. Which types of chromatography are applicable to GC?
3. Compare the analytical strategy for GC with the general analytical strategy for instrumental methods (Figure 6.5).
4. Define carrier gas, column, and vapor pressure.
5. What role does vapor pressure play in a GC separation?
6. In a GC experiment in which the liquid stationary phase is polar, which would have a shorter retention time—a nonpolar mixture component with a high vapor pressure or a polar mixture component with a low vapor pressure? Explain.

7. If a GC operator expects a given mixture component to have a very long retention time compared to other mixture components, what vapor pressure and solubility properties would this mixture component have?

8. Why is the injection port in a gas chromatograph heated to a relatively high temperature?

9. What is the general role of the detector in GC?

10. There are three heated zones in a GC instrument. Which zones are these and why does each need to be heated?

11. Why must the size of a liquid sample injected into the GC be small? What would happen if it were too large?

12. What is meant by a GC open-tubular capillary column? Why has the development of such a column been useful?

13. Contrast the packed column and the open-tubular capillary column in terms of design, diameter, length, how the stationary phase is held in place, ability to resolve complex mixtures, and amount of sample injected.

14. What is the difference between analytical GC and preparative GC?

15. What is meant by temperature programming in GC?

16. Tell how the temperature programming feature of most modern GCs can be useful in separating complex mixtures.

17. In a GC separation involving four components, it was discovered that all four components, A through D, separate cleanly at 80°C. At this temperature, A and B have fairly short retention times, but components C and D have very long retention times. It was also discovered that C and D separate cleanly at 150°C, but that A and B do not. Suggest a temperature program that would separate all four in a reasonable time, and explain why it would work.

18. What is Chromosorb™? What is its use in GC?

19. Study Table 12.3 and tell what stationary phase material would be useful for separating some low-molecular-weight alcohols.

20. Would higher carrier gas flow rates increase or decrease retention time?

21. What does it mean to say that a GC detector is universal? What does it mean to say that one GC detector is more sensitive and more selective than another?

22. Which of the different types of GC detectors
    (a) Requires the use of hydrogen gas?
    (b) Is well suited for pesticide residue analysis?
    (c) Uses the abilities of the eluting gases to conduct heat?
    (d) Is the least sensitive?
    (e) Will not work for noncombustible mixture components?
    (f) Breaks the eluting molecules into charged fragments, which are then analyzed in terms of charge and mass?
    (g) Is part of an extremely powerful (and expensive) system abbreviated GC-MS?
    (h) Uses a radioactive source?
    (i) Uses an FID with a bead of alkali metal salt positioned just above the flame?
    (j) Uses an FTIR instrument to take an IR spectrum photograph of the mixure components as they elute?
    (k) Is a detector that is specific for organic compounds of sulfur and phosphorus?
    (l) Is a detector that utilizes a UV light beam to ionize the component molecules?
    (m) Is a detector that converts eluting molecules into ions in solution so that they can then be detected by electrical conductivity measurements?

23. Which of the GC detectors we have studied
    (a) Burn(s) the mixture components in a hydrogen flame?
    (b) Is (are) not very sensitive?
    (c) Is (are) universal?
    (d) Is (are) good for pesticide residue analysis?

(e) Utilize(s) a hot filament in the flow stream?

(f) Utilize(s) a source of radioactivity?

(g) Would not detect water?

(h) Does (do) not destroy the mixture components?

(i) Convert(s) eluting molecules into ions?

(j) Make(s) a measurement like a photograph of the eluting molecules?

24. Match the detector with the items that follow: thermal conductivity detector, flame ionization detector, electron capture detector, and mass spectrometer detector.

(a) FID

(b) A GC detector good especially for pesticides

(c) A GC detector that utilizes hydrogen gas

(d) A GC detector that utilizes a radioactive source

(e) A GC detector that analyzes fragmented molecules according to their mass and charge

(f) A powerful GC detector that can perform qualitative analysis even without the use of retention times

(g) A GC detector that is based on the differences of the abilities of helium and the mixture components to conduct heat

(h) A GC detector in which noncombustible materials will not give a peak

(i) A GC detector especially good for electrophilic substances

(j) A GC detector that is part of the GC-MS assembly

(k) A universal but not very sensitive GC detector

25. Of the thermal conductivity, flame ionization, and electron capture detectors, which

(a) Requires the use of hydrogen gas?

(b) Is universally applicable, but not very sensitive?

(c) Does not destroy the sample?

(d) Is good for pesticide analysis?

(e) Can be used only for samples that are able to be burned?

26. What advantages does

(a) A thermal conductivity detector have over a flame ionization detector?

(b) A flame ionization detector have over a thermal conductivity detector?

27. What do the letters GC-MS stand for? Briefly describe how the GC-MS detector works and why it is such a useful detector.

28. Explain the principles by which qualitative analysis can be performed in GC with the use of retention times.

29. A certain sample is a mixture of four organic liquids and these liquids exhibit the following retention times in a GC experiment:

| Liquid | Retention Time |
| --- | --- |
| A | 1.6 min |
| B | 2.2 min |
| C | 4.7 min |
| D | 9.8 min |

Some known liquids were injected into the chromatograph and the following data were determined:

| Liquid | Retention Time |
| --- | --- |
| Benzene | 0.5 min |
| Toluene | 1.6 min |
| Ethylbenzene | 3.4 min |
| n-Propylbenzene | 4.7 min |
| Isopropylbenzene | 5.8 min |

(a) Can you tell which liquids from these five are definitely not present? If not, why not? If so, which liquids are they?

(b) What liquids are possibly present?

(c) Why can one not tell with certainty which liquids are present, based on the information given here.

30. A laboratory technician tests a liquid mixture with gas chromatography for the purpose of identifying the components. She injects the mixture and four peaks are displayed on the chromatogram. She then obtains four pure liquids from a stock room and injects them into the GC (same conditions) one at a time. The retention time of one of the pure liquids exactly matches one of the retention times on the mixture chromatogram. Do you think she now knows, with certainty, the identity of one of the components? Explain.

31. Fill in the blanks with terms chosen from the following list:

| | |
|---|---|
| retention time | carrier gas |
| injection port | theoretical plates |
| resolution | preparative GC |
| thermal conductivity detector | peak area |
| flame ionization detector | column |
| electron capture detector | mobile phase |
| temperature programming | stationary phase |
| open-tubular column | detector |
| column | |

(a) Two terms that describe the helium used in gas chromatography are _____ and _____.

(b) The measurement that is made when doing a qualitative analysis of a mixture in gas chromatography is _____.

(c) A type of detector in gas chromatography that is universal is the _____.

(d) A type of detector in gas chromatography that is useful for pesticide analysis is the _____.

(e) The three heated parts in a gas chromatography instrument are the _____, _____, and _____.

(f) In doing quantitative analysis with gas chromatography, the _____ is an important measurement.

(g) Increasing the length of a GC column is a way of increasing the number of _____, and therefore improving the _____.

(h) Sometimes a difficult separation in gas chromatography can be accomplished easily by changing the temperature of the column during the run. This is called _____.

(i) A gas chromatography instrument can be used to obtain pure samples of the components of a mixture to be used whenever pure samples are needed for some other experiment. This is called _____.

(j) In an experiment in which the _____ is used, one needs to have a source of hydrogen gas.

(k) The _____ has the stationary phase adsorbed directly onto the wall of the column.

32. An injection of 3.0 $\mu$L of methylene chloride (density = 1.327 g/mL) gave a peak size of 3.74 cm$^2$. The injection of 3.0 $\mu$L of an unknown sample (density = 1.174 g/mL) gave a methylene chloride peak size of 1.02 cm$^2$. Calculate a response factor for methylene chloride and the percent of methylene chloride in the sample.

33. Define internal standard. Tell why an internal standard is important in a quantitative analysis by GC. Also tell what is plotted on the x- and y-axes when plotting the standard curve in internal standard procedures.

34. Compare and differentiate between the internal standard method and the standard additions method.

35. Compare the internal standard method with the standard additions method in terms of:

(a) Solution preparation
(b) What the chromatograms look like
(c) What is plotted for the standard curve

36. Consider the quantitative gas chromatography analysis of alcohol-blended gasoline for ethyl alcohol by the internal standard method, using isopropyl alcohol as the internal standard. The peaks for these two substances are well resolved from each other and from other components. Assume there is no isopropyl alcohol in the gasoline.
    (a) Describe how you would prepare a series of standard solutions for this analysis.
    (b) What would the chromatograms of these solutions look like?
    (c) What is plotted in an analysis of this type?

37. Select one of the quantitation procedures we have discussed (response factor method, internal standard method, or standard addition method) and describe:
    (a) Experimental details (solution preparation or sample preparation)
    (b) What the raw data are
    (c) What is plotted

38. What observation on a chromatogram would lead you to conclude that you injected too much sample for the diameter of column you are using?

39. If a GC operator observes that a given mixture component gives a smaller peak than in previous work for the same sample and the same injection volume, what might be the problem?

40. What might be the cause of a drifting chromatogram baseline?

41. What will occur if a GC operator makes another sample injection before the column has been cleared of mixture components from previous injections?

42. What can a GC analyst do to solve the problem of unexpected peaks on the chromatogram?

43. The following data was obtained for Experiment 41:

| Concentration | Toluene Peak Area | Ethylbenzene Peak Area |
|---|---|---|
| 0.50 | 0.539 | 1.970 |
| 1.00 | 1.050 | 2.051 |
| 2.00 | 2.184 | 2.018 |
| 3.00 | 3.308 | 2.009 |
| Unknown | 1.382 | 2.015 |

Use the procedure outlined in this experiment to find the percent of toluene in the unknown.

44. In Experiments 42 and 45, why is there no extra calculation at the conclusion of the experiments to get the percent of the components in the samples?

45. Consider the analysis of plant material for a pesticide residue by GC. Two grams of the material is chopped up and placed in a Soxhlet extractor (Chapter 11) and the pesticide quantitatively extracted into an appropriate solvent. Following this, the solvent is evaporated to near dryness and the residue is diluted to volume in a 25-mL flask. Then 2.5 $\mu$L of this solution and standards is injected in a GC with the following results:

| Concentration (ppm) | Peak Area |
|---|---|
| 5.0 | 1168 |
| 10.0 | 2170 |
| 15.0 | 3214 |
| 20.0 | 4079 |
| 25.0 | 5392 |
| Sample | 3577 |

What is the parts per million of pesticide in the original plant material?

46. **Report**

    Find a real-world gas chromatographic analysis in a methods book (AOAC, USP, ASTM, etc.) or journal, and report on the details of the procedure according to the following scheme.

    (a) Title

    (b) General information, including type of material examined, name of the analyte, sampling procedures, and sample preparation procedures

    (c) Specifics, including whether an internal standard is used, what stationary phase and solid support are used, temperature program, mobile phase used and its flow rate, what detector is used, how the data are gathered, concentration levels for standards, how standards are prepared, and potential problems

    (d) Data handling and reporting

    (e) References

# 13

# High-Performance
# Liquid Chromatography

## 13.1 Introduction

### 13.1.1 Summary of Method

High-performance liquid chromatography (HPLC) is an instrumental chromatography configuration in which the mobile phase is a liquid. The discussion of the principles of instrumental chromatography presented in Chapter 11 (Sections 11.8.3 and 11.8.4) provides the basic background for this technique. A liquid mobile phase is made to move through a column containing the stationary phase. A mixture of compounds injected ahead of the column separates as the compounds pass through. The mixture components are then detected electronically one at a time as they elute from the column, resulting in the recording of the instrumental chromatogram by the data collection system (refer to Figures 11.6 and 11.15).

All of the types of chromatography discussed in Chapter 11 can be utilized in the HPLC configuration. Thus we have partition chromatography (liquid–liquid chromatography (LLC)), bonded phase chromatography (BPC), adsorption chromatography (liquid–solid chromatography (LSC)), ion exchange chromatography (IEC), and size exclusion chromatography (SEC), including gel permeation chromatography (GPC) and gel filtration chromatography (GFC), all as commonly used types of HPLC. More information on the specific stationary phases for each is given in Section 13.5.

The rise in popularity of HPLC is due in large part to its high-performance nature and the advantages offered over the older, noninstrumental open-column method described in Chapter 11. Separation and quantitation procedures that require hours and sometimes days with the open-column method can be completed in a matter of minutes, or even seconds, with HPLC. Modern column technology and gradient solvent elution systems, which will be described, have contributed significantly to this advantage in that extremely complex samples can be resolved with ease in a very short time.

The basic HPLC system is diagrammed in Figure 13.1. It consists of solvent reservoirs for containing the mobile phase, a special high-pressure pump for pumping the mobile phase through the column, a specially designed injection device for sample introduction, the column where the separation takes place, the detector for electronic sensing of the eluting mixture components, and a data system for acquiring and displaying the chromatogram. Besides these basic components, an HPLC unit may be equipped with a gradient programmer (Section 13.3), an auto-sampler, a guard column, and various in-line filters. The analytical strategy flow chart for HPLC is given in Figure 13.2.

### 13.1.2 Comparisons with GC

Because the HPLC mobile phase is a liquid, there are some very obvious differences between HPLC and GC. First, the mechanism of separation in HPLC involves the specific interaction of the mixture components with a specific mobile phase composition, while in GC the vapor pressure of the components,

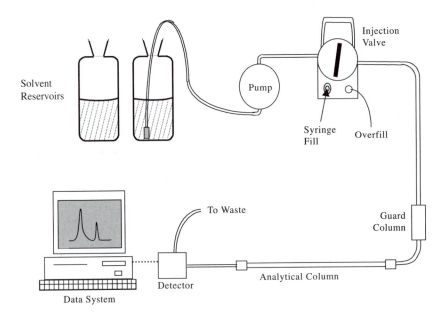

**FIGURE 13.1**   The HPLC system.

and not their interaction with a specific carrier gas, is the most important consideration (see Chapter 12). Second, the force that sustains the flow of the mobile phase is that of a high-pressure pump, rather than the regulated pressure from a compressed gas cylinder. Third, the injection device requires a totally different design due to the high pressure of the system and the possibility that a liquid mobile phase may chemically attack a rubber septum. Fourth, the detector requires a totally different design because the mobile phase is a liquid. Finally, the injector, column, and detector need not be heated as in GC, although the mode of separation occurring in the column can be affected by temperature changes, and thus sometimes elevated column temperatures are used.

## 13.2   Mobile Phase Considerations

The mobile phase reservoir is made of an inert material, usually glass. There is usually a cap on the reservoir that is vented to allow air to enter as the fluid level drops. The purpose of the cap is to prevent particulate matter from falling into the reservoir. It is very important to prevent particulates from entering the flow stream. The tip of the tube immersed in the reservoir is fitted with a coarse metal filter. It functions as a filter in the event that particulates do find their way into the reservoir. It also serves as a sinker to keep the tip well under the surface of the liquid. In addition, in specially designed mobile phase reservoirs, this sinker/filter is placed into a well on the bottom of the reservoir so that it is completely immersed in solvent, even when the reservoir is running low. This avoids drawing air into the line under those conditions. These details are shown in Figure 13.3.

The HPLC pump draws the mobile phase from the reservoir via vacuum action. In the process, air dissolved in the mobile phase may withdraw from the liquid and form bubbles in the flow stream unless such air is removed from the liquid in advance. Air in the flow stream is undesirable because it can cause a wide variety of problems, such as poor pump performance or poor detector response. Removing air from the mobile phase, called **degassing**, in advance of the chromatography is a routine matter, however, and can be done in one of several ways: 1) helium sparging, 2) ultrasonic agitation, 3) drawing a vacuum over the surface of the liquid, or 4) a combination of numbers 2 and 3.

**Helium sparging** refers to the vigorous bubbling of helium gas through the mobile phase. This can be done while it is contained in the reservoir by using a metal bubbler with tubing attached to a cylinder

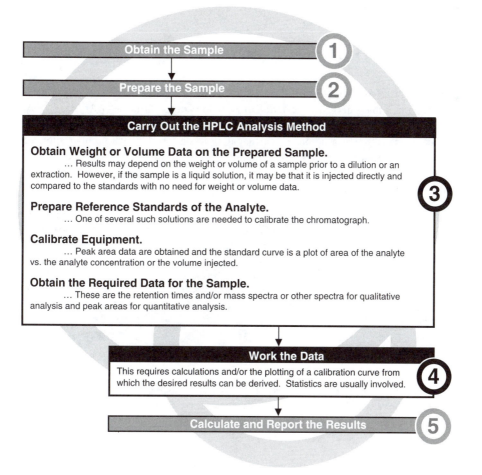

### Carry Out the HPLC Analysis Method

**Obtain Weight or Volume Data on the Prepared Sample.**
… Results may depend on the weight or volume of a sample prior to a dilution or an extraction. However, if the sample is a liquid solution, it may be that it is injected directly and compared to the standards with no need for weight or volume data.

**Prepare Reference Standards of the Analyte.**
… One of several such solutions are needed to calibrate the chromatograph.

**Calibrate Equipment.**
… Peak area data are obtained and the standard curve is a plot of area of the analyte vs. the analyte concentration or the volume injected.

**Obtain the Required Data for the Sample.**
… These are the retention times and/or mass spectra or other spectra for qualitative analysis and peak areas for quantitative analysis.

### Work the Data

This requires calculations and/or the plotting of a calibration curve from which the desired results can be derived. Statistics are usually involved.

**FIGURE 13.2** The analytical strategy flow chart for HPLC.

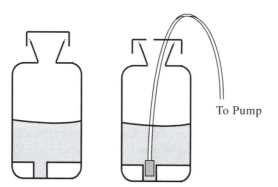

**FIGURE 13.3** Mobile phase reservoir, shown on the right, with vented cap and a coarse filter on the tip of the flow tube shown inside a well on the bottom of the reservoir.

# WORKPLACE SCENE 13.1

The Nebraska State Agriculture Laboratory participates in the annual Proficiency Testing Program of the American Association of Pesticide Control Officials (AAPCO). The AAPCO provides samples of an atrazine formulation to be assayed for atrazine by HPLC. Atrazine is popular in agriculture as a herbicide. Participating laboratories, which number approximately 40, perform the assay according to current methodology using a reverse phase system. The mobile phase is a mixture of acetonitrile and water in the ratio of 2:3 by volume, and the stationary phase is an octodecylsilane. A variable-wavelength UV absorption detector is used.

Left, Charlie Focht of the Nebraska State Agriculture Laboratory prepares the mobile phase for an atrazine assay. Note that the vacuum flask is positioned in an ultrasonic cleaner bath. Simultaneous vacuum filtration and sonication provide a more efficient means for degassing. Right, Charlie adjusts the flow rate setting on the HPLC pump.

of helium. The bubbling causes the air to be efficiently displaced from the liquid by the helium. Helium saturation of the liquid is not a problem. At a helium flow rate of 300 mL/min, complete degassing takes just a few minutes.

Drawing a **vacuum** over the surface of the liquid mobile phase has the same effect as the vacuum action of the HPLC pump—the air withdraws from the solvent. Such an action is usually sufficient for the degassing step for most applications. **Ultrasonic agitation**, by itself, is helpful in removing high levels of gases in certain samples, such as the carbonation in beverages. To remove standard levels of dissolved air, however, it must be combined with the vacuum action. Thus a laboratory analyst will often draw a vacuum over the surface while also sonicating.

If the mobile phase contains liquids that are not certified as HPLC grade solvents, they must be filtered ahead of time as well as degassed. The reason is that the packed bed of finely divided stationary phase particles through which the mobile phase percolates is itself an excellent filter. Particles in the mobile phase as small as 0.5 $\mu$m in diameter can be filtered out. The result of this is a decreased effectiveness of the column with time and possibly a blocked flow path. Unfiltered samples also may contain particles and cause this problem.

The problem is solved by **prefiltering** all mobile phases and samples before beginning the experiment. For mobile phases and large sample volumes, this involves utilizing a vacuum apparatus that draws the liquid through a 0.5-$\mu$m filter. Since such filtration involves a vacuum, the mobile phase is automatically degassed as well, so the filtration need not be a separate step. An efficient operation would be to filter a mobile phase with a vacuum apparatus while simultaneously sonicating. See Workplace Scene 13.1.

For nonaqueous solvents and their water solutions, filters made of paper are not a good choice due to the possibility of chemical incompatibility with the paper, which would then cause contamination. For these, the filter material of choice is usually nylon. A Teflon™-based material (designated PTFE) may also be used; however, if the mobile phase contains some water, the filter must be wetted first with some pure organic solvent in order to provide a reasonable filtration rate. Aqueous solutions are often impossible to filter unless the Teflon-based filter is first wetted in this manner. For filtering samples, and standard solutions as well, a small syringe-type filtering unit is used. The filter material may also be nylon, but if the Teflon-based material is used and the sample solvent contains some water, the filter must be wetted first with an organic solvent. Again, paper cannot be used if the sample or standard contains an organic liquid.

In addition to these prefiltering steps, in-line filters and a guard column are often used. The sinker in the mobile phase reservoir is a filter, as mentioned previously. Another metal in-line filter is in the flow path between the injector and the column, often immediately preceding the column. This filter would serve to remove particulates that entered via the sample injection. The **guard column** is usually placed just before the regular analytical column. Its function is to remove other contaminating substances— substances that perhaps have long retention times on the analytical column and that eventually interfere with the detection in later experiments. Guard columns are inexpensive and disposable and are changed frequently.

## 13.3   Solvent Delivery

### 13.3.1   Pumps

The pump that is used in HPLC cannot be just any pump. It must be a special pump that is capable of very high pressure (up to 5000 psi) in order to pump the mobile phase through the tightly packed stationary phase at a reasonable flow rate, usually between 0.5 and 4.0 mL/min. It also must be nearly free of pulsations so that the flow rate remains even and constant throughout. Only manufacturers of HPLC equipment manufacture such pumps.

The pump is a **reciprocating piston pump**. A diagram is shown in Figure 13.4. In this pump, a small piston (approximately $1/4$ in. in diameter) is driven back and forth drawing liquid in through an inlet check valve during its backward stroke and expelling the liquid through an outlet check valve during the forward stroke. A **check valve** is a device that allows liquid flow in one direction only. It typically consists

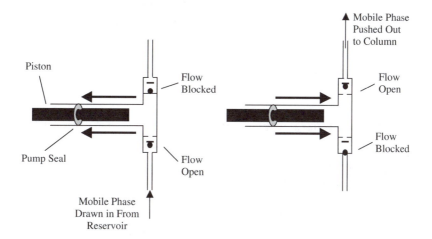

**FIGURE 13.4**   An illustration of a reciprocating piston pump with check valves. Left, the piston is in its upstroke, drawing mobile phase in from the reservoir. Right, the piston is in its downstroke, pushing mobile phase out to the column.

**FIGURE 13.5**  A photograph showing a pump piston (left), a pump seal (center), and a check valve (right). The pump seal is made to fit snugly over the piston, which is approximately 1/4 in. in diameter.

of a ruby ball that moves with the mobile phase and seals against a sapphire seat that allows flow around it in one flow direction but not in the other. See Figure 13.4. Liquid is drawn in from the reservoir when the piston is in its backstroke and pushed out toward the column when the piston is in its downstroke. The piston moves through a short, flexible sleeve called the **pump seal**, which prevents liquid from leaking through to the pump head.

This design is sometimes a twin-piston design, in which a second piston is 180° out of phase with the first. This means that when one piston is in its forward stroke, the other is in its backward stroke. The result is a flow that is free of pulsations. With the single-piston design, a pulse-damping device positioned in the flow path following the pump is used.

Modern pump designs also include a means for flushing the piston with solvent *behind* the pump seal (not shown in Figure 13.4). The solvent for this is drawn in from a separate reservoir and pumped back into this same reservoir. The purpose is to continuously rinse the piston free of mobile phase residue such that abrasive solute crystals resulting from a mobile phase that has dried out on the piston will not deposit there. These solutes, such as the salts dissolved in the buffered mobile phases used in ion exchange chromatography, may otherwise crystallize on the piston and then damage the piston or the pump seal when the piston moves back and forth. Mobile phases that contain such solutes must be flushed from the system after use so that there is also no crystallization on the front side of the seal.

Metal in-line filters and check valves may be removed and cleaned periodically, or replaced if they are damaged. This is done by dismantling the pump according to the manufacturer's instructions. In-line filters, including the sinker in the mobile phase reservoir, may be cleaned by soaking in a dilute nitric acid solution. Check valves may be cleaned by sonicating in an appropriate solvent. Photographs of a check valve, piston, and pump seal are presented in Figure 13.5.

## 13.3.2  Gradient vs. Isocratic Elution

There are two mobile phase elution methods that are used to elute mixture components from the stationary phase. These are referred to as isocratic elution and gradient elution. **Isocratic elution** is a method in which a single mobile phase composition is in use for the entire separation experiment. A different mobile phase composition can be used, but the change to a new composition can only be accomplished by stopping the flow, changing the mobile phase reservoir, and restarting the flow. **Gradient elution** is a method in which the mobile phase composition is changed, often gradually, in the middle of the run. It is analogous to temperature programming in GC.

In any liquid chromatography experiment, the composition of the mobile phase is very important in the total separation scheme. In Chapter 11, we discussed the role of a liquid mobile phase in terms of the solubility of the mixture components in both phases. Rapidly eluting components are highly soluble

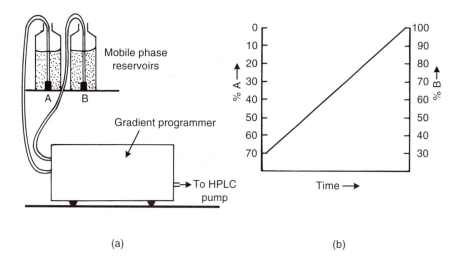

FIGURE 13.6   (a) Illustration of the gradient programmer hardware and (b) an example of a program.

in the mobile phase and insoluble in the stationary phase. Slowly eluting components are less soluble in the mobile phase and more soluble in the stationary phase. Retention times, and therefore resolution, can be altered dramatically by a change in the mobile phase composition. The chromatographer takes advantage of this by sometimes changing the mobile phase composition in the middle of the run if that helps achieve a successful separation. The gradient elution method provides a convenient and automatic method for doing this.

The **gradient programmer** is a hardware module used for gradient elution. The gradient programmer is capable of drawing from at least two mobile phase reservoirs at once and gradually, in a sequence programmed by the operator in advance, changing the composition of the mobile phase delivered to the HPLC pump. A schematic diagram of this system is shown in Figure 13.6(a), and a sample program is shown in Figure 13.6(b).

**Solvent strength** is a designation of the ability of a solvent (mobile phase) to elute mixture components. The greater the solvent strength, the shorter the retention times.

## 13.4   Sample Injection

As mentioned previously, introducing the sample to the flowing mobile phase at the head of the column is a special problem in HPLC due to the high pressure of the system and the fact that the liquid mobile phase may chemically attack a rubber septum. For these reasons, the use of the so-called loop injector is the most common method for sample introduction.

The **loop injector** is a two-position valve that directs the flow of the mobile phase along one of two different paths. One path is a sample loop, which when filled with the sample causes the sample to be swept into the column by the flowing mobile phase. The other path bypasses this loop while continuing on to the column, leaving the loop vented to the atmosphere and able to be loaded with the sample free of a pressure differential. Figure 13.7 is a diagram of this injector, showing both the load and inject positions and the path of the mobile phase in both positions.

The sample loop has a particular volume such that a careful measuring of the sample volume using the syringe is unnecessary—the sample loop is simply filled to overflowing each time (through the vent in Figure 13.7(a)) for a reproduced volume equal to the volume of the loop. If an injection volume smaller than the volume of the loop is desired, the loop may be changed to a smaller volume or a larger loop may be partially filled and the volume measured with the syringe. In some instruments, however, the dead volume between the injection port and the loop may be significant (and unknown), such that a partial filling of the loop with a known volume is impossible. In that case, an experiment that depends

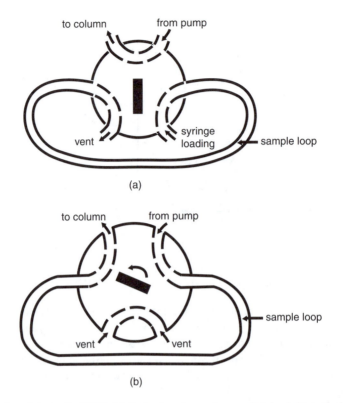

(a)

(b)

**FIGURE 13.7**    The loop injector for HPLC, (a) the load positon—the sample is loaded into the loop via a syringe at atmospheric pressure, and (b) the inject position—the mobile phase sweeps the contents of the loop onto the column.

on a reproduced injection volume (such as when injecting a series of standards in a quantitation experiment), the sample loop must be filled each time. If there is a need to inject a volume different from the volume of the sample loop installed on a given instrument, changing the loop to one of a different volume is an easy matter. See Workplace Scene 13.2.

Automated injectors are often used when large numbers of samples are to be run. Most designs involve the use of the loop injector coupled to a robotic needle that draws the samples from vials arranged in a carousel-type auto-sampler. Some designs even allow sample preparation schemes such as extraction and derivatization (chemical reactions) to occur prior to injection.

## 13.5    Column Selection

The stationary phases available for HPLC are as numerous as those available for GC. As mentioned previously, however, adsorption, partition, ion exchange, and size exclusion are all liquid chromatography methods. We can therefore classify the stationary phases according to which of these four types of chromatography they represent. Additionally, partition HPLC, which is the most common, is further classified as normal phase HPLC or reverse phase HPLC. Both of these are bonded phase chromatography, which was described in Chapter 11. Let us begin with these.

### 13.5.1    Normal Phase Columns

Normal phase HPLC consists of methods that utilize a nonpolar mobile phase in combination with a polar stationary phase. Adsorption HPLC actually fits this description, too, since the adsorbing solid stationary phase particles are very polar. (See discussion of adsorption columns in Section 13.5.3.) Normal

## WORKPLACE SCENE 13.2

Ion chromatography is used at the City of Lincoln, Nebraska, Water Treatment Plant Laboratory to analyze water samples taken from sampling sites in the distribution system around the city. The common anions determined by IC are not only nitrate, nitrite, fluoride, and sulfate, but also bromate. Bromate is found in the water because the Lincoln plant treats the water with ozone. Adding ozone to the water oxidizes any bromide to bromate. Bromate is regulated at 10 parts per billion (ppb); its concentration must be determined.

Because the expected concentration level is so low, the standard procedure for bromate using IC calls for a 250-$\mu$L sample loop on the injector, an unusually large volume for a sample loop. The procedure for the common anions listed above utilizes a 50-$\mu$L loop. It is a therefore a common task in this laboratory to change the sample loop regularly as these different anions are determined.

Eric Lee of the City of Lincoln, Nebraska, Water Treatment Plant prepares to change the sample loop on the IC injector. Notice the 250-$\mu$L sample loop (on the right) ready to be installed.

phase partition chromatography makes use of a polar liquid stationary phase chemically bonded to these polar particles, which typically consist of silica, Si–O–, bonding sites. Typical examples of normal phase bonded phases are those in which a cyano group (–CN), an amino group (–NH$_2$), or a diol group (–CHOH–CH$_2$OH) are part of the structure of the bonded phase. Typical mobile phases for normal phase HPLC are hexane, cyclohexane, carbon tetrachloride, chloroform, benzene, and toluene.

### 13.5.2 Reverse Phase Columns

Reverse phase HPLC describes methods that utilize a polar mobile phase in combination with a nonpolar stationary phase. As stated above, the nonpolar stationary phase structure is a bonded phase—a structure that is chemically bonded to the silica particles. Here, typical column names often have the carbon number designation indicating the length of a carbon chain to which the nonpolar nature is attributed. Typical designations are C8, C18 (or ODS, meaning octadecyl silane), etc. Common mobile phase liquids are water, methanol, acetonitrile (CH$_3$CN), and acetic acid buffered solutions.

### 13.5.3 Adsorption Columns

Adsorption HPLC is the classification in which the highly polar silica particles are exposed (no adsorbed or bonded liquid phase). Aluminum oxide particles fit this description too and are also readily available as the stationary phase. As mentioned earlier, this classification can also be thought of as normal phase

# WORKPLACE SCENE 13.3

SACHEM Inc., located in Cleburne, Texas, is a producer of high-purity bulk chemicals for companies that have high-purity requirements in their chemical processing. As stated in Workplace Scene 1.2, one of their products is tetramethylammonium hydroxide (TMAH), which is sold to semiconductor industries. The analysis of TMAH for trace anions such as chloride, nitrate, nitrite, and carbonate is critical for SACHEM's quality control laboratory. If these ions are present on the integrated circuit boards manufactured by one of their semiconductor customers, they may cause corrosion severe enough to affect the functionality and performance of the electronic devices in which the circuit boards are used. In SACHEM's quality control laboratory, ion chromatography procedures have been developed to measure the anion concentrations in TMAH. Because the concentration levels are trace levels, a clean room environment, like that described in Workplace Scene 1.2, is used. A special procedure for carbonate analysis is required so that the absorption of carbon dioxide from the atmosphere can be minimized.

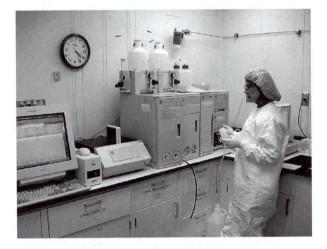

Kay Lones of SACHEM Inc. conducts analyses for trace anions in ultrapure TMAH products at the parts per billion level by using ion chromatography in a clean room like that described in Workplace Scene 1.2.

chromatography, but LSC rather than LLC. Typical normal phase mobile phases (nonpolar) are used here. The stationary phase particles can be irregular, regular, or pellicular, in which a solid core, such as a glass bead, is used to support a solid porous material.

## 13.5.4   Ion Exchange and Size Exclusion Columns

As discussed in Chapter 11, ion exchange stationary phases consist of solid resin particles that have positive or negative ionic bonding sites on their surfaces at which ions are exchanged with the mobile phase (see Figure 11.10). Cation exchange resins (SCX) have negative sites at which cations are exchanged, while anion exchange resins (SAX) have positive sites at which anions are exchanged. A popular modern name for HPLC ion exchange is simply ion chromatography (IC). Detection of ions eluting from the HPLC column has posed special problems, which are described in Section 13.6. The mobile phase for ion chromatography is always a pH-buffered water solution. See Workplace Scene 13.3.

Size exclusion columns, as discussed in Chapter 11, separate mixture components on the basis of size by the interaction of the molecules with various pore sizes on the surfaces of porous polymeric particles. Size exclusion chromatography is subdivided into two classifications: gel permeation chromatography and gel filtration chromatography. GPC utilizes nonpolar organic mobile phases, such as tetrahydrofuran (THF), trichlorobenzene, toluene, and chloroform, to analyze for organic polymers such as polystyrene. GFC utilizes mobile phases that are water-based solutions and is used to analyze for naturally occurring polymers, such as proteins and nucleic acids. GPC stationary phases are rigid gels, such as silica gel, whereas GFC stationary phases are soft gels. Neither technique utilizes gradient elution because the stationary phase pore sizes are sensitive to mobile phase changes.

## 13.5.5   Column Selection

Since each type of HPLC just discussed utilizes a different separation mechanism, the selection of a specific column packing (stationary phase) depends on whether or not the planned separation is possible or logical with a given mechanism. For example, if a given mixture consists of different molecules all of approximately the same size, then size exclusion chromatography will not work. If a mixture consists only of ions, then ion chromatography is the logical choice. While the conclusions drawn from these examples are obvious, others are less obvious and require a study of the variables and mechanisms in order for a particular stationary phase to be chosen logically.

Table 13.1 presents some guidelines for column selection. While these guidelines may prove helpful as a starting point, additional facts about the planned separation need to be determined in order to select the most appropriate chromatographic system, including facts that can be discovered only through experimentation or by searching the chemical literature. Several different mobile phase–stationary phase systems may work. Comparing reverse phase with normal phase, for example, one can see that there would only be a reversal in the order of elution. Polar components would elute first with reverse phase, whereas nonpolar components would elute first with normal phase. Experimenting with various mobile phase compositions, which may include a mixture of two or three solvents in various ratios, would be a logical starting point. Some considerations that would involve such experimentation are:

1. The mixture components should have a relatively high affinity for the stationary phase compared to the mobile phase. This would mean longer retention times and thus probably better resolution.
2. The various separation parameters should be adjusted to provide optimum resolution. These include mobile phase flow rate, stationary phase particle size, gradient elution, and column temperature (using an optional column oven).
3. Use partition chromatography for highly polar mixtures and adsorption chromatography for very nonpolar mixtures.

See Workplace Scene 13.4.

**TABLE 13.1**   Summary of Applications of the Different Types of HPLC

| Type | Useful for Components That |
| --- | --- |
| Normal and reverse phase | Have a low formula weight (<2000) |
| | Are nonionic |
| | Are either polar or nonpolar |
| | Are water or organic soluble |
| Adsorption | Have a low formula weight (<2000) |
| | Are nonpolar |
| | Are organic soluble |
| Ion exchange | Have a low formula weight (<2000) |
| | Are ionic |
| | Are water soluble |
| Size exclusion | Have a high or low formula weight |
| | Are nonionic |
| | Are water or organic soluble |

# WORKPLACE SCENE 13.4

Pharmaceutical preparations are routinely tested for quality and stability. These tests involve the analysis of these products for active ingredients and preservatives by HPLC. For example, methyl paraben, a preservative in some liquid preparations, is determined by reverse phase HPLC. Other analytes may also require reverse phase columns, but different ones in terms of column length, stationary phase particle size, and loading factors. Still other analytes may require a different type of chromatography entirely, such as ion exchange. For this reason, an analyst may need to change columns on a given instrument frequently. Changing columns involves undoing the two end fittings from the old column, pulling out the inlet and outlet tubing, and reversing these steps using the new column. It is important to maintain leak-free joints. Sometimes the replacement of the tubing ferrules, which ensure a seal between the tubing and the column, may be required.

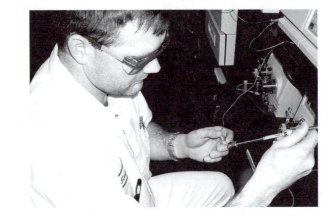

Scott Jenkins, a pharmaceutical analyst, changes HPLC columns in preparation for analyzing for a different analyte in a different pharmaceutical preparation.

## 13.6  Detectors

The function of the HPLC detector is to examine the solution that elutes from the column and output an electronic signal proportional to the concentrations of individual components present there. In Chapter 12, we discussed a number of detector designs that serve this same purpose for gas chromatography. The design of the HPLC detectors, however, is more conventional in the sense that components present in a liquid solution can be determined with conventional instruments, including spectrophotometers, fluorometers, and refractometers.

### 13.6.1  UV Absorption

The ultraviolet (UV) absorption HPLC detector is basically a UV spectrophotometer that measures a flowing solution rather than a static solution. It has a light source, a wavelength selector, and a phototube like an ordinary spectrophotometer. The cuvette is a flow cell, through which the column effluent flows. As the mobile phase elutes, the chromatogram traces a line at zero absorbance, but when a mixture

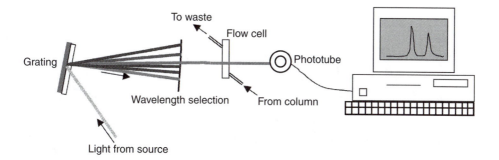

**FIGURE 13.8** The HPLC variable-wavelength UV detector—a UV spectrophotometer with a flow cell. A peak appears when a mixture component that absorbs the set wavelength elutes from the column.

component that absorbs the wavelength elutes, the absorbance changes and a peak is traced on the chromatogram. See Figure 13.8.

The wavelength selector can be a simple light filter (a so-called fixed wavelength detector) or a slit/dispersing element/slit monochromator (a variable-wavelength detector). The latter has a control for dialing in the wavelength as in a standard spectrophotometer. The fixed wavelength detector can be made to be variable in the sense that the light filter can be changed, but one cannot tune to the wavelength of maximum absorbance, and thus some sensitivity can be lost. The fixed wavelength version is no longer available from HPLC equipment vendors.

While a UV absorption detector is fairly sensitive, it is not universally applicable. The mixture components being measured must absorb light in the UV region in order for a peak to appear on the recorder. Also, the mobile phase must not absorb an appreciable amount at the selected wavelength.

## 13.6.2 Diode Array

A diode array UV detector is a diode array UV spectrophotometer with a flow cell. Refer to Section 8.2.5 for a description of an **ultraviolet–visible** (UV-VIS) diode array spectrophotometer. The light from the source passes through the flow cell and is then dispersed via a grating. The dispersed light then sprays across an array of photodiodes, each of which detects only a narrow wavelength band. With the help of the data system, the entire UV absorption spectrum can be immediately measured as each individual component elutes. See Figure 13.9. With computer banks containing a library of UV absorption spectral information, a rapid, definitive qualitative analysis is possible in a manner similar to that of gas chromatography–mass spectrometry (GC-MS) or gas chromatography–infrared spectrometry (GC-IR) (Chapter 12). In addition, the peak displayed on the chromatogram can be the result of a rapid change of the wavelength by the computer. Thus, the peaks displayed can represent the maximum possible sensitivity for each component. Finally, a diode array detector can be used to clean up a chromatogram so as to display only the peaks of interest. This is possible since we can rapidly change the wavelength giving rise to the peaks.

## 13.6.3 Fluorescence

The basic theory, principles, sensitivity, and application of fluorescence spectrometry (fluorometry) were discussed in Chapter 8. Like the UV absorption detector described above, the HPLC fluorescence detector is based on the design and application of its parent instrument, in this case the fluorometer. You should review Section 8.5 for more information about the fundamentals of the fluorescence technique.

In summary, the basic fluorometer, and thus the basic fluorescence detector, consists of a light source and a wavelength selector (usually a filter) for creating and isolating a desired wavelength, a sample compartment, and a second wavelength selector (another filter) with a phototube detector for isolating

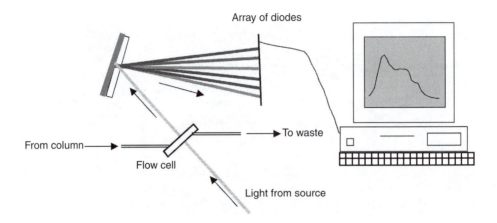

**FIGURE 13.9**   The HPLC diode array UV absorbance detector. When a mixture component elutes from the column, not only the chromatography peak but the entire UV absorption spectrum for that component can be recorded.

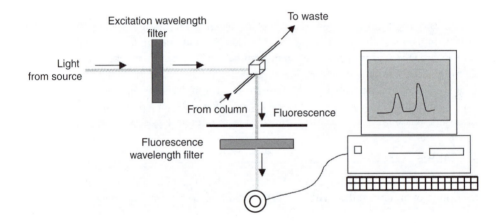

**FIGURE 13.10**   The HPLC fluorescence detector. When a mixture component that exhibits fluorescence elutes from the column, the light is detected and a peak appears on the chromatogram.

and measuring the fluorescence wavelength. The second monochromator and detector are lined up perpendicular to the light beam from the source (the so-called right-angle configuration).

As with the UV absorption detector, the sample compartment consists of a special cell for measuring a flowing, rather than static, solution. The fluorescence detector thus individually measures the fluorescence intensities of the mixture components as they elute from the column (see Figure 13.10). The electronic signal generated at the phototube is recorded on the chromatogram.

The advantages and disadvantages of the fluorometry technique in general hold true here. The fluorescence detector is not universal (it will give a peak only for fluorescing species), but it is thus very selective (almost no possibility for interference) and very sensitive.

### 13.6.4   Refractive Index

The refractive index (n) of a liquid or liquid solution is defined as the ratio of the speed of light in a vacuum ($c_{vac}$) to the speed of light in the liquid ($c_{liq}$):

$$n = c_{vac}/c_{liq} \tag{13.1}$$

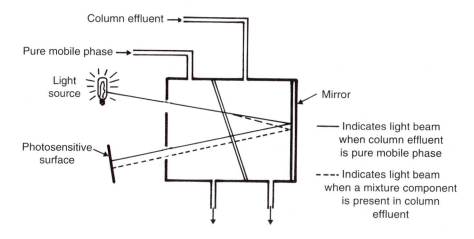

**FIGURE 13.11** An illustration of a refractive index detector.

Since the speed of light in any material medium is less than the speed of light in a vacuum, the numerical value of the refractive index for any liquid is greater than one.

An instrument known as a refractometer has been used for many years to measure the refractive index of liquids and liquid solutions for the purpose of both quantitative and qualitative analysis (see Chapter 15). A refractometer measures the degree of refraction (or bending) of a light beam passing through a thin film of the liquid. This refraction occurs when the speed of light in the sample is different from a reference liquid or air. The refractometer measures the position of the light beam relative to the reference and is calibrated directly in refractive index values. It is rare for any two liquids to have the same refractive index, and thus this instrument has been used successfully for qualitative analyses.

The refractive index detector in HPLC is a modification of this basic instrument and actually can be purchased in two different designs, depending on the manufacturer. In probably the most popular design, both the column effluent and the pure mobile phase (acting as a reference) pass through adjacent flow cells in the detector. A light beam, passing through both cells, is focused onto a photosensitive surface, the location of the beam when both cells contain pure mobile phase is taken as the reference point, and the recorder pen is zeroed. When a mixture component elutes, the refractive index in one cell changes; the light beam is bent and becomes focused onto a different point on the photosensitive surface, causing the recorder pen to deflect and trace a peak. See Figure 13.11.

The major advantage of this detector is that it is almost universal. All substances have their own characteristic refractive index (it is a physical property of the substance). Thus, the only time that a mixture component would not give a peak is when it has a refractive index equal to that of the mobile phase, a rare occurrence. The disadvantages are that it is not very sensitive and the output to the recorder is subject to temperature effects. Also, it is difficult to use this detector with the gradient elution method because it is sensitive to changes in the mobile phase composition.

## 13.6.5 Electrochemical

Various detectors that utilize electrical current or conductivity measurements for detecting eluting mixture components have been invented. These are called electrochemical detectors. Let us now examine some of the basic designs.

### 13.6.5.1 Conductivity

Perhaps the most important of all electrochemical detection schemes currently in use is the electrical conductivity detector. This detector is specifically useful for ion exchange, or ion, chromatography in which the analyte is in ionic form. Such ions elute from the column and need to be detected as peaks on the recorder trace.

A well-known fact of fundamental solution science is that the presence of ions in any solution gives the solution a low electrical resistance and the ability to conduct an electrical current. The absence of ions means that the solution would not be conductive. Thus, solutions of ionic compounds and acids, especially strong acids, have a low electrical resistance and are conductive. This means that if a pair of conductive surfaces are immersed into the solution and connected to an electrical power source, such as a simple battery, a current can be detected flowing in the circuit. Alternatively, if the resistance of the solution between the electrodes were measured (with an ohmmeter), it would be low. Conductivity cells based on this simple design are in common use in nonchromatography applications to determine the quality of deionized water, for example. Deionized water should have no ions dissolved in it and thus should have a very low conductivity. The conductivity detector is based on this simple apparatus.

For many years, the concept of the conductivity detector could not work, however. Ion chromatography experiments utilize solutions of high ion concentrations as the mobile phase. Thus, changes in conductivity due to eluting ions are not detectable above the already high conductivity of the mobile phase. This was true until the invention of so-called ion suppressors. Today, conductivity detectors are used extensively in HPLC ion chromatography instruments that also include suppressors.

A suppressor is a short column (tube) that is inserted into the flow stream just after the analytical column. It is packed with an ion exchange resin itself—a resin that removes mobile phase ions from the effluent, much like a deionizing cartridge removes the ions in laboratory tap water, and replaces them with molecular species. A popular mixed-bed ion exchange resin is used, for example, in deionizing cartridges, such that tap water ions (such as $Ca^{2+}$ and $CO_3^{2-}$) are exchanged for $H^+$ and $OH^-$ ions, which in turn react to form water. The resulting water is thus deionized. Of course, in the HPLC experiment, the analyte ions must *not* be removed in this process, and thus suppressors must be selective only for the mobile phase ions.

A typical design for a conductivity detector uses electrically isolated inlet and outlet tubes as the electrodes. This design is illustrated in Figure 13.12.

### 13.6.5.2  Amperometric

A thorough discussion of electroanalytical techniques, including polarography, voltammetry, and amperometry, is given in Chapter 14. An understanding of these would be useful for understanding the amperometric HPLC detector.

Electrochemical oxidation–reduction of eluting mixture components is the basis for amperometric electrochemical detectors. The three electrodes needed for the detection, the working (indicator) electrode, reference electrode, and auxiliary electrode, are either inserted into the flow stream or imbedded in the wall of the flow stream. See Figure 13.13. The indicator electrode is typically glassy carbon, platinum, or gold, the reference electrode a silver–silver chloride electrode, and the auxiliary a stainless steel electrode. Most often, the indicator electrode is polarized to cause oxidation of the mixture components

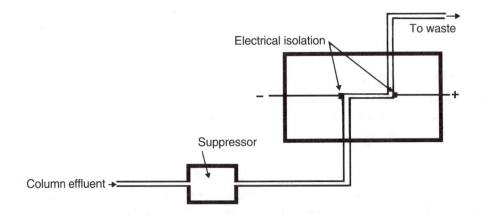

**FIGURE 13.12**  An illustration of a conductivity detector.

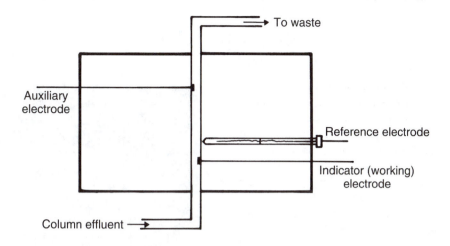

**FIGURE 13.13**   An illustration of an amperometric HPLC detector.

as they elute. The oxidation current is then measured and constitutes the signal sent to the recorder or integrator.

Advantages of this detector include broad applicability to both ionic mixture components and molecular components, as long as they are able to be oxidized (or reduced) at fairly small voltage polarizations. Selectivity can be improved by varying the potential. In addition, the sensitivity experienced with this detector is quite good—generally better than the UV detector, but not as good as the fluorescence detector. A disadvantage is that the indicator electrode can become fouled due to products of the electrochemical reaction coating the electrode surface. Thus, this detector must be able to be disassembled and cleaned with relative ease, since this may need to be done frequently.

### 13.6.6   LC-MS and LC-IR

In Chapter 12, the use of mass spectrometry and Fourier transform infrared spectrometry (FTIR) for GC detection was discussed. Details of these techniques were individually given in Chapters 10 and 12. Much of the discussion presented in Chapter 16 is applicable here. Both liquid chromatography–mass spectrometry (LC-MS) and liquid chromatography–infrared spectrometry (LC-IR) have been adapted to HPLC detection in recent years.

FTIR is a natural for HPLC in that it (FTIR) is a technique that has been used mostly for liquids. The speed introduced by the Fourier transform technique allows, as was mentioned for GC, the recording of the complete IR spectrum of mixture components as they elute, thus allowing the IR photograph to be taken and interpreted for qualitative analysis. Of course, the mobile phase and its accompanying absorptions are ever present in such a technique and water must be absent if the NaCl windows are used, but IR holds great potential, at least for nonaqueous systems, as a detector for HPLC in the future.

The mass spectrometer is also incompatible with the HPLC system, but for a different reason. The ordinary mass spectrometer operates under very low pressure (a high vacuum; see Chapter 10), and thus the liquid detection path must rapidly convert from a very high pressure and large liquid volume to a very low pressure and a gaseous state. Several approaches to this problem have been used, but probably the most popular is the thermospray (TS) technique. In this technique, the column effluent is converted to a fine mist (spray) as it passes through a small-diameter heated nozzle. The analyte molecules, which must be thermally stable, are preionized with the presence of a dissolved salt. A portion of the spray is introduced into the mass spectrometer. The analyte and mobile phase must be polar if the TS technique is used because the mobile phase must dissolve the required salt and the components must interact with the analyte molecule. See Workplace Scene 13.5.

# WORKPLACE SCENE 13.5

The progress made in interfacing HPLC instruments with mass spectrometry has been a significant development for laboratory analyses in the pharmaceutical industry. The low concentrations of test drugs in extracts of blood, plasmas, serums, and urine are no problem for this highly sensitive HPLC detector. In addition, the analysis is extremely fast. Lots of samples with very low concentrations of the test drugs can thus be analyzed in a very short time. At the MDS Pharma Services facility in Lincoln, Nebraska, for example, a very busy pharmaceutical laboratory houses over 20 LC-MS units, and they are all in heavy use daily.

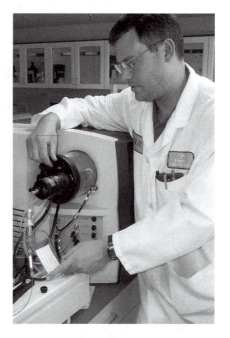

Jon Rathe, a technician in the LC-MS laboratory at MDS Pharma Services, checks out the thermospray interface on one of the LC-MS units in this laboratory.

## 13.7   Qualitative and Quantitative Analyses

Qualitative and quantitative analyses with HPLC are very similar to those with GC (Sections 12.7 and 12.8). In the absence of diode array, mass spectrometric, and FTIR detectors that give additional identification information, qualitative analysis depends solely on retention time data, $t_R$ and $t_R'$. (Remember that $t_R'$ is the time from when the solvent front is evident to the peak.) Under a given set of HPLC conditions, namely, the mobile and stationary phase compositions, mobile phase flow rate, column length, temperature (when the optional column oven is used), and instrument dead volume, the retention time is a particular value for each component. It changes only when one of the above parameters changes. Refer to Section 12.7 for further discussion of qualitative analysis.

Peak size measurement and quantitation methods were outlined in Section 11.8.5. The reproducibility of the amount injected is not nearly the problem with HPLC as it is with GC. Roughly ten times more sample is typically injected (5 to 20 $\mu$L), and there is no loss during the injection since the sample is not loaded into a higher-pressure system through a septum. In addition, the sample loop is manufactured to have a particular volume and is often the means by which a consistent amount is injected, which means reproducibility is maximized through the consistent overfilling of the loop via the injection syringe. In this way, the loop is assured of being filled at each injection and a reproducible volume is always introduced. Sometimes, however, the analyst chooses to inject varying volumes of a single standard to generate the standard curve rather than equal volumes of a series of standard solutions. In this case, the injection syringe is used to measure the volumes—a less accurate method, but better than an identical method with GC, since the sample volume is larger and there is less chance for sample loss. With this type of quantitation, the standard curve is a plot of peak size vs. amount injected, rather than concentration. As noted previously, however, some injectors have significant dead volume, which renders the method based on amount injected impossible.

The most popular quantitation method, then, is the series of standard solutions method with no internal standard or the variable injection of a single standard solution as outlined above.

# 13.8   Troubleshooting

Problems that arise with HPLC experiments are usually associated with abnormally high or low pressures, system leaks, worn injectors parts, air bubbles, or blocked in-line filters. Sometimes these manifest themselves on the chromatogram and sometimes they do not. In the following subsections, we address some of the most common problems encountered, pinpoint possible causes, and suggest methods of solving the problems. You can also refer to the troubleshooting guide in Chapter 12 for possible solutions.

## 13.8.1   Unusually High Pressure

A common cause of unusually high pressure is a plugged in-line filter. In-line filters are found at the very beginning of the flow line in the mobile phase reservoir, immediately before and/or after the injector, and just ahead of the column. With time, they can become plugged due to particles that are filtered out (particles can appear in the mobile phase and sample even if they were filtered ahead of time), and thus the pressure required to sustain a given flow rate can become quite high. The solution to this problem is to backflush the filters with solvent or clean them with a nitric acid solution in an ultrasonic bath.

Other causes of unusually high pressure are an injector blockage, mismatched mobile and stationary phases, and a flow rate that is simply too high. An injector that is left in a position between load and inject can also cause a high pressure, since the pump is pumping but there can be no flow.

## 13.8.2   Unusually Low Pressure

A sustained flow that is accompanied by low pressure may be indicative of a leak in the system. All joints should be checked for leaks (see below).

## 13.8.3   System Leaks

Leaks can occur within the pump, at the injector, at various fittings and joints, such as at the column, and in the detector. Leaks within the pump can be due to failure of pump seals and diaphragms, and loose fittings, such as at the check valves, etc. Leaky fittings should be checked for mismatched or stripped ferrules and threads, or perhaps they simply need tightening. Leaks in the injector can be due to a plugged internal line or other system blockage, gasket failure, loose connections, or use of the wrong syringe size if the leak occurs as the sample is loaded. Detector leaks are most often due to a bad gasket seal or a broken flow cell. Of course, loose or damaged fittings and a blockage in the flow line beyond the detector are possible causes.

## 13.8.4   Air Bubbles

An air pocket in the pump can cause low or no pressure or flow, erratic pressure, and changes in retention time data. It may be necessary to bleed air from the pump or prime the pump according to system start-up procedures. Air pockets in the column will mean decreased contact with the stationary phase and thus shorter retention times and decreased resolution. Tailing and peak splitting on the chromatogram may also occur due to air in the column. Air bubbles in the detector flow cell are usually manifested on the chromatogram as small spikes due to the periodic interruption of the light beam (e.g., in a UV absorbance detector). Increasing the flow rate or restricting and then releasing the postdetector flow, so as to increase the pressure, should cause such bubbles to be blown out.

### 13.8.5   Column Channeling

If the column packing becomes separated and a channel is formed in the stationary phase, the tailing and splitting of peaks will be observed on the chromatogram. In this case, the column needs to be replaced.

### 13.8.6   Decreased Retention Time

When retention times of mixture components decrease, there may be problems with either the mobile or stationary phase. It may be that the mobile phase composition was not restored after a gradient elution, or it may be that the stationary phase was altered due to irreversed adsorption of mixture components, or simply chemical decomposition. Use of guard columns may avoid stationary phase problems.

### 13.8.7   Baseline Drift

A common cause of baseline drift is a slow elution of substances previously adsorbed on the column. A column cleanup procedure may be in order, or it may need to be replaced. This problem may also be caused by temperature effects in the detector. Refractive index detectors are especially vulnerable to this. In addition, a contaminated detector can cause drift. The solution here may be to disassemble and clean the detector.

## Experiments

### Experiment 46: The Quantitative Determination of Methyl Paraben in a Prepared Sample by HPLC

**Introduction**

In this experiment, a mixture of methyl, propyl, and butyl paraben (structures shown in Figure 13.14) in methanol solvent will be separated by reverse phase HPLC. Mobile phase compositions of varying polarities will first be tested to see which one gives the optimum resolution of this mixture, and following this, a standard curve for methyl paraben will be constructed and its concentration in this solution determined.

*Remember safety glasses.*

**Part A: Determination of the Optimum Mobile Phase Composition**

1. Mobile phase compositions for this experiment are polar methanol–water mixtures in the ratios 90/10, 80/20, and 70/30 by volume. The stationary phase is C18. Prepare 200 mL of each mobile phase and then filter and degas each through 0.45-$\mu$m filters with the aid of a vacuum (instructor will demonstrate). Slowly pour each (so as to avoid reaeration by splashing) into individual mobile phase reservoirs that are labeled appropriately.
2. Obtain the sample from your instructor and filter it into a small vial using a syringe filter (instructor may choose to demonstrate). The sample is a solution of methyl, propyl, and butyl paraben in methanol. Record any identifying label in your notebook.
3. Examine the HPLC instrument to which you are assigned. Find the inlet line to the pump and place the free end of this line in the reservoir containing the mobile phase with the 90/10 composition. Trace the path of the mobile phase from the reservoir, through the pump, injection valve, column, and detector, to the waste container so that you identify and recognize all components of the flow path. Turn on the pump and detector and begin pumping the mobile phase at a rate

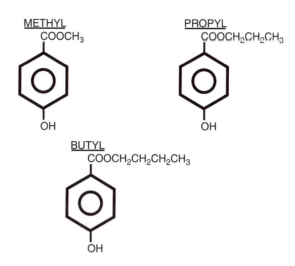

**FIGURE 13.14**   The structures of methyl, propyl, and butyl paraben.

between 1.0 and 1.5 mL/min. Allow plenty of time for the 90/10 mobile to completely flush the system before making the first injection (step 5).

4. While the mobile phase is flushing the system, locate the data system for the instrument and turn it on. Start the software for the data acquisition. Your instructor will discuss how the data system works and how to set it up.

5. Your instructor will discuss and/or demonstrate how to make an injection. With the injector valve in the load position, flush the injector's sample loop with the sample, and then make certain that the sample loop is completely filled. Turn the valve to inject and simultaneously activate the data system. The chromatogram should begin to be traced on the monitor.

6. After three peaks (perhaps poorly resolved, depending on the effect of the mobile phase used) appear in the chromatogram, stop the pump and move the free end of the inlet line to the 80/20 mobile phase. Start up the pump again and allow time to flush the system once again.

7. Repeat steps 5 and 6 to obtain second and third chromatograms of the mixture, but with the 80/20 and 70/30 mobile phases. Observe the effect of the mobile phase composition on resolution. Be sure to obtain peak area information from the data system for the experiment that gave the best resolution. This information will be used in Part B.

8. Optional: Calculate R, k, and a values according to your instructor's directions, and interpret the results.

## Part B: Methyl Paraben Quantitation

9. Select the mobile phase from Part A that gave the best resolution and, if it is not already in the system, allow plenty of time for it to purge the system.

10. Prepare a stock standard solution of methyl paraben in methyl alcohol that has a concentration of 2 mg/mL. (Your instructor may choose to prepare this so that he or she can prepare unknowns or controls from it.)

11. From this stock solution, prepare four standard solutions, using methanol as the solvent, having concentrations of 0.05, 0.1, 0.15, and 0.2 mg of methyl paraben per milliliter, in 50-mL volumetric flasks. Filter these solutions into vials as you did the sample in Part A.

12. Inject full sample loop volumes of each standard solution and the unknown and obtain a chromatogram for each. Each will show only one peak because the propyl and butyl paraben compounds are absent from the standards. Obtain the peak areas for each standard. From the retention

time of the lone peak (methyl paraben), identify the methyl paraben peak on the chromatogram from Part A and note its area also. Obtain these data for any additional unknowns or a control.

13. Plot the standard curve using the spreadsheet procedure used in Experiment 18, and obtain the correlation coefficient and the concentrations of the unknown and control.

14. Maintain the logbook for the instrument used in this experiment. Record the date, your name(s), the experiment name or number, the correlation coefficient, and the results for the control sample. Also plot the control sample results on a control chart for this experiment posted in the laboratory.

15. Dispose of all solutions as directed by your instructor.

## Experiment 47: HPLC Determination of Caffeine and Sodium Benzoate in Soda Pop

1. Prepare 500 mL of mobile phase (1.0 $M$ acetic acid in 10% acetonitrile) as follows. Dilute 50 mL of acetonitrile and 28.5 mL of glacial acetic acid to 500 mL with water. Pour into a large beaker and place on a magnetic stirrer for pH adjustment. With a pH meter and a magnetic stirrer, adjust the pH to 3.0 by adding successive small amounts of a saturated solution of sodium acetate. Filter and degas as in previous experiments or as directed by your instructor. Prepare the instrument for use by flushing the HPLC system to be used with this mobile phase for 8 min at 1.25 mL/min. The system is a reverse phase system with a C18 column.

2. Soda pop samples must be filtered and degassed. Prepare samples by vacuum filtering through paper 0.45-$\mu$m filters, and then fill small labeled vials.

3. Prepare 50 mL of a stock standard solution that is 2.0 mg/mL in caffeine and 5.0 mg/mL in sodium benzoate. Use an analytical balance to weigh the chemicals and a clean 50-mL volumetric flask for the solution. Use distilled water as the diluent. Shake well.

4. Prepare calibration standards that are 0.05, 0.10, 0.15, and 0.20 mg/mL in caffeine from the stock standard. Use 25-mL volumetric flasks and measuring pipets or pipetters suggested by your instructor. Calculate the concentration of sodium benzoate in each of these solutions.

5. Filter each calibration standard, using the syringe filtering equipment, into small labeled vials.

6. Obtain HPLC chromatograms of the calibration standards by injecting 20 $\mu$L of each. There should be two peaks. The benzoate peak will be much smaller than the caffeine peak, and the detector attenuation will need to be changed after the caffeine has eluted. Record the peak size vs. concentration data.

7. Obtain chromatograms of the samples, injecting 20 $\mu$L of each. Allow the column to completely clear before making another injection. Also obtain a chromatogram of the control sample provided. Record the peak size data.

8. When finished, flush the system with a filtered and degassed neutral pH liquid, such as pure methanol or a methanol–water mixture.

9. Create the standard curves (one for caffeine and one for benzoate) by plotting peak size vs. concentration. Use the spreadsheet procedure in Experiment 18. Obtain the concentrations of the unknowns and the control. Plot the results for the control sample on the control chart for this instrument posted in the laboratory.

10. Calculate the milligrams of caffeine and benzoate present in one 12-oz can of the soda pop. There are 0.02957 L per fluid ounce.

11. Maintain the logbook for the instrument used in this experiment. Record the date, your name(s), the experiment name or number, the correlation coefficient, and the results of the control sample.

## Experiment 48: Designing an Experiment for Determining Caffeine in Coffee and Tea

Using Experiment 47 as a base, design an experiment for determining the caffeine in various coffees and teas. Use your own ingenuity as far as brewing times, etc.

## Experiment 49: The Analysis of Mouthwash by HPLC: A Research Experiment

### Introduction

Various brands of mouthwash have a variety of components that should potentially be able to be determined by HPLC. In this experiment, you are on your own to determine what component to analyze for, what mobile and stationary phases to use, what flow rate to use, what concentrations to use, etc.

### Procedure

1. Visit your local pharmacy (outside of lab time) and examine the labels on various brands of mouthwash that are on the shelf. Select for laboratory analysis a brand that looks interesting and bring it into the laboratory.
2. Filter and degas a part of the sample. Prepare the instrument as you have done before, choosing a particular stationary and mobile phase system (such as a reverse phase system using a methanol–water mixture for the mobile phase and a nonpolar stationary phase) and flow rate that you will use as a first trial.
3. Inject the sample and observe the separation of peaks. If there is at least one peak that is resolved, proceed to step 4. Otherwise, try changing the composition of the mobile phase and the flow rate in order to achieve good resolution of at least one peak. If you are still unsuccessful, change the stationary phase and try again.
4. Prepare and filter standard solutions of several of the components shown on the label. The concentrations should be reasonable guesses of what might match what is in the sample. Inject each individually and observe the retention times for each component. Check to see if one of these retention times matches the retention time for a resolved peak from the sample.
5. If you get a match of retention times, proceed to quantitate, comparing the peak size of the sample to that of the standard.

### Optional

6. Just because the retention times match does not necessarily mean that you have identified the peak. To be sure, change the stationary phase and inject again. If a sample peak is again resolved from the others (and it is the same size), and if the retention times of the same standard peak match the resolved peak, you can be sure that you have identified the peak and your quantitation in step 5 is valid.

# Questions and Problems

1. What does HPLC stand for?
2. Define LC, LLC, LSC, BPC, IEC, IC, SEC, GPC, and GFC.
3. We use the words high performance in the name for instrumental liquid chromatography. Why?
4. Give a simple but total definition of HPLC and describe with some detail the basic HPLC system.
5. Detail the path of the mobile phase through the HPLC system from the solvent reservoir to the waste receptacle, giving brief explanatory descriptions of instrument components along the way.
6. Why is HPLC an improvement over the open-column technique?
7. Compare HPLC with GC in terms of (a) the force that moves the mobile phase through the stationary phase, (b) the nature of the mobile phase, (c) how the stationary phase is held in place, (d) what types of chromatography are applicable, (e) application of vapor pressure concepts, (f) sample injection, (g) mechanisms of separation, (h) detection systems, (i) recording systems, and (j) data obtained.
8. The relative vapor pressures of the mixture components are important in GC but not in LC. Why is that?

9. Why do air bubbles form in the flow stream between the mobile phase reservoir and the pump when the mobile is not degassed?

10. Why must HPLC mobile phases and samples be filtered prior to use in the HPLC system?

11. Explain what degassing is and how it is accomplished.

12. What is helium sparging?

13. Why must mobile phases and all samples and standards be finely filtered before an HPLC experiment?

14. Why is it important to know whether the mobile phase and samples contain an organic solvent when preparing to filter them?

15. Paper filters are available to filter LC mobile phases, but nylon filters are used more often. Why is that?

16. For what types of samples and mobile phases is Teflon-based filter material appropriate?

17. Explain the use of a guard column and in-line filters.

18. Why is a metal in-line filter placed just ahead of the column when there already is a filter in the line dipped into the mobile phase reservoir? Why do we worry about filtering the mobile phase and sample in the first place?

19. Why is it that no ordinary liquid pump can be used as the pump in an HPLC system?

20. Describe what is meant by a reciprocating piston pump.

21. What is a check valve and why is it needed in an HPLC system?

22. Distinguish between isocratic elution and gradient elution.

23. What is a gradient programmer?

24. Define the gradient elution method for HPLC, tell what instrument component is needed for it, and tell how this method is useful.

25. How is gradient elution HPLC similar to temperature programming in GC?

26. Why is it that a change in the mobile phase in the middle of the run will change the retention and resolution of the mixture components?

27. What is meant by solvent strength?

28. Give two reasons why an injection system similar to GC would not work for HPLC?

29. Describe in detail how the loop injector works and tell how it overcomes the problems that would be encountered with an injection port–septum system.

30. Show by means of a diagram the difference between the load and inject positions of the HPLC injection system.

31. What are some options that are available if an analyst wants to inject a volume that is smaller than the volume of a sample loop installed on an HPLC loop injector?

32. Distinguish normal phase HPLC from reverse phase HPLC.

33. If a given HPLC system is using a methanol–water mixture for the mobile phase and a C18 column for the stationary phase, what classification of chromatography would be in use? Explain.

34. List some typical mobile and stationary phases for: (a) reverse phase HPLC, and (b) normal phase HPLC.

35. What is meant by bonded phase chromatography? Would such a name describe normal phase, reverse phase, neither, or both? Explain.

36. Answer the following with either "polar" or "nonpolar." How would you describe the mobile phase for normal phase HPLC? How would you describe the stationary phase for reverse phase HPLC?

37. Answer the following with "normal phase" or "reverse phase." For which type of liquid chromatography is a C18 column used? Which is similar to adsorption chromatography in terms of the polarity of the stationary phase?

38. Answer the following with either "cation exchange resin" or "anion exchange resin." Which has negatively charged bonding sites to which positive ions bond? Which separates positively charged ions?

39. Distinguish between cation and anion exchange resins in terms of the nature of the charged sites on the particle surfaces and in terms of application.

40. What is ion chromatography, and what is a typical mobile phase composition for ion chromatography.

41. Distinguish between gel permeation chromatography and gel filtration chromatography in terms of mobile phases that are used and application.

42. What type of HPLC should be chosen for each of the following separation applications?
    (a) All mixture components have formula weights less than 2000, are molecular and polar, and are soluble in nonpolar organic solvents.
    (b) Mixture components have formula weights varying from very large to rather small and are nonionic.
    (c) Mixture components have formula weights less than 2000, are molecular and polar, and are water soluble.

43. Explain why the order of elution of polar and nonpolar mixture components would be reversed when switching from normal phase to reverse phase.

44. What are some HPLC system parameters that can be altered in an attempt to improve resolution?

45. Compare or differentiate between the two items in each of the following (be complete, but concise):
    (a) Isocratic elution technique and gradient elution technique
    (b) Normal phase chromatography and reverse phase chromatography
    (c) Advantages and disadvantages of the UV absorbance detector vs. the refractive index detector

46. Distinguish between the fixed-wavelength UV detector and the variable-wavelength UV detector in terms of design and use.

47. What is a diode array detector, and what are its advantages?

48. In two words or less for each, give one advantage and one disadvantage for each of the following:
    (a) UV absorbance detector
    (b) Refractive index detector
    (c) Fluorescence detector
    (d) Conductivity detector
    (e) Amperometric detector
    (f) LC-MS
    (g) LC-IR

49. Why is it that a UV absorption detector in HPLC is not universal?

50. Discuss the advantages and disadvantages of any three HPLC detectors discussed in the text.

51. Which of the four HPLC detectors (UV absorbance, refractive index, fluorescence, or conductivity)
    (a) Is the most sensitive?
    (b) Is the most universal?
    (c) Requires an ion suppressor to eliminate the ions present in the mobile phase?
    (d) Is a popular detector because it is almost universal and very sensitive?
    (e) Gives a signal based on the position of a light beam on the detector?
    (f) Has a right-angle configuration design?
    (g) Has a single monochromator that can be either a glass filter or a slit/dispersing element/slit type?
    (h) Is used frequently for ion chromatography?

52. In Chapter 12, we discussed the need to calculate response factors, specifically when a TCD detector is used (Section 12.8.2). Would response factors need to be calculated in HPLC when a UV absorbance detector is used? Explain.

53. What is a suppressor and why is one needed in an ion exchange HPLC experiment in which the mobile phase contains ions?

54. Why is FTIR considered as natural as an HPLC detector? What problem does the presence of the mobile phase pose, especially if water is present?

55. Discuss qualitative and quantitative analysis methods for HPLC and how they are different from those of GC.

56. What is the most common cause of an unusually high pressure in an HPLC flow stream and how is the problem solved?

57. What symptoms appear if air bubbles enter the HPLC flow line?

58. The splitting of a peak into two peaks is a symptom of what problem in the column? How is it solved?

59. Name some causes of baseline drift in HPLC.

60. Consider the analysis of a soda pop sample for caffeine by the standard additions method. Construct a graph from the following data and report the milligrams of caffeine in one 12-oz can of the soda pop.

| Caffeine (ppm Added) | Peak Size |
|:---:|:---:|
| 0 | 1368 |
| 5 | 1919 |
| 10 | 2431 |
| 15 | 2997 |

(Calculation hint: There are 0.02957 L per fluid ounce.)

61. The concentration of furazalidone additive in livestock feed can be determined using HPLC. The furazalidone is extracted from the feed using a water–dimethylformamide solution. The sample is then filtered and injected into the chromatograph. If 9.7186 g of the feed was weighed and 55 mL of water–dimethylformamide was added for the extraction, what is the percent furazalidone in the sample given the following:

| Concentration (ppm) | Peak Size |
|:---:|:---:|
| 2.0 | 978 |
| 10.0 | 4621 |
| 30.0 | 14017 |
| 50.0 | 21071 |
| 70.0 | 28994 |

62. **Report**

Find a real-world HPLC analysis in a methods book (AOAC, USP, ASTM, etc.) or journal and report on the details of the procedure according to the following scheme:

(a) Title

(b) General information, including type of material examined, the name of the analyte, sampling procedures, and sample preparation procedures

(c) Specifics, including quantitation procedure, type of chromatography (include the specific stationary phase used), mobile phase characteristics (composition, whether gradient or isocratic, etc.), detector, how the data are gathered, concentration levels for standards, how standards are prepared, and potential problems

(d) Data handling and reporting

(e) References

# 14

# Electroanalytical Methods

## 14.1   Introduction

**Electroanalytical methods** involve the measurement of either the electrical current flowing between a pair of electrodes immersed in the solution tested (**voltammetric** and **amperometric methods**) or an electrical potential developed between a pair of electrodes immersed in the solution tested (**potentio-metric methods**). In either case, the measured parameter (current or potential) is proportional to the concentration of analyte. See Figure 14.1.

Electroanalytical techniques are an extension of classical oxidation–reduction chemistry, and indeed oxidation and reduction processes occur at the surface of or within the two electrodes, oxidation at one and reduction at the other. Electrons are consumed by the reduction process at one electrode and generated by the oxidation process at the other. The electrode at which oxidation occurs is termed the **anode**. The electrode at which reduction occurs is termed the **cathode**. The complete system, with the anode connected to the cathode via an external conductor, is often called a **cell**. The individual oxidation and reduction reactions are called **half-reactions**.[*] The individual electrodes with their half-reactions are called **half-cells**. As we shall see in this chapter, the half-cells are often in separate containers (mostly to prevent contamination) and are themselves often referred to as electrodes because they are housed in portable glass or plastic tubes. In any case, there must be contact between the half-cells to facilitate ionic diffusion. This contact is called the **salt bridge** and may take the form of an inverted U-shaped tube filled with an electrolyte solution, as shown in Figure 14.2, or, in most cases, a small fibrous plug at the tip of the portable unit, as we will see later in this chapter.

A **galvanic cell** is one in which this current flows (and the redox reaction proceeds) spontaneously because of the strong tendency for the chemical species involved to give and take electrons. An **electrolytic cell** is one in which the current is not a spontaneous current, but rather is the result of incorporating an external power source, such as a battery, in the circuit to drive the reaction in one direction or the other. Potentiometric methods involve galvanic cells, and voltammetric and amperometric methods involve electrolytic cells.

Redox reactions, as discussed in Chapter 5, occur upon direct contact between the oxidizing agent and the reducing agent. Upon contact, the electrons jump from the reducing agent to the oxidizing agent. In the case of galvanic and electrolytic cells, the electrons transfer from one to the other by flowing through the external conductor from the anode to the cathode. The flow of electrons through the external conductor in an electrolytic cell constitutes the current that is measured in the voltammetric and amper-ometric methods. The concentration of the analyte is related to this current. When one of the electrodes is a reference electrode (to be discussed later), the potential (voltage) difference between the two half-cells in a galvanic cell is measured. These are the potentiometric methods, and the analyte concentration is related to this potential.

---

[*]The concept of oxidation and reduction half-reactions was first introduced in Chapter 5, Section 5.4.2.

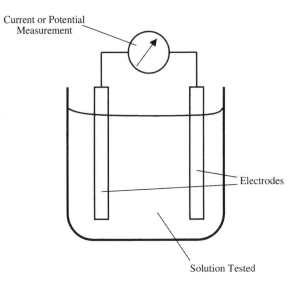

**FIGURE 14.1**   A representation of the general basis of electroanalytical methods.

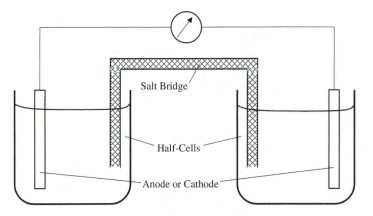

**FIGURE 14.2**   A representation of an electrochemical cell as described in the text. One electrode is the anode, the other the cathode, and electrons generated by the oxidation process at the anode flow through the external circuit to the cathode, where reduction takes place. This flow of electrons constitutes electrical current in the external circuit.

## 14.2   Transfer Tendencies: Standard Reduction Potentials

Different redox half-reactions have different tendencies to occur. Consider the three metals sodium, magnesium, and aluminum, all in the same row of the periodic table. A sodium atom, as well as atoms of other alkali metals, has a very strong tendency to give up one electron to become a sodium ion. The tendency is so strong that the reaction of sodium with ordinary water borders on explosive:

$$2\,Na + 2\,H_2O \rightarrow 2\,Na^+ + 2\,OH^- + H_2 \text{ (explosive)}$$

Magnesium, the alkaline earth metal next to sodium in the periodic table, has a relatively strong tendency to give up its two outermost electrons, but this tendency is much less than that of sodium. When a fresh surface of magnesium metal is exposed to water, the reaction occurs very slowly:

$$Mg + 2\,H_2O \rightarrow Mg^{2+} + 2\,OH^- + H_2 \text{ (slow)}$$

Aluminum metal, the next element to the right in the same row as sodium and magnesium in the periodic table, has an even lesser tendency to lose electrons, even though it does lose electrons to become the aluminum ion, $Al^{3+}$. Its reactivity with water is almost nil:

$$Al + H_2O \rightarrow \text{no reaction}$$

A similar illustration can be made with the *nonmetals* and their tendency to *take on* electrons. The point is that different half-reactions have different tendencies to occur.

To understand **potentiometric methods**, those that measure electrical potentials and determine analyte concentrations from these potentials, it is necessary that numerical values for these tendencies be known under conventional standard modes and conditions. What are these modes and conditions? First, all half-reactions must be written as either reductions or oxidations. Scientists have decided to write them as reductions. Second, the tendencies for half-reactions to proceed depend on the temperature, the concentrations of the chemical species involved,[*] and, if gases are involved, the pressure in the half-cell. Scientists have defined standard conditions to be a temperature of 25°C, a concentration of exactly 1 $M$ for all dissolved chemical species involved, and a pressure of exactly 1 atm. Third, because every cell consists of two half-cells, it is not possible to measure the value directly. However, if we were to assign the tendency of a certain half-reaction to be zero, then the tendencies of all other half-reactions can be determined relative to this reference half-reaction.

The result of all of this is what has come to be known as the table of **standard reduction potentials**. An abbreviated such table is given here as Table 14.1. In this table, the half-reactions are listed on the left and the numerical values for each are listed on the right. The heading for the right-hand column is $E^\circ$, the symbol for standard reduction potential. Notice that all half-reactions are written as reductions and that the half-reaction $2H^+ + 2e^- \leftrightarrows H_2$ is the reference half-reaction (0.0000 V).

The half-reactions at the top in the table are those that have the strongest tendency to occur as written. For example, the fluorine-to-fluoride half-reaction, the first one listed, has a very strong tendency to occur, as any student of fundamental chemistry would be able to conclude, based on the position of fluorine in the periodic table. What may be confusing is that the half-reactions involving lithium, potassium, and sodium are found at the bottom of the table, indicating that they do not have a tendency to occur. The explanation for this apparent contradiction is that the reactions are, by convention, reductions, meaning that the ions are being reduced to the metals. Thus the reductions of lithium ions, potassium ions, etc., would indeed not have a tendency to occur since the reverse reactions in each case (the metal-to-metal ion reactions) have the strong tendencies.

Another observation is that the reactions near the bottom of the table have a negative sign for their $E^\circ$, while the reactions near the top have a positive sign. The explanation for this is that, compared to the standard ($2H^+ + 2e^- \leftrightarrows H_2$), those that have a positive sign tend to occur as written, while those that have a negative sign tend to occur in the opposite direction. The more positive the $E^\circ$ value, the stronger the tendency to occur as written. The less positive (or more negative) the $E^\circ$ value, the stronger the tendency to occur in the opposite direction to what is written.

A way to peruse the table is to understand that if the reduced form in a particular half-reaction in the table comes into contact with the oxidized form in a half-reaction *above* it in the table, a redox reaction between these two forms will occur. Stated in reverse, if the oxidized form in any half-reaction in the table comes into contact with the reduced form in a half-reaction *below* it in the table, a redox reaction between these two forms will occur. Redox reactions between other species, such as a reduced form in a given reaction in contact with an oxidized form in a half-reaction *below* it in the table, will *not* occur. Mnemonic devices (devices that aid in memorizing) summarizing these statements are the arrows shown in Figure 14.3. Thus, for example, as can be seen in Figure 14.3, $Cl_2$ will react with $Fe^{2+}$ (to form $Cl^-$ and $Fe^{3+}$), but $Cl^-$ would *not* react with $Fe^{3+}$.

---

[*]To be thermodynamically correct, the tendencies for half-reactions to occur depend on the *activities* of the chemical species involved, not the concentrations. See Chapter 5 (Section 5.2.12) for a brief discussion of activity.

**TABLE 14.1**  Some Standard Reduction Potentials at 25°C

| Half-Reactions | E° (Volts) |
|---|---|
| $F_2 + 2e^- \rightleftharpoons 2F^-$ | +2.866 |
| $O_3 + 2H^+ + 2e^- \rightleftharpoons O_2 + H_2O$ | +2.076 |
| $H_2O_2 + 2H^+ + 2e^- \rightleftharpoons 2H_2O$ | +1.776 |
| $Ce^{4+} + 1e^- \rightleftharpoons Ce^{3+}$ | +1.72 |
| $MnO_4^- + 8H^+ + 5e^- \rightleftharpoons Mn^{2+} + 4H_2O$ | +1.507 |
| $Cl_2 + 2e^- \rightleftharpoons 2Cl^-$ | +1.35827 |
| $Cr_2O_7^{2-} + 14H^+ + 6e^- \rightleftharpoons 2Cr^{3+} + 7H_2O$ | +1.232 |
| $O_2 + 4H^+ + 4e^- \rightleftharpoons 2H_2O$ | +1.229 |
| $Hg^{2+} + 2e^- \rightleftharpoons Hg$ | +0.851 |
| $Ag^+ + 1e^- \rightleftharpoons Ag$ | +0.7996 |
| $Fe^{3+} + 1e^- \rightleftharpoons Fe^{+2}$ | +0.771 |
| $I_2 + 2e^- \rightleftharpoons 2I^-$ | +0.5355 |
| $Cu^{2+} + 2e^- \rightleftharpoons Cu$ | +0.3419 |
| $Hg_2Cl_2 + 2e^- \rightleftharpoons 2Hg + 2Cl^-$ (SCE) | +0.2412 |
| $AgCl + 1e^- \rightleftharpoons Ag + Cl^-$ (Ag–AgCl reference) | +0.22233 |
| $Sn^{4+} + 2e^- \rightleftharpoons Sn^{2+}$ | +0.151 |
| $2H^+ + 2e^- \rightleftharpoons H_2$ | 0.00000 |
| $Fe^{3+} + 3e^- \rightleftharpoons Fe$ | −0.037 |
| $Sn^{2+} + 2e^- \rightleftharpoons Sn$ | −0.1375 |
| $Ni^{2+} + 2e^- \rightleftharpoons Ni$ | −0.257 |
| $Fe^{2+} + 2e^- \rightleftharpoons Fe$ | −0.447 |
| $Cr^{3+} + 3e^- \rightleftharpoons Cr$ | −0.744 |
| $Zn^{2+} + 2e^- \rightleftharpoons Zn$ | −0.7618 |
| $Mg^{2+} + 2e^- \rightleftharpoons Mg$ | −2.372 |
| $Na^+ + 1e^- \rightleftharpoons Na$ | −2.71 |
| $K^+ + 1e^- \rightleftharpoons K$ | −2.931 |
| $Li^+ + 1e^- \rightleftharpoons Li$ | −3.0401 |

*Source:* From Lide, *Handbook of Chemistry and Physics,* 82nd ed., CRC Press, Boca Raton, FL, 2001–2002. With permission.

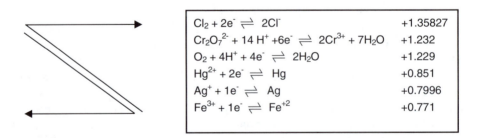

**FIGURE 14.3**  Mnemonic devices, the arrows on the left, for predicting which chemicals will participate in a redox reaction and which will not. A segment of the table of standard reduction potentials (Table 14.1) is presented on the right as a help to understand the use of the arrows. See text for an example.

# 14.3 Determination of Overall Redox Reaction Tendency: $E^o_{cell}$

The more the two half-reactions are separated in the table, the greater is the tendency for the net reaction to occur. This tendency for an overall redox reaction to occur, whether by direct contact or in an electrochemical cell, is determined from the standard reduction potentials, $E^o$ values, of the half-reactions involved, and the value of this potential are indications of the tendency of the overall redox reaction to occur. We will now present a scheme for determining this potential, which is symbolized $E^o_{cell}$.

In the following scheme, it is assumed that there is a proposed redox system given so that the half-reactions and standard reduction potentials can be found in Table 14.1, or other table of standard reduction potentials. An example follows:

Step 1: Write the equations representing the half-reactions as extracted from the overall reaction given and label as an oxidation and a reduction.

Step 2: Locate the half-reactions in a table of standard reduction potentials, such as Table 14.1, and write the $E^o$ values adjacent to the respective half-reactions. *For the oxidation half-reaction, the sign of the $E^o$ must be changed, since the reaction is written in reverse.*

Step 3: Balance charges, equalize electrons in both half-reactions, and add the two equations together (as in the scheme for equation balancing in Chapter 5), and also add the $E^o$ values together. *Do not multiply $E^o$ values by the multiplying coefficients.* The resulting E is the $E^o_{cell}$.

Step 4: If $E^o_{cell}$ is positive (+), the reaction will proceed spontaneously to the right as written. If it is negative (−), it will proceed spontaneously in the opposite direction, i.e., to the left.

## Example 14.1

What is the $E^o_{cell}$ for the following redox system and in which direction will the reaction spontaneously proceed?

$$Cu + Ag^+ \leftrightharpoons Ag + Cu^{2+}$$

### Solution 14.1

Step 1: Oxidation: $Cu \leftrightharpoons Cu^{2+} + 2e^-$

Reduction: $Ag^+ + 1e^- \leftrightharpoons Ag$

Step 2: Oxidation: $Cu \leftrightharpoons Cu^{2+} + 2e^-$     $E^o = -0.3419$ V (note sign change)

Reduction: $Ag^+ + 1e^- \leftrightharpoons Ag$     $E^o = +0.7996$ V

Step 3: Oxidation: $Cu \leftrightharpoons Cu^{2+} + 2e^-$     $E^o = -0.3419$ V

Reduction: $2(Ag^+ + 1e^- \leftrightharpoons Ag)$     $E^o = +0.7996$ V

---

$Cu + 2Ag^+ \leftrightharpoons Cu^{2+} + 2Ag$     $E^o_{cell} = +0.4577$ V

Step 4: Since $E^o_{cell}$ is positive (+), the reaction is spontaneous to the right as written.

# 14.4 The Nernst Equation

Both half- and overall reaction tendencies change with temperature, pressure (if gases are involved), and concentrations of the ions involved. Thus far, we have only been concerned with standard conditions. Standard conditions, as stated previously, are 25°C, 1 atm pressure, and 1 $M$ ion concentrations. An equation has been derived to calculate the cell potential when conditions other than standard conditions are present. This equation is called the **Nernst equation** and is used to calculate the true E (cell potential) from the $E°$, temperature, pressure, and ion concentrations. Consider the half-reaction in which "a" moles of chemical "Ox" react with n electrons ($e^-$) to give b moles of chemical "Red" (examples follow):

$$a\ Ox + ne^- \leftrightarrows b\ Red \tag{14.1}$$

The Nernst equation* is the following:

$$E = E° - \frac{0.0592}{n}\log\frac{[Red]^b}{[Ox]^a} \tag{14.2}$$

For the general overall reaction, in which "a" moles of chemical A react (with the transfer of electrons) with b moles of chemical B to give c moles of chemical C and d moles of chemical D,

$$aA + bB \leftrightarrows cC + dD \tag{14.3}$$

The Nernst equation is the following:

$$E_{cell} = E°_{cell} - \frac{0.0592}{n}\log\frac{[C]^c[D]^d}{[A]^a[B]^b} \tag{14.4}$$

If any species involved is a gas, the partial pressure of the gas is substituted for the concentration. If the temperature is different from 25°C, the constant 0.0592 changes. The number of electrons, n, in Equation (14.4) is the total number of electrons transferred, as discovered after equalizing the electrons in the two half-reactions in step 3 of the scheme for determining $E°_{cell}$ in Section 14.3.

As seen in Equations (14.2) and (14.4), the potential of cells and half-cells is dependent on the concentrations of the dissolved species involved. Clearly, the measurement of a potential can lead to the determination of the concentration of an analyte. This, therefore, is the basis for all quantitative potentiometric techniques and measurements to be discussed in this chapter.

### Example 14.2

What is the E for the $Fe^{3+}/Fe^{2+}$ half-cell if $[Fe^{3+}] = 10^{-4}$ M and $[Fe^{2+}]$ is $10^{-1}$ M at 25°C?

### *Solution 14.2*

$$Fe^{3+} + 1e^- \leftrightarrows Fe^{2+} \quad E° = +0.771\ V\ \text{(from Table 14.1)}$$

$$E = E° - \frac{0.0592}{1}\log\frac{[Fe^{2}]}{[Fe^{3+}]}$$

$$= +0.771 - \frac{0.0592}{1}\log\frac{10^{-1}}{10^{-4}}$$

$$= 0.771 - 0.0592\log 10^3$$

$$= 0.771 - 0.0592 \times 3$$

$$= 0.771 - 0.1776$$

$$= +0.593\ V$$

### Example 14.3

What is the E for the $Cu^{2+}/Cu//Ag^+/Ag$ cell if $[Cu^{2+}] = 0.10$ M and $[Ag^+] = 10^{-2}$ M at 25°C?

---

*Once again, to be thermodynamically correct, activity should be used rather than concentration. Use of concentration is an approximation.

**Solution 14.3**

One way to write the equation for the reaction involved is to use the equation in Example 14.1:

$$Cu + 2Ag^+ \rightleftharpoons Cu^{2+} + 2Ag \qquad E^o_{cell} = +0.4577 \text{ V}$$

The reaction could also be written as the reverse of this one—the way we write it is arbitrary since the sign of E tells us what direction the reaction proceeds. Using the equation above, the Nernst equation is the following. Notice that there are two electrons involved, as determined in step 3 in the solution to Example 14.1:

$$E = E^o_{cell} - \frac{0.0592}{2} \log \frac{[Cu^{+2}]}{[Ag^+]^2}$$

Note: As with equilibrium constant problems, concentrations of pure undissolved solids (in this case Ag and Cu) do not appear in the expression.

$$E = +0.4577 - \frac{0.0592}{2} \log \frac{0.10}{(0.010)^2}$$

$$= +0.4577 - \frac{0.0592}{2} \log 10^3$$

$$= +0.4577 - \frac{0.0592}{2}(3)$$

$$= +0.4577 - 0.0888 = +0.3689 \text{ V}$$

# 14.5  Potentiometry

As mentioned previously, electroanalytical techniques that measure or monitor electrode potential utilize the *galvanic* cell concept and come under the general heading of **potentiometry**. Examples include pH electrodes, ion-selective electrodes, and potentiometric titrations, each of which will be described in this section. In these techniques, a pair of electrodes are immersed, the potential (voltage) of one of the electrodes is measured relative to the other, and the concentration of an analyte in the solution into which the electrodes are dipped is determined. One of the immersed electrodes is called the **indicator electrode** and the other is called the **reference electrode**. Often, these two electrodes are housed together in one probe. Such a probe is called a **combination electrode**.

## 14.5.1  Reference Electrodes

The measurement of any voltage is a relative measurement and requires an unchanging reference point. For voltage measurements in ordinary electronic circuitry, this reference is usually ground. Ground is often a wire that is connected to the frame of the electronic unit and also to the third prong in an electrical outlet, which in turn is connected to a rod that is pushed into the earth, hence the name ground. Thus an electronics technician measures voltages relative to ground.

In electroanalytical chemistry, the unchanging reference is a half-cell that, at a given temperature, has an unchanging potential. There are two designs for this half-cell that are popular—the saturated calomel electrode (SCE) and the silver–silver chloride electrode. These are described below.

### 14.5.1.1  The Saturated Calomel Reference Electrode

The saturated calomel reference electrode is an example of a constant-potential electrode. A drawing and a photograph of a typical SCE available commercially are shown in Figure 14.4. It consists of two concentric glasses or tubes, each isolated from the other except for a small opening for electrical contact.

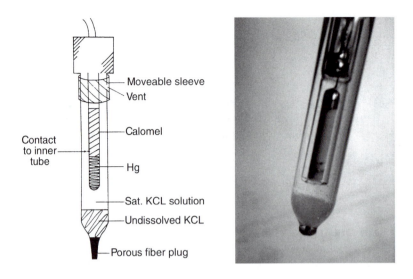

**FIGURE 14.4**   Left, a drawing of a commercial saturated calomel electrode. Right, a photograph of a commercial saturated calomel electrode.

The outer tube has a porous fiber tip, which acts as the salt bridge to the analyte solution and the other half-cell. A saturated solution of potassium chloride is in the outer tube. The saturation is evidenced by the fact that there is some undissolved KCl present. Within the inner tube is mercury metal and a paste-like material known as calomel. Calomel is made by thoroughly mixing mercury metal (Hg) with mercurous chloride ($Hg_2Cl_2$), a white solid. When in use, the following half-cell reaction occurs:

$$Hg_2Cl_2 + 2\ e^- \rightleftharpoons 2\ Hg + 2\ Cl^- \tag{14.5}$$

The Nernst equation for this reaction is

$$E = E^\circ \frac{0.0592}{2} \log [Cl^-]^2 \tag{14.6}$$

or

$$E = E^\circ - 0.0592 \log [Cl^-] \tag{14.7}$$

Obviously the only variable on which the potential depends is $[Cl^-]$. The saturated KCl present provides the $[Cl^-]$ for the reaction, and, since it is a saturated solution, $[Cl^-]$ is a constant at a given temperature represented by the solubility of KCl at that temperature. If $[Cl^-]$ is constant, the potential of this half-cell, dependent only on the $[Cl^-]$, is therefore also a constant. As long as KCl is kept saturated and the temperature kept constant, the SCE is useful as a reference against which all other potential measurements can be made. Its standard reduction potential at 25°C (see Table 14.1) is +0.2412 V.

The SCE is dipped into the analyte solution along with the indicator electrode. A voltmeter is then externally connected across the lead wires leading to the two electrodes, and the potential of the indicator electrode vs. that of the SCE is measured.

### 14.5.1.2 The Silver–Silver Chloride Electrode

The commercial silver–silver chloride electrode is similar to the SCE in that it is enclosed in glass, has nearly the same size and shape, and has a porous fiber tip for contact with the external solution. Internally, however, it is different. There is only one glass tube (unless it is a double-junction design—see Section 14.5.3) and a solution saturated in silver chloride and potassium chloride is inside. A silver wire coated at the end with a silver chloride paste extends into this solution from the external lead. See Figure 14.5. The half-reaction that occurs is

$$AgCl(s) + 1\ e^- \rightleftharpoons Ag(s) + Cl^- \qquad (14.8)$$

and the Nernst equation for this is

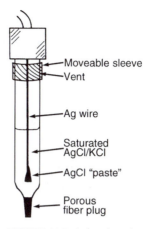

**FIGURE 14.5** A drawing of a commercial silver–silver chloride reference electrode.

$$E = E^\circ - \frac{0.0592}{1} \log[Cl^-] \qquad (14.9)$$

The standard reduction potential for this half-reaction (from Table 14.1) is +0.22233 V. The potential is dependent only on the $[Cl^-]$, as was the potential of the SCE, and once again $[Cl^-]$ is constant because the solution is saturated. Thus this electrode is also appropriate for use as a reference electrode.

## 14.5.2 Indicator Electrodes

As stated previously, the reference electrode represents half of the complete system for potentiometric measurements. The other half is the half at which the potential of analytical importance—the potential that is related to the concentration of the analyte—develops. There are a number of such **indicator electrodes** and analytical experiments that are of importance.

### 14.5.2.1 The pH Electrode

The measurement of pH is very important in many aspects of chemical analysis. Curiously, the measurement is based on the potential of a half-cell, the pH electrode.

The pH electrode consists of a closed-end glass tube that has a very thin fragile glass membrane

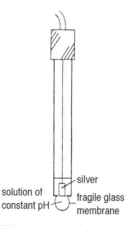

**FIGURE 14.6** A drawing of a pH electrode.

at the tip. Inside the tube is a saturated solution of silver chloride that has a particular pH. It is typically a 1 *M* solution of HCl. A silver wire coated with silver chloride is dipped into this solution to just inside the thin membrane. While this is almost the same design as that of the silver–silver chloride reference electrode, the presence of HCl and the fact that the tip is fragile glass and does not have a porous fiber plug point out the difference. See Figure 14.6.

The purpose of the silver–silver chloride combination is to prevent the potential that develops from changing due to possible changes in the interior of the electrode. The potential that develops is a **membrane potential**. Since the glass membrane at the tip is thin, a potential develops due to the fact that the chemical composition inside is different from the chemical composition outside. Specifically, it is the difference in the concentration of the hydrogen ions on opposite sides of the membrane that causes the potential—the membrane potential—to develop. There is no half-cell reaction involved. The Nernst equation is

$$E = E° - 0.0592 \log \frac{[H^+](internal)}{[H^+](external)} \tag{14.10}$$

or, since the internal $[H^+]$ is a constant, it can be combined with $E°$, which is also a constant, giving a modified $E°$, $E^*$, and eliminating $[H^+]$(internal):

$$E = E^* + 0.0592 \log [H^+](external) \tag{14.11}$$

In addition, we can recognize that $pH = -\log [H^+]$ and substitute this into the above equation:

$$E = E^* - 0.0592 \, pH \tag{14.12}$$

The beauty of this electrode is that the measured potential (measured against a reference electrode) is thus directly proportional to the pH of the solution into which it is dipped. A specially designed voltmeter, called a pH meter, is used. A pH meter displays the pH directly, rather than the value of E.

The pH meter is standardized (calibrated) with the use of buffer solutions. Usually, two buffer solutions are used for maximum accuracy. The pH values for these solutions should bracket the pH value expected for the sample. For example, if the pH of a sample to be measured is expected to be 9.0, buffers of pH = 7.0 and pH = 10.0 should be used. Buffers with pH values of 4.0, 7.0, and 10.0 are available commercially specifically for pH meter standardization. Alternatively, of course, homemade buffer solutions (see Chapter 5) may be used. In either case, when the pH electrode and reference electrode are immersed in the buffer solution being measured and the electrode leads are connected to the pH meter, the meter reading is electronically adjusted (refer to manufacturer's literature for specifics) to read the pH of this soluiton. The electrodes can then be immersed into the solution being tested and the pH directly determined.

### 14.5.2.2 The Combination pH Electrode

In order to use the pH electrode described above, two half-cells (probes) are needed—the pH electrode itself and a reference electrode, either the SCE or the silver–silver chloride electrode—and two connections are made to the pH meter. An alternative is **combination pH electrode**. This electrode incorporates both the reference probe and pH probe into a single probe and is usually made of epoxy plastic. It is by far the most popular electrode today for measuring pH. The reference portion is a silver–silver chloride reference. A drawing and a photograph of the combination pH electrode is given in Figure 14.7.

The pH electrode is found in the center of the probe as shown. It is identical to the pH electrode described above—a silver wire coated with silver chloride immersed in a solution saturated with silver chloride and having a $[H^+]$ of 1.0 $M$. This solution is in contact with a thin glass membrane at the tip. The reference electrode is in an outer tube concentric with the inner pH electrode. It has a silver wire coated with silver chloride in contact with a solution saturated with silver chloride and potassium chloride. A porous fiber strand serves as the salt bridge to the outer tube with the solution tested. The drawing in Figure 14.7 shows this strand on the side of the outer tube, but some designs have it on the bottom of the probe next to the glass membrane. Figure 14.8 shows several photographs of the porous fiber strands on various combination probes. Notice the epoxy plastic sheath surrounding the thin glass membranes in these photographs. The external end of the porous fiber strand must be in full contact with the solution being tested when the measurement is made. The connection of this electrode to the pH meter is either the bnc type or another type, depending on the pH meter used. See Figure 14.9.

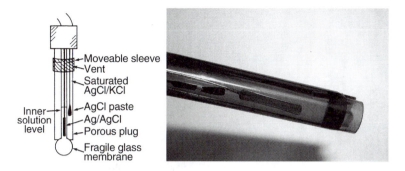

**FIGURE 14.7**    Left, a drawing of a commercial combination pH electrode. Right, a photograph of a commercial combination pH electrode. In the photograph, notice the silver electrodes in both the inner and outer tubes.

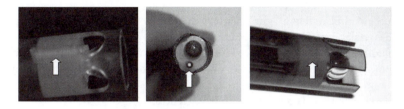

**FIGURE 14.8**    Photographs showing the porous fiber strands on various combination pH probes. The arrow points to the strand in each case. Also notice the glass membranes and the protective plastic sheaths.

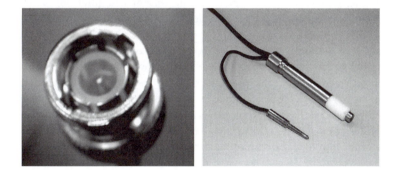

**FIGURE 14.9**    Photographs of the electrical connections to the combination pH electrode. Left, the bnc type of connection. Right, two separate connections to be made to the pH meter.

### 14.5.2.3   Ion-Selective Electrodes

The concept of the pH electrode has been extended to include other ions as well. Considerable research has gone into the development of these **ion-selective electrodes** over the years, especially in studying the composition of the membrane that separates the internal solution from the analyte solution. The internal solution must contain a constant concentration of the analyte ion, as with the pH electrode. Today we utilize electrodes with: 1) glass membranes of varying compositions, 2) crystalline membranes, 3) liquid membranes, and 4) gas-permeable membranes. In each case, the interior of the electrode has a silver–silver chloride wire immersed in a solution of the analyte ion.

Examples of electrodes that utilize a glass membrane are those for lithium ions, sodium ions, potassium ions, and silver ions. Varying percentages of $Al_2O_3$ and $B_2O_3$, along with oxides of the metal analyte, are often found in the membrane, as well as other metal oxides. The selectivity and sensitivity of these electrodes vary.

With crystalline membranes, the membrane material is most often an insoluble ionic crystal cut to a round, flat shape and having a thickness of 1 or 2 mm and a diameter of about 10 mm. This flat disk is

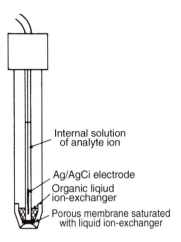

**FIGURE 14.10** A drawing of an ion-selective electrode with a liquid membrane.

Internal solution of analyte ion

Ag/AgCi electrode
Organic liqiud ion-exchanger

Porous membrane saturated with liquid ion-exchanger

mounted into the end of a Teflon™ or polyvinyl chloride (PVC) tube. The most important of the electrodes with crystalline membranes is the fluoride electrode. The membrane material for this electrode is lanthanum fluoride. The fluoride electrode is capable of accurately sensing fluoride ion concentrations over a broad range and to levels as low as $10^{-6}$ $M$. Other electrodes that utilize a crystalline membrane but with less impressive success records are chloride, bromide, iodide, cyanide, and sulfide electrodes. The main difficulty with these is problems with interferences.

Liquid membrane electrodes utilize porous polymer materials, such as PVC or other plastics. An organic liquid ion exchanger immiscible with water contacts and saturates the membrane from a reservoir around the outside of the tube containing the water solution of the analyte and the silver–silver chloride wire. See Figure 14.10. Important electrodes with this design are the calcium and nitrate ion-selective electrodes.

Finally, gas-permeable membranes are used in electrodes that are useful for dissolved gases, such as ammonia, carbon dioxide, and hydrogen cyanide. These membranes are permeated by the dissolved gases, but not by solvents or ionic solutes. Inside the electrode is a solution containing the reference wire as well as a pH probe, the latter positioned to create a thin liquid film between the glass membrane of the pH probe and the gas permeable membrane. As the gases diffuse in, the pH of the solution constituting the thin film changes, and thus the response of the pH electrode changes proportionally to the amount of gas diffusing in.

Calibration of ion-selective electrodes for use in quantitative analysis is usually done by preparing a series of standards as in most other instrumental analysis methods (see Chapter 7), since the measured potential is proportional to the logarithm of the concentration. The relationship is

$$E = E^* - \frac{0.0592}{z} \log[\text{ion}] \tag{14.13}$$

in which z is the signed charge on the ion. The analyst can measure the potential of the electrode immersed in each of the standards and the sample (vs. the SCE or silver–silver chloride reference), plot E vs. log [ion], and find the unknown concentration from the linear regression procedures (Chapter 7).

### 14.5.3 Other Details of Electrode Design

The electrodes described in this section are commercially available. The body of these electrodes may be either glass or epoxy plastic, as we have discussed. Epoxy plastic electrodes are unbreakable. Some electrodes are **gel-filled electrodes** and are sealed. This means that the KCl solution has a gelatin mixed with it. There is no vent hole, they cannot be refilled with saturated KCl, and solid KCl cannot be added.

Some electrodes are **double-junction electrodes**. Such electrodes are encased in another glass tube and therefore have two junctions, or porous plugs. The purpose of such a design is to prevent contamination—the contamination of the electrode solution with the analyte solution, the contamination of the analyte solution with the electrode solution, or both, by the diffusion of either solution through the porous tip or plug. See the next section for tips concerning these problems.

## 14.5.4 Care and Maintenance of Electrodes

While the SCE or Ag–AgCl electrode is dipped into the solution, there will be a slight leakage of the solutions through the porous tip. In order for these electrodes to be used accurately, the measurement must not be adversely affected by the slight contamination from these ions. It is a good idea to slide the moveable sleeve (Figure 14.4, for example) downward so that the outer tube is vented while the electrode is in use so that the ions do indeed freely diffuse through the porous tip. Also, the electrode, under these circumstances, should not be immersed into the solution so deep that the level of solution in the external tube is lower than the level of the solution tested. This would cause the solution to diffuse into the reference electrode rather than the reverse, and thus would contaminate the solution inside and possibly damage the electrode.

The vent hole may also be used prior to the experiment to refill the outer tube with more fill solution (often supplied by the manufacturer), as this solution is lost with time or is sometimes drained out for long-term storage or cleaning. In addition, if the undissolved KCl in the SCE disappears, more solid KCl can be added through the vent hole.

Proper storage of electrodes is important. For short-term storage (up to 1 week), the tip of the electrode should be immersed in a solution to a level above the porous plug. Some manufacturers supply a solution to be used for this. Others recommend a particular solution, such as a pH = 7 buffer solution. Other details for such storage solutions may be given in the manufacturer's literature. In any case, distilled water should *not* be used. For long-term storage (longer than 1 week), the protective plastic sleeve provided with the electrode should be placed back on the tip and a small bit of cotton moistened with the storage solution placed inside at the tip of the electrode. The manufacturer's literature usually contains the full recommendations for storage.

## 14.5.5 Potentiometric Titrations

It is possible to monitor the course of a titration using potentiometric measurements. The pH electrode, for example, is appropriate for monitoring an acid–base titration and determining an end point in lieu of an indicator, as in Experiment 10 in Chapter 5. The procedure has been called a **potentiometric titration** and the experimental setup is shown in Figure 14.11. The end point occurs when the measured pH undergoes a sharp change—when all the acid or base in the titration vessel is reacted. The same

**FIGURE 14.11**    A setup for a potentiometric titration.

# WORKPLACE SCENE 14.1

In Nebraska, state regulations require that the chemical makeup of animal feed sold in the state be accurately reflected on the labels found on the feed bags. The Nebraska State Agriculture Laboratory is charged with the task of performing the analytical laboratory work required. An example is salt (sodium chloride) content. The method used to analyze the feed for sodium chloride involves a potentiometric titration. A chloride ion-selective electrode in combination with a saturated calomel reference electrode is used. After dissolving the feed sample, the chloride is titrated with a silver nitrate standard solution. The reaction involves the formation of the insoluble precipitate silver chloride. The electrode monitors the decrease in the chloride concentration as the titration proceeds, ultimately detecting the end point (when the chloride ion concentration is zero).

Charlie Focht of the Nebraska State Agriculture Laboratory refills a saturated calomel electrode with saturated potassium chloride while preparing to analyze animal feed samples for sodium chloride via a potentiometric titration.

procedure can be used for any ion for which an ion-selective electrode has been fabricated and for which there exists an appropriate titrant. See Workplace Scene 14.1.

In addition, potentiometric titration methods exist in which an electrode other than an ion-selective electrode is used. A simple platinum wire surface can be used as the indicator electrode when an oxidation–reduction reaction occurs in the titration vessel. An example is the reaction of Ce(IV) with Fe(II):

$$Ce^{4+} + Fe^{2+} \leftrightharpoons Ce^{3+} + Fe^{3+}$$

If this reaction were to set be up as a titration, with $Ce^{4+}$ as the titrant and the $Fe^{2+}$ in the titration vessel and the potential of a platinum electrode dipped into the solution monitored (vs. a reference electrode) as the titrant is added, the potential would change with the volume of titrant added. This is because as the titrant is added, the measured E would change as the $[Fe^{2+}]$ is decreased, the $[Fe^{3+}]$ is increased, and the $[Ce^{3+}]$ is increased. At the end point and beyond, all the $Fe^{2+}$ is consumed and $[Fe^{3+}]$ and $[Ce^{3+}]$ change only by dilution; thus E is dependent mostly on the change in $[Ce^{4+}]$. At the end point, there would be a sharp change in the measured E.

Automatic titrators have been invented that are based on these principles. A sharp change in a measured potential can be used as an electrical signal to activate a solenoid and stop a titration.

# 14.6 Voltammetry and Amperometry

Electroanalytical techniques that measure or monitor the current flow between a pair of electrodes utilize the *electrolytic* cell concept. Such techniques were first mentioned in Section 13.6.5.2 in describing an electrochemical detector for HPLC. An electrolytic cell, you will recall from Section 14.1, operates with a power source in the electrode circuit to force a current to flow through the electrode system. The voltammetric and amperometric techniques usually utilize three electrodes, rather than two. The three electrodes are the **working electrode**, the **auxiliary electrode** (sometimes called the **counter electrode**), and a **reference electrode** (such as those described in Section 14.5). A special electronic circuit for carefully controlling the power that the system receives utilizes the three electrode system. The circuit applies a small, carefully controlled electrical polarization (an **applied potential**) to the working electrode, and the resulting current is due to the analyte being either reduced or oxidized at this electrode. This current is proportional to the concentration of the analyte in the solution. **Voltammetric methods** measure the current that flows as the applied potential is varied. **Amperometric methods** measure the current that flows as a result of a *constant* applied potential—a potential that causes the desired electrode reaction to take place—while stirring the solution or while causing the solution to flow across the electrode. See Workplace Scene 14.2 for a coulometric method.

## 14.6.1 Voltammetry

One classic design of a working electrode is the so-called **dropping mercury electrode**, in which small drops of mercury dropping from the end of a capillary tube constitute the working electrode. This is a good electrode for this kind of work because the surface is continuously renewed while the measurement is made and the solution is also automatically stirred. A continuously renewed surface is helpful because the surface may become fouled from the products of the reduction or oxidation forming there. The voltammetric technique utilizing this electrode is called **polarography**. There are several variations of the basic polarographic method, but these are beyond our scope.

If a stationary electrode is used, such as platinum, gold, or glassy carbon, the technique is called **voltammetry**. One useful voltammetric technique is called **stripping voltammetry**, in which the product of a reduction is deposited on the surface on purpose and then stripped off by an oxidizing potential—a potential at which the oxidation of the previously deposited material occurs. This technique can also use a mercury electrode, but one that is held stationary.

## 14.6.2 Amperometry

We have already briefly described a popular application of amperometry in Chapter 13. This was the electrochemical detector used in HPLC methods. In this application, the eluting mobile phase flows across the working electrode embedded in the wall of the detector flow cell. With a constant potential applied to the electrode (one sufficient to cause oxidation or reduction of mixture components), a current is detected when a mixture component elutes. This current translates into the chromatography peak.

The concept of amperometry can also be applied to a titration experiment, much like potential measurements were in Section 14.5 (potentiometric titration). Such an experiment is called an **amperometric titration**, a titration in which the end point is detected through the measurement of the current flowing at an electrode.

The polarization of the measuring (working) electrode, which is typically a rotating platinum disk embedded in a Teflon sheath, is held constant at some value at which the analyte reduces or oxidizes. The solution is stirred due to the rotation of the electrode. The resulting current is then measured as the titrant is added. The titrant reacts with the analyte, removing it from the solution, thus decreasing its concentration. The measured current therefore also decreases. When all of the analyte has reacted with the titrant, the decrease will stop, signaling the end point.

# WORKPLACE SCENE 14.2

Controlled-potential coulometry involves nearly complete reduction or oxidation of an analyte ion at a working electrode maintained at a constant potential and integration of the current during the elapsed time of the electrolysis. The integrated current in coulombs is related to the quantity of analyte ion by Faraday's law, where the amps per unit time (coulomb) is directly related to the number of electrons transferred, and thus to the amount of analyte electrolyzed.

Various plutonium materials are dissolved in acidic media and then fumed with sulfuric acid. In a 0.5 *M* sulfuric acid electrolyte, plutonium is reduced to Pu(III) at a platinum working electrode maintained at 0.310 V relative to a saturated calomel electrode. Plutonium (III) is oxidized to Pu(IV) at 0.670 V for the coulometric measurement. This work supports manufacturing, stock pile reduction, and pilot programs for making nuclear fuels from the stockpile.

Elmer Lujan, a technician with the Los Alamos National Laboratory in New Mexico, prepares metal samples to be analyzed for plutonium using controlled-potential coulometry. (Photo by Laurie Walker.)

## 14.7  Karl Fischer Titration

The Karl Fischer method is a titration to determine the water content in liquid and solid materials. The method utilizes a rather complex reaction in which the water in a sample is reacted with a solution of iodine, methanol, sulfur dioxide, and an organic base:

$$I_2 + CH_3OH + SO_2 + \text{organic base} + H_2O \rightarrow \text{products} \qquad (14.14)$$

There are two general ways by which the titration can take place. One is the volumetric method, in which the titrant is added to the sample via an automatic titrator. In this case, the titrant is either a mixture of all of the reactants above (a **composite** titrant) or an iodine solution (other components already in the

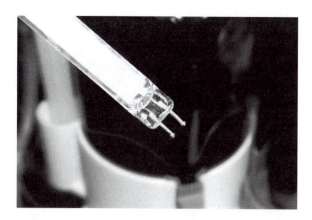

**FIGURE 14.12**    A photograph of the dual platinum electrode probe.

titration vessel). The other is the coulometric method, in which iodine is generated electrochemically. The coulometric method differs significantly from the classical titration. In either case, the end point is detected either potentiometrically or amperometrically, as described below.

## 14.7.1   End Point Detection

At the point where the last trace of water in the sample has reacted with the titrant, unreacted iodine appears in the titration vessel. The presence of unreacted iodine signals the end point. Unreacted iodine can be detected visually, since it imparts a dark yellow, or brown, color to most solutions. However, a visual detection scheme is not as reliable as the electrochemical detection scheme that now is in common use. The electrochemical scheme utilizes two platinum wire electrodes and no reference electrode. Either a constant potential or constant current is applied across these electrodes. When unreacted iodine appears in the titration vessel, either the current required to sustain the constant potential or the potential required to sustain the constant current changes sharply, signaling the end point. If the potential is monitored (constant current), it is called a **bipotentiometric method**. If the current is monitored (constant potential), it is called a **biamperometric method**. The prefix bi- is used here to distinguish the use of two platinum electrodes and no reference electrode from the use of a single indicator electrode and a reference electrode, such as in a normal potentiometric or amperometric titration. The two platinum electrode wires are available sealed in a single glass probe, as shown in Figure 14.12.

## 14.7.2   Elimination of Extraneous Water

All extraneous water present in the surrounding air and in the solvents used must be eliminated. Moisture from the air is prevented from entering the system by sealing off the cell from laboratory air. This is accomplished by the use of a rubber septum for sample injection and by the use of water-absorbing material in the other openings to the system.

The elimination of water from the solvent used is accomplished by a solvent conditioning step in which the moisture in the solvent is actually titrated with iodine prior to introducing the sample. Once the solvent moisture is eliminated, the sample can be introduced and the titration begun.

## 14.7.3   The Volumetric Method

In the volumetric method, the titrant can be a solution of iodine, methanol, sulfur dioxide, and an organic base, as described previously. Such a mixture is commonly known as the **Karl Fischer reagent** and can be purchased from any chemical vendor. It can also be a solution of iodine in methanol solvent. In that case, a **Karl Fischer solvent** containing the other required components is needed for the titration vessel.

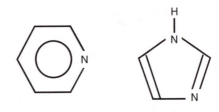

**FIGURE 14.13**    The structures of pyridine (left) and imidazole (right).

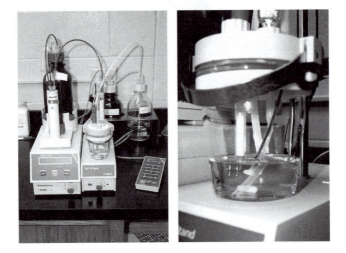

**FIGURE 14.14**    Photographs of a typical volumetric unit. The complete unit is shown on the left and a close-up of the titration vessel on the right. The dual platinum wire probe is visible in the right photograph as is the tube (dark in color) that introduces the titrant to the solution.

There are two possibilities for the organic base. One is pyridine, which is the base in the traditional Karl Fischer recipe. The structure of pyridine is similar to that of benzene. The difference is that one of the carbons in the benzene ring is replaced by a nitrogen. Pyridine is toxic, has a disagreeable odor, and does not give the optimum pH for the determination. More recently, imidazole has been used for the base. The structures of pyridine and imidazole are shown in Figure 14.13.

The complete volumetric unit typically consists of a titrant reservoir, the automatic titrator, the titration vessel with a dual-pin platinum electrode, an automatic stirrer, and an electronics module to run the detection system and display the buret reading and results. The unit may also come equipped with a system for automatically emptying the titration vessel to a waste container and introducing fresh solvent (typically methanol) for the sample. Figure 14.14 shows the complete unit and a close-up view of the titration vessel. In operation, solvent is placed in the titration vessel and, with stirring, the reagent is added so that iodine reacts with the extraneous moisture to condition the solvent. The dual-pin platinum electrode detects the excess iodine at the point when the extraneous moisture is eliminated and halts the addition of the reagent. The sample is then manually introduced and the process repeats, this time to titrate the moisture from the added sample. When the addition of titrant is again halted by the signal from the dual platinum electrode, the buret reading and results are displayed.

Like other titrants, this titrant must be standardized. Standardizaton means that the titer of the titrant must be determined. Titer is the weight of analyte (usually in milligrams) consumed by 1 ml of titrant. To determine the titer, a known weight of water is introduced and titrated (after conditioning the solvent). The titer is then calculated by dividing the weight (milligrams) of water used by the milliliters of titrant used to titrate it. The typical volumetric unit calculates the titer automatically. This occurs after the operator inputs the weight on a keypad (shown on the right of the complete unit photograph in Figure 14.14) and after the titration takes place. The percent water in a sample is then calculated as follows:

$$\% \ H_2O = \frac{\text{milliliters of titrant} \times \text{titer}}{\text{sample weight in milligrams}} \times 100$$

The volumetric method is used (rather than the coulometric method) when the water content is higher (greater than about 1%).

### 14.7.4 The Coulometric Method

As stated previously, the iodine titrant is generated electrochemically in the coulometric method. Electrochemical generation refers to the fact that a needed chemical is a product of either the oxidation half-reaction at an anode or the reduction half-reaction at a cathode. In the Karl Fischer coulometric method, iodine is generated at an anode via the oxidation of the iodide ion:

$$2 \ I^- \rightarrow I_2 + 2e^-$$

An anode–cathode assembly is required in addition to the dual-pin platinum electrode used for the end point detection. The Karl Fischer reagent in this case contains iodide and no iodine. With the reagent in the cell, the current to the anode–cathode assembly is switched on to generate the iodine needed to eliminate the extraneous moisture. When this moisture is eliminated, unreacted iodine appears and the dual-pin platinum electrode switches off the power to the anode–cathode assembly. The sample is then introduced and the current switched on again so that the iodine is again generated. When the dual-pin platinum electrode detects unreacted iodine again, the iodine generation is halted again.

The critical datum is not a buret reading, as it was in the case of the volumetric method. Rather, the amount of iodine used is determined coulometrically by computing the **coulombs** (total current over time) needed to reach the end point. The coulombs are calculated by multiplying the current applied to the anode–cathode assembly (a constant value) by the total time (seconds) required to reach the end point. The modern coulometric titrator automatically computes the amount of moisture from these data and displays it.

## Experiments

### Experiment 50: Determination of the pH of Soil Samples

Note: Safety glasses are required.

1. Obtain soil samples as directed by your instructor. These samples should be thoroughly dried and crushed to very small particles using a mortar and pestle.
2. Weigh 5 g of each soil sample into separate 50-mL beakers. An analytical balance is not necessary for this.
3. Prepare a pH meter and combination pH electrode with a pH = 7 buffer.
4. Perform this step one at a time for each sample. Pipet 5 mL of distilled water into one of the beakers. Swirl vigorously by hand for 5 sec. Let stand 10 min. Swirl lightly, dip the pH electrode in, and measure the pH.
5. Repeat step 3 for each sample. During the 10-min waiting period, the next sample(s) may be prepared, but be sure to measure each sample immediately after the 10-min period.

## Experiment 51: Red Cabbage Extract, the pH Electrode, and PowerPoint: A Group Project and Oral Presentation

### Introduction

When red cabbage is boiled in water, pigments called anthocyanins are extracted. These pigments can be used to indicate pH. In this laboratory exercise, you will observe the color of red cabbage extract as a function of pH, match each color with a color in the color palette of Microsoft PowerPoint, and, using Microsoft PowerPoint, present your findings to the class. You will work in groups of two or three for this activity.

*Remember safety glasses.*

### Part A: Determination of the Color at Various pH Values

*Procedure*

1. Obtain and clean 12 test tubes large enough to be able to dip a combination pH electrode into a solution held inside it. Also obtain one or two test tube racks to accommodate the 12 test tubes. Label ten of the test tubes as follows: pH3, pH4, pH5, pH6, pH7, pH8, pH9, pH10, pH11, and pH12. Label the remaining two as follows: 0.5 *M* HCl and 0.5 *M* NaOH. Fill these with the appropriate solutions. Place all test tubes in the rack.
2. Prepare 500 mL of red cabbage extract. To do this, obtain one large red cabbage leaf and shred it into small pieces. Place the pieces in a 1000-mL beaker filled halfway with distilled water and boil on a hot plate. Boil long enough to obtain a deep purple-colored solution. Fill each of the remaining test tubes from step 1 halfway with this red cabbage extract.
3. Obtain a pH meter and combination pH electrode. Standardize the pH meter.
4. Begin with the test tube labeled pH3. Position the pH electrode in this test tube. Since the desired pH is more acidic than the pH of the red cabbage solution, add the 0.5 *M* HCl, dropwise, and stir with a glass stirring rod until the pH is about 3. Place it back in the rack.
5. Repeat the procedure in step 4 for the remaining nine test tubes. Add 0.5 *M* HCl or 0.5 *M* NaOH as necessary to adjust the pH to the label value.

### Part B: The PowerPoint Slide Presentation

*Background*

PowerPoint is a presentation tool that is included in Microsoft Office. Just as it is easy to edit, rewrite, and print documents in Word, these features are also available when preparing slide presentations. Drawings, clip art, and other special effects are easily added to your presentation to make it more effective.

*Procedure*

1. Select "start" in the lower left-hand corner. On the pop-up menu, select Microsoft PowerPoint. On the pop-up menu, select "blank presentation." Select "OK."
2. Choose the appropriate layout for your **title** slide. Click "OK."
3. Select "new slide." Click "OK."
4. Select the appropriate layout for your **purpose** slide. Select "OK." *Briefly* state the purpose of the lab.
5. Select "new slide." Click "OK." Select the appropriate layout for your **introduction** slide. Select "OK." On this slide *briefly* summarize the **theory and background**. Use bullet format, not complete sentences. This should be a maximum of four or five lines.
6. Select "new slide." Click "OK." Select an appropriate background for the **procedure**. Here again, briefly summarize the procedure in bullet form.
7. Select "new slide." Click "OK." Select a blank background for your **experimental results**. To show the colors at different pH values, select "view" from the toolbar at the top of the page. On the

---

*This experiment was provided by Prof. Joe Rosen at New York City Technical College, Brooklyn, NY, and edited by the author.

pop-up menu select "toolbars." In the check box, be sure "drawing" is selected. You can draw shapes by left-clicking them on the drawing toolbar and then, moving the mouse to the slide, click and drag to choose the size, shape, and location of the figure. Next select "format" from the toolbar at the top of the page. On the pop-up menu, select "drawing object." Select "fill" and the color (corresponding to a test tube color) from the custom color palette. Label this colored object with the pH of the solution. Repeat for all the test tubes. You may use a new slide for each test tube if you wish.

8. Select "new slide." Click "OK." Select the appropriate layout for your **conclusion**. On this slide, summarize your conclusion.

9. On the tool bar at the top of the page, select "view." On the pop-up menu, select slide show. Press "enter" to scroll through your slides.

10. If you want to change any of your slides, select "view" from the toolbar at the top of the page and select "slide sorter" from the pop-up menu. Left-click on the slide you want to change. To delete extra slides, left click on the slide. Select "edit" from the toolbar, then "cut" or "clear."

11. When you are finished, save your work. Select "file," then "save as" in the pop-up menu. Select floppy (A:) in the "save in" text box. Type in your file's name in the "file name" text box.

***Optional Special Effects:***

12. To create transitions between slides, select "slide show" from the toolbar at the top of the page. On the pop-up menu select "slide transition." Choose the transition of choice from the text box.

13. To create animation effects, select "slide show" from the toolbar at the top of the page. On the pop-up menu select "custom animation." Choose the "effects" tab. Select the custom animation of choice from the text box.

14. Clip art can be inserted by selecting "insert" from the toolbar, "picture" from the pop-up menu, and "clip art."

## Part C: The Oral Presentation

The two or three laboratory partners should equally share the oral presentation. The main parts of an oral presentation are the introduction, the body, and the conclusion. When preparing an oral presentation always keep your listener in mind. Your presentation should be interesting and informative. It should be presented in a logical manner and include all relevant information. In the introduction, briefly describe what you plan to talk about. Here, give the **purpose**. This prepares the listener to pay attention to key points. In the body important background information is presented, and the main points are described in detail. Here, the body corresponds to the **theory, procedure**, and **observations**. What was learned is then summarized in the **conclusion**. This reminds your listener of the important points in your presentation. Practice your delivery. Speak slowly, clearly, and with enough volume during your presentation. Try to maintain eye contact with your listeners. Ask for questions after the conclusion. Be aware of your body language. Avoid rocking back and forth, fidgeting, etc. Have fun.

## Experiment 52: Potentiometric Titration of Phosphoric Acid in Soda Pop [*]

*Remember safety glasses.*

1. Prepare 500 mL of 0.10 *N* NaOH using freshly boiled water and standardize it with primary standard KHP, as done previously in Experiment 8 in Chapter 4.

2. As long as the soda pop is carbonated, the carbonic acid is present along with the phosphoric acid, although at a much smaller concentration. Citric acid may also be present, but at a smaller concentration. The carbonic acid may be eliminated by degassing to remove the carbon dioxide. Your instructor may ask you to obtain titration curves (step 4 below) for both as received samples and

---

[*] The idea for this experiment came from Ildy Boer at the County College of Morris in Randolph, New Jersey.

degassed samples to see the effect of the presence of carbonic acid. Degas about 100 mL of each soda pop sample by sonicating for 5 min or by use of a vacuum (vacuum pump and vacuum flask) with swirling. Make sure all samples are at room temperature.

3. Pipet 25.00 mL of the degassed soda pop into a 250-mL Erlenmeyer flask. Set up the titration experiment with pH probe and pH meter on an automatic stirrer. The standardized 0.10 $N$ NaOH is the titrant.

4. First obtain the complete titration curve for the sample. This will involve cautious addition of the NaOH around the inflection points so that you can carefully track the sharp change in the pH at these points. Alternatively, you may choose to use a computer for data acquisition, as in Experiment 18, Part B. Phosphoric acid ($H_3PO_4$) has three hydrogens to be neutralized. The third hydrogen is lost with only a slight change in pH, however, so you may only see two inflection points. If citric acid is also present (check the can label), there may be other inflection points as well.

5. From your titration curve, decide which inflection point you want to use for the equivalence point for the titrations in step 6, keeping in mind the possible interference from citric acid.

6. Set up the titration again with 25.00 mL of the degassed soda pop, as in step 3, and titrate to the chosen equivalence point. Repeat at least twice and average the buret readings.

7. Calculate the milligrams of phosphoric acid per milliliter of soda pop, assuming donation of one hydrogen at the first inflection point or two hydrogens at the second. Also calculate the milligrams of phosphoric acid in one can of soda pop. There are 0.02957 L per fluid oz.

## Experiment 53: Operation of Metrohm Model 701 Karl Fischer Titrator (for Liquid Samples)

1. If not already done, set up a titration unit as indicated on page 4 of the Metrohm Model 703 Ti Stand "Instructions for Use" manual, and an exchange unit as indicated on pages 72 and 73 of the Metrohm Model 701 KF Titrino "Instructions for Use" manual. Place titrant in a bottle labeled "titrant." Place methanol in a bottle labeled "solvent." Make sure both bottle caps are tight. Switch on the instrument.

2. Perform this step if the buret is not already filled. To fill the buret, press the DOS button until the piston is in the top end position. Then press the fill/stop button. Repeat so that all air bubbles are eliminated.

3. Press and hold the "IN" button on the top rear of the 703 Ti Stand in order to pump solvent into the titration vessel. Continue to add solvent until the electrode pins are completely immersed.

4. Press the start key (either on the keypad or on the exchange unit). The green "cond" light should blink as titrant is automatically added to the titration vessel to eliminate the residual water in the solvent. (The solvent is being conditioned.) When this green light stops flashing and the display reads "conditioning," the solvent is conditioned and the cell is ready for titrations.

5. Determine the titer of the titrant as follows. Tare an empty 50-$\mu$L syringe on an analytical balance, and then fill to the 40-$\mu$L line with distilled water and weigh again. This weight is the weight of the water in the syringe. Press "mode" on the keypad a few times until the display reads "titer with $H_2O$ or std." Press "enter."

6. Press "Start." Add the measured water to the titration vessel by piercing the septum in the sample inlet port with the needle of the syringe and pushing the plunger all the way in. Remove the syringe from the sample port. Type in the sample weight on the keypad and press "enter." The water will now be titrated by the automatic addition of titrant. At the completion of the titration, read and record the titer on the display.

7. Repeat steps 5 and 6 to obtain a second titer reading. The instrument will average the two readings automatically.

8. Determine the water in a liquid sample as follows. Tare an empty 50-$\mu$L syringe on an analytical balance, and then fill to the 40-$\mu$L line with the sample and weigh again. This weight is the weight of the sample. Press "mode" on the keypad a few times until the display reads "KFT." Press "enter."

9. Press "start." Add the sample to the titration vessel as you did the water in step 6, typing in the sample weight and pressing "enter" as before. When finished, the percent water in the sample should be displayed. Repeat with a second sample if desired.

10. For more determinations, proceed with step 8. When the titration vessel fills (after several runs), eliminate the solution in the titration vessel by pressing the out button on the 703 Ti Stand and holding it in. To perform more determinations after that, fresh methanol must be introduced and conditioned (steps 3 and 4). The titer of the titrant should not change over a short period of time.

## Questions and Problems

1. Differentiate between potentiometric methods and voltammetric and amperometric methods.
2. How are oxidation–reduction chemistry and electroanalytical chemistry related?
3. Define cell, half-cell, anode, cathode, electrolytic cell, and galvanic cell.
4. Distinguish between an electrolytic cell and a galvanic cell.
5. What is a salt bridge?
6. Define standard reduction potential, $E°$.
7. Without doing any calculations, look at the table of standard reduction potentials and indicate whether or not there would be a reaction between
   (a) Cu and $Ni^{2+}$
   (b) Hg and $Sn^{2+}$
   (c) Fe and $Ag^+$
8. What is the $E°_{cell}$ for a cell in which the following overall reaction occurs?

$$Ce^{3+} + Cu^{2+} \leftrightarrows Ce^{4+} + Cu$$

9. Will the reaction in question 8 proceed spontaneously to the right? Why or why not?
10. A certain voltaic cell is composed of a $Ce^{4+}/Ce^{3+}$ half-cell and a $Sn^{4+}/Sn^{2+}$ half-cell. The overall cell reaction is

$$Ce^{4+} + Sn^{2+} \leftrightarrows Sn^{4+} + Ce^{3+}$$

   (a) What is the $E°_{cell}$?
   (b) Is the reaction spontaneous as written, left to right? Why or why not?
11. What is $E°_{cell}$ for the following redox system and in which direction will the reaction spontaneously proceed? (Assume standard conditions.)

$$Zn + Cu^{2+} \leftrightarrows Zn^{2+} + Cu$$

12. What is the $E°_{cell}$ for a cell in which the following overall reaction occurs?

$$Cu^{2+} + Ni \leftrightarrows Cu + Ni^{2+}$$

13. What is the $E°_{cell}$ for a cell in which the following overall reaction occurs?

$$Ce^{3+} + Sn^{4+} \leftrightarrows Ce^{4+} + Sn^{2+}$$

14. Compare question 10 with question 13. How are the answers different and why?
15. A voltaic cell is composed of a copper electrode (Cu) dipping into a solution of copper ions ($Cu^{+2}$) and a mercury electrode in a solution of $Hg^{2+}$ ions. The cell reaction is

$$Cu + 2 Hg^{2+} \leftrightarrows Cu^{+2} + 2 Hg$$

What is the $E°$ for this cell? Is the reaction spontaneous? Explain your answer.

16. Under standard conditions, what is E for a cell in which nickel ions react with magnesium metal to form nickel metal and magnesium ions?

17. Assuming standard conditions, what is E for a cell in which iron metal reacts with manganous ion to form ferrous ion and manganese metal?

18. What is the Nernst equation and what is its usefulness in electroanalytical chemistry?

19. The $E^\circ_{cell}$ for the following reaction is +0.46 V.

$$2 \, Ag^+ + Cu \leftrightarrows 2 \, Ag + Cu^{2+}$$

If $[Ag^+] = 0.010 \, M$ and $[Cu^{2+}] = 0.0010 \, M$ initially, what is the true E for this cell at 25°C?

20. The concentration of $Ag^+$ ions in a cell is 0.010 $M$. The concentration of $Mg^{2+}$ ions is 0.0010 $M$. The cell reaction is

$$Mg + Ag^+ \leftrightarrows Mg^{2+} + Ag$$

The $E^\circ$ for the cell is +3.17 V. What is E under the above concentration conditions?

21. What is E for the $Sn^{4+}/Sn^{2+}$ half-cell if $[Sn^{4+}] = 0.10 \, M$ and $[Sn^{2+}] = 0.0010 \, M$?

22. What is E for a $Ce^{4+}/Ce^{3+}$ half-cell if $[Ce^{4+}] = 0.010 \, M$ and $[Ce^{3+}] = 0.0010 \, M$?

23. If E (under standard conditions) for the reaction of zinc ions with iron metal (to give ferric ions) is –0.72 V, what is E if the zinc ion concentration is 0.010 $M$ and the ferric ion concentration is 0.0010 $M$?

24. What is E for a cell in which tin(II) ions react with magnesium metal to give tin metal when the concentrations are as follows: magnesium ions, 0.010 $M$, and tin(II) ions, 0.0010 $M$?

25. What is E for a $Cr^{3+}/Cr//Ni^{2+}/Ni$ cell if $[Cr^{3+}] = 0.000010 \, M$ and $[Ni^{2+}] = 0.010 \, M$?

26. What is E for a $Cu^{2+}/Cu//Zn^{2+}/Zn$ cell if $[Cu^{2+}] = 0.010 \, M$ and $[Zn^{2+}] = 0.0010 \, M$?

$$Cu + Zn^{2+} \leftrightarrows Cu^{2+} + Zn$$

27. What is E for a $Sn^{2+}/Sn^{4+}//Zn^{2+}/Zn$ cell if $[Sn^{2+}] = 0.010 \, M$, $[Sn^{4+}] = 0.00010 \, M$, and $[Zn^{2+}] = 0.0010 \, M$?

$$Sn^{2+} + Zn^{2+} \leftrightarrows Sn^{4+} + Zn$$

28. Define potentiometry, reference electrode, ground, and indicator electrode.

29. Using the Nernst equation, tell how the SCE electrode works as a reference electrode.

30. Concerning the SCE, what does the fact that the KCl is saturated have to do with its usefulness as a reference electrode?

31. Tell how the concept of the salt bridge is put into practice with commercial reference electrodes.

32. Using the Nernst equation, tell how the silver–silver chloride electrode works as a reference electrode.

33. Using the Nernst equation, tell how the pH electrode works as an electrode for determining the pH of a solution.

34. Concerning the pH electrode, the following defines the potential that develops when it is immersed into a solution:

$$E = E^\circ - 0.059 \log - \frac{[H^+](\text{external})}{[H^+](\text{internal})}$$

Given this, explain how we can then say that this potential is directly proportional to the pH of the solution into which it is immersed.

35. Briefly describe how a pH meter is standardized.

36. What is a combination pH electrode? Describe its construction.

37. Discuss the relationship between the salt bridge concept illustrated in Figure 14.2 and the fiber strands in Figure 14.8.

38. What is an ion-selective electrode? Identify and describe four different types.

39. The following data were obtained using a nitrate electrode for a series of standard solutions of nitrate:

| $[NO_3^-](M)$ | E (mV) |
|---|---|
| $10^{-1}$ | 85 |
| $10^{-2}$ | 150 |
| $10^{-3}$ | 209 |
| $10^{-4}$ | 262 |

(a) Plot the calibration curve for this analysis.
(b) What is the nitrate ion concentration in a solution for which E = 184 mV?

40. What is special about gel-filled electrodes and double-junction electrodes?
41. What are two functions of the vent hole found near the top of commercial electrodes?
42. How should commercial electrodes be stored over the short term? How should they be stored over the long term?
43. Why is a platinum electrode needed in some half-cells?
44. What is a potentiometric titration?
45. Does potentiometry utilize galvanic cells or electrolytic cells? Do voltammetry and amperometry utilize galvanic cells or electrolytic cells?
46. How do voltammetry and amperometry differ from potentiometry?
47. Define working electrode, reference electrode, counter electrode, and auxiliary electrode.
48. What are two advantages of the dropping mercury electrode over a stationary electrode?
49. Differentiate between polarography and voltammetry.
50. Explain what is meant by amperometric titration.
51. What is the objective of a Karl Fischer titration?
52. What chemicals are required in a Karl Fischer reaction? What is the reaction?
53. Explain the bipotentiometric method for detecting the end point in a Karl Fischer titration. Why is it called a bipotentiometric method?
54. In a Karl Fischer experiment, why is the titration vessel sealed off from the laboratory air?
55. What is meant by conditioning the solvent in a Karl Fischer experiment?
56. Distinguish between the volumetric and coulometric methods for the Karl Fischer titration.
57. Explain how a Karl Fischer titrant is standardized in the volumetric method.
58. Define titer. If 9.38 mg of water required 17.28 mL of Karl Fischer titrant, what is the titer of this titrant?
59. A sample weighing 0.091 g required 29.22 mL of a titrant with a titer of 0.692 mg/mL. What is the percent of water in this sample?
60. Why is an anode–cathode assembly as well as a dual-pin platinum electrode needed in the coulometric Karl Fischer method?
61. There is no buret reading in the coulometric Karl Fischer method. How can the results be calculated without a buret reading?
62. **Report**
    Find a real-world electroanalytical analysis in a methods book (AOAC, USP, ASTM, etc.) or journal and report on the details of the procedure according to the following scheme:
    (a) Title
    (b) General information, including type of material examined, the name of the analyte, sampling procedures, and sample preparation procedures
    (c) Specifics, including the specific experiment (titration, series of standard solutions, Karl Fischer, voltammetric, amperometric), what the titrant is and how it is standardized or how the calibration curve is created, quantitation procedure, how the data are gathered, concentration levels for standards, how standards are prepared, and potential problems
    (d) Data handling and reporting
    (e) References

# 15 *

# Physical Testing Methods

## 15.1  Introduction

In Chapter 1 (Section 1.2), we stated that analysis methods can be classified according to the basis of the analysis: physical testing methods, wet chemicals methods, or instrumental methods. In this chapter, we examine a number of physical testing methods and equipment. The methods we will address include viscosity, thermal analysis, refractive index, optical rotation, density or specific gravity, particle sizing, impact testing, mechanical testing, tensile testing, and hardness.

## 15.2  Viscosity

### 15.2.1  Introduction

The quality of the ketchup that the American consumer uses on hamburgers and hotdogs is often judged by its resistance to flow. A ketchup that has a thick, flow-resistant consistency is generally considered to be of higher quality than one that flows readily from the bottle. However, this same consumer may become frustrated when the resistance to flow is so high that the ketchup does not flow in a timely manner from the bottle. The most desirable ketchup is thus the one that is judged to have a resistance to flow that is somewhere between the two extremes.

The same might be said of ordinary house paint. Paint that has a low resistance to flow is runny and will likely not provide satisfactory coverage of a wall with just one coat. It is therefore judged to be of poorer quality than one that does provide good one-coat coverage. On the other hand, paint that is thick and gummy may also be judged to be of poorer quality because it may not be possible to apply it uniformly with either a brush or roller.

There are dozens of other examples around the house for which we have similar quality standards in terms of their resistance to flow. Examples include other food products such as honey, pancake syrup, gravy, and ice cream toppings. Hygiene products include shampoo, hand lotion, and liquid soap. We might also cite pharmaceutical formulations, such as cough medicines, milk of magnesia, and liquid dietary supplements, as well as home and car care products such as caulks, glues, and motor oils.

Just as a consumer may judge the quality of a product by its resistance to flow, so also quality assurance technicians in the chemical process industries may judge the quality of a fluid material or product by its resistance to flow. Examples include solutions of polymers (where a solution's resistance to flow is indicative of the quality of the undissolved polymer), asphalt formulations for roads and parking lots, lubricating oils, etc.

---

*Kirk Hunter of Texas State Technical College in Waco, Texas, authored Sections 15.4 and 15.6–15.10.

## 15.2.2  Definitions

The science that deals with the deformation and flow of matter is called **rheology**. An important rheological concept is the **shear force**, sometimes called the **shear stress**, or the force that causes a layer of a fluid material to flow over a layer of stationary material. The rate at which a layer of a fluid material flows over a layer of stationary material is called the **shear rate**. A fluid flowing through a tube, for example, would be the fluid material, while the tube wall would be the stationary material. An important rheological measurement that is closely related to the resistance to flow is called **viscosity**. The technical definition of viscosity is the ratio of shear stress to shear rate:

$$\text{viscosity} = \frac{\text{shear stress}}{\text{shear rate}} \tag{15.1}$$

If increasing (or decreasing) the shear stress increases (or decreases) the shear rate proportionally at a given temperature (such that the ratio does not change), the fluid is said to be a **Newtonian** fluid. If the shear rate does not increase proportionally with the shear stress, the fluid is said to be a **non-Newtonian** fluid. In the above example, if the flow rate of a fluid in a tube increases due to an increase in the force pushing it through, and the ratio of the force to the rate does not change, then the fluid is a Newtonian fluid. If it does change, then it is a non-Newtonian fluid. Some fluids exhibit Newtonian behavior and some do not. This has practical significance in that if a parameter relating to force or rate changes in the course of laboratory measurements of viscosity, then the results will vary.

Viscosity, as it was defined in Equation (15.1), is often called the **dynamic viscosity**. The most common unit of dynamic viscosity is the **centipoise**, a unit based on force per rate. The base unit, the **poise** (100 centipoise per poise) is seldom used. Of more practical significance is **kinematic viscosity**, which is the dynamic viscosity divided by the density of the fluid:

$$\text{kinematic viscosity} = \frac{\text{dynamic viscosity}}{\text{density}} \tag{15.2}$$

The most common unit of kinematic viscosity is the **centistokes** (cS), the centipoise per density unit. Again, the base unit, the **stokes** (S) (100 cS/S), is seldom used.

The laboratory technique for measuring viscosity is called **viscometry** (sometimes **viscosimetry**), and the device in the laboratory used to measure viscosity is called a **viscometer** (sometimes **viscosimeter**).

## 15.2.3  Temperature Dependence

Viscosity depends on temperature. The higher the temperature, the lower the viscosity. Pancake syrup, for example, flows more freely when heated. For reasonable accuracy when measuring viscosity, the temperature must be very carefully controlled. This means that the viscometer and sample must be immersed in a constant temperature bath and the temperature given time to equilibrate before the measurement is recorded. A calibrated thermometer must be used to measure the temperature.

## 15.2.4  Capillary Viscometry

**Capillary viscometry** measures viscosity by measuring the time it takes for the fluid test material to pass through a very small diameter tube—a capillary tube. While any capillary tube can conceivably be used, there are tubes that are commercially produced for this purpose. These tubes are called **capillary viscometers**. The basic design consists of a U-shaped glass tube with capillary portions in one arm of the U. This portion has a large central bulb and two etched calibration lines. It opens up near the top so that suction can be conveniently applied. The other side has a larger opening at the top into which the test material is poured when preparing for a measurement. A simplified representation is given in Figure 15.1.

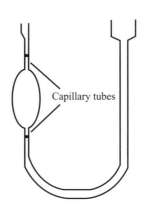

**FIGURE 15.1**   A simplified representation of a capillary viscometer.

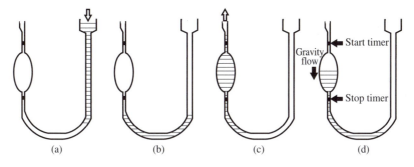

**FIGURE 15.2**   The concept of capillary viscosity measurement. See text for full description.

To measure the time of flow, the test fluid is poured into the appropriate side (Figure 15.2(a)) so that it occupies the bottom of the U and so that the meniscus on the capillary side is below the lower calibration mark (Figure 15.2(b)). The suction is then applied to the capillary side so that the fluid is drawn up to above the upper calibration mark (Figure 15.2(c)). When the fluid is released, it flows by gravity through the capillary (Figure 15.2(d)) so that the levels in the two sides will once again be at the same height (Figure 15.2(b)). The time it takes for the meniscus on the capillary side to pass the two calibration marks is the time that is measured (with a timer).

Capillary viscometers that have this design are called Ostwald viscometers. There are many specific designs of Ostwald viscometers. The most frequently used are the Cannon–Fenske viscometer and the Ubbelohde viscometer; these are the two that will be described here.

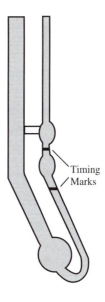

**FIGURE 15.3**   The Cannon–Fenske capillary viscometer.

The Cannon–Fenske viscometer is pictured in Figure 15.3. The U-tube has a bend in the center. There are several bulbs in the U that allow a greater volume of fluid to be tested and a long and precise time measurement. The timing marks are pointed out in Figure 15.3. The Ubbelohde viscometer is pictured in Figure 15.4. In this design, the U is completely vertical and, like the Cannon–Fenske viscometer, also has several bulbs for containing the fluid. The timing marks are also pointed out in Figure 15.4.

The Cannon–Fenske viscometer is used for measuring the kinematic viscosity of transparent Newtonian liquids, especially petroleum products and lubricants. The Ubbelohde viscometer is also used for the measurement of kinematic viscosity of transparent Newtonian liquids, but by the suspended level principle.

The time of flow is proportional to the kinematic viscosity as follows:

$$\text{kinematic viscosity} = \text{calibration constant} \times \text{time} \quad (15.3)$$

In order to calculate the kinematic viscosity, the calibration constant for the viscometer in question must be known. This calibration constant is often determined by the vendor before it is shipped, but it is also often checked by the user. In the calibration procedure, a fluid of known viscosity is tested so that the calibration constant can then be calculated:

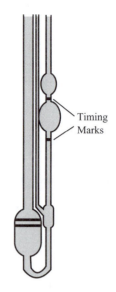

Timing Marks

**FIGURE 15.4** The Ubbelohde capillary viscometer.

$$\text{calibration constant} = \frac{\text{kinematic viscosity}}{\text{time}} \quad (15.4)$$

### Example 15.1

A given calibration liquid is known to have a kinematic viscosity of 15.61 cS at 25°C. Testing this liquid in a capillary viscometer gave a time of 139 sec. An unknown liquid was then tested with the same viscometer and found to give a time of 238 sec. What is the kinematic viscosity of the unknown liquid?

### *Solution 15.1*

First, the calibration constant of the viscometer is determined from the calibration data:

$$\text{calibration constant} = \frac{15.61 \text{ cS}}{139 \text{ sec}} = 0.112 \text{ cS sec}^{-1}$$

Then the viscosity of the unknown liquid can be calculated:

$$\text{kinematic viscosity} = \text{calibration constant} \times \text{time}$$

$$= 0.112 \text{ cS sec}^{-1} \times 238 \text{ sec} = 26.7 \text{ cS}$$

### 15.2.5  Rotational Viscometry

Rotational viscosity methods measure viscosity by measuring the torque required to rotate a spindle immersed in the fluid sample. A motor rotates the spindle and the torque is proportional to the resulting deflection of a spring. This deflection is indicated by either a pointer or dial mechanism or a digital display. In the digital model, the viscosity is displayed directly. A wide range of viscosities can be measured using interchangeable spindles and multiple rotation speeds, as well as different motors (different rotational units). For a given spindle geometry and speed, an increase in viscosity is indicated by an increase in the deflection of the spring and thus the readout. A diagram of the rotational viscometer is shown in Figure 15.5. See Workplace Scene 15.1.

# WORKPLACE SCENE 15.1

When designing mixing equipment for a manufacturing plant, it is vital that accurate viscosity data be known for all conditions to which a fluid will be subjected during the manufacturing process. For example, viscosity data are needed to fully characterize the mixing performance of an impeller. An impeller system that yields an ideal flow pattern in a mixing tank with a low-viscosity fluid may not generate the same flow pattern with a high-viscosity fluid.

In a research and development laboratory at the Dow Chemical Company in Midland, Michigan, rotational viscometry experiments on various dilutions of a test fluid, such as corn syrup, can generate the required data. Once various challenges are overcome, such as obtaining a uniform and constant temperature throughout the fluid and dealing with unusual physical behaviors of the test fluid, accurate viscosity measurements can be made and the project to optimize mixing performance can move forward.

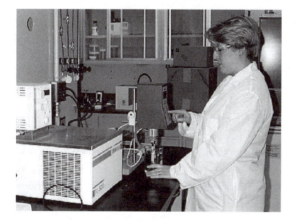

Amy Betz of the Dow Chemical Company collects viscosity data on a nonhazardous sample.

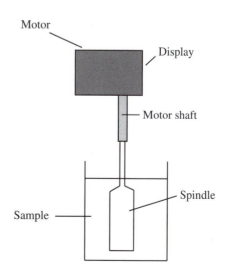

**FIGURE 15.5** A diagram of a rotational viscometer.

## 15.3   Thermal Analysis

### 15.3.1   Introduction

**Thermal analysis** is the characterization of a sample of matter based on the properties it displays while subjected to a controlled temperature program. These properties include any property related to temperature changes (especially temperature increases). Examples include heat-related phase changes and degradations, crystallizations, heat capacities, heats of reaction, glass transitions, curing rates for adhesives, and weight changes. The properties are observed either by monitoring temperature or heat flow in and out of the sample or by monitoring the sample weight during the process.

**Differential thermal analysis** (DTA) is a technique in which the temperature difference between the sample tested and a reference material is measured while both are subjected to the controlled temperature program. **Differential scanning calorimetry** (DSC) is a technique in which the heat flow difference between the sample and reference material is monitored while both are subjected to the controlled temperature program. **Thermogravimetric analysis** (TGA) is a technique in which the weight of a sample is monitored during the controlled temperature program.

### 15.3.2   DTA and DSC

As a sample of a pure substance is heated, its temperature increases (Figure 15.6(a)). When a phase change begins to occur, however, such as when a solid sample begins to melt, the temperature does not increase, even though the heat continues to be added. This is because the heat added is used to change the phase of the sample (heat of fusion or heat of vaporization), rather than to raise its temperature. Once the phase transition is completed, its temperature catches up to the surroundings and increases again (Figure 15.6(b)).

Now consider the use of a reference material that does not melt in the temperature range used in the above scenario. Its temperature would match the temperature of the surroundings ($T_E$) for the entire temperature program. Consider plotting the difference ($\Delta T = T_S - T_R$) between the temperature of the sample ($T_S$) and the temperature of the reference material ($T_R$) vs. the temperature of the surroundings. Initially, there would be no difference, $\Delta T$ is zero, since the sample and surroundings are heated equally. However, when the sample melts, $T_S$ lags behind $T_R$ temporarily, making $\Delta T$ negative. After melting is complete, the sample catches up such that the two temperatures are again equal, $\Delta T = 0$. A plot of $\Delta T$ vs. $T_E$ then results in the DTA curve in Figure 15.7. A negative peak occurs when the sample melts.

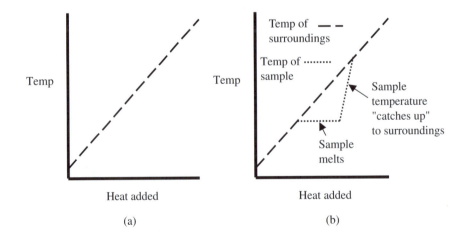

**FIGURE 15.6**   (a) A linear temperature program. (b) Applying the temperature program in (a) to a sample that melts at a particular temperature.

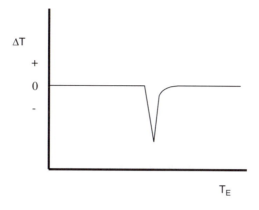

**FIGURE 15.7** A representation of the results of the differential thermal analysis (a DTA curve) of a sample that melts at a particular temperature.

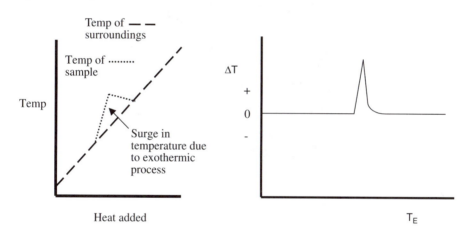

**FIGURE 15.8** Left, applying the temperature program in Figure 15.6(a) to a sample that undergoes an exothermic process. Right, a representation of the DTA curve of the exothermic process.

The negative peak shown in Figure 15.7 is a result of an endothermic process (a process that absorbs heat) such as melting. Other endothermic processes, other than melting, would also produce a negative peak. Examples include a chemical reaction or a decomposition. The particular characteristics of this peak (shape, width, sharpness, smoothness, etc.) provide clues concerning the sample composition and properties that are the object of a DTA.

Exothermic processes (processes that evolve heat) may also occur during the experiment. This would produce a surge in the sample temperature, rather than the flattening observed in Figure 15.6(b), and would produce a positive peak in the $\Delta T$ vs. $T_E$ plot. See Figure 15.8. Again, the characteristics of this positive peak provide clues concerning the sample. Exothermic processes include crystallization as well as some chemical and decomposition reactions.

Differential scanning calorimetry is similar to DTA. However, rather than monitoring the temperature difference between the sample and reference as the temperature of the surroundings is increased, as in DTA, the energy required to keep the sample temperature equal to the reference temperature is monitored. This energy is monitored according to either the oven power required or the actual heat flow that occurs. In any case, separate electrical heating elements heat the sample and reference. When the sample temperature lags behind the reference temperature (when an endothermic process occurs), the sample heating element is given more power and there is greater heat flow to the sample. When the sample temperature surges ahead of the reference temperature (when the exothermic process occurs), the reference heating

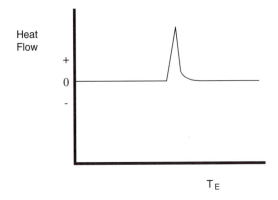

**FIGURE 15.9**   A representation of a DSC curve for an exothermic process. Note that the y-axis is heat flow.

**FIGURE 15.10**   A photograph of the interior of a small oven used for differential scanning calorimetry. The crucibles for the sample (S) and reference material (R) are placed in the small circles.

element is given more power and there is greater heat flow to the reference. The difference in either the power or the heat flow (whichever is being monitored) is then plotted vs. the temperature of the surroundings, and a DSC curve that appears very similar to the DTA curve is obtained. See Figure 15.9.

Some scientists describe DSC techniques as a subset of DTA. DTA can be considered a more global term, covering all differential thermal techniques, while DSC is a DTA technique that gives calorimetric (heat transfer) information. This is the reason that DSC has calorimetry as part of its name. Most thermal analysis work is DSC, and Sections 15.3.3 and 15.3.4 provide information about the instrumentation and applications of this technique.

### 15.3.3   DSC Instrumentation

The instrumentation used for heat flow monitoring (rather than oven power monitoring) is described here. The DSC oven has a cylindrical shape and is approximately 2.5 cm in diameter and 3 cm high. Two tiny aluminum crucibles (approximately 20 $\mu$L capacity) are used, one to contain the sample and one to contain the reference material. These crucibles are placed on a platform that is positioned approximately 1 cm above the floor of the oven. The specific locations for the crucibles are designated (such as S for sample and R for reference). See Figure 15.10. The platform is made of a special material to facilitate thermoelectric heating and monitoring via thermocouples on the underside. Immediately below the sample and reference locations under the platform are small metal discs (not visible in Figure 15.10) that form the junction with the platform for the thermocouple.

The reference material is usually elemental indium, a soft metal (atomic number 49 in group IIIA in the periodic table). One small plug of indium is placed in an aluminum crucible, which is then positioned on the reference location in the oven. A few milligrams of the sample is placed in another aluminum crucible, which in turn is positioned on the sample location in the oven. The oven is then closed and the temperature program begun.

It is common to provide a particular gaseous environment in the oven, i.e., to purge the oven with a gas such as air, helium, oxygen, nitrogen, etc. The oven module has an inlet for the purge gas. The purpose of a purge gas is to sweep out gases that form during the program and thus provide a constant environment for the sample and reference so that effects of environment changes are not measured. Also, a particular gas may create favorable conditions for a particular effect to be observed. For example, a chemical reaction caused by the presence of oxygen will be observed in the DSC curve if oxygen is used as the purge gas. This reaction will not be observed if helium or another inert gas is used. In addition, different gases have different thermal conductivities and this will affect the results.

The rate of heating is variable. The slower this rate the better, because fast heating will shift temperatures higher and cause the size of the peak to increase. The rate of heating during calibration should equal the rate of heating during sample analysis.

### 15.3.4   Applications of DSC

Two important applications of DSC are in the pharmaceutical industry and in the polymer industry. In the pharmaceutical industry, the purity of formulations and raw materials can be measured. Various levels of purity give different melting points and melting ranges. Very pure materials melt sharply (within 1 to 2°) and melt at expected temperatures. Impure materials have broader melting ranges and melt at lower temperatures than pure materials. Such phenomena can easily be detected with DSC.

In the polymer industry, the melting or degradation behaviors of polymers are important to determine. For example, when a polymer is extruded (i.e., when polymer pellets are converted to film by heating them and drawing them through an extruder), the thermal analysis of the polymer material determines the amount of heat needed in the extruder to make the material pliable. Given the large number of polymer formulations that have been developed and continue to be developed, thermal analysis procedures can be quite important.

## 15.4   Refractive Index

Another important physical property of liquids is the **refractive index**. Since the refractive index is a constant for a particular liquid at a given temperature, it can be used to help identify substances, check for purity, and measure concentrations. One type of detector found in some liquid chromatograph instruments (Chapter 13) uses refractive index.

Small amounts of impurities have a significant effect on the refractive index. In fact, the refractive index for many binary mixtures changes linearly with concentration over a wide range of concentrations. A calibration curve of refractive index vs. concentration along with the refractive index of a sample can be used to find the concentration of a species in the sample. For example, the food and beverage industry uses the refractive index to find the concentration of sugar solutions. Table 15.1 lists several additional applications for refractive index.

**TABLE 15.1**   Applications of Refractive Index Measurement

| Industry | Application |
|---|---|
| Chemicals | Solvents, distillation products, organic solutions, organic polymers |
| Petrochemicals | Oils, fats, waxes, naphthalenes, paints |
| Food | Jam, fruit extract, syrup, honey, milk, chocolate, baby food |
| Beverage | Soft drinks, fruit juice, wine, beer |
| Pharmaceutical | Aromas, essential oils, solutions |
| Medical | Blood, serum, urine |

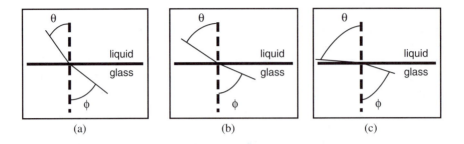

**FIGURE 15.11**  As the angle of the light passing through the liquid becomes smaller (a → c), the change of the angle of the light entering the glass also becomes smaller.

Newer techniques for measuring the refractive index allow for instantaneous, real-time measurement in process streams, or alternatively, a special continuous-flow sample well can be installed on bench top instruments. Small, pocket-sized refractometers also make field measurement very simple and reliable. Fiber optic sensors find uses in biomedical applications.

How is refractive index measured? As discussed in Chapter 7, the speed of light in a vacuum is a constant, approximately $3 \times 10^{10}$ cm/sec. In dilute gases and gas mixtures, such as air, the speed of light is only slightly slower and the difference is essentially negligible. However, in liquids and solids the speed of light is significantly slower.

The refractive index is defined as the ratio of the speed of light in a vacuum to the speed of light in the medium:

$$\eta = \frac{c}{v} \tag{15.5}$$

where $\eta$ (Greek letter eta) is the refractive index, c is the speed of light in a vacuum, and $v$ is the speed of light in the medium. For example, the refractive index of water is 1.33. This means that light travels 1.33 times faster in a vacuum than in water.

As a beam of parallel light travels from one medium to another that has a different density, the light beam bends or changes direction due to a change in the speed of light. In Figure 15.11, a liquid sample is placed in contact with a glass surface. A beam of light passes through the liquid and into the glass at an angle of $\theta$ (Greek letter theta) with respect to the perpendicular, as shown. Since the glass and the liquid do not have the same index of refraction, the direction of the light beam changes as it enters the glass. This is called refraction and is the reason objects in water appear distorted when viewed from above the surface.

As the angle in the liquid ($\theta$) increases, the angle in the glass ($\phi$) also increases (Figure 15.11, a → b → c). However, the changes in angles $\theta$ and $\phi$ are not the same. As the angle $\theta$ approaches 90°, the corresponding *change* in angle $\phi$ becomes smaller and smaller and angle $\phi$ approaches a limiting value called the critical angle, $\phi_{crit}$ (Figure 15.11(b) and (c)). The critical angle depends only on the difference in the refractive index between the glass and the liquid.

The mathematical relationship between the angle of the light passing through the liquid and the angle of the light passing through the glass is expressed in **Snell's law**:

$$\frac{\eta_{liquid}}{\eta_{glass}} = \frac{\sin \phi}{\sin \theta} \tag{15.6}$$

In this equaiton, $\phi$ is the angle of the light entering the glass measured from the perpendicular and $\theta$ is the angle of the light in the liquid. When $\phi$ is 90°, $\sin \theta = 1$ and $\phi$ has its critical angle, $\phi_{crit}$. By rearranging

Equation (15.6), the refractive index of the liquid is given by:

$$\eta_{\text{liquid}} = \eta_{\text{glass}} \times \sin\phi_{\text{crit}} \tag{15.7}$$

The instrument that measures the refractive index is called a **refractometer**. In the **Abbe refractometer**, a common type of refractometer, a small liquid sample is placed on the glass surface, or prism, using either a dropper or pipet. Light of a specific wavelength is then passed through the liquid film. Since the liquid and the glass do not have the same index of refraction, the light beam changes direction as it leaves the liquid and enters the glass prism. The optics of the refractometer are then adjusted to find the critical angle. Since the refractive index of the glass prism is known, the refractive index is read directly.

The refractive index is usually reported to four decimal places and has no units. Since small amounts of impurities can significantly affect the refractive index, substances must be very carefully purified in order to match the established reported refractive indices. An example of a refractive index is given as $\eta_{\text{D}}^{20} = 1.4567$.

The superscript 20 in the refractive index indicates the temperature at which the measurement was made. The subscript D indicates that the sodium D line (wavelength = 589 nm) was used.

The temperature is important in refractive index measurements since the refractive index generally decreases with temperature. To maintain the sample at a constant temperature, temperature-controlled water is circulated around the prisms.

The refractive index also varies with the wavelength, and the wavelength at which it is measured must also be specified. Yellow sodium light for illumination is not very handy. For that reason, commercial, low-cost refractometers have built-in compensators and filters that enable the use of ordinary white light. The resulting refractive index is that which would have been obtained had a sodium light source been used.

Newer techniques for measuring the refractive index include the automatic, or digital, refractometer and the fiber optic refractive index sensor. The **digital refractometer** measures the critical angle by detecting the position of the bright and dark borderlines with a linear diode array. This solid-state detector consists of 1024 photodiodes arranged at intervals of 25 $\mu$m. The light sources for these newer instruments range from filtered white light to light emitting diodes (LEDs) having a wavelength of 589 nm.

In the digital refractometer the sample is placed in a sample well where a portion of the prism surface is exposed. The prism in these instruments is made of sapphire, and not glass, because it has superior properties. It is scratch-proof, corrosion resistant, and has an excellent thermal conductivity. The sidewalls of the sapphire prism are totally reflecting mirrors. As light is directed against the interface between the prism and the sample solution, the multiple rays strike the interface surface at different angles. They are then reflected to the diode array detector to form an image with a bright area and a dark area. Those rays that strike the surface at too shallow of an angle are totally reflected to detector area A. Light rays that are at too steep of an angle are partially reflected to detector area B. The position of the borderline, C, is the critical angle and the refractive index. See Figure 15.12.

Interferences such as bubbles, undissolved particles, and color do not affect the resulting refractive index, making this method quite suitable for opaque, highly viscous, and highly colored substances such as milk and pulp liquor.

The **fiber optic refractive index sensor** finds use in biomedical applications. It uses a silicon chip with optical waveguides forming ring resonators. When the laser wavelength is scanned, the resonators cause dips in the power transmitted through the device. The wavelength at which these dips occur is a measure of the refractive index of the substance in contact with the chip surface.

The relationship between the refractive index and the amount of dry substance content is well known for sucrose and is the basis for the degree Brix (°Brix) scale. It is arbitrarily set such that 1° Brix is equal to a concentration of 1% sucrose. In other words, the °Brix scale indicates the number of grams of sucrose per 100 g of solution. This relationship also holds for a large number of similar substances and finds extensive use in the food industry. For example, a reading of 40° Brix would mean that the sample contained 40 g of solid per 100 g of solution.

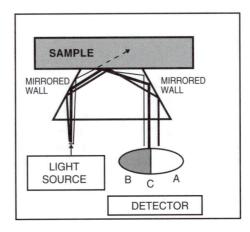

**FIGURE 15.12**  A digital refractometer with a diode array detector.

Just like refractive index, the °Brix scale is quite dependent on the temperature. Manual Abbe refractometers do not compensate for this temperature effect. Special correlation tables are used to adjust the readings to a standard temperature, 20°C. Digital refractometers, on the other hand, can operate over a fairly wide range of sample temperatures (+15 to +40°C) and automatically apply these temperature corrections. See Workplace Scene 15.2.

## 15.5   Optical Rotation

Ordinary white light, such as has been dealt with in previous chapters, does not move in just one plane. In actuality, the light waves exist in all planes around the line of travel. There are an infinite number of such planes. Figure 15.13(a) is a simplified view of this idea, showing a head-on view of a beam of light with the planes depicted as double arrows. It is possible to polarize light so as to block all planes of light except for one (Figure 15.13(b)). This is done with the use of a polarizing filter. Light consisting of just a single plane of light such as this is called **plane-polarized light.**

Some compounds exhibit the property of being able to rotate the plane of polarized light. In other words, when a beam of plane-polarized light passes through a sample of such a compound, the plane is rotated to another position around the line of travel (Figure 15.13(c)). The property is called **optical rotation,** or **optical activity.** In order to be optically active, a compound must possess an asymmetric carbon atom in its molecular structure. An **asymmetric carbon atom** is one that has four different structural groups attached to it. An example of such a compound is 2-butanol:

OH<br>
|<br>
$CH_3-CH-CH_2-CH_3$           2-butanol

The second carbon from the left is an asymmetric carbon atom because it has four different groups attached to it—a methyl group, a hydrogen, a hydroxyl group, and an ethyl group.

Having an asymmetric carbon atom means that a structure will have a nonsuperimposable mirror image (like a right hand and a left hand), which in turn means that in drawing a structure such as 2-butanol, we have actually represented two compounds with one drawing. There are actually two different structures in one and they are mirror images of each other. A pair of structures that are nonsuperimposable mirror images are called **enantiomers.** It turns out that one enantiomer rotates plane-polarized light in one direction, and the other rotates it through exactly the same angle, but in the opposite direction. Two compounds representing a pair of enantiomers are exactly identical in all other physical properties (color, odor, boiling point, solubility in water, refractive index, viscosity, etc.). They even have identical infrared spectra. The only property by which they differ is in the direction in which they rotate plane-polarized light.

# WORKPLACE SCENE 15.2

One ingredient in pharmaceutical cough medicine preparations is a flavor oil. Pharmaceutical companies that produce such preparations purchase the flavor oil in bulk quantities and then perform a laboratory analysis on a sample of it to determine its quality. The method of choice for this is refractive index. The flavor oil raw material must be within the refractive index specification limits established for it before it can be used in the cough medicine production process.

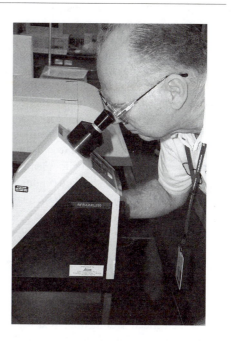

John Hannon, an analyst for a pharmaceutical company, uses a refractometer for the determination of the quality of flavor oil used in cough medicine preparations.

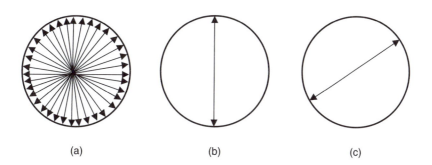

(a)          (b)          (c)

**FIGURE 15.13** (a) A representation of a light beam with all wave planes shown as double arrows around the line of travel. (b) A representation of planed-polarized light—all planes are blocked but one. (c) A representation of plane-polarized light that has been rotated—the plane in (b) has been rotated as a result of passing through a solution of an optically active compound.

Most of the time, enantiomers are found equally mixed together. Equally mixed enantiomers are *not* optically active because the rotation in one direction by one structure is canceled by the rotation in the opposite direction by the other structure. Hence, a sample of 2-butanol, for example, as normally obtained from a chemical vendor, is *not* optically active. An equimolar mixture of two enantiomers is called a **racemic mixture** and is optically inactive. Separation of a racemic mixture is not possible by conventional methods because the enantiomers are identical with respect to properties that are used to effect the separation. However, it may be possible to separate them by *chemical* methods, meaning that one may undergo a chemical reaction that the other does not. Some biological reactions are such reactions, and hence a single enantiomeric structure is sometimes found in nature.

The rotation of plane-polarized light can be measured with a laboratory instrument called a **polarimeter**. A simple polarimeter consists of a light source, a polarizing filter, a sample tube for containing the sample, and a device to observe or measure the rotation. The exact degree of rotation depends on how much of the enantiomer is in the path of the light. Thus one can measure the concentration of an enantiomer with a polarimeter. The degree of rotation would also depend on the length of the sample tube. There is also an effect of temperature and wavelength. A temperature of 25°C and a wavelength of 589.3 nm (the primary sodium emission line) are the standard conditions for reporting this property in tables of physical properties. When reporting the optical rotation in such tables, it is referred to as the **specific rotation**. The measured angle of rotation, usually symbolized by the Greek letter theta ($\theta$), is divided by the concentration (in grams per milliliter) and the length of the sample tube (in decimeters) to calculate the specific rotation.

## 15.6 Density and Specific Gravity

### 15.6.1 Introduction to Density

Matter can be defined as something that has mass and occupies space. If something occupies space, it has volume. Therefore, matter may be defined as any substance that has density. As an intrinsic physical property of matter, density can be used to help identify and differentiate substances. The method that is used to determine the density depends on the substance being measured. Is it a solid, liquid, or gas? If it is a solid, is it regularly shaped? Is it porous? What level of precision is required?

In the laboratory density is one of the most important characteristics that we have to describe various substances. It is also one of the most important tasks that a chemical technician performs. For example, to confirm the identity of an incoming raw material one of the tests technicians may perform is density. Is it sodium chloride or magnesium sulfate or sucrose?

In the design of parts for a particular device, density can be an important factor in the selection of a material, especially when deciding whether to use metal or plastic. For example, some specially designed plastics may cost $5/lb compared to $1/lb for a steel part. If the plastic has a density of 0.04 lb/in.$^3$ and the steel has a density of 0.28 lb/in.$^3$, the difference in price will be small if the part has a small volume. Using the figures above, the cost of 1 in.$^3$ would be $0.20 for the plastic part and $0.28 for the steel part.

Density is defined as mass per unit volume:

$$\text{Density} = \frac{\text{Mass}}{\text{Volume}} = \frac{m}{v} \tag{15.8}$$

The density of a substance may be obtained simply by measuring its mass and volume and making the necessary calculation. The units for mass are typically given in grams, while the units for volume may be milliliters (cubic centimeters) or liters, depending on the physical state of the substance. The volume of solids and liquids is given in cubic centimeters or milliliters.

The volume of a substance changes with the temperature, thus affecting its density. When determining the density of a substance, the temperature should also be measured and recorded. It is generally reported along with the density value in one of the following formats:

D = 0.9897 g/mL @ 25°C

or

$$D^{23°C} = 0.978 \text{ g/mL}$$

## 15.6.2 The Density of Regular Solids

If we wanted to measure the density of a rectangularly shaped solid, we would measure the mass in grams and then measure the dimensions of the solid—the length, width, and height—in centimeters to find the volume. The density would be calculated from these measurements. The tools needed are a balance and a ruler or some other device that measures length. The type of balance and the type of measuring device that we would select would depend on the level of precision desired. Similarly, the volume of other regular geometric shapes, such as spheres, cylinders, and cones, could be found by making the appropriate measurements and using the volume formulas.

## 15.6.3 The Density of Irregularly Shaped Solids

The density of an irregularly shaped solid is usually determined by measuring the mass and then measuring the volume of liquid that it displaces. The volume of liquid in a graduated cylinder is measured before the object is submerged and then measured again with the object submerged. The difference in the volume equals the volume of the object.

Care must be taken to ensure that the object is completely wetted by the liquid. If any air bubbles cling to the object or if any voids are not filled by the liquid, the resulting density will be low. The volume of the object will appear to be larger than it actually is. For this reason, liquids with low surface tensions, such as kerosene or alcohol, are frequently used for this type of measurement. However, for most substances, water is fine. The most accurate method for determining the volume of an irregularly shaped object is to weigh it both in air and in a liquid of known density (Figure 15.14). The apparent loss in weight of the object when submerged is equal to the mass of the displaced liquid (Archimedes' principle). If the density of the liquid is known, the volume of the substance is easily calculated.

### Example 15.2

A chunk of metal has a mass of 29.7 g when weighed in air and 18.7 g when weighed in water. What is the density of the metal?

### Solution 15.2

| | |
|---|---|
| Mass$_{air}$ | 29.7 g |
| Mass$_{water}$ | 18.7 g |
| Mass$_{"lost"}$ | 11.0 g |

The "loss" in mass is equal to the volume of water displaced, which is equal to the volume of the object. Since the density of water is 1.0 g/mL, the volume of the water displaced is 11.0 mL. This is also the volume

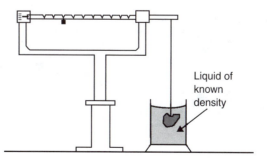

**FIGURE 15.14** Using a balance to find the volume of an irregularly shaped object.

of the object.

$$volume = 11.0 \text{ mL}$$

Then, calculating density, we substitute into Equation (15.8):

$$Density = \frac{m}{v} = \frac{29.7 \text{ grams}}{11.0 \text{ mL}} = 2.70 \frac{g}{mL}$$

If the object being measured should float, a weight is attached such that the object is submerged and then both the mass and the volume of the weight are subtracted from the combined weight.

## 15.6.4   The Density of Liquids

To determine the density of a liquid, the mass in grams of a measured volume of liquid in milliliters is determined. The density is calculated from these measurements. The volume may be measured by a variety of devices, such as a graduated cylinder, pipet, or buret. Very precise determinations of volume are measured with pycnometers. These devices hold a specified volume. A cap or stopper with a capillary overflow tube ensures repeatability in the measurement. See Figure 15.15.

The volume of viscous materials, such as grease or ointments, can be measured with a special pycnometer made of aluminum. The chamber is milled to ensure precise volume. A cap with a capillary opening is screwed into the body such that any air is expelled. See Figure 15.16 and Workplace Scene 15.3.

### Example 15.3

An empty and dry pycnometer weighs 23.4532 g and has a volume of 10.00 mL at 20°C. When filled with hexane it weighs 30.0532 g. What is the density of the hexane?

**FIGURE 15.15**   Two glass pycnometers. The capacity of the one on the left is 5 mL. The capacity of the one on the right is 10 mL.

**FIGURE 15.16**   An aluminum pycnometer assembled (left) and disassembled (right). The quarter in the photos is to indicate the size of the pycnometer.

# WORKPLACE SCENE 15.3

R esearch and development technologists at the Dow Chemical Company can characterize materials in a variety of ways. One material property that is especially critical in polymer foaming and processing technology is density. A tool used for measuring the density of a material is called a pycnometer. There are many different manual and automatic types to choose from. For extremely accurate and precise density measurements, an easy-to-use, fully automatic gas displacement pycnometer is utilized. Analyses are commenced with a single keystroke. Once an analysis is initiated, data are collected, calculations performed, and results displayed without further operator intervention.

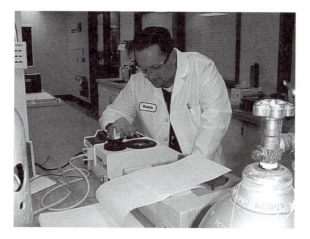

Dow technologist Kenn Bouchard utilizes an automatic gas displacement pycnometer for measuring density in materials characterization.

### Solution 15.3

The mass of the hexane is determined by subtracting the mass of the empty pyconometer from the mass of the filled pycnometer.

| | |
|---|---|
| Mass filled | 30.0532 g |
| Mass empty | 23.4532 g |
| Mass hexane | 6.6000 g |

The density is then calculated by substituting into Equation (15.8):

$$\text{Density} = \frac{m}{v} = \frac{6.6000 \text{ grams}}{10.00 \text{ mL}} = 0.6600 \frac{g}{mL}$$

## 15.6.5  Bulk Density

Bulk density, or the apparent density, is the total mass per unit of total volume. It is not an intrinsic property of a material since it varies with the size distribution of the particles and their environment. The porosity of the solid and the material with which the pores, or voids, are filled also affect the bulk density. For a single nonporous particle, the true density equals the bulk density.

### Example 15.4

A sample of quartz sand weighs 2.65 g and occupies a volume of 2.0 cm$^3$. What is its bulk density?

*Solution 15.4*

$$mass = 2.65 \text{ g}$$

$$volume = 2.0 \text{ cm}^3$$

Density is calculated by substituting into Equation (15.8):

$$Density = \frac{m}{v} = \frac{2.65 \text{ grams}}{2.0 \text{ cm}^3} = 1.3 \frac{g}{cm^3}$$

Bulk density can give information about a material's porosity. Ceramics and powder metals that are made by compaction and sintering have varying degrees of porosity. In structural parts, porosity is undesirable. However, in powdered metal wear parts, porosity is desired for retention of lubrication.

To find the percentage void space, we use the following formula:

$$\%void = \left(1 - \frac{bulk \ density}{actual \ density}\right) \times 100 \tag{15.9}$$

In soils, the bulk density is an indication of the degree of compaction and also the capacity for holding water, air, and nutrients. Highly compacted soils with low porosity (voids) are desirable for roadbeds and dams, but are not suitable for plant growth. The actual density, or particle density, of soils is determined by the displacement of water of a given mass of soil.

Factors that affect the bulk density are the fill rate—how fast the substance is transferred to the container—and the size of particles. As the fill rate increases, the particles have less time to arrange or settle and the resulting apparent volume is larger. We see this when we purchase cereal at the grocery store and the box is only two thirds full.

---

# DEMONSTRATION

Very slowly (~1 mL/sec) pour ice cream salt (rock salt) to the 100-mL level in a graduated cylinder. Quickly pour the contents through a funnel into another 100-mL graduated cylinder. Repeat using both fast and slow fill rates. With the level greater than 100 mL, cover the top of the cylinder with your hand and shake vigorously until the level falls to 100 mL.

---

## 15.6.6  Specific Gravity

Closely related to density is specific gravity. Specific gravity compares the density of a substance to the density of a reference substance. Typically, for solids and liquids, this reference substance is water at 4°C, which has a density of 1.00 g/mL. It is defined by the following relationship:

$$\text{Specific gravity} = \frac{\text{density of substance}}{\text{density of reference substance}} \qquad (15.10)$$

$$= \frac{\text{density of substance}}{\text{density of water @ 4°C}} \qquad (15.11)$$

$$= \frac{\text{density of substance}}{1.0 \text{ g mL}^{-1}} \qquad (15.12)$$

Note that the specific gravity has no units. When reporting the specific gravity, the temperature of the sample and the reference substance are also noted. The typical format is with a superscript and a subscript. The superscript is the temperature of the substance, while the subscript is the temperature of the reference substance.

**Example 15.5**

The density of gold at 20°C is 19.3 g/mL. What is the specific gravity of gold?

*Solution 15.5*

$$\text{Specific gravity Au}_{20°C} = \frac{19.3 \dfrac{\text{gram}}{\text{mL}} @ 20°C}{1.00 \dfrac{\text{gram}}{\text{mL}} @ 4°C} = 19.3^{20°C}_{4°C}$$

## 15.6.7 Hydrometers

Hydrometers utilize Archimedes' principle (see Section 15.6.3). The body of the instrument is weighted so that it is partially submerged when placed in the liquid. The stem is only partially submerged. Since the stem is thin, small changes in the density of the liquid will cause a considerable change in the amount of the stem that is submerged. The stem is calibrated to read the specific gravity directly. See Figure 15.17.

Hydrometers are calibrated in density, specific gravity, and several arbitrary units such as degrees Baume (°Be), degrees American Petroleum Institute (°API), or °Brix. These units are used for specialized purposes in various industries. The relationship between the specific gravity (sp. gr.) of a liquid with a density less than water and these specialized units is given by the following equations:

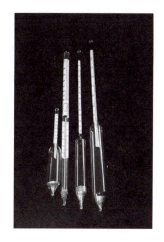

**FIGURE 15.17**  Examples of hydrometers.

$$°Be = 145 - \frac{145}{\text{sp.gr.}^{60°F}_{60°F}} \qquad (15.13)$$

$$°A.P.I. = \frac{141.5}{\text{sp.gr.}^{70°F}_{60°F}} - 131.5 \qquad (15.14)$$

$$°Brix = \frac{400}{\text{sp.gr.}^{60°F}_{60°F}} - 400 \qquad (15.15)$$

To convert hydrometer readings or specific gravities to °Baume for liquids heavier than water, the following formula is used:

$$°\text{Be} = \frac{140}{\text{sp.gr.}_{60°F}^{60°F}} - 130 \tag{15.16}$$

The Brix scale is arbitrarily graduated such that 1° Brix = 1% sugar. °Brix measurements are mostly carried out in the food and beverage industry for the quality control of sucrose solution (sugar syrup). The Brix scale measures the percentage by weight of sucrose in pure water solutions and is only valid for pure sucrose solutions. However, many industries use the Brix scale to refer to any sweet solids in a product, e.g., malt sugar, glucose, or honey. The values obtained are approximate and not true Brix values. They are referred to as apparent Brix values. Sugar syrup used by the food and beverage industry is sold by weight and concentration. Here, the determination of °Brix is important for both quality control and product cost. Additionally, Brix determinations are used to determine the proper harvest time and the maturity of fruits. Brix determinations can also be made with a refractometer, as described in Section 15.4.

The petroleum industry uses API gravity standards to describe hydrocarbons. Baume measurements are used by industrial manufacturers for nonhydrocarbons.

## 15.6.8  The Westphal Specific Gravity Balance

The Westphal specific gravity balance is a device that will measure the specific gravity directly if water is one of the liquids. It applies two concepts:

1. The mass of a floating object is equal to the mass of the liquid it displaces.
2. If a body of constant mass is immersed in different liquids, then the corresponding apparent loss in weight of the body is proportional to the masses of the equal volumes of liquid that the body displaced.

A plummet of known mass is placed in the liquid. The beam is balanced by placing riders on the beam. The specific gavity is read directly from the position of the riders on the beam. See Figure 15.18.

## 15.6.9  Density Gradient Columns

The density of some materials may be found using a density gradient column. This is a cylinder containing liquid layers of varying densities, with the most dense at the bottom of the column and the least dense at the top. A sample to be determined is placed in the column and falls to the level that is similar to its density.

These devices are somewhat complex to set up and maintain. A simple density column can be made from relatively ordinary materials found around the house or lab. Some recipes are given in Table 15.2.

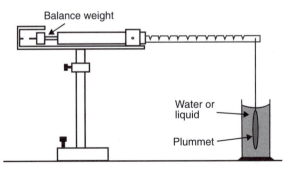

**FIGURE 15.18**   The Westphal specific gravity balance.

**TABLE 15.2**  Examples of Density Column Components

| | Column A | Column B | Column C | Column D |
|---|---|---|---|---|
| | Karo™ syrup | Water | Colored water | Salt water (approximately saturated) |
| | Glycerin | Ethanol | Cooking oil | Joy™ or Dawn™ dishwashing liquid |
| | Water | Cooking oil | Dawn™ liquid | Water |
| | Salad oil | Motor oil | Karo™ syrup | Isopropyl alcohol |
| | Rubbing alcohol | Molasses | | Mineral oil |
| | Lamp oil | Antifreeze | | |
| | | Ivory™ liquid (white) | | |
| | | Palmolive™ liquid (green) | | |
| | | Joy™ | | |
| | | Dawn™ (blue) | | |
| Note: | Color the lamp oil and rubbing alcohol with food dyes | | | All layers, except the mineral oil, can be colored with a drop or two of food coloring |

However, a simpler approach to a single column is to create several solutions of known density, each in their own container. For solutions with a density greater than water, varying concentrations of calcium nitrate in deionized water are used. Solutions having a density less than water are made with varying concentrations of isopropanol and deionized water. The specific gravity or the density of each solution is measured using a hydrometer, Westphal balance, pycnometer, or other appropriate method (Reference ASTM D1505).

The sample is placed in each solution in series. When the sample sinks, the density of the sample is said to lie between a range of densities corresponding to the one it sank in and the closest solution in which it floated. The precision of this method may be improved by increasing the number of solutions within a given density range.

# 15.7 Particle Sizing

## 15.7.1 Introduction

Solids can be separated on the basis of their particle size. When crushing and grinding equipment is used, a determination of the distribution of the various particle sizes is important in evaluating the performance of the equipment. The size of particles in soil gives an indication of the porosity and the ability for the soil to hold moisture and nutrients. It affects manufacturing processes and the formulation of products. The size of particles in dry feed for farm animals, such as swine, is related to its nutritional value. Many times the size of particles in a process may be specified in a standard method, such as the USP or FDA.

Several methods are used to determine the size of particles. They include **sieves, sedimentation, electrozone sensing, microscopy**, and **light scattering**. The method selected will depend on the application.

## 15.7.2 Sieves and Screen Analysis

**Sieves** are used in **screen**, or **sieve, analysis**. This is an inexpensive and reliable method for classifying larger particles based on size and mass. Screen analyses are generally used for quality control and analytical work and find use in a wide variety of industries. They are not suitable for emulsions, sprays, or very fine powders (<0.05 mm in diameter).

To make a screen analysis, the individual screens (see Figure 3.4) are cleaned and tapped free of any adhering particles. The screens are nested together vertically with the coarsest mesh at the top and the finest mesh at the bottom. A bottom pan and a top cover complete the set. A known amount of material is then placed on the top screen, the cover is replaced, and the stack is rotated, shaken, and bumped at intervals. Although this process could be done by hand, it is usually done mechanically to avoid operator

**TABLE 15.3**  U.S. Sieve Series and Tyler Equivalents for Selected Screens

| Sieve Designation | | | Nominal Wire | Tyler Equivalent |
|---|---|---|---|---|
| Standard | Alternate | Sieve Opening (mm) | Diameter (mm) | Designation (mesh) |
| 5.66 mm | No. 3 | 5.66 | 1.68 | 3 |
| 4.00 mm | No. 5 | 4.00 | 1.37 | 5 |
| 2.83 mm | No. 7 | 2.83 | 1.10 | 7 |
| 2.00 mm | No. 10 | 2.00 | 0.900 | 9 |
| 1.41 mm | No. 14 | 1.41 | 0.725 | 12 |
| 1.00 mm | No. 18 | 1.00 | 0.582 | 16 |
| 707 micron | No. 25 | 0.707 | 0.450 | 24 |
| 500 micron | No. 35 | 0.500 | 0.340 | 32 |
| 354 micron | No. 45 | 0.354 | 0.247 | 42 |
| 250 micron | No. 60 | 0.250 | 0.180 | 60 |
| 177 micron | No. 80 | 0.177 | 0.131 | 80 |
| 125 micron | No. 120 | 0.125 | 0.091 | 115 |
| 88 micron | No. 170 | 0.088 | 0.064 | 170 |
| 63 micron | No. 230 | 0.063 | 0.044 | 250 |
| 44 micron | No. 325 | 0.044 | 0.030 | 325 |
| 37 micron | No. 400 | 0.037 | 0.025 | 400 |

bias. After a period of time the **residuals**, or **fines**, are removed from the bottom pan. The pan is replaced and the shaking is continued until no fines appear on the pan. When completed, the mass of the particles on each of the screens is measured.

The sieve is made of a wire mesh cloth of standard sizes held in place by a circular metal frame. The diameter of the sieve ranges from 4 to 18 in., and the height may vary from 1 to 2 in. Three popular sieve series are used: the Tyler Series, the first commercial laboratory screens; the U.S. Standard Series; and the U.S. Alternative Series. Both the U.S. Alternative Series and the Tyler Series designations use the approximate number of opening per linear inch. However, the U.S. Standard Series is the preferred designation and is recommended by the International Standards Organization (ISO). Table 15.3 shows the U.S. Sieve Series and Tyler Series equivalents for a selected number of screens.

The clear space between the individual wires of the screen is termed the **screen aperture**. The number of apertures per linear inch is called the **mesh**. For example, a 10-mesh screen will have ten openings per linear inch. The actual size of the opening is found by subtracting the diameter of the wire. The material passing through the 80-mesh screen but not through the 100-mesh screen is designated 80/100 or the −80/+100 fraction. This is frequently seen in the description of gas chromatograph column packing material.

### 15.7.3  Data Handling and Analysis

Particles come in all shapes and sizes and in large numbers. Data are presented graphically using **histograms, fractional plots**, or **cumulative plots**. These graphs are primarily useful as pictures of the size distribution of the mixture. Table 15.4 gives a typical screen analysis for a 900-g sample. The measured experimental data are the mesh sizes, and the masses of the particles on each of the sieves are the masses of the residuals or fines. The other quantities are calculated.

The **mass fraction** is calculated by taking the mass of the particles on each screen and dividing by the total mass of the sample, as shown in Equation (15.17):

$$\text{Mass Fraction} = \frac{0.036 \text{ kg}}{0.900 \text{ kg}} = 0.04 \tag{15.17}$$

For example, the mass fraction on sieve 1 would be:

$$\text{Mass Fraction} = \frac{\text{mass on sieve}}{\text{total mass of sample}} \tag{15.18}$$

**TABLE 15.4**   A Typical Screen Analysis for a Sample Size of 900 g

| Sieve No. | Mesh Size (mm) | Mass (kg) | Mass Fraction | Relative Frequency |
|---|---|---|---|---|
| 1 | 0.71 | 0.036 | 0.04 | |
| 2 | 0.500 | 0.144 | 0.16 | 0.7619048 |
| 3 | 0.355 | 0.261 | 0.29 | 1.9310345 |
| 4 | 0.25 | 0.18 | 0.20 | 1.9047619 |
| 5 | 0.18 | 0.126 | 0.14 | 2 |
| 6 | 0.125 | 0.081 | 0.09 | 1.6363636 |
| 7 | 0.071 | 0.054 | 0.06 | 1.1111111 |
| 8 | 0.045 | 0.018 | 0.02 | 0.7692308 |
| Fines (residual) | 0 | | | 0.2222222 |
| Total | | 0.900 | 1.00 | |

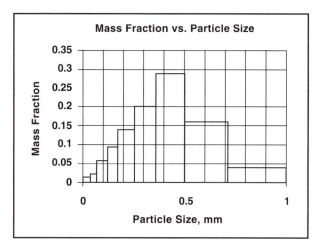

**FIGURE 15.19**   Mass fraction vs. particle size for Table 15.4 data.

The mass fraction may also be reported as the weight percent by multiplying the mass fraction by 100, as in the following relationship:

$$\text{Weight \%} = \frac{\text{mass on sieve}}{\text{total mass of sample}} \times 100 \tag{15.19}$$

The weight percent on sieve 1 would then be 4.0%.

## 15.7.4   Histogram Representation

Mass fractions are easily reported using a **histogram**. Figure 15.19 is a histogram of the mass fraction vs. the particle size for the data in Table 15.4.

Using this method makes comparisons difficult. If we divide each mass fraction by the screen interval, we obtain the **relative frequency** for each class:

$$\text{Relative Frequency} = \frac{\text{mass fraction}}{\text{screen interval}}$$

$$\text{Relative Frequency} = \frac{0.16}{(0.71-0.50)\,\text{mm}} = 0.7619\ \text{mm}^{-1} \tag{15.20}$$

**TABLE 15.5** A Representative Sieve Analysis for a Sample Size of 900 g

| Sieve No. | Mesh Size (mm) | Average Particle Size | Mass (kg) | Mass Fraction | Cumulative Mass Fraction Oversize | Cumulative Mass Fraction Undersize |
|---|---|---|---|---|---|---|
| 1 | 0.71 | | 0.036 | 0.04 | 0.04 | 0.96 |
| 2 | 0.500 | 0.605 | 0.144 | 0.16 | 0.2 | 0.8 |
| 3 | 0.355 | 0.4275 | 0.261 | 0.29 | 0.49 | 0.52 |
| 4 | 0.25 | 0.3025 | 0.18 | 0.20 | 0.68 | 0.32 |
| 5 | 0.18 | 0.215 | 0.126 | 0.14 | 0.82 | 0.18 |
| 6 | 0.125 | 0.1525 | 0.081 | 0.09 | 0.91 | 0.09 |
| 7 | 0.071 | 0.098 | 0.054 | 0.06 | 0.97 | 0.03 |
| 8 | 0.045 | 0.058 | 0.018 | 0.02 | 0.99 | 0.01 |
| Fines (residual) | 0 | 0.0225 | | | 1 | 0 |
| Total | | | 0.900 | 1.00 | | |

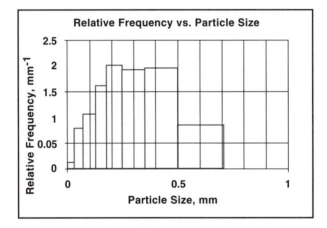

**FIGURE 15.20** Relative frequency vs. particle size.

The screen interval for sieves 1 and 2 is 0.71 mm–0.50 mm, or 0.21 mm, and the relative frequency would be 0.7619 mm$^{-1}$.

When the relative frequency vs. the particle size is plotted (Figure 15.20), the area under each rectangle is equal to the mass fraction of particles in that interval. The total area, like the sum of the mass fractions, is equal to 1.

## 15.7.5 Fractional and Cumulative Representations

Screen analysis data may also be presented by **fractional plots** or **cumulative plots**. Fractional plots show the mass fraction retained on each screen. Cumulative plots show either the sum of the mass fractions passing through each screen (undersize) or the sum of the mass fractions retained on each screen (oversize). The data are presented in Table 15.5. Figure 15.21 shows the mass fraction vs. the average aperture for the data in Table 15.5. Since the data for the smaller screen openings are crowded together, the logarithm of the average screen opening is used to spread these out, as shown in Figure 15.22. Additionally, the logarithm of both the mass fraction and the particle size may be used as in Figure 15.23. The average particle size is calculated by simply averaging the aperture of a sieve and the aperture of the next larger sieve.

$$\text{Average Particle Size} = \frac{(\text{Aperture}_1 + \text{Aperture}_2)}{2}$$

$$\text{Average Particle Size} = \frac{(0.71 + 0.500)\,\text{mm}}{2} = 0.605\,\text{mm}$$

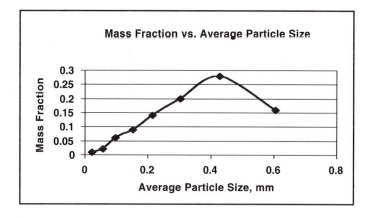

**FIGURE 15.21**  Plot of mass fraction vs. average particle size for the data in Table 15.5.

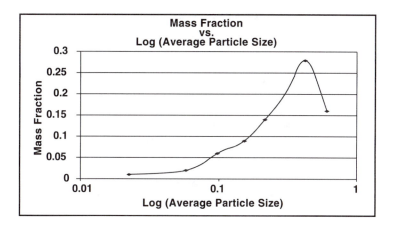

**FIGURE 15.22**  Mass fraction vs. logarithm of the average screen apertures.

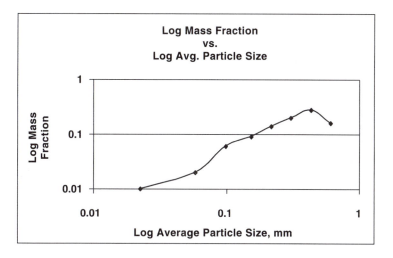

**FIGURE 15.23**  Plot of the logarithm of both the mass fraction and the particle size.

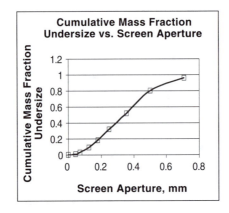

**FIGURE 15.24**    Cumulative mass fraction undersize vs. screen aperture.

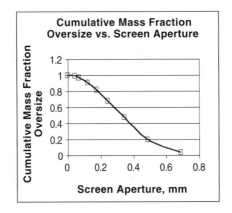

**FIGURE 15.25**    Cumulative mass fraction oversize vs. screen aperture.

The **cumulative mass fraction plot** is another method of representing the data. The cumulative mass fraction oversize is the relative mass of particles retained on each sieve. Similarly, the cumulative mass fraction undersize is the relative mass of particles passing through each sieve. Figures 15.24 and 15.25 show these cumulative plots. The cumulative mass fraction oversize is calculated by adding the mass fraction on a sieve ($m_i$) to the sum of the mass fractions on all preceding sieves ($m_1 + m_2 + \cdots + m_{i-1}$). Each mass fraction is that which is retained on the sieve. For example, the cumulative mass fraction oversize for sieve 3 would be 0.49:

$$\text{Cumulative Mass Fraction Oversize} = m_i + (m_1 + m_2 + \cdots + m_{i-1})$$
$$\text{Cumulative Mass Fraction Oversize} = 0.29 + (0.04 + 0.16) = 0.49$$

(15.21)

The cumulative mass fraction undersize (particles that pass through the sieve) can be calculated by subtracting the cumulative mass fraction oversize from 1. For example, the cumulative mass fraction undersize for sieve 3 would be 0.51:

$$\text{Cumulative Mass Fraction Undersize} = 1 - \text{Cumulative Mass Fraction Oversize}$$
$$\text{Cumulative Mass Fraction Undersize} = 1 - 0.49 = 0.51$$

(15.22)

The sieve analysis only gives an approximation of the particle distribution. The geometric shape of the particles is a factor in its moving to the proper-size sieve. For many process operations, more detail about the shape and surface area of the particles is important for the proper design and operation of equipment.

## 15.7.6  Sedimentation Analysis

Sedimentation techniques such as gravity and centrifugal settling are fairly simple methods of determining the size of particles in soils and in matrices, such as paints and ceramics, where screening is not practical. For some analysis, particularly of soils, the time required to perform the analysis may take as long as 24 h.

Knowing the particle size distribution for soils provides information about many of the soil properties, such as how much heat, water, and nutrients the soil will hold, how fast they will move through the soil, and what kind of structure, bulk density, and consistency the soil will have. The texture of the soil, how it feels, is based on the relative amounts of sand, silt, and clay present. Particles larger than 2.0 mm are called **stones** or **gravels** and are not considered soil material. Sand varies in size from 2.0 to 0.05 mm. Silt varies from 0.05 to 0.002 mm. Clays are less than 0.002 mm.

The settling method is based on Stoke's law, which says that denser, and usually larger, particles sink farther than less dense or smaller particles. This method assumes that particles all have the same density and that they are spherical.

In gravity settling the sample is dispersed in a liquid and then allowed to settle in a sedimentation cell. The height of the particles in the cell is then measured, or the specific gravity of the dispersing liquid is measured at different time intervals, which gives an indication of the size distribution. The **hydrometer**, or **Bouyoucos**, method is frequently used to classify the amount of sand, silt, and clay in soils.

To perform a **hydrometer test** for a soil sample, a known mass of sieved (<2 mm) soil is thoroughly mixed with a dispersing agent. It is then placed in a graduated cylinder, water is added, and the mixture is carefully agitated. A hydrometer is carefully placed in the mixture and read at 2 min, 12 min, and after 24 h.

According to the U.S. Department of Agriculture (USDA) method, the sand particles will settle to the bottom of the cylinder in 2 min, leaving silt and clay in suspension. The International Soil Science Society (ISSS) uses the 12-min time period for sand to settle. After 24 h, all of the silt particles have settled, leaving only clay in suspension. The hydrometer reading at each of these intervals is converted to grams of soils per liter using a correlation chart. See Workplace Scene 15.4.

Another classification method is the **pipet method**, in which the weight percent solids for each fraction are measured directly. A known mass of soil is dispersed in a liquid and placed in a graduated cylinder. A sample of the suspension is withdrawn by pipet from various depths at specified time intervals. The solvent is removed by drying, and the remaining solids are weighed. A final hydrometer reading determines the amount of clay remaining in the suspension. The weight percent of each fraction is calculated. This method allows for more detailed determination and classification of the various soil fractions.

## 15.7.7  Electrozone Sensing

The **electrozone sensing** technique, also called the Coulter principle, was originally developed for biomedical applications for counting blood cells. This method counts and sizes particle based on changes in the electrical resistance caused by nonconductive particles suspended in an electrolyte. It presently finds uses in a wide variety of industries, including the food, environmental, coatings, ceramics, and metals industries.

Figure 15.26 shows a schematic of an electrozone sensing cell. The suspended particles are placed in a container with two electrodes. One of the electrodes is within a tube that has a small hole, or aperture, in it. If there is no obstruction in the orifice, a steady current flows between the electrodes. As the particle moves near the sensing zone, it displaces a volume of the conductive electrolyte equal to the volume of

# WORKPLACE SCENE 15.4

Particle sizing is important in agriculture laboratories, especially soil laboratories, to determine the percentages of sand, clay, and silt. Besides the dry and wet sieving methods mentioned in the text, particle sizing of soil samples is sometimes done by preparing a soil dispersion, a slurry of the soil in a water solution of sodium pyrophosphate, and measuring the density of the dispersion at 40 sec and again at 6 h. The sand-sized particles settle to the bottom in approximately 40 sec, while only clay-sized particles are still in suspension after 6 h. The densities are measured with a hydrometer. The percentages of sand, clay, and silt can be calculated from this data and the temperature of the dispersions.

Larry Arnold, a technician with the National Soil Survey Laboratory in Lincoln, Nebraska, prepares soil samples for particle size analysis.

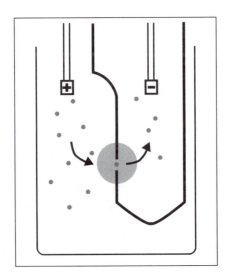

**FIGURE 15.26**   The electrozone or Coulter sensor.

the particle. This causes a temporary change in the current that is seen as a pulse. The number of pulses reflect the number of particles. The size of the pulse determines the volume, or size, of the particle.

Drawing a partial vacuum on the sensing cell carefully controls the quantity of the suspension drawn through the aperture. This allows for repeatable measurements.

### 15.7.8 Microscopy

Microscopy allows for direct examination of the particle, but it is not useful for determining size distributions. An ordinary optical microscope can be used for some determinations. An electron microscope will allow more detailed evaluation of the particle.

### 15.7.9 Light Scattering

One of the newest particle sizing techniques is **light scattering**. This technique is used to measure particle size distribution, colloid behavior, particle size growth, aerosol research, clean room monitoring, and pollution monitoring.

Particles in suspension have Brownian movement, the random motion resulting from thermal effects. A **laser** light source is used with the measurement point, the intersection of two laser beams. The measurements are performed on single particles as they move through the sample area. When the laser beam strikes a suspended particle, the light is scattered in many directions and is measured by photodetectors. The movements of the particle cause short-term fluctuations in the intensity of the scattered light. Since two laser beams are used, the scattered light generates an optical interference pattern. The signal response has a frequency proportional to the particle velocity. The responses from each photodetector are collected and electronically manipulated to provide a direct measure of the particle diameter.

## 15.8 Mechanical Testing

The mechanical properties of a material describe how it responds to the application of either a force or a load. When this is compared to an area, it is called **stress**, another term for pressure. Three types of **mechanical stress** can affect a material: **tension** (pulling), **compression** (pushing), and **shear** (tearing). Figure 15.27 shows the direction of the forces for these stresses. The mechanical tests consider each of these forces individually or in some combination. For example, tensile, compression, and shear tests only measure those individual forces. Flexural, impact, and hardness tests involve two or more forces simultaneously.

### 15.8.1 Impact Testing

When a material is loaded quickly or shocked or impacted, its behavior is, in many cases, different from the behavior exhibited in the tensile test, where the load is applied slowly. Whether shocks occur as part of

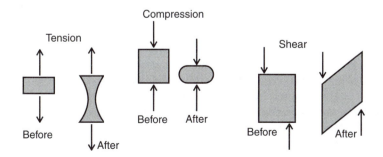

**FIGURE 15.27** Mechanical stresses.

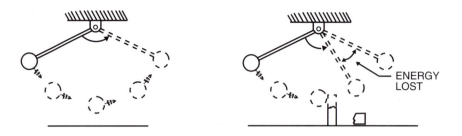

**FIGURE 15.28**    Left, the swinging pendulum. Right, the pendulum after striking an object.

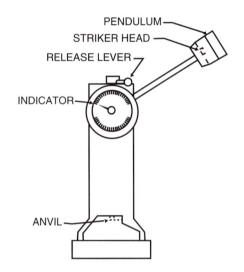

**FIGURE 15.29**    The impact testing machine.

normal service or by accident, the **impact strength** of the material will determine its suitability for a particular use. Any tendency of a material toward brittle behavior can be demonstrated through the impact test.

The **impact test** measures the amount of energy absorbed by the material before it fractures. A tough material will absorb more energy than a brittle one. These tests can be classified as **falling mass tests** and **pendulum tests**. Falling mass tests drop a ball-shaped mass from a given height onto the material to be tested. Helmets, containers, and eyewear are often tested in this manner. On the other hand, pendulum tests use a swinging hammer to strike the material and the amount of energy absorbed by the sample is measured.

The principle of the pendulum-type impact test involves the application of simple physics. If a pendulum is allowed to swing freely, it will swing through a known arc or angle as shown on the left in Figure 15.28. If the swinging pendulum strikes an object, it loses energy after the strike and the resulting arc becomes smaller. This is shown on the right in Figure 15.28.

The impact testing machine shown in Figure 15.29 measures the arc, or angle, through which the pendulum swings and converts this measurement to the energy needed to break the specimen. This energy is reported in either foot-pounds (ft·lbs) or Newton-meters (N·m.), also called joules.

The **Charpy test** and the **Izod test** are both pendulum-type impact tests. The difference between these two tests is essentially the orientation of the sample. In the Charpy test, or simple-beam method, the sample is supported at both ends, but is not held down, as shown on the left in Figure 15.30. In the Izod test, or cantilever beam method, the sample is supported on one end in a vice, as shown on the right in Figure 15.30. In both tests the sample is at the bottom of the pendulum arc and the amount of energy absorbed by the sample is measured.

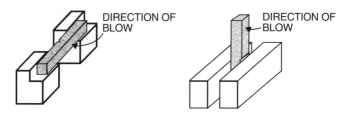

**FIGURE 15.30**   Left, Charpy test. Right, Izod test.

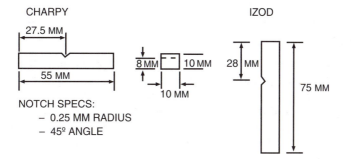

**FIGURE 15.31**   The dimensions and specifications of a metal impact specimen.

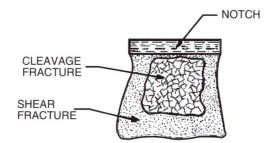

**FIGURE 15.32**   A fracture analysis of a metal impact test specimen.

The test method may specify either a notched sample or an unnotched sample. If a notched specimen is specified, the dimensions of the notch are also given. Figure 15.31 is an example of these dimensions for a metal sample. In the Charpy test the notched sample is struck from behind the notch. In the Izod test the notch is facing the hammer. The notch, called a stress raiser, concentrates the stresses applied by the impact load. It makes the material brittle and increases the elastic limit in the notch area. Without the notch, many materials will bend without fracture, and the capacity to absorb energy will not be accurately determined.

The radius of the notch is quite important, particularly for plastics. For example, polyvinyl chloride (PVC) is a notch-sensitive material. If the notch is blunt (2-mm radius), the impact strength is higher than that for ABS. If the notch is sharp (0.25-mm radius), the impact strength of PVC is lower than that for ABS. Other polymers that are notch brittle are high-density polyethylene (HDPE), polypropylene (PP), polyethylene teraphthalate (PET), dry polyamides (PAs), and acetals.

For metals, the failure will be either ductile, brittle, or a combination of both. Fine-grained, ductile steel will have a smooth, silky appearance. A coarse, granular appearance shows brittleness and lack of toughness. An examination of a fractured metal impact test specimen will typically show two distinctly defined regions—a smooth, silky region where the ductile (shear) fracture was initiated, and a coarse, granular region where the final brittle (cleavage) fracture occurred. Figure 15.32 shows these regions.

The percentage of each type of fracture is reported and is an indication of the ductility. The impact tests are used to indicate the effectiveness of heat treatment processes for metals. For example, overheating a metal before quenching will likely reduce the toughness even though the hardness is not affected. Likewise, some tempering processes following hardening can make the metal brittle and less tough.

Temperature has a great affect on the impact test results. The notch-bar impact strength of ordinary mild steel decreases drastically over the range of atmospheric temperatures (−40 to 100°F). In addition to temperature, moisture can also affect the impact strength of plastics. For example, polyamides (nylons) exhibit a wide range of impact strengths with moisture content. Impact strengths can be as high as 20 kJ/m$^2$ when containing moisture, and as low as 5 kJ/m$^2$ when thoroughly dried.

## 15.9   Tensile Test

### 15.9.1   Introduction

The **tensile test** is used to determine the ductility and the toughness of a material. **Ductility** refers to the ability of a material to deform plastically, or permanently, without breaking when stresses are applied. The amount of deformation that occurs before fracture is a measure of the material's ductility. It is quantified by the **percent elongation** and the **percent reduction in area**. **Toughness** is the ability to withstand stresses before breaking and is quantified by measuring the area under the **stress–strain curve**. **Brittle** materials tend to be relatively strong, but do not stretch. **Ductile** materials will stretch, but are relatively weak. The ultimate tensile strength, breaking strength, and yield strength are all determined from the tensile test.

The tensile test can give an indication of the chemical structure of a polymer, the conditions of sample preparation, the molecular weight, the molecular weight distribution, crystallinity, and the extent of cross-linking or branching. For metals, the tensile test provides information about heat treatment, alloying conditions, and certain manufacturing processes such as forging.

The tensile test is performed by placing a specially shaped specimen in the heads of the testing machine. The specimen is pulled apart through a hydraulic or mechanical loading system (Figure 15.33). Most ordinary tensile tests are conducted at room temperature and the tensile load is applied slowly. The unit measure of tensile strength is the pascal (Pa), or newtons per square meter (N·m$^{-2}$), and is defined by the following equation:

$$\text{Tensile Strength (Pa)} = \frac{\text{Pulling force (N)}}{\text{Cross-sectional area (m}^2)} \qquad (15.23)$$

As the tensile stress is applied to the specimen, it usually deforms and becomes thinner and longer. The change in length relative to the sample's original length is called **strain**. It is calculated as the change in length of a fixed, or gauge, length divided by the original length and is expressed as percent elongation.

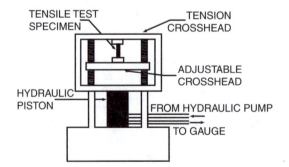

**FIGURE 15.33**   A schematic of a hydraulic tensile testing machine.

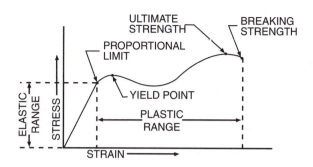

**FIGURE 15.34**  A representative stress–strain diagram for a ductile material.

The gauge length is specified in the testing method.

$$\% \text{ Elongation} = \frac{\text{Final length} - \text{Original length}}{\text{Original length}} \times 100 \qquad (15.24)$$

For metals, the percent reduction in area provides additional information about the material's deformation or its ductility. It is calculated by dividing the change in the cross-sectional area by the original cross-sectional area:

$$\% \text{ Reduction of Area} = \frac{\text{Original cross sectional area} - \text{ Final cross sectional area}}{\text{Original cross sectional area}} \times 100 \qquad (15.25)$$

## 15.9.2  The Stress–Strain Diagram

If the amount of force applied to the specimen is monitored along with the resulting strain, a **stress–strain diagram** may be obtained. Figure 15.34 is a representative example of a stress–strain diagram for a ductile material showing various regions. Understanding a few of these terms will assist in interpreting the diagram. Notice that at the lower stress values the diagram is a straight line. In this region stress is directly proportional to strain. This is the **elastic range**. If a stress within this range is applied to a sample, it will spring back to its original shape when the stress is removed.

**Young's modulus of elasticity**, or the **tensile modulus**, is the ratio of the stress applied to the strain within this linear region. It provides an indication of stiffness or how much a material or part will stretch under a given load. For example, a material that has a high tensile modulus is rigid and resists stretching.

The tensile modulus can be determined from the slope of the linear portion of this stress–strain curve. If the relationship between stress and strain is linear to the **yield point**, where deformation continues without an increased load, the modulus of elasticity can be calculated by dividing the yield strength (pascals) by the elongation to yield:

$$\text{Modulus of Elasticity} = \frac{\text{Stress (Pa)}}{\text{Strain}} \qquad (15.26)$$

When a linear relationship between the stress and strain is no longer present, the **proportional limit** is reached. On the diagram this is the highest point on the linear portion of the graph or where the curve no longer is a straight line. The material at this point is still elastic. The proportional limit is sometimes called the **yield point**.

At some stress level above the proportional limit the material will no longer return to its original shape; it will be permanently deformed. This region beyond the yield point is called the **plastic range**.

For many polymers and metals the yield point is not clearly defined. In these cases the **offset yield strength** is used. To obtain this value, a line parallel to the linear portion of the curve is drawn such that

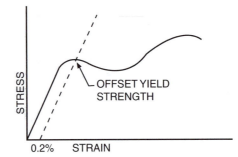

**FIGURE 15.35**   Offset yield strength.

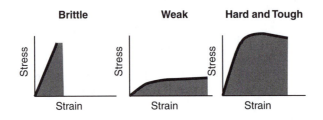

**FIGURE 15.36**   Toughness.

it intercepts the x-axis (strain) at a specified offset strain. For example, a 0.10 or 0.20% offset is used for some steels. Figure 15.35 shows the determination of the offset yield strength for a representative ductile material.

The **ultimate tensile strength** is the maximum stress sustained by the specimen. It is calculated by dividing the maximum load by the original cross-sectional area. A brittle material will break at its ultimate strength, while a ductile material will continue to stretch. At the maximum load a ductile material will also experience localized deformation or necking. This necking is not uniform and occurs rapidly to the point of rupture, or the **breaking point**. The **breaking strength** can be calculated by dividing the breaking load by the original cross-sectional area. This value is always less than the ultimate tensile strength.

The ability of a material to absorb energy without breaking is a function of ductility and is a measure of toughness. The area under the stress–strain curve represents the amount of energy needed to break the sample and an estimation of its toughness. The larger the area under the curve, as shown in Figure 15.36, the greater the material's toughness. See Workplace Scene 15.5.

# 15.10   Hardness

## 15.10.1   Introduction

**Hardness** can be defined as a measure of the resistance to abrasion, to deformation, or to penetration by a much harder body. In plastics the resistance to scratching, marring, and abrasion is related to the hardness.

Because the hardness of a material may be easily determined, it has become the most common method for the inspection of metals. For ferrous alloys there is a definite relationship, within limits, between hardness and tensile strength. However, tensile strength values obtained from hardness measurements are only approximations, and the true tensile strength of a metal can only be determined by the tensile test. Hardness testing is used frequently as a quality control test for estimating tensile strength because it is cheaper than machining tensile specimens and performing the tests.

# WORKPLACE SCENE 15.5

Physical characterization of polymers is a common activity that research and development technologists at the Dow Chemical Company perform. A material property evaluation that is critical for most polymer systems is a tensile test. Many instruments such as an Instron® test frame can perform a tensile test and, by using specialized software, can acquire and process data. Use of an extensometer eliminates calibration errors and allows the console to display strain and deformation in engineering units. Some common results from a tensile test are modulus, percent elongation, stress at break, and strain at yield. These data are then used to better understand the capabilities of the polymer system and in what end-use applications it may be used.

In order to have better accuracy, technologist Sarah Kushon of the Dow Chemical Company utilizes an extensometer during a tensile test. With permission.

**TABLE 15.6** Moh's Hardness Scale

| Substance | Moh's Scale | |
|---|---|---|
| Talc | *Softest* | 1.0 |
| Rock salt or gypsum | | 2.0 |
| Calcite | | 3.0 |
| Fluorite | | 4.0 |
| Apatite | | 5.0 |
| Feldspar | | 6.0 |
| Quartz | | 7.0 |
| Topaz | | 8.0 |
| Corundum | | 9.0 |
| Diamond | *Hardest* | 10.0 |

## 15.10.2 Simple Hardness Tests

One simple hardness test is the **Moh hardness test**; it is based on the fact that a harder material will scratch a softer material. Geologists and mineralogists frequently use this test. The **Moh scale** is an arbitrary scale of hardness based on the ability of ten selected minerals to scratch each other. The relative Moh hardness for several substances is given in Table 15.6.

**TABLE 15.7**    Shore Units

| Height of First Rebound (in.) | Shore Units |
| --- | --- |
| 6_ | 100 |
| 4 11/16 | 75 |
| 3 1/8 | 50 |
| 1 9/16 | 24 |
| No bounce | 0 |

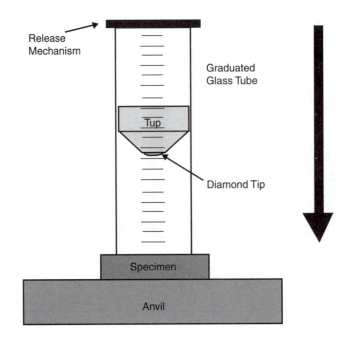

**FIGURE 15.37**    Scleroscope.

Another simple hardness test is the Shore scleroscope. This instrument measures the rebound height of a free-falling hammer called a tup. Figure 15.37 is a diagram of a scleroscope. The hammer drops from a predetermined height of 10 in., strikes the test sample, and rebounds. The higher the hammer rebounds, the harder the sample. The height of the first rebound is monitored and converted into "Shore" units. Table 15.7 shows the relationship between the rebound height and the Shore units.

The surface of the material to be tested should be free of imperfections since they could affect the rebound height.

## 15.10.3 Indentation Hardness Tests

Quantitative hardness tests slowly apply a fixed load to an indentor that is forced into the smooth surface of the specimen. After the load is removed, either the diameter across the impression or the depth of the impression is measured. The size of the penetration is proportional to the material's hardness. Rockwell, Brinnell, Vickers, and Knoop are well-known indentation hardness testing instruments.

Because the penetrator is applied to the surface of the material, the surface must be prepared such that it is representative of the structure of the material. The surface should be free from surface imperfections, such as scales and pits, and the surface should be flat. Some tests may require the surface to be highly polished.

Rockwell hardness tests apply a **minor load** and a **major load**. The minor load essentially sets the indentor on the surface of the material and through any slight imperfections. The major load pushes the indentor into the material to be tested. The minor and major loads vary depending on the test.

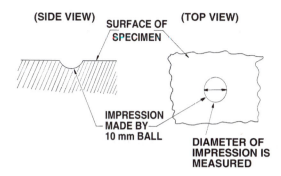

**FIGURE 15.38**   The Brinnell hardness test.

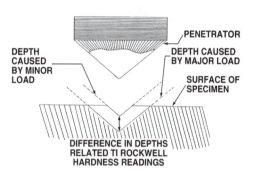

**FIGURE 15.39**   The Rockwell hardness test.

The hardness tests are classified according to the magnitude of the load. **Macrohardness tests** apply major loads that are greater than 1000 g (1 kg). **Microhardness tests** employ loads less than 1000 g and are commonly used to measure the hardness of a single grain or a very small area. Brinnell, Rockwell, and Vickers are examples of macrohardness tests, and Knoop is an example of a microhardness test.

## 15.10.4   The Brinnell Hardness Test

The **Brinnell hardness test** applies a major load of 3000 kg through a 10-mm ball. The diameter of the resulting impression is measured with a Brinnell microscope (Figure 15.38). Because of the fairly large indentation area, the Brinnell test is not suitable for materials that exhibit high creep factors—the tendency for plastic flow under stress. Materials with high creep factors would have a ridge around the edge of the indentation.

## 15.10.5   Rockwell Hardness Tests

The **Rockwell hardness tests** (**superficial** and **standard**) measure the depth of the impression made by either a 1/8-in. ball, a 1/16-in. ball, or a brale, diamond-point penetrator under major loads of 15, 30, 45, 60, 100, or 150 kg, depending on the type of test (Figure 15.39). The minor load for the Rockwell superficial test is 3 kg, and for the Rockwell standard test 10 kg.

## 15.10.6   The Knoop Microhardness Test

The **Knoop microhardness test** measures the length of the longest diagonal of the elongated diamond-shaped impression made by a load of less than 1 kg. The length is measured by setting the reticules in the microscope view and reading a vernier dial (Figure 15.40).

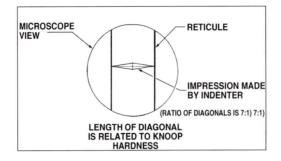

**FIGURE 15.40**    The Knoop hardness test.

**TABLE 15.8**    Comparison of Selected Hardness Tests

| Method | Basis | Scale | Major Load (kg) | Indentor | Symbol |
|---|---|---|---|---|---|
| Brinnell | Area of penetration | | 500–3000 | 10-mm ball | BHN |
| Rockwell | Depth of penetration | C, A, D | 150, 100, 60 | Brale (diamond cone) | $R_C$, $R_A$, etc. |
| | | E, K, H | 150, 100, 60 | 1/8-in. ball | |
| | | B, F, G | 150, 100, 60 | 1/16-in. ball | |
| | | L, M | 100, 60 | $\pi$-in. ball | |
| | | R | 60 | 1/2-in. ball | |
| Rockwell superficial | Depth of penetration | N | 45, 30, 15 | Brale (diamond cone) | 15N, etc. |
| | | T | 45, 30, 15 | 1/16-in. ball | 15T, etc. |
| Vickers | Area of penetration | | 5–120 | Pyramidal diamond | DPH |
| Knoop | Area of penetration | | 0.025–35 | Pyramidal diamond | Units Knoop |
| Moh | Appearance of scratch | | Manual | 10 stones | Units Moh |
| Shore | Height of bounce | | Gravity | 40 grain weight | Units Shore |

Table 15.8 compares selected hardness tests, showing scales, the indentor, and the major loads.

# Experiments

## Experiment 54: Capillary Viscometry

*Remember safety glasses.*

1. Set up a constant temperature bath for capillary viscometry at a temperature of 25°C.
2. Obtain a capillary viscometer with a known calibration constant. Also obtain samples of the following alcohols: methanol, ethanol, 1-propanol, 2-propanol, 1-butanol, and any others your instructor may suggest.
3. Pour the first liquid to be measured into the viscometer tube and place the tube in the constant temperature bath. After allowing plenty of time for the temperature to equilibrate, measure the time of flow in the manner discussed in Section 15.2.4. Using the known calibration constant, calculate the kinematic viscosity at 25°C. Repeat with each of the other alcohols.
4. Obtain an unknown alcohol from your instructor. Perform step 3 with this alcohol and identify it.
5. Maintain the logbook for the equipment used in this experiment, recording your name(s), the date, and the experiment number and name.

## Experiment 55: Rotational Viscometry

*Remember safety glasses.*
References: U.S. Pharmacopeia, 2000, p. 313; ASTM #D 1439.

### Part A: Dependence of Viscosity on Temperature

1. Prepare 500 mL of a 2% solution of carboxymethylcellulose, sodium salt, in water in the manner described in the U.S. Pharmacopeia reference above. Since the solution preparation is time-consuming, your instructor may prepare it ahead of time. Using a rotational viscometer with an appropriate spindle and a constant temperature bath, measure the viscosity of this solution at various temperatures. Plot viscosity vs. temperature.

### Part B: Determination of the Concentration of Carboxymethylcellulose, Sodium Salt, in an Unknown

1. Prepare a carboxymethylcellulose solution as in Part A. Prepare dilutions to bracket the suspected concentration in the unknown. Measure the viscosities of the solutions and the sample with a rotational viscometer and in a constant temperature bath set at 25°C. Plot viscosity vs. concentration and determine the concentration in the unknown.
2. Maintain the logbook for the equipment used in this experiment (Parts A and B), recording your name(s), the date, and the experiment number and name.

## Experiment 56: Measuring Refractive Index

Note: At the conclusion of the experimental work with refractive index, maintain the logbook for the equipment used, recording the date, your name(s), the experiment name and number (the parts you performed), and calibration information, if applicable.
*Remember safety glasses.*

### Part A: Measuring the Refractive Index of a Liquid

Equipment: Nondigital Abbe refractometer

1. If the refractometer uses a temperature-controlled water bath, ensure that the temperature is set properly and the pumps are circulating the temperature-controlled water through the sampling platform of the refractometer.
2. Carefully clean the prism with lint-free lens paper.
3. Place a drop of the liquid on the lower prism and carefully lower the upper prism.
4. Look through the eyepiece and turn the compensator knob until the colored indistinct boundary between the light and dark fields becomes a sharp line.
5. Still looking through the eyepiece, adjust the large knob until the sharp line intersects the midpoint of the crosshairs.
6. Depress the button on the side of the refractometer to read the refractive index from the scale.
7. Open the prisms and clean them with lens paper.

Note: If a digital refractometer is used, follow the manufacturer's operating instructions.

### Part B: Use of Refractive Index to Determine the Concentration of an Unknown Solution

1. Prepare standard water solutions of isopropanol of 1, 5, 10, 20, and 30%.
2. Measure the refractive index of each of the solutions, the unknowns, and a control, if one is provided.
3. Plot the standard curve using the spreadsheet procedure from Experiment 18 and obtain the correlation coefficient and the concentrations of the unknowns.

**Part C: Demonstration of a Nonlinear Refractive Index Calibration Curve**

1. Prepare standard solutions of 0, 25, 50, 75, and 100% methanol in deionized water. (Note: 0% is only deionized water and 100% is only methanol.) Optional: Additional standard solutions may be prepared to provide more data for the calibration curve, e.g., 10, 20, 30%, etc.
2. Measure the refractive index of each of the solutions.
3. Manually prepare a calibration curve, on graph paper, of refractive index vs. concentration. (A French curve may be helpful in drawing the curve.)
4. Measure the refractive index of an unknown solution.
5. Using the calibration curve, find the concentration of the methanol in the unknown.

Note: An additional calibration curve of specific gravity vs. concentration of each of the solutions will be necessary to give an estimate of the concentration.

**Part D: Percent Sugar in Soft Drinks by Refractive Index Measurements**

1. Prepare solutions of sucrose in water that are 4, 8, 12, and 16% sucrose by diluting the available 20%. Use 25-mL volumetric flasks.
2. Using the Abbe refractometer, read and record the refractive indexes of these solutions, the soft drink unknowns, and a control sample, if one is provided.
3. Plot the standard curve using the spreadsheet procedure from Experiment 18 and obtain the correlation coefficient and the concentrations of the unknowns and control.

# Experiment 57: Particle Size Analysis

*Remember safety glasses.*

## Part A: Particle Size Distribution (Screen Analysis)*

### Introduction

In this experiment, a solid material, such as pecan hulls, are crushed, ground, and separated into various sizes to observe the effects of the variation of size distribution with screening time and the variation of size distribution on rate of vibration. The size and distribution of particles may be determined by several methods. Screening is commonly used for this purpose. In this method a known mass of material of various sizes is passed over a series of standard screens and the amount of material collected on each screen is determined. The rate of vibrating the screen and the time allowed for vibrating have definite effects on the distribution of particles.

### Procedure

1. Obtain approximately 2 lb of material to be crushed. Determine accurately the amount obtained.
2. If necessary, use a hammer or mallet to break the material into pieces no larger than 1/2 in. (Note: *Wear safety glasses.*)
3. Place the material into the grinder and collect in hopper bag.
4. Determine the weight of material obtained. Calculate the amount of material lost in the grinding process.
5. Screen this material through weighed, U.S. Standard screens (eight screens to be specified) at a rate of 500 vibrations per minute.
6. Obtain the weight of material on each screen and pan at the end of 2, 4, 8, and 10 min.
7. Determine the amount of material lost in this process. (Perform a material balance.)
8. Repeat steps 4 to 7 using the same material for 700 and 900 vibrations per minute. At 900 vibrations per minute, continue the classification until it is complete. Check every 3 to 4 min until no fines are added to the bottom pan. (Some shakers may not have variable vibration rates.)
9. Record the grinding screen(s) used and the U.S. Standard sieves.

---

*References: ASTM E276-98 and ASTM D422.

10. Determine the screen size, screen openings, mass fraction, and cumulative mass fraction through each screen for all runs.
11. Plot:
    A. Histogram of relative frequency vs. particle size
    B. Mass fraction (or weight percent) retained (oversize) vs. average particle diameter
    C. Cumulative mass fraction (or weight percent) (undersize) through each screen vs. average screen opening

## Part B: Particle Size of Soils (Jar Method)

### Introduction

The particle size and distribution in soils can be easily determined by the sedimentation method. It is based on the fact that large particles will settle faster than smaller ones. No special equipment is needed.

### Procedure

1. Fill a quart jar about two thirds full with water. Into this jar, place approximately a 1/2 cup of soil that has been finely crushed. Any large clumps can be broken up with your fingers.
2. Add 5 tablespoons of 8% Calgon™ solution.
3. Replace the cap and shake for 5 min.
4. Set the jar aside and let stand for 24 h.
5. At the end of 24 h, measure the depth of the settled soil. This represents the total depth of soil.
6. Shake the jar for 5 min.
7. Set the jar on the table and let stand for 40 sec.
8. Measure the depth of the soil. This is the sand layer.
9. At the end of 30 min, measure the depth of the soil. The difference between this depth and the depth of the sand layer is the depth of the silt layer.
10. The remaining unsettled part is the clay fraction.
11. Convert each of the measurements into percentages for each of the fractions—sand, silt, and clay.

## Part C: Particle Size of Soils (Bouyoucos Method)*

### Procedure

1. Prepare the dispersing solution by mixing 50 g of sodium hexametaphosphate in 1 L of distilled water.
2. Weigh 25 g of dried, sieved soil and pour it into a 250-mL beaker. Add 100 mL of the dispersing agent and mix thoroughly.
3. After the soil and the dispersing agent are mixed, set the beaker aside for 24 h.
4. Measure the distance between the 500-mL mark and the base of the cylinder by placing a measuring stick inside the cylinder.
5. Record the calibration temperature of the hydrometer that is found on the body.
6. After the soil-dispersing agent has set for 24 h, stir the suspension and transfer it to the 500-mL graduated cylinder. Rinse any remaining soil particles out of the beaker with distilled water.
7. Add enough distilled water to fill the cylinder to the 500-mL mark.
8. Cover the cylinder with plastic wrap or other cover and mix vigorously by rotating the cylinder at least ten times. Be sure that the soil is thoroughly mixed.
9. Set the cylinder on the bench and begin timing with a stopwatch.
10. After 1 min, carefully place the hydrometer in the suspension.
11. At exactly 2 min after the cylinder was set down, record the hydrometer reading. (According to the USDA definition, sand will settle in this time period.)

*References: USDA; ISSS (International Soil Science Society).

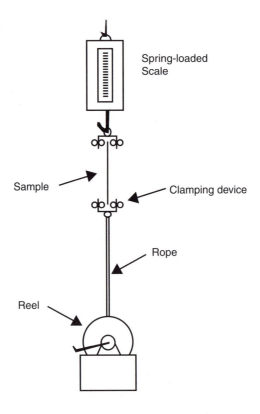

**FIGURE 15.41**   A homemade tensile tester.

12. Remove the hydrometer and place a thermometer in the suspension. After 1 min, record the temperature.
13. At 12 min, take another hydrometer reading and temperature reading. (According to the ISSS definition, sand will settle in this time period.)
14. Leave the cylinder undisturbed for 24 h.
15. At 24 h take another hydrometer reading and temperature reading. (The silt particles have settled in this time period, leaving the clay in suspension.)
16. Discard the soil suspension in a special pail. (The soil may be placed outside in a designated location.) Do not pour down the sink.

## Experiment 58: Tensile Testing of Polymers Using a Homemade Tester*

Purpose: To measure the breaking strength of various household polymer samples using a homemade tensile tester.

Note: A homemade tensile tester (see Figure 15.41) is used in this experiment. The device can be constructed as follows:

Attach the scale to a secure location.
Use nylon or other strong fiber rope that will not break before the sample.
The clamping devices can be made from a 1/8-in. metal strip, chair gliders, and wing nuts.

The samples can be from a variety of plastic materials, e.g., margarine tubs, plastic bags, etc.
*Remember safety glasses.*

---

*Adapted from *Polymer Chemistry*, National Science Teachers Association, Washington, D.C., 1989, pp. 169–170. Reference: ASTM D-638.

*Procedure*

1. Cut standard-size strips (1 × 3 cm) from different types of plastics to be tested.
2. Clamp the strips in place.
3. Slowly crank the wrench until the sample breaks.
4. Record the reading on the spring scale at the breaking point.
5. Compare tensile strength values of the various samples.

Caution: Although many types of materials can be used, samples should be limited to safe breaks and within the limits of the scale.

## Questions and Problems

1. Other than those mentioned in Section 15.2.1, think of other formulations for which the quality may be judged by their resistance to flow.
2. Define rheology, shear force, shear stress, shear rate, Newtonian fluid, dynamic viscosity, centipoise, kinematic viscosity, centistokes, viscometry, and viscometer.
3. How does capillary viscometry measure viscosity?
4. In capillary viscometry, the fluid being measured flows through a capillary tube. How is this helpful in measuring viscosity?
5. What is an Ostwald viscometer?
6. Is a Cannon–Fenske viscometer a capillary type or a rotational type? Explain.
7. Distinguish between a Cannon–Fenske viscometer and a Ubbelohde viscometer.
8. A given calibration liquid is known to have a kinematic viscosity of 12.72 cS at 25°C. What is the calibration constant for a viscometer if it takes this liquid 109 sec to pass from the upper mark to the lower mark of a capillary viscometer?
9. The calibration constant of a capillary viscometer is 0.133 cS sec$^{-1}$. If an unknown liquid flows between the marks in 98 sec, what is the viscosity of the liquid?
10. Explain how a rotational viscometer measures viscosity.
11. Define thermal analysis, differential thermal analysis, differential scanning calorimetry, and thermogravimetric analysis.
12. What do the following stand for: DTA, DSC, TGA?
13. When you heat a sample that is melting, why does its temperature not increase?
14. How do the temperature of a sample and the temperature of its surroundings compare as the temperature of the surroundings is increased?
15. Explain the negative peak in Figure 15.7.
16. In a plot of the difference in temperature between a sample and a reference material vs. the temperature of the surroundings, why does an endothermic process produce a negative peak while an exothermic process produces a positive peak.
17. How does DSC differ from DTA? How are they similar? Why is DSC often considered a subset of DTA?
18. In Figure 15.10, what do the circles designated S and R refer to? Why must there be two locations like these in the oven?
19. For what purpose is indium often used in DSC?
20. What reactions other than melting produce results in DSC?
21. What are some applications of thermal analysis in the pharmaceutical and polymer industries?
22. The refractive index of methanol is given as $\eta_D^{20} = 1.3292$. What does the D mean? What does the 20 mean?
23. Why would a technician measure the refractive index of a distillation product?
24. Why do refractometers utilize compensators and filters for the light source?
25. Why is the temperature surrounding the prisms maintained at a constant temperature?
26. The sugar concentration, measured as sucrose, of a soft drink was determined using a refractometer with a °Brix scale. The refractometer reading was 10.5° Brix. What is the sugar concentration?

27. Define plane-polarized light, optical rotation, optical activity, asymmetric carbon atom, enantiomers, racemic mixture, polarimeter, and specific rotation.
28. What exactly is the physical property known as optical rotation?
29. Write down the structures of at least five compounds that have an asymmetric carbon atom.
30. The compound 2-butanol has an asymmetric carbon atom, but a sample of 2-butanol out of the bottle is not optically active. Why is that?
31. If two structures are nonsuperimposable mirror images, what can you say about the direction that each will rotate the plane of polarized light? Would a mixture of the two be optically active? Explain.
32. Describe how a polarimeter measures optical rotation.
33. The mass of an empty, dry graduated cylinder is 55.556 g; 25.00 mL of a liquid was placed in the graduated cylinder and reweighed. The mass of the cylinder and liquid was found to be 75.081 g. What is the density of the liquid?
34. The density of a solid, nonporous material is needed whose specific gravity is known to be greater than 1.0. The mass of the material in air was determined to be 49.833 g. The mass of the material in water was determined to be 43.533 g. What is the density of the material?
35. A 2-ft-long copper tube having a 1.0-in. inside diameter is filled with steel balls having a diameter of 1.0 in. The space between the balls is filled with water. What is the bulk density of the tube in grams per cubic centimeter? (The specific gravity of steel is 7.8. The volume of a sphere is calculated by the relationship $V = 1.333\pi r^3$.)
36. The bulk density of a sample of rock salt (sodium chloride) is to be determined. An empty 100-mL graduated cylinder had a mass of 123.248 g. Rock salt was dispensed into the cylinder at a rate of 1 mL/sec until it reached 100 mL. The graduated cylinder was reweighed and had a mass of 286.587 g. What is the bulk density of the rock salt?
37. Explain why the bulk density differs from the actual density.
38. Explain why the fill rate is an important factor when determining the bulk density?
39. Given the following data, calculate the average bulk density for the substance.

| Trial | Time (sec) | Volume (mL) | Mass$_{initial}$ (g) | Mass$_{final}$ (g) |
|-------|-----------|-------------|---------------------|--------------------|
| 1 | 98 | 100.0 | 78.26 | 211.47 |
| 2 | 102 | 100.0 | 78.34 | 210.82 |
| 3 | 97 | 100.0 | 78.31 | 209.55 |

40. A screen used for a sieve analysis has a designation of 60 mesh. What does this indicate?
41. One type of packing material for a column for a gas chromatograph is Chromosorb W −80/+100. What does −80/+100 indicate?
42. Given the following data for a screen analysis, calculate the mass fraction, cumulative mass fraction, and relative frequency.
    Sample Size: 1253 g

| Sieve No. | Mesh Size (mm) | Mass (kg) | Mass Fraction | Cumulative Mass Fraction Oversize | Relative Frequency |
|-----------|----------------|-----------|---------------|-----------------------------------|--------------------|
| 1 | 0.71 | 0.094 | | | |
| 2 | 0.500 | 0.063 | | | |
| 3 | 0.355 | 0.078 | | | |
| 4 | 0.25 | 0.167 | | | |
| 5 | 0.18 | 0.398 | | | |
| 6 | 0.125 | 0.288 | | | |
| 7 | 0.071 | 0.096 | | | |
| 8 | 0.045 | 0.069 | | | |
| Fines (residual) | 0 | | | | |
| Total | | 1.253 | | | |

43. Describe the difference among the following stresses: tension, compression, and shear.
44. List three mechanical tests that involve two or more forces simultaneously.
45. What is the difference between a ductile material and a brittle material?
46. What does the modulus of elasticity, or Young's modulus, indicate?
47. On the stress–strain diagram, what does the elastic range indicate? The plastic range?
48. A material's toughness, or its ability to absorb energy without breaking, is determined by what mechanical property?
49. Define hardness.
50. Which is harder, quartz or calcite?
51. Which hardness tests measure the depth of penetration?
52. Which hardness tests measure the diameter of a penetration?
53. What is the purpose of the minor load in the Rockwall Hardness test?
54. Should the thickness of the material be considered when conducting a Rockwell Hardness test? Explain.

# 16 *

# Bioanalysis

## 16.1  Introduction

**Bioanalysis** may be defined as laboratory analysis of biomolecules. **Biomolecules**, in turn, are organic compounds with biological activity, generally important *only* in biological systems, or cells. **Biochemistry** is the study of structure and function of biomolecules. **Biotechnology**, a related concept, concerns the industrial applications of biochemical techniques. Thus bioanalysis, biochemistry, and biotechnology are closely related concepts, all concerned primarily with biomolecules.

## 16.2  Biomolecules

The four groups of biomolecules are carbohydrates, lipids, proteins, and nucleic acids. All four are essential to life and are therefore found in living cells. All are organic compounds, based on the element carbon. In addition to carbon, all four also contain hydrogen and oxygen. Proteins, nucleic acids, and a few lipids contain nitrogen. Of the four, proteins alone contain sulfur, while nucleic acids and some lipids also contain phosphorus. Notably, few elements are found in biomolecules, and all of these are relatively small and simple, with the highest atomic number, 16, being sulfur. This is reflected in the fact that over 96% of a living cell consists of just four elements: carbon, oxygen, hydrogen, and nitrogen. The two most important biomolecules in modern bioanalysis are proteins and nucleic acids.

### 16.2.1  Carbohydrates

Carbohydrates are literally hydrates of carbon, containing only the elements carbon, oxygen, and hydrogen. In the human diet, they are considered macronutrients, along with proteins and fats (triacylglycerols). The three types of carbohydrates are monosaccharides, disaccharides, and polysaccharides.

Monosaccharides are the smallest carbohydrates, usually possessing molecular weights between 90 and 200. Chemically, they are either aldehydes (the aldoses) or ketones (the ketoses). Common examples of aldoses are glucose, galactose, and ribose. Ribulose and fructose are ketoses. Monosaccharides are commonly classified according to the number of carbon atoms in their chemical backbones. A three-carbon monosaccharide is called a triose, five-carbon a pentose, and six-carbon a hexose. Glucose, galactose, and fructose are hexoses, while ribose and ribulose are pentoses. Most monosaccharides have hydroxyl functional groups totaling one less than the number of carbon atoms in the skeleton. Thus, hexoses have five hydroxyl groups and pentoses have four. The presence of these very polar functional groups ensures that all monosaccharides are hydrophilic (water loving) and therefore water soluble. Examples of monosaccharides are shown in Figure 16.1. In nature, monosaccharides exist in two forms, rings and open-chain carbon skeletons (see Figures 16.1 and 16.2).

A disaccharide consists of any two monosaccharides covalently linked together, as shown below and in Figure 16.2:

---

*This chapter was written by Don Mumm of Southeast Community College, Lincoln, Nebraska.

NUMBER OF CARBON ATOMS

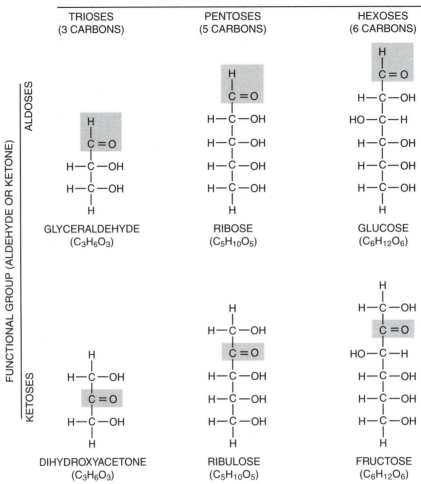

FIGURE 16.1    Monosaccharide structures drawn as open-chain carbon skeletons.

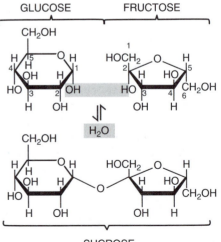

FIGURE 16.2    Monosaccharide structures drawn as rings. Disaccharides form by linking two rings through the loss of water.

$$\text{sucrose} = \text{glucose} + \text{fructose}$$

$$\text{lactose} = \text{glucose} + \text{galactose}$$

$$\text{maltose} = \text{glucose} + \text{glucose}$$

Like monosaccharides, disaccharides are hydrophilic. In nature, they tend to be exclusively either plant or animal products. Sucrose (table sugar) and maltose (malt or grain sugar) are produced only by plants, while lactose (milk sugar) is exclusively an animal product.

The largest and most complex carbohydrates are the polysaccharides. They are polymers, long chains of repeating chemical units. Each individual unit is called a monomer. The monomer unit of polysaccharides is the monosaccharide, normally glucose. A typical polysaccharide contains several hundred individual monomers. Examples of common polysaccharides are starches, plant products that are major macronutrients in the human diet, and cellulose, found in plant cell walls. In the human diet, cellulose is referred to as fiber, indigestible but beneficial for normal intestinal motility. More than half of the Earth's total carbon is stored in these two polysaccharides.

The most important biological function of carbohydrates is to provide energy. Most current dietary recommendations suggest that about 55% of total caloric requirements be provided by carbohydrates.

## 16.2.2   Lipids

The one and only distinctive feature of lipids is that they are all totally hydrophobic (water fearing), or nearly so. They generally will not chemically interact with water and therefore will not dissolve in water. Chemically, lipids fit into several categories, each of which is structurally unique. Common types of lipids include triacylglycerols, phospholipids, and steroids.

Triacylglycerols, commonly refered to as fats and oils, consist of three fatty acids linked to a molecule of glycerol, a three-carbon alcohol. Fatty acids are long-carbon-chain molecules, each with a single carboxyl functional group. Common examples are stearic acid and palmitic acid, shown in Figure 16.3.

Most biologically important fatty acids have even numbers of carbon atoms in their skeletons. Palmitic acid, for example, has 16. Since palmitic acid has no carbon–carbon double bonds, it is considered **saturated**. An **unsaturated** fatty acid, such as oleic acid (also shown in Figure 16.3), possesses one or more carbon–carbon double bonds (C=C).

Free fatty acids all have a hydrophilic end, the carboxyl group. However, when they link to glycerol to form triacylglycerols, water is chemically removed (as shown in Figure 16.3), resulting in molecules that exhibit totally hydrophobic behavior. The three types of triacylglycerols are saturated, with no carbon–carbon double bond (C=C); monounsaturated, with one C=C bond; and polyunsaturated, with more than one C=C bond. The triacylglycerol forming in Figure 16.3 would be classified as monounsaturated, since it has only the one C=C bond provided by oleic acid. The degree of unsaturation is reflected in physical appearance, with polyunsaturated molecules generally appearing as yellow liquids (oils), and saturated molecules as white solids (fats). In nature, triacylglycerols are widespread and are crucial energy storage

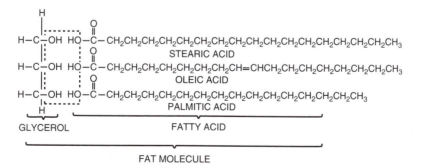

**FIGURE 16.3**   Examples of fatty acids linked to glycerol.

# WORKPLACE SCENE 16.1

Antibodies are complex proteins directly involved in immune responses against foreign agents such as viruses and bacteria. They are produced by the immune system in response to the entry of anything foreign into our bodies. The laboratory measurement of antibodies in biological samples is frequently used to diagnose various bacterial and viral diseases. The general laboratory method, called **immunoassay**, usually utilizes a UV-VIS absorption technique. Since proteins, including antibodies, and the products of the reaction of antibodies with drugs absorb UV and visible light, their presence in blood, urine, semen, and other body fluids may be detected by this method. In one procedure, a 96-well plate is coated with the drug and plasma samples are placed in the wells, where antibodies are subsequently produced in order to bind to the drug molecules. The product absorbs UV light. The wells are then scanned by a UV absorbance unit designed specifically for the analysis of the plates for the antibodies.

Mary Peterson of MDS Pharma Services in Lincoln, Nebraska, prepares to insert a 96-well plate into the UV-VIS scanner used to detect antibodies that have been bound to a seed drug.

molecules (and function as insulation in mammals). They are one of the macronutrients in the human diet, along with carbohydrates and proteins, providing on average about 30% of the total calories in a high-quality diet.

Phospholipids consist of two fatty acids linked to glycerol, with an additional chemical head that contains a phosphate functional group. Due to the polarity of the phosphate group, all phospholipids possess a small hydrophilic region. This property is critically involved in phospholipid function, which is to provide the backbone of biological membranes. In a membrane, phospholipids align in a double layer, the phospholipid bilayer, with the hydrophilic heads outside (top and bottom) and the long hydrophobic tails (fatty acids) inside. The vast majority of a biological membrane, therefore, is hydrophobic. An example of a phospolipid is shown in Figure 16.4.

Steroids are the only common lipids without fatty acids. Chemically, they consist of a fused carbon ring structure consisting of four rings, three rings of which are six membered and one ring of which is five membered, as shown in Figure 16.5. Some steroids, such as cortisone, function as hormones. Cholesterol is an important component of membranes in animal cells.

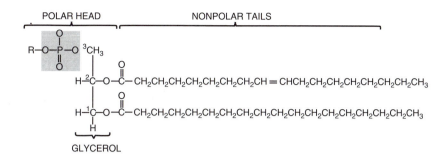

**FIGURE 16.4** An example of a phospholipid.

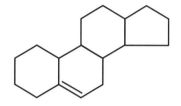

**FIGURE 16.5** An example of the fused ring structure found in steroids.

## 16.2.3 Proteins

Proteins are almost certainly the most structurally complex molecules found in nature. They are large polymers (many have molecular weights in excess of 100,000) with intricate three-dimensional structures and tremendous diversity. See Workplace Scene 16.1. In nature there are at least 10,000 different proteins, each possessing its own structural identity. Amino acids are the monomer units. There are only 20 common amino acids found in these 10,000 proteins. True proteins contain only the elements carbon, hydrogen, oxygen, nitrogen, and sulfur. Historically, amino acid sequences have been used to trace possible evolutionary relationships among organisms, since they are coded by DNA base sequences. Mutations therefore are manifested structurally by different amino acid sequences of proteins.

### 16.2.3.1 Amino Acids

The 20 common amino acids are the monomers for proteins. All 20 are similar in structure, but not identical. (Structures of all 20 amino acids are shown in Figure 16.6.) Each amino acid has a central carbon atom, the alpha carbon, to which is attached a carboxyl group, an amino group, a hydrogen atom, and a unique R group. The 20 amino acids differ *only* in their R groups. These R groups collectively determine the properties of proteins. For example, a water-soluble protein is likely to have many amino acids with hydrophilic R groups. Techniques commonly used to analyze amino acids include paper chromatography, thin-layer chromatography (TLC), and electrophoresis, as discussed in Section 16.3. Analysis and separations are possible due to the different R groups found in all amino acids. These techniques take advantage of the unique chemistry of R groups.

### 16.2.3.2 Peptides

Peptides are short chains of amino acids linked by peptide bonds. Most biologically active peptides contain two to ten amino acids. Peptide bonds are formed between the carboxyl carbon of one amino acid and the amino nitrogen of another. Since water is released, this is an example of dehydration synthesis. The bond forms as illustrated in Figure 16.7.

Examples of biologically active peptides include:

1. Aspartame, a dipeptide (aspartic acid + phenylalanine) artificial sweetener marketed under the trade name Nutrasweet
2. Glutathione, a tripeptide (glutamic acid + cysteine + glycine) antioxidant believed to be an anticancer agent

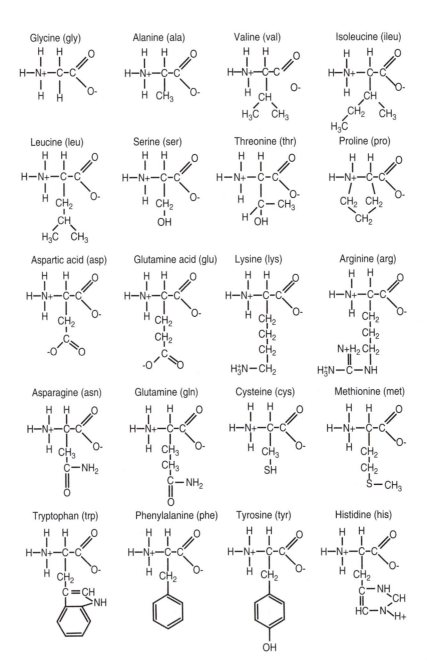

**FIGURE 16.6**   The amino acids.

3. Enkephalins, pentapeptides found in the central nervous system that are natural analgesics
4. Oxytocin, a hormone composed of nine amino acids, which induces labor by stimulating contraction of uterine muscles

### 16.2.3.3   Proteins: Primary Structure

The primary structure of a protein is the linear sequence of amino acids linked by peptide bonds. The structural heirarchy of proteins is as follows.

amino acids $\rightarrow$ peptides $\rightarrow$ polypeptide chains $\rightarrow$ proteins

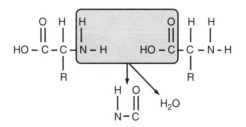

**FIGURE 16.7**   The formation of a peptide bond.

The polymer level is reached with the polypeptide chain, a long (usually 20+) chain of amino acids. A protein consists of one or more polypeptide chains. For proteins with only one chain, the chain *is* the protein. For multiple-chain proteins, known as oligomeric proteins, the polypeptide chain and protein levels must be viewed separately. Each different polypeptide chain has its own unique primary structure.

Historically, data on primary structures of proteins were used to help establish evolutionary relationships between organisms by reversing the following sequence:

DNA base sequence → RNA base sequence → protein amino acid sequence

This process starts by comparing amino acid sequences of a protein in two different organisms to determine the number of amino acids in common. The more they have in common, the more their genes that code for the protein must have in common. Similar genes imply closer evolutionary relationships. The need for this approach to obtaining genetic information, however, decreased dramatically when DNA sequencing techniques were developed in the late 1970s. There was no longer the need to work backwards, since direct data on DNA nucleotide sequences were obtainable. The advent of industrial applications of recombinant DNA technology breathed new life into amino acid sequencing. Many genetic engineering companies employ protein chemists to sequence proteins produced by genetically altered organisms to verify that their primary structures are identical to those produced by humans.

### 16.2.3.4   Proteins: Secondary Structure

At this level of structure, the three-dimensionality of proteins begins. Secondary structures are spatial arrangements of polypeptide chains stabilized by backbone interactions. Because R groups are not involved, and because all polypeptide chains have fundamentally the same backbone, the total number of secondary structures in nature is limited. The most common is the alpha helix (a **helix** is a spiral-shaped structure), discovered by Linus Pauling and Robert Corey in the 1940s. The alpha helix is stabilized by hydrogen bonds between NH and C=O groups in the backbone of a chain. Other examples of secondary structures include the beta pleated sheet found in silk and the collagen triple helix. The beta pleated sheet is a flat, open structure whose end view resembles pleats. Many fibrous proteins have only one secondary structure. For example, the alpha keratins, found in body coverings (fingernails, toenails, horns, hooves), are entirely alpha helix, and collagen is entirely triple helix. Globular proteins often possess multiple secondary structures. The enzyme ribonuclease, for example, has significant regions of both alpha helix and beta pleated sheet.

### 16.2.3.5   Proteins: Tertiary Structure

Each polypeptide chain in nature has its own unique tertiary structure. For example, if 25,000 polypeptide chains exist, there are 25,000 distinct tertiary structures. A number this large means necessarily that detailed tertiary structures for many of the polypeptide chains are unknown at this time. Determining

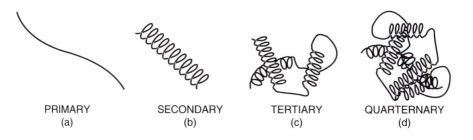

**FIGURE 16.8**   Conceptual illustrations of the four levels of proteins.

tertiary structures in the laboratory is slow, tedious, and in many cases presently impossible. At present, detailed information on tertiary structures exists for at most 5% of all proteins.

This tremendous diversity and complexity results from the nearly infinite number of total possible combinations of interactions stabilizing tertiary structures. R group interactions are entirely responsible. The four R group interactions are: 1) hydrophobic interactions, 2) ionic (charge) attractions, 3) hydrogen bonding, and 4) disulfide bonds between cysteine residues (amino acids in a chain). Since disulfide bonds are the only covalent linkages, their presence is deemed crucial for protein stability. The complexity of tertiary structures is derived from the sheer number of possible interactions. For example, consider a hypothetical polypeptide chain with 200 amino acids, 100 of which have hydrophobic R groups. Theoretically, any one of these 100 may form a hydrophobic interaction with any of the remaining 99. To determine the complete tertiary structure, one must prove which amino acids actually do form these interactions (e.g., position 18 with position 95).

#### 16.2.3.6   Proteins: Quaternary Structure

This level of structure is found only in oligomeric proteins, those with multiple polypeptide chains. Myoglobin, for example, has only one polypeptide chain, and therefore has no quaternary structure. Myoglobin is the red pigmented protein in muscles that stores oxygen gas. Hemoglobin, the protein that carries oxygen from the lungs to the tissues, has four polypeptide chains, and therefore possesses a quaternary structure. This level is concerned with the interactions among chains that stabilize the three-dimensional structures needed for functional molecules. Quaternary structures result from R group interactions, frequently hydrogen bonding. A conceptual illustration of all four levels is shown in Figure 16.8.

### 16.2.4   Nucleic Acids

Only two nucleic acids exist. They are DNA (deoxyribonucleic acid) and RNA (ribonucleic acid). The structural complexity of nucleic acids falls far short of that of proteins. Like proteins, however, nucleic acids are polymers, with nucleotides being the monomer units.

Each nucleotide consists of three structural components: a nitrogenous base, a pentose monosaccharide, and a phosphate functional group. The bases, which are heterocyclic and aromatic, are of two types, purines and pyrimidines. Purine bases consist of a six-atom ring fused to a five-atom ring. Of the nine atoms in the fused rings, five are carbon and four are nitrogen. The common purines in DNA and RNA are adenine (A) and guanine (G). Less common purines include uric acid, xanthine, and hypoxanthine. Pyrimidine bases possess a single six-atom ring, four atoms of carbon and two of nitrogen. Pyrimidines found in DNA are cytosine (C) and thymine(T), and RNA has cytosine and uracil (U). DNA nucleotides include the pentose deoxyribose, and RNA has ribose. The linkage of any base to either pentose forms a nucleoside. Covalently linking a phosphate functional group to the pentose results in a complete nucleotide.

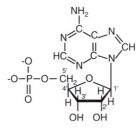

Adenosine 5'-monophosphate
AMP

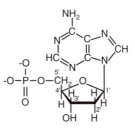

Deoxyadenosine 5'-monophosphate
dAMP

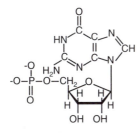

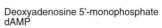

Guanosine 5'-monophosphate
GMP

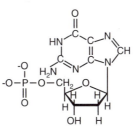

Deoxyguanosine 5'-monophosphate
dGMP

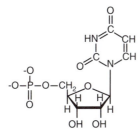

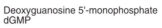

Uridine 5'-monophosphate
UMP

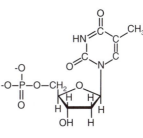

Deoxythymidine 5'-monophosphate
dTMP

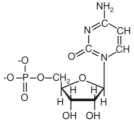

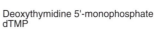

Cytidine 5'-monophosphate
CMP

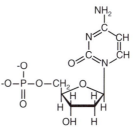

Deoxycytidine 5'-monophosphate
dCMP

**FIGURE 16.9** Common nucleotides.

In nature, eight common nucleotides exist, four found in DNA and four in RNA. In the standard abbreviations for DNA nucleotides, a lowercase d specifies the presence of deoxyribose. RNA nucleotides lack this designation. Nucleosides have names of one word (e.g., deoxyadenosine, cytidine, and uridine). The ending "monophosphate" completes the nucleotide names. Table 16.1 lists correct names for all common nucleotides and nucleosides, and Figure 16.9 shows linkages and structures for all eight nucleotides.

**TABLE 16.1**    Common Nucleosides and Nucleotides

| Nucleosides | Nucleotides |
|---|---|
| Adenosine | AMP = adenosine monophosphate |
| Guanosine | GMP = guanosine monophosphate |
| Cytidine | CMP = cytidine monophosphate |
| Uridine | UMP = uridine monophosphate |
| Deoxyadenosine | dAMP = deoxyadenosine monophosphate |
| Deoxyguanosine | dGMP = deoxyguanosine monophosphate |
| Deoxycytidine | dCMP = deoxycytidine monophosphate |
| Deoxythymidine | dTMP = deoxythymidine monophosphate |

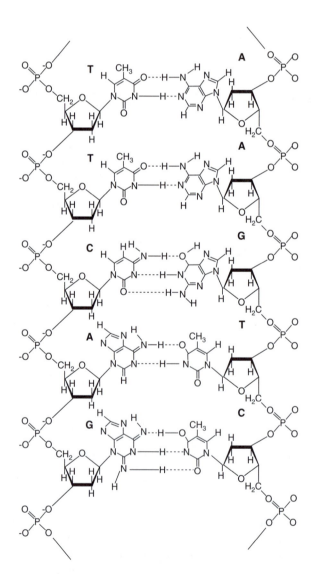

**FIGURE 16.10**    Details of the A–T and C–C pairing.

To form a polymer, nucleotides covalently link in long chains called polynucleotides. The linkage is always between the monosaccharide of one nucleotide and the phosphate group of the next. This linkage results in a repetitive monosaccharide–phosphate backbone in all polynucleotides. DNA consists of two polynucleotides wrapped around each other in a double helix. The total number of nucleotides in each strand often exceeds 2 million. DNA, therefore, has a very high molecular weight. Some bacterial DNAs, for example, have molecular weights in excess of 1 billion. Despite the large size, the structure is very repetitive and lacks the complexity found in proteins. DNA is the hereditary material in all cellular organisms. The hereditary or genetic information is carried in the base sequence found in the interior of the double helix. The bases are attracted to each other by hydrogen bonds, two between A and T, and three between G and C. These hydrogen bonds hold the two strands on the double helix together. Details of the A–T and G–C pairing are shown in Figure 16.10, and the double helix itself is shown in Figure 16.11. RNA has more structural variability than DNA, existing in a variety of three-dimensional configurations, mostly single stranded. The common types of RNA, messenger (mRNA), transfer (tRNA), and ribosomal (rRNA), all participate in the expression of hereditary information in living cells.

**FIGURE 16.11** An illustration of the DNA double helix.

## 16.3 Laboratory Analysis of Biomolecules

### 16.3.1 Introduction

Laboratory methods for the analysis of biomolecules and the analysis of biological samples for biomolecules are at the same time old and new technologies. Many common laboratory techniques for determining protein structure are over 50 years old. As discussed in Section 16.2.3 and the subsections, information on protein structure was crucial to our understanding of evolutionary relationships among organisms. Testing food samples for carbohydrates, triacylglycerols, and proteins is also an older technology. Processed foods are routinely analyzed for these molecules. With recent focus on nutrition, this information is crucial for consumers.

The newer applications involve the field of biotechnology. Proteins produced by genetically altered organisms such as bacteria must be examined to verify that they are identical to the same proteins produced by humans. Also, analysis of DNA from crime scenes is relatively recent. Indeed, DNA analysis and fingerprinting are powerful tools in modern forensics.

### 16.3.2 Electrophoresis

As discussed in Chapter 11, electrophoresis refers to a group of techniques used to separate and study molecules with electrical charges. Based upon these charges, both sign and magnitude, biomolecules such as proteins, peptides, amino acids, nucleic acids, and fragmented nucleic acids migrate in an electrical field. Amino acids, peptides, and proteins all possess ionizable R groups, resulting in net charges when these molecules are in solution. Examples of ionizable R groups include amino ($NH_2$), which ionizes to $NH_3^+$; carboxyl (COOH), which ionizes to $COO^-$; and hydroxyl (OH), which ionizes to $O^-$. Nucleic acid fragments and nucleotides are all anionic due to the phosphate groups. The direction of migration is determined by the sign of the charge. The speed and distance of migration are determined by the magnitude of the charge. For example, molecules with charges of ++++ and + will both move toward the cathode (negative electrode), but the ++++ sample will move faster and therefore farther.

Matrices for separation include paper, various gels, and liquids. We mentioned in Chapter 11 that in paper electrophoresis, a piece of absorbent filter paper soaked in a current-carrying buffer is the matrix for separation. This type of electrophoresis is most commonly applied to amino acids and peptides. The two most common types of gels are polyacrylamide (polyacrylamide gel electrophoresis (PAGE)) and agarose. The essential component in all PAGE gels is acrylamide, a potential neurotoxin. Caution is therefore required when handling acrylamide, especially in raw powder form. Agarose is a polysaccharide that is much safer to handle. PAGE is usually run vertically, while agarose gels are horizontal. Both techniques are performed in special apparatus. PAGE is most commonly used with proteins, while agarose gel electrophoresis is the favored method for nucleic acids and their fragments, but is also effective for proteins. In capillary electrophoresis, separation occurs in a glass capillary tube containing buffer only, as we discussed in Chapter 11.

A variety of buffers is used in electrophoresis. The selected buffer must contain ions to carry the current. Other than current-carrying capacity, the most critical criterion for buffer selection is the stability of the sample to be analyzed. Many proteins are unstable in acidic pHs, so alkaline buffers are frequently employed. Tris-(hydroxymethyl)amino methane (TRIS or THAM), sodium acetate, and ethylenedi-aminetetraacetate (EDTA) are common solutes in buffers, with pHs between 7.9 and 8.9 typical. (Refer to Chapter 5 for a discussion of buffers.) These buffers also work well with nucleic acid fragments. In addition, phosphate buffers, e.g., 10 m$M$ $K_3PO_4$, are often used with nucleic acid fragments (1.0 m$M$ = 0.0010 $M$).

Since by definition all types of electrophoresis involve current flow, a power supply is required. Most applications also require some apparatus for containing the separation matrix. Thus electrophoresis may be considered instrumental analysis. Most apparatus and power supplies are relatively inexpensive. The one exception is capillary electrophoresis, the most recent development in the field. Capillary electrophoresis (CE) is a highly sophisticated and expensive instrumental analysis that, other than charge-based separation, has little in common with other types of electrophoresis. In modern analytical laboratories, CE most frequently replaces or supplements high-performance liquid chromatography (HPLC). CE has two major advantages over HPLC. First, it usually yields higher resolution. Second, unlike HPLC, CE generates virtually no volatile solvents for disposal, since there is no mobile phase pumped through the instrument. Rather, samples are separated as they move through a fixed arc of glass capillary that contains buffer. There is no net flow of buffer. Each end of the capillary tube is dipped into a vial of buffer. Samples start at one end and move toward the other based on magnitude of charge (as shown in Figure 16.12), with higher magnitude components moving faster. An ultraviolet detector located near the end of the capillary monitors the movement, showing separate peaks for each component of the sample. A common CE buffer is 10 m$M$ $K_3PO_4$. Theoretically, any buffer that carries current is suitable.

### 16.3.3  Chromatography

Chapters 11 to 13 introduce the subject of chromatography and also provide specific examples of uses of the technology in analytical chemistry. Chromatographic techniques are also powerful tools for the analysis of biomolecules. Various types of chromatography are useful, especially for analyzing proteins and nucleic acids. In general, methods tend to be complex, due to the complex nature of these molecules. Paper, thin-layer, and high-performance liquid chromatography are frequently employed in the analysis of proteins and nucleic acids in biological materials such as food.

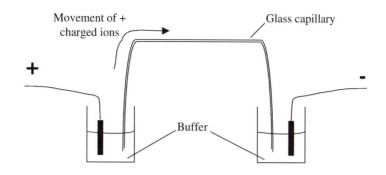

**FIGURE 16.12**   An illustration of capillary electrophoresis.

### 16.3.3.1  Paper Chromatography

This method is very useful for separating amino acids found in food samples. The most effective matrix for separation is an absorbent cellulose-based filter paper. A very effective mobile phase is 70% isopropyl alcohol in water. Although the 20 amino acids are chemically very similar, they may be successfully separated by this method. Amino acids interact with the stationary phase to different extents, thus moving at different speeds. Chemical differences among amino acids that determine migration speed include molecular weight, charge, and polarity.

### 16.3.3.2  Thin-Layer Chromatography

TLC has similar applications to paper chromatography. The stationary phase is a coating, such as silica gel, on a glass or plastic plate. Depending on the TLC plate used, components may be separated based on differences in molecular weight, charge, or polarity (see Chapter 11). TLC with a 70% isopropyl alcohol mobile phase and a silica gel plate is an effective substitute for paper chromatography separation of amino acids. Nucleotides may be separated on a special silica gel plate and a 20% ethanol (in water) mobile phase.

### 16.3.3.3  High-Performance Liquid Chromatography

HPLC is frequently employed in the analysis of amino acids, peptides, proteins, nucleic acids, and nucleotides. HPLC is also often used to analyze for drugs in biological samples (see Workplace Scene 16.2). Due to the complex nature of the molecules to be analyzed, these techniques tend to be more complex than HPLC applications in other areas of analytical chemistry. For example, separation of nucleotides or amino acids is more difficult than testing for caffeine in beverages, even though the same instrument and same general methods would be employed. A variety of columns and mobile phases are regularly employed.

#### 16.3.3.3.1  Columns

Three common chemical parameters used to separate molecules via HPLC are molecular weight, charge, and polarity. Molecular weight separation, using a size exclusion column such as gel permeation (GPC-gel permeation chromatography) requires a minimum molecular weight difference of 10%. The order of sample elution is from largest to smallest. This method may be useful for separating a mixture of proteins, but is generally ineffective for amino acids, peptides, or nucleotides, since they do not differ by the required minimum of 10% in molecular weight. Separation by charge is ion exchange chromatography. A common column, strong anion exchange (SAX), may be employed for amino acids, peptides, and nucleotides. The column packing in this case is studded with positive charges, facilitating the separation of anions based on magnitude of charge (see Chapter 11). Using this column, the most cationic component will elute first and the most anionic will elute last. Although all nucleotides are anionic due to their phosphate groups, they differ in *magnitude* of charge, due to their unique nitrogenous bases. This difference may be exploited to separate them. The HPLC column used must be an anion exchanger (see Figure 16.13), such as SAX. The stronger the charge on the nucleotide, the stronger the interaction with

# WORKPLACE SCENE 16.2

It is common for pharmaceutical companies and their contract laboratories to perform laboratory work on biological samples to track drugs in the body and to profile body fluids and excretions to determine exactly where the drugs end up in the body and at what concentrations. The drugs include over-the-counter medications as well as proprietary drugs that are still being researched. Samples analyzed in these kinds of studies include whole blood, plasma, and serum, as well as urine and feces. Samples may also be parts of an animal's body, since the research is often performed on animals. In any case, the laboratory work begins by extracting the analyte from the sample and concludes with one of any number of methods to arrive at the concentration of the analyte drug. The analyst often uses HPLC, since there can be many other components of the sample that need to be separated before the analyte drug can be measured.

John Howe, an analyst in the laboratory of MDS Pharma Services in Lincoln, Nebraska, checks an auto-sampler tray used with an HPLC instrument in the analysis of biological samples for drugs. A given run may include a large number of sample extracts to run alongside quality controls and standards, hence the need for an auto-sampler.

the stationary phase. Therefore, the nucleotide with the lowest magnitude charge will elute first, and that with the strongest charge will elute last. Differences in charge magnitude, however, are subtle, so elution peaks will be close together.

Amino acids and peptides also may be separated via an SAX column, providing that the correct pH is used. Amino acids and peptides bear a charge at every pH except their isoelectric point, where they are electrically neutral. At a pH above their isoelectric point, the net charge is negative, with positive charges existing at pHs below their isoelectric points. To insure that all amino acids and peptides have negative charges, a pH of 9.0 or higher must be used. A pH of 3.5 or less may also be used. At this low pH, amino acids and peptides have positive charges, and therefore a cation exchanger must be used.

Nucleotides, peptides, and amino acids also differ subtly in their polarities Some are more hydrophobic than others. Thus, separation via reverse phase HPLC is possible. A reverse phase column, such as C18 or C8, has a low- to medium-polarity stationary phase. The more hydrophobic sample components interact to a greater degree with the stationary phase, and therefore elute more slowly than the more hydrophilic components. The sample elution order is from most hydrophilic to most hydrophobic.

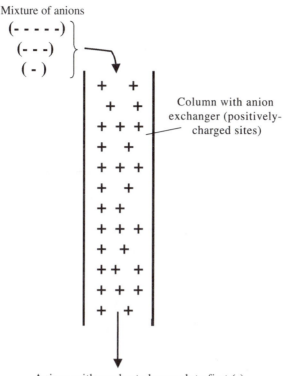

Mixture of anions

(- - - - -)
(- - -)
(- )

Column with anion exchanger (positively-charged sites)

Anions with weakest charge elute first (-)
Anions with strongest charge elute last (-----)

**FIGURE 16.13** The principle of anion exchange chromatography.

Since amino acids and nucleotides are all polar and hydrophilic, they will be eluted quickly by the column. The mobile phase (see below) is also selected on the basis of polarity, with a medium- to high-polarity solvent required. The opposite of reverse phase chromatography is normal phase, where the column packing is medium to high polarity and the mobile phase is nonpolar. This technology is generally not applied to the analysis of polar molecules such as amino acids or nucleotides. Some peptides are more hydrophobic, making this method potentially more useful for peptides than for amino acids or nucleotides.

### 16.3.3.3.2 Mobile Phases

Most HPLC applications involving biomolecules utilize aqueous mobile phases. Critical parameters include both ionic strength and pH. Common solutes include TRIS, sodium phosphate, sodium acetate, and sodium chloride. Slightly alkaline pHs are preferable, for stability reasons. Specific examples of mobile phases include 50 m$M$ TRIS, 25 m$M$ KCl, and 5 m$M$ MgCl$_2$ (pH 7.2) for nucleotides, and 50 m$M$ NaH$_2$PO$_4$ (pH 7.0) and 20 m$M$ TRIS and 0.1 $M$ sodium acetate (pH 7.5) for both peptides and amino acids. All of these mobile phases are suitable for reverse phase or ion exchange applications.

### 16.3.3.3.3 Other Considerations

Most HPLC instruments monitor sample elution via ultraviolet (UV) light absorption, so the technique is most useful for molecules that absorb UV. Pure amino acids generally do not absorb UV; therefore, they normally must be chemically derivatized (structurally altered) before HPLC analysis is possible. The need to derivatize increases the complexity of the methods. Examples of derivatizing agents include o-phthaldehyde, dansyl chloride, and phenylisothiocyanate. Peptides, proteins, amino acids cleaved from polypeptide chains, nucleotides, and nucleic acid fragments all absorb UV, so derivatization is not required for these molecules.

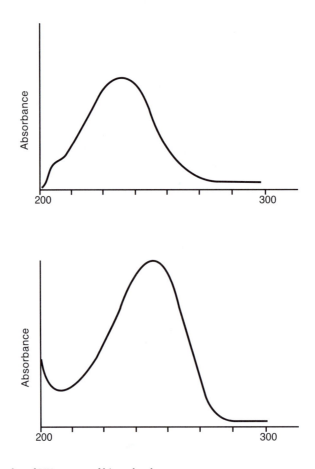

FIGURE 16.14    Examples of UV spectra of biomolecules.

### 16.3.3.4  Spectroscopy

Various spectroscopic methods are also useful in the analysis of biomolecules. The most notable is UV–visible (UV-VIS) spectroscopy, discussed in Chapter 8. Proteins and nucleic acids possess a conjugated pi bond skeleton, and therefore absorb UV light, with absorption maxima between 200 and 300 nm. Nucleic acid solutions are colorless, with no absorption of visible light. Some proteins, called conjugated proteins, have a chromophore that absorbs visible light as well as UV. Examples of such proteins include hemoglobin and cytochrome C, both red in color, with absorption maxima in the 520 to 580 nm region. All proteins have some UV absorption at 280 nm, and all nucleic acids and nucleotides absorb at 260 nm. These wavelengths are therefore useful in monitoring the presence of these molecules in various solutions.

UV or visible absorption spectra of proteins and nucleic acids are very simple, with either one or two absorption peaks. Examples are shown in Figure 16.14.

## Experiments

## Experiment 59: Qualitative Testing of Food Products for Carbohydrates

### Introduction

Benedict's reagent is commonly used to detect the presence of monosaccharides and disaccharides. Iodine detects the presence of starches.

*Remember safety glasses.*

## Procedure

1. Prepare a hot water bath (near the boiling point of water) by placing a 250-mL beaker filled halfway with distilled water on a hot plate.
2. Obtain various food samples. Place small amounts of each into two separate small test tubes — two tubes for each sample.
3. Add 80 drops of Benedict's reagent to one tube of each food sample. Heat each of these tubes in a boiling water bath for 5 min. If the color changes to red, orange, yellow, or green, the presence of monosaccharides or disaccharides is indicated.
4. Add 20 drops iodine solution to one tube of each food sample. Shake well. A color change to blue or black indicates the presence of starches.
5. In your laboratory notebook, list all food samples analyzed and indicate the presence or absence of monosaccharides/disaccharides and starches in each.

## Experiment 60: Fat Extraction and Determination

### Introduction

Since all triacyglycerols (fats) are hydrophobic, they may be extracted from food samples using hydrophobic solvents such as various ethers. Also, since fat content in processed foods is of critical concern to consumers, this technique has considerable real-world significance. Rubber gloves should be worn, and it is very important that there be no open flames in the laboratory.

*Remember safety glasses.*

### Procedure

1. Assemble a boiling water bath in a fume hood using a hot plate and a 250-mL beaker filled halfway with distilled water.
2. Obtain samples of various foods and weigh approximately 1 g of each on a top-loading balance. Record and label each weight (as to the type of food) in your notebook. Also in your notebook, give each an identifying number. Place each sample in a 100-mL round-bottom flask labeled with the identifying number.
3. Place the flask in the water bath without the stopper and add 10.0 mL of concentrated hydrochloric acid to the flask. Heat for 45 min at a gentle boil and shake the flask at 10-min intervals. The HCl–heat combination breaks up the food sample, releasing free fatty acids, monosaccharides, amino acids, etc., thus making extraction easier.
4. After 45 min, remove the flask from the water bath and add 10 mL of distilled water. Then cool the flask by running water over its exterior. Stopper the flask and place it in a freezer for 30 min. This aids in the separation of the fat.
5. After 30 min, remove the flask from the freezer and, in the fume hood, add 25 mL of ethyl ether. Stopper the flask and, continuing to work in the fume hood, shake vigorously for 30 sec, releasing pressure every 10 sec by removing the stopper. Then add 25 mL of petroleum ether and repeat the shaking and venting procedure for another 30 sec.
6. Carefully decant the ether layer (the top layer) into a tared 600-mL beaker. Set the beaker on a hot plate at the lowest setting to evaporate the ether. Then repeat the extraction with both ethers two more times, decanting the ether layer into the beaker and evaporating the ether layer both times. The combination of the two ethers and the repetitive extractions will easily extract the fat from the food sample. All other food components will remain in the bottom, aqueous layer.
7. Leave the beaker on the hot plate until all the ether is evaporated, such that only the fat remains. Weigh the beaker and fat and record the weight. Calculate percentage fat in the sample as follows:

$$\% \text{ fat} = \frac{\text{weight of fat}}{\text{weight of sample}} \times 100$$

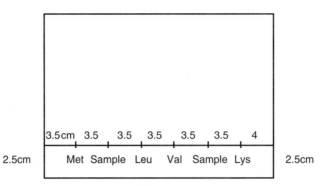

**FIGURE 16.15**  Suggested pattern for paper chromatography in Experiment 61, step 1.

8. The weight of fat is determined by subtracting the weight of the beaker from the weight of the beaker and fat. Record percent fat of all food samples analyzed.

## Experiment 61: Identification of Amino Acids in Food by Paper Chromatography

### Introduction

The purpose of this experiment is to determine whether selected food samples contain certain amino acids. The 20 amino acids are chemically very similar. Therefore, separating them in the laboratory is often difficult. One effective method of separating and studying amino acids is paper chromatography. Refer to Chapter 11 for the details of this technique.

*Remember safety glasses.*

### Procedure

1. Obtain a 25 × 16-cm piece of Whatman #1 chromatography paper or filter paper. Take precautions to keep the paper clean. Only touch the edges with your fingers, and lay it on a clean paper towel on the bench top. Using a pencil, draw a horizontal line across the paper at a distance of 2.5 cm from one edge, as shown in Figure 16.15. Also draw short vertical lines at 3.5-cm intervals across the horizontal lines, as shown. The intersections of these lines are where the sample spots will be applied. Also label each intersection as shown.

2. Prepare 0.01 *M* solutions of leucine, methionine, valine, and lysine using a solvent that is 70% isopropyl alcohol and 30% water. Liquid food samples, such as fruit juice samples, can be used directly. Solid food samples must be soluble in the isopropyl alcohol–water solvent. Prepare solutions of solid samples as concentrated as possible.

3. Using separate capillary tubes for each standard and sample, spot the paper by placing a *small* drop of each from the capillary tube at the appropriate points (where the lines cross). The smaller the spot the better. *Perform this step carefully.* Allow the spots to dry. Sample solutions should be spotted a second time over the original spot to make certain that there is sufficient amino acid in each spot. A single spotting is adequate for the four amino acid solutions.

4. Obtain a paper chromatography developing chamber (similar to that shown in Figure 11.12). Pour 70% isopropyl alcohol (the mobile phase) into the chamber to a depth of 2 cm. Then, place the paper sheet in the chamber as shown in Figure 11.12. Cover the chamber.

5. After 2 to 3 h (the longer the better), remove the paper, quickly mark the solvent front with a pencil, and hang the paper in a fume hood to dry. After drying for 5 to 10 min, spray the developed portion with ninhydrin solution (0.1% ninhydrin in butanol or isopropyl alcohol). Allow to dry for 10 min. Then place it in a drying oven at 100°C for 5 min. Amino acid spots will be violet to blue in color.

6. Compare the spots for the standards with those of the food samples and determine whether any of the standard amino acids are in the samples. Your instructor may ask you to calculate $R_f$ factors as part of the identification process. For a review of $R_f$ factors, see Section 11.8.1.

# Experiment 62: Separation of Hemoglobin and Cytochrome C by Horizontal Agarose Gel Electrophoresis

## Introduction

In this exercise, hemoglobin and cytochrome C, both proteins, are separated in an agarose gel. Agarose is a porous polysaccharide. The porosity of agarose allows a sample to move through it based on both sign and magnitude of charge. For example, samples with charges of ++++ and + will both move toward the cathode (negative electrode), but the ++++ sample will move faster and therefore farther in a given time period. Every protein is neutral or uncharged at a characteristic pH, the isoelectric point or isoelectric pH. At a pH above the isoelectric point, the protein will have a net negative charge, and at a pH below the isoelectric point a net positive charge.

This experiment requires a horizontal electrophoresis apparatus, power supply, and a gel-pouring tray with a comb to form sample wells. The FB 1001 apparatus available from Fisher Biotech and an apparatus available from EdVotek both are appropriate. The gel-pouring tray with combs is also available from EdVotek. Directions for preparing the tray accompany the unit. The Fisher apparatus includes a cooling unit that allows running at a constant temperature. In the EdVotek apparatus, the gel is under buffer. Since samples are loaded after the gel is immersed in buffer, sucrose must be added to increase sample density.

*Remember safety glasses.*

## Procedure

1. Prepare 100 mL (500 mL if the gel is to be run under the buffer) of an electrophoresis buffer that is 20 m$M$ tris-(hydroxymethyl)amino methane (TRIS or THAM), 6 m$M$ sodium acetate, and 1 m$M$ disodium EDTA. Adjust the pH of this solution to 7.9 using concentrated HCl. Also prepare small volumes of solutions of hemoglobin and cytochrome C in the buffer (the concentration is not important) and also a mixture solution of these two solutes. Add a quantity of sucrose to each.

2. Weigh 0.48 g of DNA grade agarose into a 150-mL beaker and add 60 mL of the buffer prepared in step 1. Heat this solution to boiling. While this solution is heating, prepare a gel-pouring tray (next step).

3. While the agarose solution is heating, prepare a gel-pouring tray by taping both ends. As the agarose solution on the hot plate begins to boil, it will become clear. When it begins to boil, run a bead along each end of the tray to seal it, using an eyedropper or pipet. When the bead solidifies, carefully pour the rest of the solution into the tray, insert the comb into the center slot, and allow gel to solidify (10 to 15 min).

4. Remove the comb carefully, and then remove the tape or sealing apparatus from both ends of the tray. Carefully let the gel slide out of the tray onto the electrophoresis apparatus.

5. Using a micropipettor, spot 50-$\mu$L samples of the hemoglobin and cytochrome solutions into the preformed slots as follows: slots 1 and 4 = hemoglobin, slots 2 and 5 = cytochrome C, and slots 3 and 6 = mixture.

6. Run the experiment for 30 min at 125 V. The separated bands should be clearly visible in the gel.

7. In your notebook, make a drawing of the observed banding pattern, labeling both the hemoglobin and cytchrome C. Explain, in relevant detail, what you can determine about the isoelectric points of these proteins and what might happen in the gel if the buffer pH were changed from 7.9 to 3.9.

# Experiment 63: HPLC Separation of Nucleotides

*Remember safety glasses and wear gloves.*

## Procedure

1. Prepare 500 mL of a mobile phase that is 25 m$M$ KCl, 5 m$M$ MgCl$_2$, and 50 m$M$ tris-(hydroxymethyl)amino methane (TRIS or THAM). Adjust the pH to 7.2 using concentrated HCl. Filter and degas this mobile phase using a vacuum filtration apparatus equipped with 0.45-$\mu$L filters.

2. The first part of this experiment uses an HPLC instrument with an anion exchange column installed. Flush the flow path with the mobile phase at a flow rate of 1.25 mL/min for 10 min.

3. Prepare separate solutions of four nucleotides (concentration not important). Either DNA or RNA nucleotides may be used. Filter approximately 2-mL quantities of each nucleotide solution into small vials. Also prepare and filter a mixture of all four by combining small quantities of each in one vial.

4. Inject 20 $\mu$L of each sample and observe the retention times. Then inject 20 $\mu$L of the mixture. Record all experiment parameters, e.g., flow rate, attenuation, pressure (if pressure gauge is installed), etc. Be sure to print hard copies of all chromatograms. When finished, flush the flow path for 10 min with filtered and degassed methanol.

5. Install a reverse phase column and repeat. *Do not change the parameters recorded in step 4.* When finished, again flush the flow path with filtered and degassed methanol for 10 min.

6. On the mixture chromatograms, identify each peak (as well as possible) based on retention time data from the individual chromatograms. Note these identities on the mixture chromatograms. The chromatograms should be placed in the data section of your notebook. Also in your notebook indicate which column provides the best separation and explain this observation. Also explain how the resolution might be improved.

## Experiment 64: Ultraviolet Spectra of Nucleotides

### Introduction

All nucleotides absorb electromagnetic radiation in the ultraviolet region, specifically between 200 and 300 nm. Spectra of nucleotides obtained from a spectrophotometer may be useful in identifying nucleotides.

### Procedure

1. Prepare a 10 m$M$ solution of $K_3PO_4$ buffer. Use this solution as a blank and for diluting solutions of the standards. Prepare standard solutions of selected nucleotides and nucleoside triphosphates, 0.10 m$M$ in 10 m$M$ $K_3PO_4$ buffer. Obtain unknown nucleotide or nucleoside triphosphate solutions from your instructor.

2. Zero the spectrophotometer at 300 nm, using the blank. Obtain molecular absorption spectra of all nucleotide and nucleoside solutions in the range of 200 to 300 nm.

3. Be sure to note on each spectrum the name of the nucleoside or nucleotide and the wavelength of maximun absorption.

4. Identify your unknown by comparing with the known spectra. Comment on whether it is possible to distinguish between a nucleotide and a nucleoside triphosphate or between two nucleotides.

## Experiment 65: Restriction Endonuclease Cleavage of DNA

### Introduction

This activity is intended to be performed in conjunction with Experiment 66. Restriction endonucleases, or restriction enzymes, cleave DNA at specific base sequences, fragmenting the DNA into smaller pieces. The two strands of a DNA double helix are cleaved at different places, resulting in uneven fragments called sticky ends. Cleavage of DNA by restriction enzymes is a required first step in various types of DNA analysis, including DNA fingerprinting and recombinant DNA technology.

### Procedure

1. Obtain samples of selected restriction enzymes and reaction buffer (supplied with enzymes). Also obtain a DNA sample, either a viral (lambda) or bacterial plasmid.

2. Using a micropipettor, add 10 $\mu$L of the reaction buffer to a small reaction vial or microcentrifuge tube. Add 15 $\mu$L of DNA. Then, with a fresh micropipet tip, add 15 $\mu$L of one enzyme.

3. Repeat step 2 using separate vials for each additional enzyme.

4. Mix all tubes by microcentrifuging for 5 min at 15,000 rpm, or by tapping tubes on any flat surface at least 50 times. Then incubate in a water bath at 37°C for 50 min. Carry these tubes over to Experiment 66.

5. In your notebook, invent your own restriction enzyme. Show the DNA base sequence where it cleaves. Then construct a double stranded section of DNA with at least 25 base pairs and mark the points of cleavage by your enzyme. Finally, show all fragments formed.

## Experiment 66: Separation of Restriction Enzyme Digestion Fragments via Horizontal Agarose Gel Electrophoresis

### Introduction

Fragmented DNA from restriction enzyme digestion is very useful in modern biotechnology. One increasingly common technique that requires DNA fragments is DNA fingerprinting. In this technique, the fragments are separated by movement through an electrical field in a gel matrix composed of agarose, a polysaccharide. Due to the phosphate groups in DNA, all fragments bear a negative charge and will migrate toward the positive pole (anode), but based on their different sizes, they will migrate at different speeds. This creates a distinct banding pattern of the fragments, visible clearly after staining the gels with methylene blue. Since each restriction enzyme forms different size fragments, one would expect different electrophoretic banding patterns from the same DNA digested by two different enzymes. In modern forensic science, DNA from a crime scene is cleaved by several different enzymes to obtain several different banding patterns. These fingerprints are then compared with those obtained from suspects' DNA, which was cleaved with the same enzymes. If the fingerprints match perfectly, the guilty individual is identified. The degree of certainty of such information exceeds that of conventional fingerprints.

The discovery of restriction enzymes was a necessary first step in modern recombinant DNA technology. In order to recombine DNA fragments from two different organisms, short, single-strand sequences must be available. The sticky ends formed by restriction enzymes provide the necessary single-strand sequences. When DNAs from two different organisms, such as the bacterium *Escherichia coli* and human, are cleaved by the same enzyme, DNA base sequences complementary to each other are always formed, allowing for recombination when the DNAs are mixed. Thus, for every TTGCAG single-strand sequence in *E. coli* sticky ends, a complementary AACGTC sticky end will be formed from human DNA. This means that a human gene, such as that coding for insulin, may be routinely incorporated into a bacterial cell. The recombinant cell will now synthesize human insulin. Other human products produced in this manner include human growth hormones and interferons, antitumor agents.

This experiment utilizes the restriction enzyme digests from Experiment 65.

*Remember safety glasses and wear gloves.*

### Procedure

1. Prepare a buffer solution that is 20 m$M$ tris-(hydroxymethyl)amino methane (TRIS or THAM), 6 m$M$ sodium acetate, and 1 m$M$ disodium EDTA. Adjust to pH = 7.9.
2. Prepare 0.8% agarose gel by adding 0.48 g of agarose to 60 mL of the buffer prepared in step 1. Heat to boiling.
3. While the agarose solution is heating, prepare a gel-pouring tray by taping both ends. When the agarose solution begins to boil, run a bead along each end of the tray to seal it, using an eyedropper or pipet. When the bead solidifies, carefully pour the rest of the solution into the tray and insert the comb into the slot closest to one end. When the gel is solid (10 to 15 min), carefully remove the comb and then the tape. Slide the gel out of the tray and onto the electrophoresis apparatus.
4. Using a micropipettor, add 40 $\mu$L of each restriction enzyme digest from Experiment 65 to separate wells.
5. Add buffer to electrode chambers of the electrophoresis apparatus. Set up the electrodes so that the positive pole is opposite the loaded wells. Run the electrophoresis at 60 V for 40 min.
6. Remove the gel, carefully place it in a tray or large watch glass, and cover it with methylene blue. *Wear gloves when handling stain.* Stain for at least 30 min. Overnight is fine.
7. *Wearing gloves,* rinse the gels in tap water until all excess methylene blue has been removed. Blue DNA bands should be clearly visible.

8. In your notebook, analyze the results as follows:
   (a) Examine and draw DNA banding patterns for all samples. Show all differences found with different enzymes.
   (b) What do you think would happen if the agarose concentration was increased? Decreased?
   (c) Why are all DNA fragments anionic?
   (d) What would happen if the buffer were replaced with distilled water?
   (e) Explain how this technique might be used to determine guilt or innocence of a crime. Explain how this technique might be used to establish paternity.

# Questions and Problems

1. Name the monomer units of: (a) cellulose, (b) protein, (c) DNA, (d) RNA, (e) hemoglobin, (f) starch, and (g) collagen.
2. Name two polymers with helical structures.
3. What are the major biological functions of carbohydrates?
4. What disaccharide is formed from the linkage of two molecules of glucose?
5. Name a type of lipid composed entirely of glycerol and fatty acids.
6. Draw examples of both saturated and unsaturated fatty acids.
7. What type of lipid lacks fatty acids?
8. Why are triacylglycerols difficult to transport through human blood?
9. Show how peptide bonds form.
10. Describe the four levels of protein structure.
11. How would hydrophobic R groups affect the three-dimensional structure of a protein?
12. Protein X is abnormal and lacks biological function. The problem is in the quaternary structure. Examine the following statements, determine whether each is true of false, and in either case explain why:
    (a) X must contain at least two polypeptide chains.
    (b) R groups are responsible for making X nonfunctional.
    (c) A primary structural alteration would account for X being nonfunctional.
    (d) X probably will not form an alpha helix.
13. Name the three structural components of a nucleotide.
14. Why is it essential to prevent damage to DNA bases?
15. List the three major types of RNA.
16. What chemical interactions hold together the two strands of a DNA double helix?
17. Explain the structural and functional differences between DNA and RNA.
18. Why is agarose gel electrophoresis generally preferable to polyacrylamide gel electrophoresis?
19. What are some requirements for an effective electrophoresis buffer?
20. List three examples of instrumental analysis used to study biomolecules.
21. What charge would a peptide with an isoelectric point of 7.2 have at pH 8.9?
22. Name two types of HPLC columns that might be used to separate nucleotides.
23. Why are both nucleotides and derivatized amino acids difficult to separate via reverse phase chromatography?
24. Explain the principle of ion exchange chromatography.
25. Of alkane chains and conjugated pi bond structures, which chemical structure do molecules that absorb UV light have?

# Appendix 1

## Good Laboratory Practices

In Chapter 1, we mentioned that various laboratories are under federal regulation in relation to the activities and the quality of work performed in the laboratory. These regulations have come to be known as good laboratory practices (GLPs). There are two sets of GLP regulations, one for the Food and Drug Administration (FDA) and one for the Environmental Protection Agency (EPA). They differ only in statements relating to these agencies' objectives and purposes. The FDA regulations may be accessed at http://www.access.gpo.gov/nara/cfr/waisidx_01/21cfr58_01.html. The EPA regulations may be accessed at http://www.access.gpo.gov/nara/cfr/waisidx_99/40cfr160_99.html.

The major subparts of the GLP regulations are listed below:

Subpart A: General Provisions
Subpart B: Organization and Personnel
Subpart C: Facilities
Subpart D: Equipment
Subpart E: Testing Facilities Operations
Subpart F: Test, Control, and Reference Substances
Subpart G: Protocols for and Conduct of a Study
Subparts H and I: Reserved
Subpart J: Records and Reports
Subpart K (FDA only): Disqualification of Testing Facilities

We now take a look at each subpart individually.

## A1.1   General Provisions

The first subpart in each set is titled "General Provisions." In this subpart, in separate numbered paragraphs, the scope of the regulations is laid out, a number of definitions are listed, and the applicability of the regulations to studies performed under grants and contracts is covered. Also presented is a paragraph that indicates that the inspection of a testing facility is permitted. Besides these, the EPA GLPs include a statement of compliance or noncompliance, as well as a statement on the effects of noncompliance.

The scope paragraph of the EPA GLPs mentions that the research and marketing of pesticide products are regulated. In addition, all data relating to certain sections of the Federal Insecticide, Fungicide, and Rodenticide Act (FIFRA) are regulated. The scope paragraph of the FDA GLPs mentions that the research and marketing of products regulated by the FDA are regulated. Specific products are listed. In addition, under FDA GLPs, all data relating to certain sections of the Federal Food, Drug and Cosmetic Act (FFDCA) and the Public Health Service Act (PHSA) are regulated.

The definitions are for those words and phrases that the reader encounters in the regulations. Examples are given below.

**Quality assurance unit**: Any person or organizational element, except the study director, designated by testing facility management to perform the duties relating to quality assurance of the (non-clinical laboratory) studies.

**Raw data**: Any laboratory worksheets, records, memoranda, notes, or exact copies thereof, that are the result of original observations and activities of a (nonclinical laboratory) study and are necessary for the reconstruction and evaluation of the report of that study.

**Reference substance** (EPA only): Any chemical substance or mixture, or analytical standard, or material other than a test substance, feed, or water, that is administered to or used in analyzing the test system in the course of a study for the purposes of establishing a basis for comparison with the test substance for known chemical or biological measurements.

**Person**: An individual, partnership, corporation, association, scientific or academic establishment, government agency, or organizational unit thereof, and any other legal entity.

**Sponsor**: 1) A person who initiates and supports, by provision of financial or other resources, a (nonclinical laboratory) study; 2) a person who submits a (nonclinical laboratory) study to the EPA (FDA) in support of an application for a research or marketing permit; or 3) a testing facility, if it both initiates and actually conducts the study.

**Study** (EPA): An experiment in which a test substance is studied in a test system under laboratory conditions or in the environment to determine or help predict its effects.

**Nonclinical laboratory study** (FDA): An experiment in which test articles are studied prospectively in test systems under laboratory conditions to determine their safety.

**Study director**: The individual responsible for the overall conduct of a (nonclinical laboratory) study.

**Test substance** (EPA): A substance or mixture administered or added to a test system in a study, in which the substance or mixture: 1) is the subject, or contemplated subject, of an application for a research or marketing permit, or 2) is an ingredient or product of a substance as defined above.

**Test article** (FDA): Any food additive, color additive, drug, biological product, electronic product, medical device for human use, or any other article subject to regulation under the FFDCA or PHSA.

**Test system** (EPA): Any animal, plant, microorganism, chemical or physical matrix, including but not limited to soil or water, or subparts thereof, to which the test, control, or reference substance is administered or added for the study.

**Test system** (FDA): Any animal, plant, microorganism, or subparts thereof, to which the test or control article is administered or added for study.

**Testing facility** (both EPA and FDA): A person who actually conducts a (nonclinical laboratory) study, i.e., who actually uses the test substance (article) in a test system.

In the paragraph entitled "Applicability to Studies Performed Under Grants and Contracts," the regulations state that whenever a sponsor utilizes the services of a consulting laboratory, contractor, or grantee to perform an analysis or other service, it shall notify this laboratory, contractor, or grantee that the service must be conducted in compliance with GLPs. Hence, essentially all phases of the work are covered by the regulations whether performed by an outside organization or the sponsor's own laboratory.

The paragraph that discusses the "Inspection of a Testing Facility" is of significant importance. It is found in both the EPA and FDA GLPs and, in essence, allows government representatives to enter the facilities and inspect and copy all records within the scope of the work at reasonable times and in a reasonable manner. A key statement here is that the EPA (or FDA) will not consider a study in support of an application for a permit if the testing facility refuses to permit inspection. Thus a facility must permit inspection. This inspection, or audit as it is often called, is therefore taken very seriously.

The EPA regulations include a statement of compliance or noncompliance (§160.12) and a statement indicating the effects of noncompliance (§160.17), while the FDA regulations address these issues in their Subpart K. The EPA statements say that any person who submits an application for a research or marketing permit shall submit a statement signed by the applicant, the sponsor, and the study director that the study was either conducted in accordance with GLPs, conducted in part in accordance with GLPs (with those parts not so conducted described), or conducted such that it was not known whether it was in accordance with GLPs. As to noncompliance consequences, the EPA regulations say that the EPA may refuse to consider reliable any data from a study that was not conducted in accordance with GLPs and that severe penalties may result from submission of a compliance statement that is false.

The FDA statements in Subpart K are much more explicit and detailed. While the EPA statements are described in just the two paragraphs mentioned above, the FDA statements are written in a total of eight paragraphs.

## A1.2   Organization and Personnel

Subparts with this title appear under both the FDA and EPA regulations and are essentially identical. Included are requirements for personnel (FDA §58.29 and EPA §160.29), requirements for testing facility management (FDA §58.31 and EPA §160.31), requirements for the study director (FDA §58.33 and EPA §160.33), and requirements for the quality assurance unit (FDA §58.35 and EPA §160.35).

The paragraphs on personnel state that each individual engaged in the work must have the proper education, training, and experience; that each testing facility must maintain a file summarizing the education, training, and experience of the workers; and that there shall be a sufficient number of workers to complete the study in a timely manner. In addition, these paragraphs state that the workers follow sanitation and health procedures; that the workers must wear appropriate clothing and that this clothing be changed as necessary to prevent contamination; and that any worker found to be ill be appropriately excluded from contact with any operation or function that may adversely affect the study.

The paragraphs on testing facility management specifically state the duties of the management. These include: 1) designating the study director before the study is initiated, 2) replacing the study director if necessary, 3) assuring that there is a quality assurance unit, 4) assuring that all test substances (articles) involved in the study be tested for identity, strength, purity, stability, and uniformity, as applicable, 5) assuring that personnel, resources, facilities, equipment, materials, and methodologies are available as scheduled, 6) assuring that personnel clearly understand their functions, and 7) assuring that any reported deviations from regulations are communicated to the study director and corrective actions are taken and documented.

The paragraphs outlining the requirements of the study director, the individual responsible for the overall conduct of the study, state that he or she be a scientist or other professional of appropriate education, training, and experience. Additionally, this individual is responsible for the interpretation, analysis, documentation, and reporting of results, and represents the single point of study control. He or she assures that: 1) the protocol, including any change, is approved and followed, 2) all experimental data are accurately recorded and verified, 3) unforeseen circumstances affecting the quality and integrity of the study are noted and corrective action taken and documented, 4) test systems are identified in the protocol, 5) all applicable GLPs are followed, and 6) all raw data, documentation, protocols, specimens, and final reports are transferred to the archives during or at the close of the study.

Finally, the paragraphs outlining the requirements of the quality assurance unit state that such a unit is required; the duties of this unit; that the specific characteristics of the unit, including responsibilities, applicable procedures, and records, must be maintained in writing; and that the EPA or FDA shall have access to the documents. The quality assurance unit is responsible for monitoring each study to assure management that the facilities, equipment, personnel, methods, practices, records, and controls are in conformance with the regulations. The quality assurance unit must be completely separate from and independent of the personnel engaged in the direction and conduct of the study.

## A1.3   Facilities

The facilities for the study are described in Subpart C for both the EPA and the FDA regulations. There are six paragraphs under Subpart C in each case, and the titles of these paragraphs are very similar. The key points here are: 1) facilities must be of suitable size and construction and must be designed so that they are separate from other activities so as to prohibit adverse effects on the study from the other activities, 2) proper separation of test systems, including waste disposal systems, must be maintained, 3) storage areas for feed, nutrients, soils, bedding, supplies, and equipment shall be provided, and the areas for storing feed, nutrients, soils, and bedding shall be separated from the test systems, and, in addition, these facilities shall be provided as required by the written protocol, 4) facilities for handling test articles, control articles, and reference substances, including receipt and storage, mixing, and storage of prepared mixtures, shall be organized so as to prevent mix-ups and to preserve the identity, strength, purity, and stability of the articles and mixtures, 5) special laboratory space shall be provided as needed for all procedures, and 6) there shall be archives, limited to access by authorized personnel, for storage of study records and specimens.

## A1.4   Equipment

Subpart D presents the regulations for the equipment used in a study. There are two paragraphs in both sets, the FDA and the EPA GLPs, one for equipment design and one for maintenance and calibration. The texts of both sets are virtually identical, stating that all equipment must perform as required by the protocol and that this equipment must be maintained and calibrated. Formal standard operating procedures (SOPs) must be on file for this and must set forth in sufficient detail the methods, materials, and schedules to be used in the inspection, cleaning, maintenance, testing, calibration, and standardization of equipment and also name the person responsible. In addition, written records shall be maintained for this activity, including records of any malfunctions, how these were discovered, and what corrective action was taken. We mentioned this subpart previously in our discussion of calibration of equipment.

## A1.5   Testing Facility Operations

Subpart E of both the FDA and EPA GLP regulations addresses the operations aspects of the work. This is the subpart that specifies the use and design of standard operating procedures. This is also the subpart that specifies the labeling of reagents and solutions. Some important points in the paragraph describing standard operating procedures are: 1) SOPs that satisfy management of the quality and integrity of the data are required, 2) the study director may authorize deviations from an SOP, 3) significant changes must be authorized by management, 4) SOPs must be immediately available to personnel, and 5) a historical file of SOPs must be maintained. An important point regarding labeling is that *every* bottle in the laboratory must have a label with the required information. The required information, as delineated in this subpart, includes identity, titer or concentration, storage requirements, and expiration date.

There is also a paragraph in Subpart E providing considerable detail regarding the care of test systems that are animals or plants.

## A1.6   Test, Control, and Reference Substances

Test substances (or articles), control substances (or articles), and reference substances are covered by Subpart F in both the FDA and EPA regulations. Basically this subpart covers all substances under investigation and all known substances used in the investigation in terms of their characterization, handling, and mixing.

First, in terms of their characterization, the regulations state that the identity, strength, and other characteristics shall be determined and documented before use. The regulations also state that their stability shall be determined before the experimental start date or concurrently with the study. In addition, test substance storage containers must be labeled and must be kept for the duration of the study, and for studies lasting more that 4 weeks, reserve samples must be retained according to paragraphs 58.195 (FDA) and 160.195 (EPA). Finally, the EPA (but not the FDA) regulations state that the stability of the substances under storage conditions at the test site shall be known for all studies.

Second, in terms of handling, the regulations state that procedures must be established to ensure that: 1) there is proper storage, 2) contamination or deterioration is avoided during handling, 3) proper identification is maintained throughout the study, and 4) the receipt and distribution is documented for each batch. The concept of the chain-of-custody documentation is thus covered in the regulations.

Third, for each substance (or article) that is mixed with a carrier (solvent or other medium): 1) the uniformity or concentration shall be determined, 2) the solubility shall be determined (EPA only), and 3) the stability in the mixture shall be determined. Expiration dates shall be clearly shown and (EPA only) the device used to formulate the mixture shall not interfere with the integrity of the test.

## A1.7 Protocols for and Conduct of a (Nonclinical Laboratory) Study

The term protocol has a specific meaning in GLPs. It is defined in Subpart G of both the FDA and EPA GLPs as an official written document that clearly indicates the objectives and all methods for the conduct of the study. In the case of the EPA GLPs, it contains a set of 15 specific items. In the case of the FDA GLPs, it contains a set of 12 specific items. Like SOPs, an approved protocol can be changed or revised, but the changes and revisions must be documented, signed by the study director, dated, and maintained with the original document.

A specific path for conducting a study is also outlined in Subpart G. Besides stating that the study shall be conducted and test systems monitored in accordance with the protocol, and that specimens shall be properly identified and results made available to a pathologist, an important statement regarding the data generated is presented.

The EPA GLPs include a paragraph under Subpart G (§160.135) that deals with physical and chemical characterization studies. This paragraph states that all provisions of GLP standards apply to certain specific physical and chemical characterization studies of test, control, and reference substances. These studies are listed. It also states that certain specified paragraphs do not apply to studies other than those listed.

## A1.8 Records and Reports

Subpart J in both the FDA and EPA GLPs deals with the records and reports generated by a study. Specifically, this subpart in both the EPA and FDA GLPs deals with the reporting of the study results, the storage and retrieval of records and data, and the retention of records. Under "Report of the Study Results" (EPA §160.185 and FDA §58.185), the items to be included in the final report are listed. The list includes 13 enumerated items: names, dates, objectives and procedures, statistical methods, identities of and data concerning substances used, laboratory methods, test systems, dosages, data integrity issues, specific data handling procedures, reports of scientists involved, data storage locations, and the quality assurance unit statement. Corrections or additions to the final report are handled via amendments.

All raw data, documentation, records, protocols, specimens, final reports, and correspondence relating to data interpretation must be retained and archived according to the "Storage and Retrieval of Records and Data" paragraphs (EPA §160.190 and FDA §58.190) found in this subpart, and these records must be available for expedient retrieval. The archived records shall be carefully protected and indexed.

Additional statements regarding records retention, which do not supersede the previous statements, are presented in the "Retention of Records" paragraph (EPA §160.195 and FDA §58.195). The period of time specific records must be retained is indicated here.

## A1.9 Disqualification of Testing Facilities

Subpart K of the FDA regulations discusses the grounds for the disqualification of testing facilities, the procedures and actions associated with disqualification, and the procedure for the reinstatement of disqualified facilities.

# Appendix 2

## Significant Figure Rules

### Rules for Determining the Number of Significant Figures in a Given Number

1. Any nonzero digit is significant.

   Example:     916.3     four significant figures

2. Any zero located between two significant figures is significant.

   Example:     1208.4     five significant figures

3. Any zero to the left of nonzero digits is not significant unless it is also covered by rule 2.

   Example:     0.00345     three significant figures

4. Any zero to the right of nonzero digits and also to the right of a decimal point is significant.

   Example:     34.10     four significant figures

5. Any zero to the right of nonzero digits and to the left of a decimal point and not covered by rule 2 may or may not be significant, depending on whether the zero is a placeholder or actually part of the measurement. Such a number should be expressed in scientific notation to avoid any confusion.

   Example:     430     don't know

   ($4.3 \times 10^2$ or $4.30 \times 10^2$ would be better ways to express this number, depending on whether there are two or three significant figures in the number.)

### Rules for Determining Significant Figures in the Answer to a Calculation

1. The answer to a multiplication or division problem has the same total number of significant figures as the number with the least significant figures used in the calculation. Rounding to decrease the count of digits, or the addition of zeros to increase the count of digits, may be necessary.

   Example 1:     $4.3 \times 0.882 = 3.7926$ (calculator answer)

   $= 3.8$ (answer with correct number of significant figures)

Example 2:     $\dfrac{0.900}{0.2250}$ = 4 (calculator answer)

   = 4.00 (answer with correct number of significant figures)

2. The correct answer to an addition or subtraction problem has the same number of digits to the right of the decimal point as the number with the least such digits that is used in the calculation. Once again, rounding to decrease the count of digits, or addition of zeros to increase the count of digits, may be necessary.

Example 1:     24.992 + 3.2 = 28.192 (calculator answer)

   = 28.2 (answer with correct number of significant figures)

Example 2:     772.2490 – 0.049 = 772.2 (calculator answer)

   = 772.200 (answer with correct number of significant figures)

3. When several calculation steps are required, no rounding is done until the final answer is determined.

Example:     $\dfrac{3.026 \times 4.7}{7.23}$ = 1.9363762 (calculator answer with premature rounding)

   = 1.9 (incorrect answer due to premature rounding)

   = 1.9671093 (calculator answer correctly determined)

   = 2.0 (correct answer)

4. When both rules 1 and 2 apply in the same calculation, follow rules 1 and 2 in the order they are needed while also keeping rule 3 in mind.

Example:     (3.22 – 3.034) × 5.61 = 1.0659 (calculator answer with premature rounding)

   = 1.1 (incorrect answer due to premature rounding)

   = 1.04346 (calculator answer correctly determined)

   = 1.0 (correct answer)

5. Conversion factors that are exact numbers have an infinite number of significant figures.

Example:     There are exactly 3 ft/yd. How many feet are there in 2.7 yd?

   2.7 × 3 = 8.1 (two significant figures, not one, in correct answer)

6. In cases in which the logarithm of a number needs to be determined, such as in converting $[H^+]$ to pH or transmittance to absorbance, the number of digits in the mantissa of the logarithm (the series of digits to the right of the decimal point) must equal the number of significant figures in the original number.

Example:     $[H^+] = 4.9 \times 10^{-6}\ M$

   $pH = -\log[H^+] = 5.31$

# Appendix 3

## Stoichiometric Basis for Gravimetric Factors

In a stoichiometry calculation, the weight of one substance involved in a chemical reaction (reactant or product) is converted to the weight of another substance (reactant or product) appearing in the same reaction. The balanced equation is the basis for the calculation, and the formula weights of the reactant and product involved are needed. In the following general example,

$$aA + bB \rightarrow cC + dD \tag{A3.1}$$

if the weight of D is known and the weight of A is needed, this weight of A can be calculated using a stoichiometry calculation of the form

$$\text{weight of A} = \text{weight of D} \times \frac{1}{FW_D} \times \frac{a \text{ moles of A}}{d \text{ moles of D}} \times FW_A \tag{A3.2}$$

in which FW represents formula weight.

Dimensional analysis shows that the units in this equation cancel appropriately:

$$\cancel{\text{grams of D}} \times \frac{\cancel{\text{moles of D}}}{\cancel{\text{grams of D}}} \times \frac{\cancel{\text{moles of A}}}{\cancel{\text{moles of D}}} \times \frac{\text{grams of A}}{\cancel{\text{moles of A}}} = \text{grams of A} \tag{A3.3}$$

A stoichiometry calculation is thus essentially a three-step procedure in which: 1) the weight of D is divided by its formula weight to get moles of D, 2) the moles of D are converted to the moles of A by multiplying by the mole ratio a/d, as found in the chemical equation, and 3) the moles of A are converted to grams of A by multiplying by the formula weight of A.

The calculation process just described can be simplified by observing that the three steps can be combined as follows:

$$\text{weight of A} = \text{weight of D} \times \frac{FW_A}{FW_D} \times \text{mole ratio } \frac{a}{d} \tag{A3.4}$$

Consider the following portion of this equation:

$$\frac{FW_A}{FW_D} \times \text{mole ratio } \frac{a}{d} \tag{A3.5}$$

The same mole ratio, a/d, can also be found by simply balancing the common element in the formulas of A and D. Thus, the ratio, a/d, is the same as the ratio $Q_S/Q_K$ seen in the equation for a gravimetric factor derived in Section 3.6.3 (Equation (3.12)) and is used to convert the weight of one substance to the weight of another, just as we described in Chapter 3 was the purpose of the gravimetric factor.

Thus the concept of a gravimetric factor is based on stoichiometry.

# Appendix 4

## Solution and Titrimetric Analysis Calculation Formulas

Refer to Chapters 4 and 5 for the meaning of any symbols.

## Calculation of the Molarity of a Solution from Solution Preparation Data

$$\text{molarity} = \frac{\text{moles of solute}}{\text{liters of solution}} \tag{A4.1}$$

$$\text{molarity} = \frac{\text{grams/FW}}{\text{liters of solution}} \tag{A4.2}$$

$$C_A = \frac{C_B \times V_B}{V_A} \tag{A4.3}$$

## Calculation of the Normality of a Solution from Solution Preparation Data

$$\text{normality} = \frac{\text{equivalents of solute}}{\text{liters of solution}} \tag{A4.4}$$

$$\text{normality} = \text{molarity} \times \text{equivalents per mole} \tag{A4.5}$$

$$\text{normality} = \frac{\text{grams/EW}}{\text{liters of solution}} \tag{A4.6}$$

$$C_A = \frac{C_B \times V_B}{V_A} \tag{A4.7}$$

# Calculation of Parts Per Million of a Solution from Solution Preparation Data

$$\text{ppm} = \frac{\text{milligrams dissolved}}{\text{liters of solution}} \tag{A4.8}$$

## Preparation of Molar Solutions

If solute is a pure solid (or liquid) that will be weighed,

$$\text{grams to weigh} = L_D \times M_D \times FW_{SOL} \tag{A4.9}$$

If solute is already dissolved, but the solution is to be diluted,

$$V_B = \frac{C_A \times V_A}{C_B} \tag{A4.10}$$

## Preparaton of Normal Solutions

If solute is a pure solid (or liquid) that will be weighed,

$$\text{grams to weigh} = L_D \times N_D \times EW_{SOL} \tag{A4.11}$$

If solute is already dissolved, but the solution is to be diluted,

$$V_B = \frac{C_A \times V_A}{C_B} \tag{A4.12}$$

## Preparation of Parts Per Million Solutions

If the substance the parts per million is expressed as is the same substance as that to be weighed,

$$\text{ppm}_{desired} \times L_{desired} = \text{milligrams to be weighed} \tag{A4.13}$$

If the substance the parts per million is expressed as is not the same substance as that to be weighed,

$$\text{ppm}_{desired} \times L_{desired} \times \text{gravimetric factor} = \text{milligrams to be weighed} \tag{A4.14}$$

If a more concentrated solution is to be diluted,

$$V_B = \frac{C_A \times V_A}{C_B} \tag{A4.15}$$

## Standardization Using a Standard Solution

If working with molarities,

$$V_T \times M_T \times \text{mole ratio (ST/T)} = V_{ST} \times M_{ST} \tag{A4.16}$$

If working with normalities,

$$V_T \times N_T = V_{ST} \times N_{ST} \tag{A4.17}$$

## Standardization Using a Primary Standard

If molarity is to be calculated,

$$L_T \times M_T \times \text{mole ratio(PS/T)} = \frac{\text{grams}_{PS}}{FW_{PS}} \tag{A4.18}$$

If normality is to be calculated,

$$L_T \times N_T = \frac{\text{grams}_{PS}}{EW_{PS}} \tag{A4.19}$$

## Percent Analyte Calculations

If the molarity of the titrant is to be used,

$$\% \text{ analyte} = \frac{L_T \times M_T \times \text{mole ratio (ST/T)} \times FW_{\text{analyte}}}{\text{weight of sample}} \times 100 \tag{A4.20}$$

If the normality of the titrant is to be used,

$$\% \text{ analyte} = \frac{L_T \times N_T \times EW_{\text{analyte}}}{\text{weight of sample}} \times 100 \tag{A4.21}$$

Back titration, if normalities are used:

$$\% \text{ analyte} = \frac{(L_T \times N_T - L_{BT} \times N_{BT}) \times EW_{\text{analyte}}}{\text{sample weight}} \times 100 \tag{A4.22}$$

# Appendix 5
## Answers to Questions and Problems

## Chapter 1

1. See Section 1.1.
2. An assay is an analysis in which a named material is analyzed for that named material. For example, the assay of a tablet of ibuprofen is an analysis of the tablet for ibuprofen content. An analysis is not an assay when the analyte is some material other than the named material. An analysis of a calcium supplement for a color additive is not an assay.
3. Qualitative analysis is identification—analysis for what a substance is or what is in a sample. Quantitative analysis is the analysis for quantity, or how much of a given analyte is in a sample. Examples of qualitative analysis: 1) identify the contaminant giving an off-color to a manufactured product, 2) identify the byproduct of a chemical reaction, 3) identify an organic substance leaching into ground water from a hazardous waste site. Examples of quantitative analysis: 1) the determination of the nitrate concentration in a drinking water sample, 2) the determination of the concentration of the active ingredient in a pharmaceutical preparation, and 3) the determination of the concentration of an additive in a solvent manufactured at a chemical plant.
4. It is a qualitative analysis because the task is to identify the material, not to determine quantity.
5. See Section 1.2.
6. The activities carried out in a wet lab would probably include sample preparation and wet chemical analysis procedures (for example, extractions, solution preparations, and titrations)—activities that do not utilize sophisticated electronic instrumentation.
7. A wet chemical analysis would likely be chosen when a more precise result is needed or when the analyte is a major, rather than a minor, constituent. An instrumental analysis procedure would likely be chosen when a minor constituent present at a low level is to be determined and when a faster method with a greater scope or practicality is needed.
8. 1) Obtain the sample, 2) prepare the sample, 3) carry out the analysis method, 4) work the data, and 5) calculate and report the results.
9. A sample is a portion of a larger bulk system that has all the characteristics of the bulk system and is acquired for the purpose of testing it in a laboratory. To obtain a sample means to carry out a process by which a sample is acquired from the bulk system such that the sample truly represents the bulk system and is not altered in any way before the analysis takes place in the laboratory.
10. To prepare a sample means to carry out a laboratory procedure by which a sample is appropriately readied for the analysis method chosen. Such procedures include (but are not limited to) drying, dissolving, extracting, crushing, etc.
11. An analytical method is an operation (possibly involving many steps) by which data on a prepared sample and associated standard(s) are obtained and recorded so that the quality or quantity of the analyte in the sample can be ultimately determined. It is the heart of an analysis in the sense that it is the step following sample preparation in which the crucial data is acquired.

12. To carry out the analytical method means to perform the series of steps in which the crucial data for qualitative or quantitative analysis is directly obtained.

13. The raw data (measurements obtained directly from the laboratory equipment used without undergoing any additional process) usually need to be used in calculations, graphing, etc., in order to be put into a form that is most useful. Once the data are worked up in this way, they are ready for final calculation such that results are obtained and able to be reported.

14. A laboratory worker that has good analytical technique is one that takes special care in carrying out manual tasks to ensure that all data are obtained in a careful manner so that errors involved in handling samples and standards are eliminated or at least minimized.

15. The stirring rod is wet with your sample solution and rinsing it back into the beaker ensures that all of your sample is present in the beaker for subsequent reactions or measurement.

16. GLP stands for good laboratory practices. These are federal regulations governing FDA- and EPA-affiliated laboratories and pertain to proper procedures in the laboratory to ensure that results are obtained in as trustworthy a manner as possible. SOP stands for standard operating procedure. These are step-by-step written procedures that are specially approved by laboratory directors for carrying out certain specific tasks.

17. GLP regulations address such things as labeling, record keeping and storage, documentation and updating of SOPs and laboratory protocols, identifying who has authority to change SOPs, the processes by which they are changed, and audits.

18. The laboratory notebook is a legal document, and good record keeping is central to good analytical science. Not only must data be recorded, but they must be recorded with diligence, with considerable thought being given to integrity and purpose.

19. Data and results consist of sample descriptions, numerical measurements made on samples and standards (sometimes in the form of tables, chart recordings, and computer printouts), and mathematical formulas and calculations. Results consist of the final answer to an analysis, i.e., the result that is sought by doing the analysis.

20. Even if a piece of data is perceived to be bad, such that the analyst is tempted to erase it, it may happen later that it is valuable in some way.

21. There is a certain amount of error associated with all measurements regardless of the care with which the device was calibrated. This error is the uncertainty inherent in the measurement and the human error, either determinate or indeterminate, that can creep into an experiment. If the analyst and his or her client are to rely on the results, these must be taken into account.

22. Determinate errors are avoidable blunders that are known to have occurred. Indeterminate errors are errors that are assumed to have occurred but there is no direct knowledge of such or errors that are inherent in measurement in general. They are the errors that are dealt with by statistical treatment of the data or results.

23. A bias is an error that is known to occur each time a given procedure is carried out. Its effect is usually known such that a correction factor can be applied.

24. This is a determinate error because it is an error that is known to have occurred. Indeterminate errors are either errors inherent in a measurement or human errors that are not known to have occurred.

25. This is an indeterminate error because it is inherent in the use of the analytical balance.

26. Yes, if all determinate errors are eliminated and if indeterminate errors are taken into account by statistics.

27. Accuracy refers to how close a measurement or result is to being correct. Precision refers to the repeatability of a measurement. A precise measurement or result is not necessarily accurate.

28. They are very precise because they are within the uncertainty in the last significant figure expressed. It is not known with certainty if the results are accurate, although good precision usually indicates good accuracy.

29. The measurements are not precise because they differ widely. We cannot say anything about accuracy because we do not know the correct answer.

30. The results are still quite precise (see the answer to number 28 above), but given the additional information, we now know that they are not accurate.

31. Standard deviation is 0.0013. Relative standard deviaton is 0.000075.

32. The percent standard deviation is 1.3. The stated precision is exceeded.

33. Because it relates the value of the standard deviation to the value of the mean.

34. This means that if the experiment is carried out properly the precision of the data or result is such that a relative standard deviation of 1% or less can be achieved.

35. If the result of the analytical testing of this batch shows that it is out of statistical control (by being plotted outside the action limits on a control chart, for example), then this batch must be at least quarantined for further testing. If further testing indicates the same result, then the batch must be rejected.

36. A quality control chart is a visual aid for determining whether a given analytical result is outside the action limits determined for the results for that procedure. If it is outside the action limits, the cause may be a problem with the procedure, among other things.

# Chapter 2

1. A representative sample must be obtained and transported to the laboratory safely without alteration and the sample must be appropriately prepared (such as dried, dissolved, extracted, etc.) for the particular method to be used.

2. If a sample does not represent the bulk system in the manner intended, if the sample's integrity is not maintained during transportation, or if the preparation schemes go awry and do not provide the intended product, then the analytical results to be reported will be incorrect regardless of whether the analytical method was performed without error.

3. A representative sample is a sample that has all the characteristics in exactly the same proportions as the bulk system from which it came.

4. If a particular part of the bulk system under investigation is known to have an analyte concentration very different from the rest of the system, then taking a selective sample taken from that part of the system and analyzing it independently would make sense.

5. (a) At a given site, sample at different lateral positions, flowing water and semistagnant water, different depths, etc., then combine all samples into one composite sample. (b) The film is likely produced in different lots and on different production lines. Take random samples manufactured on a given day on a given production line (from a given lot) and combine. (c) Randomly sample several tablets from the bottle and combine. (d) Sample at different locations, at different depths, etc., and combine. (e) Take samples from all sides of the building, at different heights, under the eaves, close to the ground, etc., and combine. (f) Tissue from a particular organ may be more advantageous than others. Obtain samples from whatever organ is targeted and combine.

6. A bulk sample is the original undivided sample that was taken directly from the bulk system being characterized. A primary sample is the same as a bulk sample. A secondary sample (also subsample) is a part of the primary sample taken to the next step. A laboratory sample is a sample taken to a laboratory for analysis. The laboratory sample could be a primary sample or a subsample. A test sample is that part of a laboratory sample actually measured out for the method used.

7. The act of obtaining samples from a bulk system is subject to errors that can be neither detected nor compensated due to the bulk system often being nonhomogeneous and the sample therefore possibly not being exactly representative. Such errors are indeterminate and must be dealt with by statistics.

8. If the integrity of a sample is called into question, a paper trail, or chain of custody, must be examined in order to discover errors. This chain of custody documents who had custody of the sample and what actions were performed so that the sample integrity can be verified.

9. Glass containers can leach trace levels of metals and contaminate the sample.

10. It can be refrigerated to slow down the bacterial action.

11. Answer must be discovered in a reference book or website.

12. Particle size reduction makes dissolving procedures more efficient and extracting procedures more accurate because of the improved contact of the solvent with the sample. If accompanied by thorough mixing, particle size reduction also results in more homogeneous samples, which are more representative.

13. If the sample is not homogeneous, its division into smaller samples may result in a test sample that has a different overall composition than the original sample.

14. Extraction is an incomplete dissolution. As such, certain parts of the sample are dissolved in this procedure while other parts are not. The parts that are dissolved are thought of as having been extracted, or removed, from the sample.

15. A solid–liquid extraction can be as simple as mixing the solid material in a flask with solvent followed by filtration, or a Soxhlet extraction. See Section 2.5.3.

16. A supercritical fluid is a state of matter achieved by high temperature and extremely high pressure, exceeding the so-called critical temperature and pressure for that substance. The solvent properties of a supercritical fluid are much improved over the normal solvent properties of that fluid.

17. (a) water, (b) $HNO_3$, (c) HF, $HClO_4$, or aqua regia, (d) HCl, (e) aqua regia, (f) HF, (g) $H_2SO_4$

18. (a), (g), (l), (m)—sulfuric acid
    (b), (h), (i)—hydrofluoric acid
    (c), (e), (f)—nitric acid
    (d), (j)—hydrochloric acid
    (k)—perchloric acid

19. A mixture of concentrated $HNO_3$ and concentrated HCl in the ratio 1:3 by volume.

20. Refer to Table 2.2 and accompanying discussion.

21. (a) nitric acid, (b) hydrochloric acid, (c) hydrochloric acid, (d) sulfuric acid, (e) hydrofluoric acid, (f) sulfuric acid, (g) sulfuric acid

22. HCl—(b)
    $HNO_3$—(e)

23. Fusion is the dissolving of a sample with the use of a molten inorganic salt called a flux.

24. Refer to the answer to number 23 above.

25. Such material must be insoluble in the flux. Examples are platinum, gold, nickel, and porcelain.

26. A solid–liquid extraction is an extraction in which an analyte in a solid sample is extracted into a liquid solvent. A liquid–liquid extraction is an extraction in which an analyte dissolved in a liquid sample is extracted into a second liquid solvent that is immiscible with the first. A solid phase extraction is an extraction in which an analyte or its liquid solvent is extracted via contact with a solid material (sorbent) as the solution passes through a cartridge containing the sorbent.

27. 1) Analysis of water for pesticide residue, and 2) analysis of soil for metals.

28. A separatory funnel is a glass container that is half bottle (top) and half funnel (bottom) with a stopcock in the stem of the funnel at the bottom. See Figures 2.2 and 11.2. It is used for liquid–liquid extractions. The analyte solution and extracting solvent are placed together in the separatory funnel and shaken vigorously. After allowing the two immiscible layers to separate, each can be drained one at a time through the stopcock.

29. Extraction is the selective dissolution, as opposed to total sample dissolution, of the analyte from a sample. The sample can be either a solid material or a liquid, such as a water solution.

30. They are nonpolar and, as such, do not mix with water and will extract nonpolar solutes from water.

31. Diethyl ether is highly flammable and also highly volatile, a combination that can result in an explosion due to a flame source igniting the vapors in a lab in which it is being used. In addition, decomposition over time results in the formation of highly unstable and explosive peroxides in the stored containers. Precautions include avoiding open flames, working in a fume hood, storing the containers in explosion-proof refrigerators, the use of metal containers, which slow the peroxide formation, and disposal after about 9 months of storage.

32. N-hexane, benzene, toluene, and diethyl ether are less dense than water, and thus would be the top layer in an extraction experiment. Liquids with a higher density than water would sink to the bottom of the separatory funnel.

33. (a) All the organic liquids mentioned are toxic to a certain extent, and certainly they should never be ingested. However, chloroform is more toxic than the others and requires special handling.
    (b) n-hexane, toluene, and diethyl ether are flammable.

34. A purge-and-trap procedure is one in which a volatile analyte is purged from solvent by helium sparging and trapped on a sorbent held in a cartridge through which the helium then passes.

35. See Section 2.6.2.

36. Solvent exchange is a process in which the analyte solution is evaporated to dryness and the residue reconstituted with a different solvent.

37. The quality of the reagent must be assured in order for the analyst to have the confidence that the sample and prepared reagents will not be contaminated and give inaccurate results.

38. ACS certified means that a reagent meets or exceeds the specifications of purity set by the American Chemical Society.

39. No, because they generally are not pure enough.

40. Spectro grade or spectroanalyzed for spectrophotometric analysis. HPLC grade for liquid chromatographic analysis.

41. See Section 2.8.

# Chapter 3

1. Gravimetric analysis is a wet chemical method of analysis in which the measurement of weight is the primary measurement, and most of the time the only measurement, that is made on the analyte and its matrices.

2. Gravimetric analyses do not usually require standard solutions of any kind and do not require the calibration of any equipment beyond a balance.

3. Mass is a measure of the amount of a sample. Weight is the measure of the gravitational pull on this amount of sample. On the surface of the Earth, a weight measurement is taken to be the same as a mass measurement, utilizing the same units, although technically they are not the same.

4. A weighing device is called a balance because a weight is often determined by balancing the object to be weighed with a series of known weights across a fulcrum. Modern electronic balances, however, do not operate this way, but they are still called balances.

5. A single-pan balance is a balance with a single pan, which is used only for the object to be weighed. In the older styles, a constant counterbalancing weight is used while a number of removable weights on the same side as the object are added or removed in order to obtain the weight of the object. The more modern single-pan balances are of the torsion variety.

6. To say that a weight measurement was made to the nearest 0.01 g means that the second digit past the decimal point is the last digit available on the balance used. To say that the precision of a balance is ±0.1 mg (±0.0001 g) means that the fourth place past the decimal point is the last digit that can be obtained by that balance.

7. A top-loading balance is an electronic balance with the sample pan on the top. It is not enclosed, which means that it is capable of measuring only to the nearest hundredth of a gram. See Figure 3.3.

8. To say a balance has the tare feature means that the balance can be zeroed with an object, such as a piece of weighing paper, on the pan. This helps to obtain the weight of a chemical directly without having to obtain and subtract the weight of the weighing paper.

9. Analytical balances are balances that measure to 0.1 or 0.01 mg.

10. A number of considerations are important. For example, the analytical balance must be level, the pan must be protected from air currents, the object to be weighed must be at room temperature, the object must be protected from fingerprints, etc.

11. A desiccator is a storage container used either to dry samples or, more commonly, to keep samples and crucibles dry and protected from the laboratory environment once they have been dried by other means. An indicating desiccant is a desiccant that changes color when saturated with adsorbed water. Drierite is an example of a desiccant. It is anhydrous calcium sulfate.

12. To check the calibration of a balance is to weigh a known weight to verify that the balance gives the correct weight.

13. It depends on how heavy the sample must be. If it is less than a gram, then an analytical balance must be used because only such a balance would give four significant figures. However, if the sample is greater than 10 g, then an ordinary balance is satisfactory since it would give four significant figures.

14. A physical separation is a separation the results from some nonchemical operation, such as evaporating a solvent, filtering, etc. A chemical separation uses a chemical reaction to effect the separation.

15. (a) Loss on drying is the determination of the percentage weight loss that occurs on a sample as a result of heating at or below the boiling point of water or as a result of desiccation.
    (b) 50.594%

16. 27.239%

17. Heat to constant weight refers to repeating heating, cooling, and weighing operations performed until there is no longer any loss in weight.

18. Loss on ignition is the determination of the percentage weight loss that occurs on a sample as a result of heating to ignition temperatures. Residue on ignition is the determination of the percentage residue weight remaining after heating to ignition temperatures.

19. Volatile organics, 48.301%; residue, 51.699%.

20. Volatile organics, 55.992%; residue, 44.008%.

21. Insoluble matter in a general analysis report on the label of a chemical container is the percentage of the material that does not dissolve in water.

22. See Section 3.6.1.5, "Solids in Water and Wastewater."

23. 1024 mg/L

24. $1.098 \times 10^4$ mg/L total solids; $1.048 \times 10^4$ mg/L settleable solids.

25. Sand, 78.565%; salt, 21.435%.

26. 58%

27. (a) 0.5745        (b) 0.3430        (c) 0.8084        (d) 1.473
    (e) 0.7071        (f) 0.6665        (g) 0.9679        (h) 1.103

28. (a) 1.15730       (b) 1.29963       (c) 1.1768

29. (a) 0.36219       (b) 0.93089       (c) 0.966222

30. 0.0984 g

31. 0.8131 g

32. 1.513 g

33. 49.17%

34. 0.1840 g

35. 34.98%

36. 18.18%

37. 43.30%

38. 26.905%

39. 26.31%

40. There is no chance for the weighed sample to be anywhere but in your beaker. Also eliminates the chance of contamination.

41. To get the precipitate particles to clump together to make them more filterable.

42. From the time you weigh the item the first time until you weigh it the second time. It is during that time that the weight of a sample would be inaccurate due to fingerprints, since such added weight would alter the calculated weight of the sample.

43. This precipitate is AgCl, not BaSO₄. Its presence only signals the presence of chloride in the rinsings, thus indicating that more rinsing is needed.

44. Even though they are heat stable, the weight of such a label will change during heating. If there is a weight change for this reason, then the weight of the crucible contents calculated later will be incorrect.

45. The precipitate and filter paper are sopping wet at this point and would saturate the desiccant. Desiccators are mostly used to keep things dry once they are already dry, not to dry things that are wet.

46. A filter paper is ashless if it is completely volatilized when burned, leaving no ash residue. In Experiments 6 and 7, if the filter paper were not ashless, a residue would remain in the crucible and add weight to the precipitate. This in turn would increase the calculated percent, and the results reported would be high.

47. The final results would be lower than the correct answer because some of the wax from the pencil would volatilize during the second heating, causing the crucible with the precipitate to weigh less than it should, thus causing the analyst to report a lower precipitate weight than he or she would otherwise.

48. The percent would be lower than the true percent because the weight of the precipitate (the numerator in the percent calculation) would be lower.

49. An ordinary balance would suffice because this solution is used only for a qualitative test—the identification of chloride in the rinsings.

# Chapter 4

1. Gravimetric analysis utilizes primarily weight measurements and may or may not involve chemical reactions. Titrimetric analysis utilizes both weight and volume measurements and always involves solution chemistry and stoichiometry.

2. Titrimetric analysis is sometimes called volumetric analysis because it is characterized by the frequent measurement of solution volume utilizing precision glassware.

3. For gravimetric analysis, a solution *may* be needed to react with the analyte. Otherwise it consists of just physical separation operations and usually initial and final weight measurements. For titrimetric analysis, solutions are always needed to react with the analyte and these solutions must be standardized. Also, a critical measurement is a volume measurement (buret reading).

4. See Section 4.2 for the definition of terms.

5. (a) $0.195\ M$      (b) $3.13\ M$      (c) $0.01041\ M$      (d) $0.01607\ M$
   (e) $1.8\ M$      (f) $2.40\ M$

6. (a) NaOH, 39.997; HCl, 36.461      (b) NaOH, 39.997; $H_2SO_4$, 49.040
   (c) HCl, 36.461; $Ba(OH)_2$, 85.6710      (d) NaOH, 39.997; $H_3PO_4$, 32.665
   (e) HCl, 36.461; $Mg(OH)_2$, 29.160      (f) NaOH, 39.997; $H_3PO_4$, 48.998
   (g) NaOH, 39.997; $Na_2HPO_4$, 141.959      (h) NaOH, 39.997; $H_3PO_4$, 97.995
   (i) $Na_2CO_3$, 52.9945; HCl, 36.461

7. (a) $0.159\ N$      (b) $5.16\ N$      (c) $10.8\ N$      (d) $9.92\ N$
   (e) $0.0492\ N$      (f) $0.0121\ N$      (g) $0.0428\ N$

8. $0.2411\ M$

9. (a) Dissolve 2.8 g of KOH in water, dilute to 500.0 mL, and shake.
   (b) Dissolve 2.2 g of NaCl in water, dilute to 250.0 mL, and shake.
   (c) Dissolve 36 g of glucose in water, dilute to 100.0 mL, and shake.
   (d) Dilute 4.2 mL of 12.0 $M$ HCl with water to 500.0 mL and shake.
   (e) Dilute 12 mL of 2.0 $M$ NaOH with water to 100.0 mL and shake.
   (f) Dilute 56 mL of 18.0 $M$ $H_2SO_4$ with water to 2.0 L and shake.

10. (a) Dissolve 6.8 g of $KH_2PO_4$ in water, dilute to 500.0 mL, and shake.
   (b) Dilute 1.5 mL of 18.0 $M$ $H_2SO_4$ with water to 500.0 mL and shake.

(c) Dissolve 7.1 g of $Ba(OH)_2$ in water, dilute to 750.0 mL, and shake.

(d) Dissolve 1.6 g of $Na_2CO_3$ in water, dilute to 200.0 mL, and shake.

(e) Dissolve 15 g of $NaHCO_3$ in water, dilute to 700 mL, and shake.

(f) Dilute 14 mL of 15.0 $N$ $Ba(OH)_2$ with water to 700.0 mL and shake.

(g) Dilute 1.0 mL of 15 $M$ $H_3PO_4$ with water to 300.0 mL and shake.

11. (a) Dissolve 3.5 g of NaOH in water, dilute to 250.0 mL, and shake.

    (b) Dilute 15 mL of 6.0 $M$ NaOH with water to 250.0 mL and shake.

12. 29 mL

13. 93 mL

14. (a) Dissolve 18 g of $KNO_3$ in water, dilute to 500.0 mL, and shake.

    (b) Dilute 39 mL of 4.5 $M$ $KNO_3$ with water to 500.0 mL and shake.

15. Use 5.0 mL of 17 $M$ acetic acid and 12 g of sodium acetate. Dissolve the sodium acetate in water, add the 17 $M$ acetic acid, dilute to 500.0 mL, and shake.

16. $1.56 \times 10^3$ mL

17. 0.05586 $M$

18. 0.2671 $M$

19. 0.1056 $M$

20. 0.2250 $N$

21. 0.08798 $N$

22. 0.1733 $N$

23. No. The equivalent weight of sulfuric acid is different from that of hydrochloric acid, but it does not enter into the calculation.

24. See Section 4.6.2.

25. See Section 4.6.2.

26. SRM stands for standard reference material. It is a standard chemical manufactured and certified by the National Institute of Standards and Technology (NIST) as being exactly as labeled. It is the ultimate standard. CRM stands for certified reference material. It is a standard chemical manufactured and certified by a vendor as being exactly as labeled. It is traceable to a SRM, meaning it has been compared to and certified with the use of an ultimate standard.

27. Titer is an alternate concentration expression for a titrant that is specific to a particular substance titrated. It is the weight of the substance titrated that is consumed by 1 mL of the titrant.

28. 2.22 mg/mL

29. 75.87%

30. 21.22%

31. 38.36%

32. 36.94%

33. (a) false           (b) false           (c) true           (d) true

    (e) false           (f) false           (g) false          (h) false

34. (a) Measuring pipet.

    (b) The last bit of solution in the tip should *not* be blown out into the receiving vessel.

    (c) Otherwise, the solution would be diluted with the film of water on the inside surface.

35. A volumetric pipet, because the diameter of the tube where the calibration line is located is narrower.

36. No. A volumetric flask is calibrated TC, not TD.

37. A frosted ring near the top means the pipet is calibrated for blowout.

38. A volumetric flask is a precision piece of glassware with a single calibration line in a narrow neck. An Erlenmeyer is not a precision piece of glassware and has graduation lines on it only as a rough indication of volume.

39. Rinsing a pipet with the solution to be transferred assures that the liquid film adhering to the inside surface is the solution to be transferred and not water, which would contaminate or dilute the solution to be transferred.

40. The serological pipet is calibrated through the tip and may be used by adjusting the meniscus to the line that would deliver the desired volume by letting it drain completely (and blowing it out). The Mohr pipet is calibrated only to the lowest calibration line and cannot be drained below this line.

41. (a) A volumetric flask is not calibrated to deliver a volume, only to contain. There is a difference because the thin liquid film adhering to the inner wall is contained but not delivered.

    (b) I would tell him to use a 50-mL volumetric pipet.

42. Because the liquid film adhering to the inner wall would contaminate or dilute the solution being transferred if it were not rinsed first with this solution.

43. By dilution: measure out 6.22 mL of the 4.021 *M* solution as precisely as possible (measuring pipet), dilute it to 100.0 mL (volumetric flask), and shake. By weight of the pure solid: weigh out 2.650 g on an analytical balance, place in a 100-mL volumetric flask, dissolve in water, dilute to 100.0 mL, and shake.

44. If I chose a Mohr pipet, it would not be calibrated for blowout. If I chose a serological pipet, it would be calibrated for blowout. A blowout pipet would have one or two frosted rings completely circumscribing the top of the pipet. See Figure 4.12 for details on how to make the delivery.

45. (a) Use a volumetric flask and an analytical balance.

    (b) Standardization is an experiment by which a solution concentration is determined with good precision.

46. (a) Volumetric pipet, one calibration line; serological pipet, many graduation lines.

    (b) Volumetric pipet, not calibrated for blowout; serological pipet, calibrated for blowout.

    (c) Serological pipet, because since volumetric pipets have just one calibration line, they are not manufactured for odd volumes such as 3.72 mL.

    (d) Both are calibrated TD.

47. Pipets calibrated TC are used for viscous solutions that do not drain well. For these, a pipet contains the solution and then the contained volume is rinsed out into the receiving vessel.

48. A graduated cylinder is sufficient because this volume does not affect the result and does not enter into the calculation.

# Chapter 5

1. The equivalence point must be easily and accurately detected; the reaction involved must be fast; the reaction must be quantitative.

2. Monoprotic acid—An acid that has just one hydrogen ion to donate per molecule.
   Polyprotic acid—An acid that has more than one hydrogen ion to donate per molecule.
   Monobasic base—A base that will accept just one hydrogen ion per formula unit.
   Polybasic base—A base that will accept more than one hydrogen per formula unit.
   Titration curve—A plot of pH vs. milliliter of titrant showing the manner in which pH changes vs. milliliter of titrant during an acid–base titration.
   Inflection point—A point in a titration curve at which the slope of the curve is a maximum.

3. These curves are shown in Figure 5.1(a) and (c). Both curves show the same pattern after the inflection point because in both cases the acids have been neutralized at that point and the pH depends only on the added NaOH. The acetic acid curve shows a higher pH level leading up to the equivalence point than the HCl curve. Also, the inflection point of the HCl curve covers a broader pH range than that of the acetic acid. The reason for these differences is that acetic acid is a weak acid, meaning fewer hydrogen ions in solution for the same total concentration. Fewer indicators will work for the acetic acid titration because of this narrower range.

4. These curves are shown in Figure 5.3. Both curves show the same pattern after the inflection point because in both cases the bases have been neutralized at that point and the pH depends only on the added HCl. The ammonium hydroxide curve shows a lower pH level leading up to the equivalence point than the NaOH curve. Also, the inflection point of the NaOH curve covers a

broader pH range than that of the ammonium hydroxide. The reason for these differences is that ammonium hydroxide is a weak base, meaning more hydrogen ions in solution for the same total concentration. Fewer indicators will work for the ammonium hydroxide titration because of this narrower range.

5. These curves are shown in Figures 5.6 and 5.7. The only similarity is the final pH level near the end of the titration. At that point, both acids have been neutralized, and in each case, the pH depends only on the added NaOH. The phosphoric curve shows a higher pH level at the beginning. This is because it is a weaker acid and there are fewer hydrogen ions in solution. The phosphoric acid curve shows two inflection points, while the sulfuric acid curve shows essentially one inflection point. This is because phosphoric acid has two weakly acidic hydrogen ions (See Section 5.2.5) to be neutralized per molecule, whereas sulfuric has just two hydrogen ions that are strongly acidic and are neutralized together. The inflection point of the sulfuric acid covers a broader pH range and many indicators would work. With phosphoric acid, the analyst has a choice of which inflection point to use and the indicator choice depends on which is chosen. Once a range is chosen, there would be fewer choices for the indicator because of the narrower pH range of each inflection point.

6. (a) See Figure 5.1(c).
   (b) See the two lower curves in Figure 5.4.

7. (a) See Figure 5.1(c).
   (b) See the NaOH curve in Figure 5.3.
   (c) See the $NH_4OH$ curve in Figure 5.3.

8. No. See Figure 5.11. For Experiment 11, the second inflection point was chosen and is at a pH range that is too low to utilize the phenolphthalein end point.

9. No. See Figure 5.1(c). The inflection point is at a pH that is too high range for the bromcresol green end point.

10. The indicator must change color in the range of approximately 8.5 to 10.5. Phenolphthalein would be best.

11. The P in KHP refers to phthalate. The phthalate ion is a benzene ring with two carboxyl groups on adjacent carbons. See Figure 5.8. Potassium hydrogen phthalate (KHP) is useful as a primary standard because it possesses all the qualities sought in a primary standard. See Section 4.5.2.

12. THAM, or TRIS, is tris-(hydroxymethyl)amino methane. Its structure is $(HOCH_2)_3CNH_2$. It is a base because it contains the $-NH_2$ group, which accepts a hydrogen ion to form $-NH_3^+$. It is useful as a primary standard because it possesses all the qualities sought in a primary standard, as discussed in Section 4.6.2.

13. Sodium carbonate accepts two hydrogen ions per formula unit, and each is accepted separately during a titration, hence two inflection points. In the region leading up to the second inflection point, carbonic acid is a product of the reaction. Because carbonic acid is in equilibrium with $CO_2$ and water, the reaction is sluggish and does not go to completion unless the $CO_2$ is eliminated by boiling. The second inflection point is at a pH range that is too low for phenolphthalein to work, but is just right for bromcresol green.

14. Total alkalinity is the capacity for a volume of water to neutralize an added standard acid solution, usually to obtain a pH of 4.5. It is expressed as the millimoles of $H^+$ per liter of water.

15. 6.14 mmol/L

16. A back titration is a titration in which the titrant is added in excess and the excess amount titrated with a second titrant, the so-called back titrant, so as to come back to the end point.

17. Calcium carbonate is not soluble in water. The addition of the standard acid to a tablet immersed in water causes it to dissolve, but rather slowly. A direct titration is not appropriate because the reaction is not fast, and it would be difficult to detect the end point. A back titration is therefore useful because the standard acid can be added in excess so that the calcium carbonate completely dissolves. The excess acid can then be titrated with a standard base.

18. 85.53%, 1105 mg

19. (a) Concentrated sulfuric acid
    (b) Concentrated sodium hydroxide
    (c) Ammonia
    (d) Acid
    (e) Acid
    (f) Base

20. (a) It is used to dissolve or digest the sample.
    (b) It is one option for the acid in the receiving flask to react with the ammonia.

21. We must subtract the excess equivalents of titrant (which is equal to the equivalents of back titrant used) from the total equivalents of titrant added so that we have the equivalents of titrant that actually reacted with the analyte.

22. 4.832

23. 5.052

24. Buffer solution—A solution that resists changes in pH even when a strong acid or base is added or when it is diluted with water.
    Conjugate acid—The product of the neutralization of a base that is an acid because it can lose the hydrogen ion that it gained during the neutralization.
    Conjugate base—The product of the neutralization of an acid that is a base because it can gain back the hydrogen ion that it lost during the neutralization.
    Conjugate acid–base pair—A pair of compounds consisting of a base and its conjugate acid or an acid and its conjugate base.
    Buffer capacity—The capacity of a buffer solution to resist pH changes.
    Buffer region—The region of a titration curve leading up to the inflection point.

25. Sodium acetate is a conjugate base when it results from the neutralization of acetic acid. The acetate ion can gain back the hydrogen ion that it lost and become acetic acid again. Ammonium chloride is a conjugate acid when it results from the neutralization of ammonia. The ammonium ion can lose the hydrogen ion it gained and become ammonia again.

26. The Henderson–Hasselbalch equation is an equation expressing the relationship between pH, $pK_a$, and the log of the ratio of the concentrations of the base to its conjugate acid or an acid to its conjugate base. It is derived from the $K_a$ or $K_b$ expression. See Equations (5.26) to (5.30) in the text. They are each a form of this equation.

27. pH = 2.77

28. pH = 3.25

29. pH = 7.65

30. 5.7

31. A certain combination of chloroacetic acid and sodium chloroacetate would give a pH of 3.00 since the pH range for this combination is 1.8 to 3.8. From question 27, the $K_a$ of chloroacetic acid is $1.36 \times 10^{-3}$. The ratio is 0.74.

32. This is the same question as number 15 in Chapter 4. Use 5.0 mL of 17 $M$ acetic acid and 12 g of sodium acetate. Dissolve the sodium acetate in water, add the 17 $M$ acetic acid, dilute to 500.0 mL, and shake.

33. Weigh 9.1 g of THAM and 32 g of THAM hydrochloride. Place both in the same 500-mL container. Add water to dissolve. Add water to the 500-mL mark and shake.

34. pH = 10.06

35. Monodentate, bidentate, hexadentate—These are adjectives that describe a ligand in terms of the number of bonding sites (pairs of electrons) available for bonding to the metal ion. Mono = 1, bi = 2, hexa = 6.
    Ligand—The charged or uncharged chemical species that reacts with a metal ion forming a complex ion.
    Complex ion—A charged aggregate consisting of a metal ion in combination with one or more ligands.

Coordinate covalent bond—A covalent bond in which the two shared electrons are contributed to the bond by only one of the two atoms involved.

Water hardness—A term used to denote the $Ca^{+2}$, $Mg^{+2}$, and $Fe^{+3}$ content of water both quantitatively and qualitatively.

Aliquot—A portion of a larger volume of a solution, usually a transferred or pipetted volume.

36. Ligand—The charged or uncharged chemical species that reacts with a metal ion forming a complex ion. Examples: See Table 5.3.

    Complex ion—A charged aggregate consisting of a metal ion in combination with one or more ligands. Examples: See Table 5.3.

37. $CoCl_4^{2-}$ = complex ion; $Cl^-$ = ligand.

38. Ammonia is an example of a monodentate ligand. See Table 5.3 for other examples. EDTA is a hexadentate ligand. A monodentate ligand is one that has just one bonding site for bonding to a metal ion.

39. Bidentate, because there are two sites (hence "bi") at which a metal ion will bond.

40. (a) The ligand is the reactant (left side of the equation) consisting of the three aromatic rings and two nitrogens.
    (b) The complex ion is the product of the reaction (right side of the equation).
    (c) The ligand is bidentate because it has two bonding sites that bond to the metal ion.
    (d) The complex ion is a chelate because the ligand involved has two more bonding sites.

41. (a) 6                        (b) 1                        (c) hexadentate

42. A basic pH is needed in order to expose all bonding sites on the EDTA (i.e., remove all acidic hydrogens) so that it can react completely with the metal ions. A pH of 8 is not basic enough for this (see Figure 5.21), and at a pH of 12, magnesium ions precipitate as the hydroxide and are lost to the analysis. A pH of 10 will work.

43. (a) A complex ion formed between calcium ions and the Eriochrome Black T.
    (b) The free Eriochrome Black T ligand.
    (c) As a masking agent, cyanide acts as a ligand and forms a complex ion with a metal ion, in effect preventing the metal ion from interfering in an analysis. We say the metal ion is masked and that the cyanide is a masking agent.

44. See answer to number 42.

45. (a) 6.2 mg of Mg is dissolved and the solution diluted to 250.0 mL.
    (b) 22.5 mg of Ag is dissolved and the solution diluted to 750.0 mL.
    (c) 24.0 mg of Al is dissolved and the solution diluted to 600.0 mL.
    (d) 7.5 mg of Mg is dissolved and the solution diluted to 500.0 mL.
    (e) 7.50 mg of Fe is dissolved and the solution diluted to 250.0 mL.
    (f) 12.5 mg of Cu is dissolved and the solution diluted to 100.0 mL.

46. (a) 12.5 mL
    (b) 9.0 mL
    (c) 6.2 mL

47. (a) 63.5 mg NaCl is dissolved and diluted to 500.0 mL.
    (b) 25.0 mL of the 1000.0 ppm is diluted to 500 mL.

48. (a) 7.50 mg of Fe is dissolved and diluted to 250.0 mL.
    (b) 54.3 mg is dissolved and diluted to 250.0 mL.
    (c) 7.50 mL of the 1000.0 ppm is diluted to 250 mL.

49. (a) 19.6 mg is dissolved and diluted to 100.0 mL.
    (b) 5.00 mg of copper metal is dissolved and diluted to 100.0 mL.
    (c) 5.00 mL of the 1000.0 ppm is diluted to 100.0 mL.

50. (a) 12.5 mL
    (b) 0.0125 g
    (c) 0.0222 g

51. (a) 35.7 mg

(b) 25.3 mg

(c) 36.1 mg

52. 4.70 g of disodium dihydrogen EDTA dihydrate is dissolved and diluted to 500.0 mL.

53. 3.7 g

54. (a) 0.0103 $M$

    (b) 0.0196 $M$

    (c) 0.03688 $M$

    (d) 0.05755 $M$

    (e) 0.00775 $M$

    (f) 0.006418 $M$

    (g) 0.006578 $M$

    (h) 0.008402 $M$

    (i) 0.009371 $M$

55. (a) 404.2 ppm $CaCO_3$

    (b) 536.4 ppm $CaCO_3$

    (c) 271.7 ppm $CaCO_3$

    (d) 160.7 ppm $CaCO_3$

    (e) 110.2 ppm $CaCO_3$

    (f) 314.5 ppm $CaCO_3$

    (g) 283.5 ppm $CaCO_3$

56. Oxidation—The loss of electrons or the increase in oxidation number.

Reduction—The gain of electrons or the decrease in oxidation number.

Oxidation number—A number indicating the state an element is in with respect to bonding.

Oxidizing agent—A chemical that causes something to be oxidized while being reduced itself.

Reducing agent—A chemical that causes something to be reduced while being oxidized itself.

57. (a) +5      (b) +3      (c) +6      (d) +5

    (e) +7      (f) +1      (g) +4      (h) +6

    (i) +3      (j) +3

58. (a) +1      (b) −1      (c) +5      (d) 0

    (e) +3      (f) +5

59. (a) +3      (b) 0      (c) +6      (d) +6

    (e) +6

60. (a) +7      (b) −1      (c) 0      (d) +3

    (e) +1

61. (a) +4      (b) −2      (c) 0      (d) +6

    (e) +4      (f) −2      (g) +6      (h) +4

    (i) +4      (j) +6

62. (a) +5      (b) +3      (c) +5      (d) +3

    (e) +3      (f) +5      (g) −3      (h) +5

    (i) 0      (j) +5

63. (a) CuO (or Cu) was reduced (+2 to 0), and $NH_3$ (or N) was oxidized (−3 to 0).

    (b) $Cl_2$ was reduced (0 to −1), and KBr (Br) was oxidized (−1 to 0).

64. (a) Mg is the reducing agent because it was oxidized from 0 to +2. HBr is the oxidizing agent because H is reduced from +1 to 0.

    (b) Fe is the reducing agent because it was oxidized from 0 to +3. $O_2$ is the oxidizing agent because it was reduced from 0 to −2.

65. (a) No—This is neutralization, no change in oxidation number.

    (b) Yes—S and N have changed oxidation number.

    (c) Yes—Both Na and H have changed oxidation number.

    (d) No—This is neutralization, no change in oxidation number.

    (e) No—No oxidation number changes.

(f) Yes—Both K and Br have changed oxidation number.

(g) Yes—Both Cl and O have changed oxidation number.

(h) No—This is neutralization, no change in oxidation number.

(i) No—No change in oxidation number.

(j) Yes—Both Mg and H have changed oxidation number.

66. (b) is redox, Cu is oxidized, $HNO_3$ (N) is reduced. Cu is reducing agent, $HNO_3$ the oxidizing agent.

67. (a) (2) is redox

(b) HCl (or H)

(c) Zn

(d) Lose

68. (a) $3 H_2O + 4 Cl^- + 3 NO_3^- \rightarrow 4 ClO_3^- + 6 H^+ + 3 N^{3-}$

(b) $2 H^+ + 2 Cl^- + 2 NO_3^- \rightarrow 2 ClO_2^- + N_2O + H_2O$

(c) $ClO^- + 2 NO_3^- \rightarrow ClO_3^- + 2 NO_2^-$

(d) $3 ClO^- + 2 H^+ + 2 NO_3^- \rightarrow 3 ClO_2^- + 2 NO + H_2O$

(e) $4 ClO_3^- + SO_4^{2-} \rightarrow 4 ClO_4^- + S^{2-}$

(f) Balanced as is

(g) $H_2O + 4 IO_3^- + SO_3 \rightarrow 4IO_4^- + 2 H^+ + S^{2-}$

(h) $Cl^- + 2 H^+ + SO_4^{2-} \rightarrow ClO^- + SO_2 + H_2O$

(i) $3 Cl^- + 8 H^+ 4 SO_4^{2-} \rightarrow 3 ClO_4^- + 4 S + 4 H_2O$

(j) $6 Br^- + 6 H^+ + SO_3 \rightarrow 3 Br_2 + S + 3 H_2O$

(k) $5 I^- + 4 H^+ + 4 NO_3^- \rightarrow 5 IO_2^- + 2 N_2 + 2 H_2O$

(l) $12 H_2O + 8 P + 5 IO_4^- \rightarrow 8 PO_4^{3-} + 24 H^+ + 5 I^-$

(m) $3 H_2O + 3 SO_2 + BrO_3^- \rightarrow 3 SO_4^{2-} + 6 H^+ + Br^-$

(n) $10 Fe + 30 H^+ + 3 P_2O_5 \rightarrow 10 Fe^{3+} + 6 P + 15 H_2O$

(o) $2 Cr + 6 H^+ + 3 PO_4^{3-} \rightarrow 2 Cr^{3+} + 3 PO_3^- + 3 H_2O$

(p) $5 Ni + 16 H^+ + 2 PO_4^{3-} \rightarrow 5 Ni^{2+} + 2 P + 8 H_2O$

(q) $6 H^+ + 2 MnO_4^- + 5 H_2C_2O_4 \rightarrow 2 Mn^{2+} + 8 H_2O + 10 CO_2$

(r) $6 I^- + 14 H^+ + Cr_2O_7^{2-} \rightarrow 3 I_2 + 2 Cr^{3+} + 7 H_2O$

(s) $Cl_2 + H_2O + NO_2^- \rightarrow 2 Cl^- + NO_3^- + 2 H^+$

(t) $3 S^{2-} + 8 H^+ + 2 NO_3^- \rightarrow 3 S + 2 NO + 4 H_2O$

(u) Balanced as is

69. (a) $6 S_2O_3^{2-} + 14 H^+ + Cr_2O_7^{2-} \rightarrow 3 S_4O_6^{2-} + 2 Cr^{3+} + 7 H_2O$

(b) $0.4475 \, M$

70. $0.03438 \, M$

71. $0.02497 \, M$

72. $0.007842 \, M$

73. 19.76%

74. 70.37%

75. 55.76%

76. 42.09%

77. The intensely purple-colored $MnO_4^-$ solution becomes easily visible when there is no longer any ST available to react.

78. (a) It takes on electrons readily.

(b) If potassium permanganate contacts oxidizable substances, a reaction will take place, and the concentration of the permanganate will change to an unknown value.

79. An indirect titration is one in which the ST is determined indirectly by titrating a second species that is proportional to the ST. A back titration is one in which the end point is intentionally overshot and the excess back titrated.

80. Iodometry is a titrimetric method involving iodine. It is an indirect method because the product of the reaction of ST with $I^-$ ($I_2$) is titrated.

81. KI—The titrant from which the $I_2$ is liberated.
    $Na_2S_2O_3$—The titrant for the liberated $I_2$.
    $K_2Cr_2O_7$—The primary standard for the $Na_2S_2O_3$.

# Chapter 6

1. 1) Obtain the sample; 2) prepare the sample; 3) carry out the analysis method; 4) work the data; and 5) calculate and report results.
2. Wet methods are those that involve physical separation and classical chemical reaction stoichiometry, but no instrumentation beyond an analytical balance. Instrumental methods are those that involve additional high-tech electronic instrumentation, often complex hardware and software. Common analytical strategy operations include sampling, sampling preparation, data analysis, and calculations. Also, weight or volume data are required for almost all methods as part of the analysis method itself.
3. Sampling activities are important because if the sample does not represent what it is intended to represent, all operations that follow, whether they are wet or instrumental methods, will not give a reliable result. Sample preparation schemes are important because if the sample is not prepared properly for the method chosen, the method, whether wet or instrumental, again will not give a reliable result.
4. In an instrumental method, a series of reference solutions are usually prepared for the purpose of calibrating the equipment. In addition, a chemical reaction, often at the heart of a wet method, is not necessarily required for instrumental methods.
5. The general principle of analysis with analytical instrumentation is depicted in Figure 6.4. Some property of the standards and sample solution is detected and measured by the instrument. An electronic signal is generated proportional to this property and read on a readout device.
6. The three major classifications, with instrument names in parentheses, are spectroscopy (spectrometer), chromatography (chromatograph), and electroanalytical chemistry (no specific name).
7. Spectroscopic methods involve the use of light and measure either the amount of light absorbed (absorbance) or the amount of light emitted by solutions of the analyte under certain conditions. Chromatographic methods involve more complex samples in which the analyte is separated from interfering substances using specific instrument components and electronically detected, with the electrical signal generated by any one of a number of detection devices.
8. Calibration is a procedure by which any instrument or measuring device is tested with a standard in order to determine its response for an analyte in a sample for which the true response is either already known or needs to be established.
9. In the case of an analytical balance, the standard is a known weight. This weight is measured and if this result and the known value are the same, the balance is calibrated. If they are not the same, the balance is taken out of service and repaired. In the case of a pH meter, the standard is a buffer solution. The pH of this buffer solution is measured and if the known pH and the measured pH are the same, the meter is calibrated. If they are not the same, the readout is electronically tweaked until it gives the correct result. It is then said to be calibrated.
10. The standard curve is the plot of an instrument's readout vs. concentration, the data for which are the results of measuring a series of standard solutions prepared for the experiment.
11. Preparing and measuring a series of standard solutions and plotting the standard curve is the usual process of calibration when the response of a device to a standard is not known in advance (see answer to number 8). In other words, a series of standards establishes the result that is not known in advance. With such a calibration in effect, the operator can then measure samples by comparing the measured result for the sample to the standards results.
12. Samples solutions can be held in place in a small container inside the instrument, injected into the instrument with the use of a syringe, aspirated into an instrument with a sucking device, or not placed inside the instrument at all, but externally tested by dipping a probe into it.

13. A sensor is a kind of translator. It receives specific information about the system under investigation and transmits this information in the form of an electrical signal.

14. Sensor, signal processor, power supply, and readout device.

15. See Section 6.3.2 for the definitions.

16. 0.431 mA, assuming the resistance is known to three significant figures.

17. An instrumentation amplifier provides precise voltage amplification for sensors that produce a voltage output. There is usually an offset control for expanded scale measurements. It is a difference amplifier, meaning that it amplifies the difference between two input signals. An operational amplifier converts current output from sensors to a voltage and performs mathematical operations, such as comparison, summing, and logarithm calculations. It may also serve as a constant current source.

18. See Figure 6.10.

19. A thermocouple is a junction of two metals that produces a voltage proportional to temperature and is therefore a device for measuring temperature. A thermocouple can be calibrated with the use of known temperatures, such as the freezing point of water (slushy ice–water mixture) or temperatures known because they are measured with a calibrated thermometer.

20. The advantages of a single standard are that it takes less time and is less involved. The series of standards method is preferred because we do not rely on just one data point (for which an error may go undetected) and we can establish the response over a range of concentrations rather than at just one concentration.

21. 0.0170

22. 0.514

23. 7.3 ppm

24. 0.655 ppm

25. $C_u$ = 6.55 ppm

26. The linearity (or lack of linearity) of the readings is known only for the range of concentrations of the standards prepared. Without testing other standards, it is not known if the linearity extends beyond this range, and so the answer to the unknown cannot be reliably determined.

27. 1.8

28. 0.140

29. 4.84

30. The interpolation of the unknown results relies on the establishment of the linear relationship between the readout and the concentration.

31. The method of least squares is a procedure by which the best straight line through a series of data points is mathematically determined. More details are given in Section 6.4.4. It is useful because it eliminates guesswork as to the exact placement of the line and provides the slope and y-intercept of the line.

32. The best straight line fit is obtained when the sum of the squares of the individual y-axis value deviations (deviations between the plotted y values and the values on the proposed line) is at a minimum.

33. Linear regression is another name for the process of determining the straight line for a series of data points via the method of least squares.

34. Slope, y-intercept, correlation coefficient, and concentrations of samples.

35. The term perfectly linear data refers to data in which the instrument readout is exactly the same multiple of the concentration at all concentrations measured and all the points lie exactly on the line. A value of exactly 1 for the correlation coefficient would indicate perfect linear data.

36. Values that have at least two nines (0.99) are satisfactory for some devices, but values with up to three nines, and sometimes four nines, are attainable in some cases.

37. Serial dilution is the preparation of a series of solutions by always diluting the solution just prepared to make the next one. For example, to make solutions with concentrations 10, 20, 30, 40, and 50 ppm, serial dilution would mean to prepare the 50 ppm first, then to prepare the 40 ppm from the 50 ppm, the 30 ppm from the 40 pmm, the 20 ppm from the 30 ppm, etc.

38. The series of standard solutions does not work well when the instrument readout is dependent on some other variable factor in addition to the concentration, such as variable injection volume in gas chromatography or the variable solution viscosity in atomic absorption.

39.

| Concentration | Milliliters of 1000 ppm Needed |
|---|---|
| 1.00 | 0.0500 |
| 2.00 | 0.100 |
| 3.00 | 0.150 |
| 4.00 | 0.200 |
| 5.00 | 0.250 |

The milliliters of 1000 ppm needed is pipetted into separate 50-mL volumetric flasks and water is added to each to the 50-mL mark. Each flask is then shaken to make the solutions homogeneous. The pipet needed would be a small serological pipet, perhaps 0.50 mL in capacity. Alternatively, a micropipet, such as described in Chapter 4, can be used.

40. Assuming the need to achieve three significant figures in the concentrations, the following volumes of the 100.0 ppm stock would need to be pipetted for 25.00 mL each of 2.00, 4.00, 6.00, and 8.00 ppm solutions, respectively: 0.500, 1.00, 1.50, and 2.00 mL.

41. The blank (reagent blank) is a solution that contains all the substances present in the standards and the unknown (if possible) except for the analyte. A sample blank takes into account any chemical changes that may take place as the sample is taken or prepared. See Section 6.6.2 for an example.

42. See Section 6.6.3.

43. Most of the time the unknown sample requires some pretreatment, such as dilution, extraction, or, if it is a solid, dissolving. The analytical concentration of the analyte in the untreated sample must usually be reported, rather than the concentration in the sample solution. Thus, a calculation is usually required to obtain the final answer.

44. 0.0131 mg

45. 0.462 mg

46. $1.86 \times 10^3$ ppm

47. $4.58 \times 10^4$ ppm

48. Assuming three significant figures in the volume measurements, 535 ppm.

49. 0.002562 g, 52.06 ppm

50. $7.50 \times 10^2$ ppm

51. Assuming three significant figures in the volume measurements, $2.09 \times 10^3$ ppm.

52. Assuming three significant figures in the volume measurements, 0.650 ppm.

53. Data acquisition by computer refers to the use of a computer to obtain data (instrument readout values) directly by interfacing to the instrument. These data are then found in the computer's memory, or on disk, and are not necessarily recorded independently on a recorder or in a notebook.

54. The volume of data acquired in modern laboratories is such that computer storage is most efficient and eliminates the need to provide the space for hard copies.

55. For plotting standard curves.

56. Laboratory information management system. See Section 6.8.2.

# Chapter 7

1. See Section 7.1.

2. The strategy for spectrochemical analysis specifies the type of instrument readings made on the standards and samples: light absorption and emission measurements.

3. Some qualities of light are best explained if we describe it as consisting of moving particles, often called photons or quanta (called the particle theory of light). Other qualities are best explained if

we describe it as consisting of moving electromagnetic disturbances referred to as electromagnetic waves (called the wave theory of light). Thus light has a dual nature.

4. Mechanical waves require matter to exist. Electromagnetic waves do not. Since outer space is relatively free of matter, electromagnetic waves can exist there, while mechanical waves cannot.

5. See Section 7.2.1.

6. (a) $8.31 \times 10^{-5}$ cm
   (b) $7.49 \times 10^{-5}$ cm
   (c) $4.297 \times 10^{-5}$ cm
   (d) $3.826 \times 10^{-5}$ cm

7. (a) 0.317 nm
   (b) $5.11 \times 10^3$ nm

8. (a) $4.79 \times 10^{-2}$ cm
   (b) $3.84 \times 10^{-3}$ cm
   (c) $7.61 \times 10^{-4}$ cm

9. (a) $7.04 \times 10^{13}$ sec$^{-1}$
   (b) $4.13 \times 10^{14}$ sec$^{-1}$
   (c) $4.59 \times 10^{14}$ sec$^{-1}$

10. (a) $2.00 \times 10^{-20}$ J
    (b) $2.47 \times 10^{-22}$ J
    (c) $5.07 \times 10^{-23}$ J

11. (a) $1.04 \times 10^{11}$ sec$^{-1}$
    (b) $4.10 \times 10^8$ sec$^{-1}$

12. (a) $4.31 \times 10^{-26}$ J
    (b) $4.24 \times 10^{-18}$ J

13. (a) 4.64 cm
    (b) 0.0220 cm

14. (a) 176 cm$^{-1}$
    (b) $1.06 \times 10^3$ cm$^{-1}$

15. (a) $1.70 \times 10^{15}$ sec$^{-1}$
    (b) $9.45 \times 10^{15}$ sec$^{-1}$

16. (a) $1.26 \times 10^4$ cm$^{-1}$
    (b) $2.99 \times 10^4$ cm$^{-1}$

17. (a) $2.00 \times 10^{-8}$ cm
    (b) $1.62 \times 10^{-6}$ cm

18. (a) $2.54 \times 10^{-5}$ cm
    (b) $1.18 \times 10^{15}$ sec$^{-1}$
    (c) $7.83 \times 10^{-19}$ J
    (d) $3.94 \times 10^4$ cm$^{-1}$

19. Light with wavelength 627 Å

20. Light with frequency $7.84 \times 10^{13}$ sec$^{-1}$

21. Light with wavelength 591 nm

22. Light with energy $5.23 \times 10^{-14}$ J

23. Light with frequency $7.14 \times 10^{13}$ sec$^{-1}$

24. (a) A                (b) B                (c) A

25. (a) decreased        (b) decreased        (c) decreased

26. IR light has a longer wavelength and lower frequency and wavenumber than UV light.

27. Energy and frequency: radio waves < infrared light < visible light < UV light < x-rays.
    Wavelength: x-rays < UV light < visible light < infrared light < radio waves.

28. Lower limit, approximately 350 nm; upper limit, approximately 750 nm.

29. (a) UV has more energy than IR.
    (b) UV causes electronic transitions; IR causes vibrational and rotational transitions.

30. Yellow-colored objects appear to be yellow because yellow wavelengths are reflected and not absorbed like the other wavelengths.

31. See Figures 7.7, 7.10, and 7.11, for example.

32. If the energy of light striking the atoms or molecules exactly matches an energy transition possible within the atoms or molecules, i.e., the transition of an electron from a lower state to a higher state, the transition will occur and the energy that once was light will be possessed by the atoms or molecules, i.e., absorption.

33. An electronic transition is one in which an electron in an atom or molecule is moved from one electronic state to another with the absorption of the equivalent energy, such as from UV or visible light. A vibrational transition is one in which a molecular bond's vibrational state changes due to the movement from one vibrational energy level to another because of the absorption of the equivalent energy, such as from short-wavelength infrared light. A rotational transition is one in which a molecule's rotational state changes due to the movement from one rotational energy level to another because of the absorption of the equivalent energy, such as from longer-wavelength infrared light. Electronic transitions require the most energy—the energy of visible or UV light. Rotational transitions require the least energy—longer-wavelength IR light.

34. See Figure 7.12.

35. UV-VIS spectrophotometry is a technique that measures molecules and complex ions using light in the ultraviolet regions of the electromagnetic spectrum. IR spectrometry is a technique that measures molecules using the infrared region of the electromagnetic spectrum. Atomic spectroscopy is a technique that measures atoms using light in the UV and visible regions of the electromagnetic spectrum.

36. An absorption spectrum is the plot of absorbance vs. wavelength—the unique pattern of absorption useful for qualitative analysis. A molecular absorption spectrum is of the continuous variety, while an atomic absorption spectrum is a line spectrum. Only specific wavelengths get absorbed by atoms because only specific energy transitions are possible (no vibrational transitions—only electronic). Both vibrational and electronic are possible with molecules, and thus all wavelengths get absorbed to some degree.

37. A line spectrum is an absorption or emission spectrum that displays a series of vertical lines indicating that only certain narrow wavelength bands (lines) are absorbed or emitted. A line spectrum results when atoms are measured. This is the case because there are no vibrational levels in atoms, and therefore only very few transitions are allowed.

38. It is because of the presence or absence of vibrational energy levels in the species in question. Atoms do not have vibrational levels and so very few transitions are possible. Molecules have vibrational levels superimposed on the electronic levels, so many transitions are allowed.

39. Because no two chemical substances display identical spectra.

40. A transmission spectrum is a plot of transmittance or percent transmittance vs. wavelength. An emission spectrum is a plot of emission intensity vs. wavelength.

41. An energy level diagram meant to depict atomic absorption will have arrows pointing upward to indicate that energy is being absorbed, resulting in a transition from lower to higher energy levels. An energy level diagram meant to depict atomic emission will have arrows pointing downward to indicate that energy is being emitted, resulting in a transition from higher to lower energy levels.

42. $T = I/I_o$. $I_o$ = intensity of light striking the detector with the blank in the path of the light. $I$ = intensity of light striking the detector with a sample in the path of the light.

43. (a) 0.0857     (b) 0.308     (c) 0.613

44. (a) 0.331     (b) 0.539     (c) 0.166

45. (a) 0.239     (b) 0.465     (c) 0.315

46. (a) 40.6%     (b) 13.1%     (c) 32.4%

47. Beer's law is $A = abc$. A is absorbance, a is absorptivity, b is pathlength, and c is concentration.

48. 1.04

49. $3.72 \times 10^{-5} M$

50. $9.27 \times 10^{-6}\ M$
51. (a) 0.118                     (b) $0.981\ \text{L mol}^{-1}\ \text{cm}^{-1}$ (c) $7.87 \times 10^{-6}\ M$
52. T = 0.400. In order to calculate the molar absorptivity, the molarity of the solution would be needed.
53. $1.46 \times 10^{5}\ \text{L mol}^{-1}\ \text{cm}^{-1}$
54. $1.84 \times 10^{-5}\ M$
55. $1.8 \times 10^{-5}\ M$
56. 0.405 cm
57. 0.0234 cm
58. 0.886
59. $1.67 \times 10^{3}\ \text{L mol}^{-1}\ \text{cm}^{-1}$
60. (a) 0.259                     (b) $8.38 \times 10^{3}\ \text{L mol}^{-1}\ \text{cm}^{-1}$                     (c) 0.505
    (d) 0.428                     (e) 90.2%
61. (a) 0.186                     (b) $5.98 \times 10^{3}\ \text{L mol}^{-1}\ \text{cm}^{-1}$                     (c) 0.635
    (d) 0.628                     (e) 89.6%
62. 3.79 ppm
63. 3.31 ppm
64. In order, from top to bottom: e, f, a, g, h, d, b, i, c
65. (a) 522 nm, because this is the wavelength of maximum absorbance.
    (b) Estimating percent transmittance at 23%, A = 0.64.
    (c) 13 L/mol cm
66. See Figure 7.20.
67. The wavelength giving the most absorbance is the wavelength giving the best sensitivity.

# Chapter 8

1. See Figure 7.12.
2. See Section 8.2.1.
3. The tungsten filament source is a visible light source. It is a light bulb with a tungsten filament and emits the visible wavelengths with significant intensity, but the intensity varies with wavelength, as it does with all light sources. The deuterium lamp is a UV light source. It contains deuterium gas at a low pressure and emits UV light when electricity is applied across a pair of electrodes. The xenon arc lamp is a source for both visible and UV light. It contains xenon at a high pressure, and UV and visible is generated via a discharge across a pair of electrodes.
4. No, the intensity varies with wavelength. See Figure 8.1 for the tungsten filament intensity profile.
5. Either a radiation filter (glass) or monochromator is used. A light filter absorbs all wavelengths except for a somewhat narrow band of wavelengths (see Figure 8.2). A monochromator utilizes a dispersing element in combination with a slit to select a very narrow wavelength band (see Figure 8.3).
6. Bandwidth is the width of the wavelength band that is allowed to exit a monochromator. The narrowness of this band is called the resolution. High resolution corresponds to a very narrow bandwidth and vice versa.
7. An absorption filter is a wavelength selector consisting of a piece of glass that transmits only a certain rather narrow band of wavelengths from a light source and absorbs the rest. A monochromator is a wavelength selector consisting of a dispersing element–slit combination that selects a very narrow band of wavelengths by sliding the spray of wavelength coming from the dispersing element across the exit slit, as shown in Figure 8.4.
8. A monochromator is a wavelength selector. It consists of two slits and a dispersing element in combination. The dispersing element splits the light from a source into the wavelengths of which

it is composed. Upon rotating this element, different wavelengths pass through the exit slit, and thus the position of the dispersing element dictates what wavelength is selected.

9. Turning the knob on the exterior of the instrument rotates the dispersing element of a mono-chromator on the interior of the instrument such that the spray of wavelengths coming from the dispersing element slides across the exit slit. At 728 nm, only a very narrow wavelength band emerges from the exit slit.

10. A diffraction grating is a dispersing element consisting of a highly polished mirror with a large number of regular, narrowly spaced lines scribed on the surface.

11. The sample compartment must consist of a light–tight box so that no stray light—only light from the instrument's light source—reaches the light sensor.

12. A single-beam spectrophotometer is one in which a single, continuous light beam shines from the light source through the monochromator and sample compartment to the detector. A double-beam spectrophotometer is one in which a single beam of light from the source is split into two beams in order to provide certain advantages. A single-beam instrument would be used in situations in which the expense of the double-beam is not warranted, such as in less precise work.

13. The precalibration of a single-beam spectrophotometer consists of tweaking the readout to read 100% T when the blank is in the sample compartment.

14. It is slow and tedious for an experiment in which a molecular absorption spectrum is measured. It is slow and tedious because the wavelengths are manually scanned in small increments, and each time the wavelength is changed, the calibration step with the blank needs to be performed due to the variability of light intensity from the light source at the different wavelengths.

15. Rapid-scanning single-beam instruments exist. The absorption spectrum of the blank is first obtained, followed by that of the sample. The sample scan is then adjusted to its proper measure-ment by using the spectrum of the blank.

16. Since the time between reading the blank and reading the sample can be significant, the instrument may lose its calibration due to minor electrical fluctuations in either the source or detector.

17. A light chopper is a rotating circular partial mirror used to split a light beam into two beams. A beamsplitter is a mirror with slots also used to split a light beam into two beams. See Figure 8.5. A chopper creates a double beam in time, while a beamsplitter creates a double beam in space.

18. A double-beam spectrophotometer is one in which either a beamsplitter or light chopper is used to create two beams of light in order to deal with the problem of variable light intensity of the different wavelengths emitted by the source.

19. The two designs are those shown in Figures 8.6 and 8.7. In one case, the second beam passes through the blank, while in the other case, the second beam goes directly to a detector.

20. 1) Do not have to continually replace sample with blank when obtaining a molecular absorption spectrum; 2) errors due to light source and detector fluctuations are minimized; and 3) accurate rapid scanning of wavelengths is possible.

21. A double-beam instrument is preferred for rapid scanning because adjustments for intensity changes after each wavelength can be made immediately before a sample is read. With a single-beam instrument, there is a delay.

22. A double-beam instrument is preferred for the reasons expressed in the answer to question 20.

23. A photomultiplier tube is a light sensor combined with a signal amplifier. See Section 8.2.4.

24. See Section 8.2.4.

25. A diode array spectrophotometer is one that utilizes a series of photodiodes to detect the light intensity of all wavelengths after the light has passed through the sample. See Figure 8.9. The advantage is that an absorption spectrum can be measured in a matter of seconds.

26. Cuvettes for visible spectrophotometry can be made of clear, colorless plastic. The only require-ment is that none of the visible light wavelengths be absorbed, which is the reason that they must be colorless.

27. Cuvettes are matched if they are identical in terms of pathlength and reflective–refractive properties. Cuvettes used for calibration and the analysis of samples after calibration must be matched so that absorption readings are due solely to concentration effects and not cuvette differences.

28. Things that may be different include pathlength and the reflective–refractive properties at the interior and exterior interfaces.

29. An interfering substance absorbs the same wavelength as the analyte or otherwise inhibits the accurate reading of the analyte.

30. A deviation from Beer's law refers to the linear relationship between absorbance and concentration becoming nonlinear, as in Figure 8.10.

31. The optimum working range for percent transmittance (to avoid instrumental deviations from Beer's law) is between 15 and 80%, which corresponds to an absorbance range of 0.10 to 0.82.

32. See Section 8.4.3.

33. The wavelength calibration can be checked by comparing the maximum absorbance wavelength for a known substance as measured by the instrument to what it is reported to be otherwise.

34. See Section 8.4.4.

35. The wavelength of fluorescence is longer than the wavelength of absorption. This is because vibrational relaxation occurs, causing the energy drop to be less between the excited and ground states. See Figure 8.11.

36. See the answer to question 35.

37. One is needed to select the wavelength of absorption and one to select the wavelength of fluorescence.

38. The fluorescence is measured (with a monochromator and phototube) at right angles to the incoming light beam.

39. (a) Fluorometers have two monochromators and a right-angle configuration.

    (b) Two monochromators are needed, one to select the wavelength of absorption and one to select the wavelength of fluorescence. A right-angle configuration is important in order to avoid measuring the light from the source while trying to measure the fluorescence.

40. The differences are as mentioned in the answer to question 39(a). See Figure 8.12 for the instrument diagram.

41. Fluorescence intensity.

42. The electrons in benzene ring systems are the kinds of electrons that undergo the energy changes required for fluorescence.

43. To say fluorometry is more selective means that there are fewer interferences. To say that it is more sensitive means that it can detect smaller concentrations.

44. A fluorometric procedure would be used when analyzing for a substance that is able to fluoresce and when the absorption procedure is prone to interferences and does not give a satisfactory sensitivity.

45. More sensitive, more selective.

46. Advantage: virtually no interferences exist.
    Disadvantage: not useful for very many analytes.

47. (a) Fluorometers have two monochromators and a right-angle confuration. Absorption spectrophotometers do not.

    (b) Fluorometry is more sensitive

    (c) Absorption spectrophotometry is more highly applicable.

48. Infrared absorption causes vibrational transitions. UV-VIS absorption causes electronic transitions.

49. UV, quartz glass; IR, inorganic salt crystals.

50. Similar: Both are molecular fingerprints and therefore useful for qualitative analysis, and both display absorption behavior over a wavelength range.
    Different: IR displays sharper absorption bands that have greater specificity. IR spectra are usually displayed as transmission spectra rather than absorption spectra. IR spectra utilize wavenumber, rather than wavelength, on the x-axis.

51. Infrared absorption patterns can be more directly assigned to more specific structural features.
52. Fourier transform infrared spectrometry.
53. (a) FTIR                                   (b) FTIR
    (c) Double-beam dispersive            (d) FTIR
    (e) Double-beam dispersive            (f) FTIR
54. An interferometer is a device that utilizes a moveable and a fixed mirror to manipulate the wave patterns of a split light beam to create constructive and destructive interference in this beam.
55. (a) FTIR                                   (b) FTIR
    (c) Double-beam dispersive            (d) FTIR
56. Inorganic compounds consist of ionic bonds that do not absorb infrared light and therefore would present no interfering absorption bands.
57. Describing a liquid as neat means that it is pure and not in solution.
58. 1) Sealed cell, 2) demountable cell, and 3) sealed demountable cell.
59. See Section 8.8.1.
60. The spacer between the NaCl windows.
61. Refer to Figure 8.20. When filling a cell with a liquid that has an unusually high viscosity, two syringes may be used—one to push the liquid into the cell by pushing down on the plunger of the syringe containing the liquid in the inlet port, and one to pull the liquid into the cell by pulling up on the plunger of an empty syringe in the outlet port.
62. 1) The windows may fog or become disfigured since water dissolves them; and 2) water will interfere with the detection of alcohols in the spectrum.
63. See number 1 in the answer to question 62.
64. The solvent will exhibit absorption bands that may interfere with those of the analyte. Carbon tetrachloride has only one kind of bond, the C–Cl bond, and this bond will not absorb IR light at wavelengths that are usually important.
65. 1) Dissolve in a solvent and then use a liquid sampling cell to measure the solution; 2) dissolve in a solvent, place several drops of the solution on a salt plate, evaporate the solvent, and measure the residue; 3) KBr pellet; 4) Nujol mull; and 5) reflectance.
66. A KBr pellet is a thin wafer of a mixture of KBr and the sample made by first thoroughly mixing the sample with dry, powdered KBr and then compressing a quantity of this mixture in a laboratory press.
67. 1) The best pellets are made from dry KBr; and 2) the presence of water will result in absorption patterns that may cause the analyst to make erroneous conclusions.
68. A hydraulic press may be used to help make a higher-quality KBr pellet.
69. A Nujol mull is a mixture of a solid sample with Nujol, or mineral oil, for the purpose of more conveniently obtaining an infrared spectrum of the solid sample.
70. Neither utilize solvents or other materials that would present possibly interfering absorption bands in the spectrum.
71. The mineral oil has a rather simple spectrum (only carbon–hydrogen bonds). Additionally, as with solvents, a computer maybe used to subtract the mineral oil bands from the spectrum of the solid.
72. See Section 8.9.5.
73. The diffuse reflectance method is a method for solids in which the powdered solid, held in a cup, is irradiated with the IR beam. The scattered reflected light is captured by the detector and the spectrum displayed.
74. Yes, gases may be measured by IR spectrometry. See Section 8.9.6.
75. Figure 8.32: A broad, fairly strong absorption band at 3300 cm$^{-1}$ indicating an alcohol. No benzene ring absorptions or carbonyl group. It is an aliphatic alcohol.
    Figure 8.33: Strong, sharp band at 1700 cm$^{-1}$ indicating a carbonyl group. Absorption bands on the high side of 3000 cm$^{-1}$ and a series of weak bands between 1700 and 2000 cm$^{-1}$ indicating a benzene ring. Possibly benzaldehyde, or a similar compound.

Figure 8.34: Strong, sharp absorption at 1700 cm$^{-1}$ indicating a carbonyl group. No other significant patterns except the C–H pattern on the low side of 3000 cm$^{-1}$. It is an aliphatic aldehyde or ketone. Figure 8.35: A benzene ring is indicated because of the band on the high side of 3000 cm$^{-1}$ and the series of weak peaks between 1700 and 2000 cm$^{-1}$. Aliphatic C–H bonds are also indicated (absorption bands on the low side of 3000 cm$^{-1}$). Possibly ethylbenzene, or a similar structure.

76. (a) It appears that Figure 8.37 is the spectrum of an alcohol because of the broad absorption in the 3500 cm$^{-1}$ range.

   (b) Yes, Figures 8.36 and 8.37 are both spectra of compounds that have a benzene ring. The three patterns to observe for benzene rings are sharp bands on the high side of 3000 cm$^{-1}$, a series of weak peaks between 1600 and 2000 cm$^{-1}$, and two peaks between 1500 and 1600 cm$^{-1}$.

   (c) Yes, Figure 8.38 is probably the spectrum of a compound with just C–H bonds. It does not have a benzene ring, however, because the spectrum does not display the absorption patterns of a benzene ring (see answer to (b) above).

   (d) Yes, Figure 8.36 is the spectrum of a compound that has a carbonyl group because it displays the strong sharp peak at 1700 cm$^{-1}$.

   (e) Figure 8.36: benzophenone, Figure 8.37: 3-methylphenol, Figure 8.38: n-pentane.

77. Calibrate the instrument by preparing a series of standard solutions of the analyte in a solvent, measuring the absorbance of each at an appropriate wavelength and plotting absorbance vs. concentration. Obtain the concentration of the unknown from this graph. See Section 8.11 for specific details.

# Chapter 9

1. Spectral lines for atoms are absorption or emission bands that are so narrow that they appear as lines rather than bands.

2. (a) See Figure 9.1. The sources of the lines are the very specific energy transitions that are allowed in atoms.

   (b) The two most common sources of atomic emission spectra are flames and the ICP torch.

3. A given emission line is caused by a transition between the same two energy levels as the corresponding absorption line. Since both therefore represent the same energy difference, they occur at the same wavelength.

4. See Section 9.1.

5. An atomizer is a device that forms atoms from ions. Examples include flames, a graphite furnace, an ICP source, vapor generators, etc.

6. Most common atomic absorption techniques: flame atomic absorption and graphite furnace atomic absorption. Most common atomic emission techniques: ICP and flame emission.

7. 1) solvent evaporates, 2) ions atomize, 3) atoms are raised to excited states, and 4) excited atoms drop back to ground state and emit light.

8. (a) Flame AA is therefore useful because there is a large percentage of unexcited atoms present that can absorb light from the light source.

   (b) Resonance is the continuous movement of atoms back and forth between the ground state and excited states.

9. (a) 1800 K              (b) 2300 K              (c) 2900 K              (d) 3100 K

10. Air–acetylene and N$_2$O–acetylene flames are the most commonly used because their temperatures are high enough to provide sufficient atomization for most metals while not burning at too fast a rate.

11. An oxygen–acetylene flame has a high burning velocity, which decreases the completeness of atomization and thus lowers the sensitivity.

12. The element to be analyzed is contained in the cathode of the hollow cathode lamp, since its atoms become excited and emit light and this light is what is needed for absorption in the flame. No monochromator is needed because the wavelength is already specific for the atoms in the flame.

13. Unless the analyte metal is contained in the cathode, the lamp will not emit the required wavelength.

14. Many hollow cathode lamps are needed because each element, since it must be contained in the cathode, requires a different lamp. Some lamps, however, are multielement. The element analyzed must be contained in the cathode so that its line spectrum will be generated and absorbed by the same in the flame.

15. EDL stands for electrodeless discharge lamp. It is an alternative to the hollow cathode lamp as a light source in atomic absorption spectroscopy.

16. A nebulizer is a device that converts a flowing liquid to a fine mist or cloud. A nebulizer is needed in conjunction with a premix burner so that the analyte solution can be sufficiently mixed with the fuel and oxidant gases prior to reaching the flame.

17. Refer to Figure 9.8 and the accompanying discussion.

18. Since the solution, air, and fuel are premixed, some solution droplets will not make it all the way to the flame. These collect in the bottom of the mixing chamber unless they are allowed to drain out.

19. The light chopper serves to allow the detector to differentiate between the light emitted by the flame and the light originating from the light source. This, in turn, allows the detector to measure an absorbance that is free of the interfering light emitted by the flame.

20. The single-beam instrument uses a single light beam, albeit modulated, from the light source, through the flame and monochromator, to the detector. In a double-beam instrument, a second beam is created to bypass the flame to be rejoined after the first beam has passed through the flame. See Figures 9.10 and 9.11. The advantage of the single beam is a less expensive and less complicated instrument. The advantage of the double beam is that it eliminates problems due to source drift and noise.

21. The reference beam in the atomic absorption instrument does not pass through the blank, but merely bypasses the flame. Thus the fluctuations in light intensity are accounted for, but the blank adjustment must be made at a separate time.

22. Room light is mostly eliminated by the monochromator positioned between the flame and the detector. Room light that is the same wavelength as the light being measured is eliminated along with flame emissions by modulating or chopping the light from the source such that the detector electronics is able to generate a signal based only on the light from the source.

23. The primary line is the line in the line spectrum that is used most often for analyzing for that element because it is the most sensitive and useful. Secondary lines are other lines that are sometimes used for various reasons.

24. Controls that need to be optimized are wavelength, slit width, lamp current, lamp alignment, aspiration rate, burner head position, and fuel and oxidant flow rates. See Section 9.3.5 for details.

25. Yes, Beer's law applies. The width of the flame is the pathlength, and each analyte has an absorptivity for the conditions chosen.

26. A chemical interference is one in which the sample matrix affects the chemical behavior of the analyte. A spectral interference is one that interferes with accurate measurement of the desired spectral line.

27. Matrix matching refers to the preparation of standards in such a way that their matrices match that of the sample as closely as possible. It is important because there may be a component in the sample that affects the reading in some way, and unless that component is present in the standards at the same concentration level, it may affect the results in a negative way.

28. The standard additions method is one in which the standards prepared for the standard curve consist of the sample to which varying amounts of a standard solution have been added. It helps with the problem of chemical interferences because the sample is the matrix for the standards and the interference occurs in all measurements such that its effect is negated.

29. 1) It is not possible to prepare a blank with the same matrix, and 2) the sample concentration is outside the range of the standards, and the linearity of the standard curve cannot be verified.

30. To release the calcium from the sample matrix so that it can be atomized in the flame.

31. Background absorption is light absorption due to molecular substances or particles in the flame. Background correction refers to the technique in which the background absorption is isolated and subtracted out. The absorption of a continuum light beam passing through the flame along with the light from the source allows the required subtraction to take place.

32. See Section 9.3.7.

33. The graphite furnace method of atomization utilizes a small graphite tube furnace to electrically heat rapidly a small volume of the analyte solution contained inside to a temperature that eventually causes atomization.

34. The analyte solution is placed (injected) into the furnace with a micropipet or auto-sampler. Following this, a temperature program is initiated in which the furnace heats rapidly to: 1) evaporate the solvent, 2) char the solid residue, and finally 3) atomize the analyte, creating the atomic vapor.

35. Compare Figure 9.10 with Figure 9.14. The difference is that the light beam passes through the furnace tube rather than the flame.

36. Argon gas is needed to provide an inert atmosphere for the graphite surface so that it is not oxidized (and damaged) by exposure to air at the high temperature. Cold water is needed to provide rapid cooldown between measurements. A source of high voltage is needed to accommodate the rapid heating to the very high temperature required.

37. See Figure 9.17.

38. It must be protected from air because air, at the high temperature achieved during the experiment, can oxidize and damage and disintegrate the graphite surface.

39. The absorbance signal originates from a very small volume of solution placed in the furnace, and since the furnace is continuously flushed with an inert gas, the vapors from this volume are swept out of the furnace after a short time.

40. Advantages: 1) high sensitivity, 2) only small volumes of sample needed, and 3) better detection limit. Disadvantages: 1) matrix effects, and 2) poor precision.

41. The Zeeman background correction is a background correction procedure for graphite furnace AA in which a powerful pulsing magnetic field is used to shift the energy levels of atoms and molecules and thereby shift the wavelengths that are absorbed and allow subtraction of the background.

42. Inductively coupled plasma. Flame photometry, because emission is measured and not absorption.

43. See Section 9.5.

44. Advantages: 1) more sensitive, 2) broader concentration range measurable, and 3) multielement analysis possible. Disadvantages: cost.

45. (a) See answer to number 40.        (b) See answer to number 44.

46. Mercury is the only metal that is a liquid at room temperature and is therefore the only metal that has a significant vapor pressure such that an atomic vapor can be created without heat.

47. The hydride generation technique is a technique in which volatile metal hydrides are formed by chemical reaction of the analyte solutions with sodium borohydride. The hydrides are guided to the path of the light, heated to relatively low temperatures, and atomized. It is useful because it provides an improved method for arsenic, bismuth, germanium, lead, antimony, selenium, tin, and tellurium.

48. See Table 9.3 and discussions in the text.

49. (a) ICP
    (b) Atomic fluorescence
    (c) Flame AA, graphite furnace AA, and atomic fluorescence
    (d) Spark emission
    (e) Flame photometry, atomic fluorescence, ICP, and spark emission
    (f) Graphite furnace AA
    (g) ICP

50. (a) T          (b) T          (c) F          (d) T
    (e) F          (f) F          (g) T          (h) F

| (i) T | (j) T | (k) T | (l) F |
|-------|-------|-------|-------|
| (m) T | (n) F | (o) F | (p) T |
| (q) F | (r) T | (s) T | (t) T |

51. Sensitivity is the concentration of analyte that will produce an absorption of 1%. Detection limit is the concentration that gives a readout level that is double the noise level in the baseline. See Table 9.2.

# Chapter 10

1. X-ray diffraction—Measurement of atomic and ionic spacings in crystal structures and thicknesses of thin metal films.
   X-ray absorption—Quantitative analysis of heavy metals in matrices of lighter metals.
   X-ray fluorescence—Quantitative analysis for elements in a wide variety of materials.
2. X-ray tube—A metal anode is bombarded with high-energy electrons causing inner-shell electrons to be ejected and replaced by higher shell electrons. The loss in energy of these electrons as they drop to the lower levels is on the order of the energy of x-rays, and x-rays are emitted.
   X-ray fluorescence—X-rays emitted by an x-ray tube irradiate an elemental material causing inner-shell electrons to be ejected. X-ray emission then follows, as with the x-ray tube.
3. $K_\alpha$ emission—An x-ray emission resulting from an electron dropping from the L shell to the K shell.
   $L_\beta$ emission—An x-ray emission resulting from an electron dropping from the N shell to the L shell.
   $K_{\alpha 2}$ emission—An x-ray emission resulting from an electron dropping from the next to lowest sublevel in the L shell to the K shell.
4. Diffraction is the scattering of light that occurs when the light is directed at a structure that has regularly spaced lines or points in it, and these spacings are similar to the wavelength of the light. The scattering takes place because of internal reflections that occur and the destructive and constructive interferences that then also occur.
5. Bragg's law is $n\lambda = 2d \sin\theta$. n is an integer, the order of the diffraction, $\lambda$ is the wavelength in angstroms, d is the interplanar spacing, and $\theta$ is the diffraction angle in degrees (see Figure 10.4).
6. 2.04 Å
7. 65.75°
8. It is fluorescence because a longer wavelength (lower energy) is emitted than the wavelength absorbed.
9. It is atomic fluorescence because it occurs with unbound atoms rather than with molecules.
10. Use of the analyzer crystal (wavelength-dispersive system) means better resolution, but it is slower and less sensitive.
11. See Section 10.3.3.
12. Nuclear magnetic resonance
13. The nuclear energy transitions that occur with the absorption of radio frequency light occur only in a strong magnetic field.
14. Radio frequency wavelengths, because the energy required to cause the nuclear energy transitions in a magnetic field is on the order of radio frequency energy.
15. The two states are: 1) when the small magnetic field due to the spinning–precessing nucleus aligns with the applied magnetic field, and (2) when it opposes the applied magnetic field. The latter is a slightly higher energy state.
16. Most of the time, the nucleus measured is the $^1$H nucleus, which is essentially a proton. Thus the P in PMR stands for proton.
17. Fourier transform nuclear magnetic resonance
18. Hertz, cycles per second; megahertz, million cycles per second.

19. A 60-MHz instrument requires a field strength of 14,092 G, while a 100-MHz instrument requires 23,486 G.

20. The gauss is a unit of magnetic field strength.

21. Seven components: magnet, sample holder, RF generator, RF detector, sweep generator, sweep coils, and data system. See Section 10.4.2 for the function of each.

22. In FTNMR experiments, very brief repeating pulses of RF energy are applied to the sample while holding the magnetic field constant. The resulting detector signal contains the absorption information for all the nuclear energy transitions of all $^1H$ nuclei. The Fourier transformation sorts out and plots the absorption as a function of frequency, giving the same result, the NMR spectrum, as in a traditional experiment but in much less time.

23. y-axis, absorption; x-axis, chemical shift (in parts per million).

24. The chemical shift is the effect that the environment of a nucleus has on the position of an absorption peak in an NMR spectrum. Such peaks are shifted to an extent dependent on this environment.

25. The structural features near each hydrogen in the molecule, since they display a small and variable magnetic field of their own, impact the magnetic field that these hydrogens see and thus the apparent magnetic field values at which the molecule absorbs.

26. (a) 1                    (b) 2                    (c) 3                    (d) 1
    (e) 2

27. There are two kinds of hydrogen, the methyl group hydrogen and the hydrogen of the alcohol functional group, thus two peaks.

28. TMS provides a reference for displaying peaks of each type of hydrogen. The chemical shifts observed for each type of hydrogen are so slight that they would be difficult to measure without this reference.

29. Since there are three separate peaks, there are probably three different kinds of hydrogens in the structure.

30. The integrator trace tells us that the number of hydrogens represented by the peak at 5 ppm is one half the number at 4.2 ppm and one third the number at 1.8 ppm.

31. Three peaks mean three different kinds of hydrogens. There are three different kinds of hydrogens in ethyl alcohol, but only two different kinds in diethyl ether.

32. The integration trace shows that the three peaks are in the ratio 1:2:3, left to right. This is what is expected with ethyl alcohol since there are one of one kind of hydrogen, two of another, and three of another. For ethyl methyl ether, a ratio of 3:2:3 would be expected.

33. One quartet is expected because the carbon with two hydrogens is adjacent to a carbon with three (n) equivalent hydrogens (n + 1 = 4). One triplet is expected because the carbon with three hydrogens is adjacent to a carbon with two (n) equivalent hydrogens (n + 1 = 3).

34. The spectrum is that of isopropyl alcohol. The septet is due to the splitting of the peak for the hydrogen on the center carbon because of the six (n) equivalent hydrogens on either side. The doublet is due to the splitting of the peak for the six hydrogens on the two methyl groups because of the one hydrogen on the adjacent carbon. The singlet is due to the hydrogen on the OH group.

35. The integrator trace would show that the singlet is due to one hydrogen, the septet due to one hydrogen, and the doublet due to six hydrogens.

36. A given peak will be split into a number of peaks equal to the number of equivalent hydrogens on the adjacent carbons plus 1 (the n + 1 rule). This represents a clue about the structure that can help lead to identification of the compound.

37. The high-energy electron beam is used to fragment the molecules of the analyte into particles of mass-to-charge ratios characteristic of that molecule.

38. Particles of different mass-to-charge ratios will exhibit different properties in the various mass analyzer designs and be detected accordingly. This is useful for qualitative analysis because the different mass-to-charge ratios represent molecular fragments that can be pieced together such that the compound can be identified.

39. A high-vacuum system is needed so that the components of air will not be fragmented and cause an interference.

40. See Sections 10.5.2 to 10.5.5.

41. y-axis, fragment count; x-axis, mass-to-charge ratio.

42. Mass spectroscopy is useful for qualitative analysis because the signals due to the different mass-to-charge ratios, as seen in the mass spectrum, represent molecular fragments that can be pieced together such that the compound can be identified.

43. The three largest peaks: 43—$CH_3CO^+$, 57—$CH_3COCH_2^+$, 15—$CH_3^+$.

44. Methanol is Figure 10.25, ethanol is Figure 10.26, benzene is Figure 10.27, and 2-butanone is Figure 10.28.

45. Different isotopes are detected because the differing count of neutrons that characterizes atoms of different isotopes produces ions of different mass-to-charge ratios in the mass spectrometer. Their relative abundances are related to the intensity of the respective signals in the mass spectrum.

46. In ICP-MS we measure metal ions produced in the plasma. Since these are already particles of a particular mass and charge, there is no need to use an electron beam to create them.

47. ICP-MS offers a very high sensitivity. See Table 10.1.

48. GC-MS, gas chromatography–mass spectrometry; LC-MS, liquid chromatography–mass spectrometry.

49. Mass spectrometry can serve as the detection system for mixture components as they elute from instrumental chromatographic columns. The specific advantage is that the mass spectrum of each mixture component is produced, and this greatly aids in identification.

# Chapter 11

1. It is not uncommon for real-world analysis samples to be very complex in terms of the number of chemical substances present. The study of modern separation science is thus very important in analytical chemistry from the standpoint that many potentially interfering substances must be identified and eliminated.

2. Recrystallization is the process by which an impure crystalline substance is purified by the dissolving and subsequent reconstitution of the crystals such that both soluble and insoluble impurities are removed. See the text for more details.

Distillation is the process by which two or more liquids, or a solution of liquids and solids, are separated as a result of differences in vapor pressure and boiling points through the evaporation and condensation of the mixture components above the boiling liquid mixture.

Fractional distillation is the same as distillation, except that a fractionating column (a tube containing a high-surface-area material) is positioned above the boiling liquid mixture such that continuous evaporation and condensation occur with time, resulting in a cleaner separation of all components.

Extraction is the separation of a particular mixture component from a sample through contact with a solvent in which the component is soluble.

Liquid–liquid extraction is the separation of a solute from a liquid sample with the use of a liquid extracting solvent that is immiscible with the sample.

Solvent extraction is another name for liquid–liquid extraction.

Countercurrent distribution is a liquid–liquid extraction process in which actually dozens of extractions are performed when extracting solvent and sample solvent are contacted in the manner depicted in Figure 11.4. It is a process involving a series of separatory funnels in a way that approximates partition chromatography.

Liquid–solid extraction is the separation of a component of a solid sample through contact with a liquid extractant.

Chromatography is the separation of mixture components as a result of the varying degrees of interaction that the mixture components have with a mobile phase and a stationary phase.

3. Insoluble impurities are removed by filtering at the elevated temperature—they are filtered out while all other mixture components pass through the filter. Soluble impurities are removed after cooling again by filtration—the purified crystals are captured by the filter while the soluble impurities pass through with the filtrate.

4. The minimum amount of solvent is important so as to maximize the yield of crystals upon cooling. Using too much solvent would cause more solid to stay dissolved when cooled.

5. Vapor pressure is a measure of the tendency of a substance to be in the gas phase at a given temperature. At the boiling point of a liquid mixture, the component with the higher vapor pressure will have a higher concentration in the condensing vapors and thus will be purer than it was before.

6. A high vapor pressure compared to the hardness minerals means that water has a much stronger tendency to evaporate such that upon distillation, water will vaporize free of contamination from the minerals and, when the vapors are condensed, will be much purer than before.

7. The difference in vapor pressure and boiling point of water and dissolved hardness minerals is substantial, such that one simple distillation will result in significant purification. In the case of two liquids, however, these differences are much narrower, such that many distillations or a fractionating column is required.

8. Two liquids in a mixture typically have a significant vapor pressure and similar boiling points. Thus, a clean separation does not occur with only a simple distillation. A fractional distillation is usually required.

9. A theoretical plate consists of one evaporation–condensation step in the distillation process. The height equivalent to a theoretical plate is the length of fractioning column in which is contained one theoretical plate.

10. In a liquid–liquid extraction experiment, two immiscible liquids, one pure and one with the analyte dissolved in it, are brought into intimate contact by vigorous shaking in a common container (separatory funnel). If the analyte is more soluble in the pure liquid than in the original liquid, its molecules move from the original to the pure during the shaking. Following the shaking, the immiscible liquids separate into two liquid layers in the container and most of the analyte is found in the extracting solvent.

11. The two liquid phases must be immiscible, and the analyte must be more soluble in the extracting solvent.

12. The separatory funnel is the glass or plastic container that is used for liquid–liquid extractions. It is shaped like an inverted teardrop and has a stopper at the top and a stopcock at the bottom (See Figure 11.2). It is used as described in number 10 above in a repeated shaking and venting procedure, as pictured in Figure 11.3. Following the extraction, the liquid layers are allowed to separate and the bottom layer is drawn off through the stopcock.

13. If the water containing the iodine is placed in the same container as pure hexane and the container shaken, extraction will be demonstrated because when the iodine transfers to the hexane layer, this layer will turn pink.

14. (a) 4.6 $M$ (b) 0.0447 $M$

15. If we assume that the concentration of analyte in the original solvent is 0.060 $M$ *before* extraction, then, after rounding the answer of the calculation according to significant figure rules, the concentration in the extraction solvent *after* extraction is also 0.060 $M$ (a very large distribution coefficient, as in this problem, would indicate that virtually all the analyte is extracted).

16. Distribution coefficient = 0.731%, extracted = 19.6%.

17. As in number 15 above, the distribution coefficient is very large and the calculation shows, after rounding according to significant figure rules, that virtually all of the analyte is extracted, 0.037 g.

18. These two quantities are the same, except that the distribution ratio takes into account all dissolved forms of the analyte, while the distribution coefficient takes into account only one form. If only one form exists, then the two are identical.

19. An evaporator is a device that eliminates excess solvent from a liquid solution and concentrates the solutes by vaporizing a portion or all of the solvent. The modern way to accomplish this is by using a stream of inert gas blowing over the surface of the solution while applying gentle heat.

20. A solid–liquid extraction is the selective dissolution of a component of a solid material through contact with a liquid solvent. Two methods of accomplishing this are: 1) shaking the solid with the solvent in the same container for a specified period of time, and 2) using a Soxhlet extractor, in which fresh solvent is continually cycled through the solid for a specified period of time.

21. No. Besides the convenient use of the separatory funnel for the actual extraction, they are also designed for easy separation of two immiscible liquids after the extraction through the stopcock. It would not be easy to separate a liquid from a solid through the stopcock.

22. Potassium, iron, phosphorus, and other elements in soil; formaldehyde residue in insulation, cellophane, and other materials; pesticide and herbicide residue in plants.

23. Chromatography is a separate technique in which mixture components are separated based on the differences in the extent of their interaction with two phases, a mobile phase and a stationary phase.

24. 1) Adsorption—Mixture components are separated based on varying degrees of adsorption forces between them and the stationary phase.
    2) Partition—Mixture components are separated based on solubility differences in the liquid stationary phase.
    3) Ion exchange—Ionic mixture components are separated based on the varying strength of ionic bonds formed with sites on the stationary phase.
    4) Size exclusion—Mixture components are separated based on the varying abilities to penetrate the pores on the stationary phase.

25. Partition chromatography utilizes the varying solubilities of the mixture components in a liquid stationary phase. Absorption chromatography utilizes the varying tendencies for mixture components to adhere to the surface of a solid stationary phase.

26. The apparatus consists of a series of separatory funnels by which the countercurrent distribution discussed in the text can be accomplished. The sample is introduced into the first separatory funnel along with the extracting solvent. Following this first extraction, each layer moves into other separatory funnels in which fresh sample solvent and fresh extracting solvent are introduced and shaken together. Each of the fours layers then move into other separatory funnels, introducing fresh solvents again, etc. The result is a separation that approximates chromatography, as discussed in the text and as pictured in Figure 11.4 of the text.

27. (a) B would emerge first because polar mixture components tend to dissolve more in the polar mobile phase and thus will come through the column with the mobile phase and emerge first. A, being nonpolar, will tend to remain behind in the stationary phase. (b) A, since it is nonpolar, will emerge first since it will tend to dissolve more in the nonpolar mobile phase.

28. (a) no         (b) yes         (c) yes         (d) no
    (e) yes        (f) no

29. The four types are partition, adsorption, ion exchange, and size exclusion.
    (a) Partition only     (b) None         (c) All four
    The five configurations are paper, thin layer, open column, instrumental gas, and instrumental liquid.
    (d) All five           (e) Thin layer, open column, and instrumental liquid

30. GC, gas chromatography; LC, liquid chromatography; GSC, gas–solid chromatography; LSC, liquid–solid chromatography; GLC, gas–liquid chromatography; LLC, liquid–liquid chromatography; BPC, bonded phase chromatography; IEC, ion exchange chromatography; SEC, size exclusion chromatography; GPC, gel permeation chromatography; and GFC, gel filtration chromatography.

31. Size exclusion chromatography. The separation occurs because the stationary phase particles are porous and the small molecules enter the pores and are slowed from passing through the column, while the large molecules pass through more quickly since they do not enter the pores.

32. With a cation exchange resin, the bonding sites are negatively charged and cations are exchanged. With an anion exchange resin, the bonding sites are positively charged and anions are exchanged.

33. The five configurations studied in this chapter are 1) paper chromatography, 2) thin-layer chromatography, 3) open-column chromatography, 4) gas chromatography, and 5) high-performance liquid chromatography.

34. (a) In open-column chromatography, the stationary phase consists of solid particles or a thin film of liquid bonded to solid particles, and is contained in a vertical tube. In HPLC, the stationary phase also consists of solid particles or a thin layer of liquid bonded to solid particles, but it is contained in a metal tube that does not have to be in any particular spatial orientation.

   (b) In open-column chromatography, the column is held vertically, so the force moving the mobile phase is gravity. In HPLC, the force that moves the mobile phase is the pressure from a high-pressure pump.

   (c) In the open-column procedure, fractions of eluate are collected in tubes and analyzed later by some specified method. In HPLC, an electronic sensor (detector) on the eluate end of the column senses the mixture components as they elute and displays peaks, the sizes of which are related to quantity.

35. (a) In thin-layer chromatography, the mobile phase moves by capillary action. In GC, the mobile phase moves by gas pressure.

   (b) Thin-layer chromatography is a planar procedure. GC is a column procedure.

   (c) In thin-layer chromatography, $R_f$ factors are used. In GC, retention times are used.

36. LSC is liquid–solid chromatography. Strictly speaking, any type that utilizes a solid stationary phase, namely, adsorption, ion exchange, and size exclusion, can be referred to as LSC. However, adsorption chromatography is the only one of this group that is routinely referred to as LSC.

37. The sensor sends an electronic signal to the data system and this signal increases and then decreases again as the individual mixtures elute, thus creating the appearance of a peak on the screen.

38. a-k, b-n, c-l, d-o, e-p, f-j, g-i, h-m

39. (a) Stationary phase, partition        (b) Size exclusion
    (c) Electrophoresis                     (d) HPLC
    (e) Adsorption                          (f) Thin layer

40. Blanks, left to right, starting upper left: water, liquids or dissolved solids, size exclusion, porous polymer beads, any liquid type, polymer beads with ionic sites; ions, gas, thin liquid film, thin layer, liquids or dissolved solids.

41. (a) F          (b) T          (c) T          (d) F
    (e) T          (f) F          (g) T          (h) T
    (i) F          (j) T          (k) T          (l) T
    (m) F          (n) F          (o) F          (p) T
    (q) T

42. (a) Thin-layer chromatography
    (b) Partition chromatography
    (c) Ion exchange chromatography
    (d) High-performance liquid chromatography
    (e) Partition chromatography and adsorption chromatography
    (f) Open-column chromatography
    (g) Partition chromatography
    (h) Paper chromatography and thin-layer chromatography

43. The retention time of a mixture component is the time from when the sample is first injected until the component's peak is at its apex. Adjusted retention time is the difference between the retention time of the mixture component and the retention time of air. See Figure 11.17.

44. See Figure 11.17.

45. R = 1.5. The two peaks are considered to be satisfactorily resolved since the resolution is 1.5 or greater.

46. N = $2.2 \times 10^2$ theoretical plates. H = 0.33 in.
47. Longer column, lower temperature, slower flow rate, different stationary phase.
48. The capacity factor is 3.7. It is within the optimum range, which is 2 to 6.
49. Selectivity = 1.82. This is a good separation based on the criterion that a value of 1.2 or better is considered good.
50. The capacity factor for methyl paraben is 2.7. The capacity factor for propyl paraben is 4.9.
51. See Figure 11.20.
52. See Section 11.9 for these descriptions.
53. See introduction under Section 11.9.
54. Capillary electrophoresis is an electrophoresis technique in which the mixture components are separated in a capillary tube and detected with an on-line detector after the separation occurs. The advantages include smaller quantity of sample and qualitative and quantitative analysis in a much shorter time.

# Chapter 12

1. GC, gas chromatography; GLC, gas–liquid chromatography; GSC, gas–solid chromatography.
2. Partition and adsorption
3. In the strategy for GC, it is noted that there may be no need for weight or volume data for the sample because the sample itself may be injected directly and quantitation performed solely from the chromatographic information. It is also noted that the internal standard method is common, and the solution preparation and calibration procedure are altered accordingly.
4. Carrier gas—A name for the mobile phase in gas chromatgraphy.
   Column—The designation for the tube in which the stationary phase in instrumental chromatography is contained.
   Vapor pressure—The tendency of a liquid substance to escape the liquid phase and become a gas, or the pressure exerted by the gas molecules of a substance above a liquid containing that substance.
5. Vapor pressure is a measure of the tendency of a substance to be in the gas phase at a given temperature. Since different mixture components will have different vapor pressures, the separation occurs in part due to the different tendencies of the components to be in the mobile gas phase.
6. A nonpolar mixture component with a high vapor pressure would have the shorter retention time because both the high vapor pressure and the fact that it has a polarity different from the stationary phase means that it would likely be found in the mobile phase most of the time.
7. Such a mixture of components would have a low vapor pressure and a high solubility in the stationary phase.
8. To flash vaporize the liquid samples.
9. The role of the detector is to generate the electronic signal when a mixture component elutes from the stationary phase.
10. The injection port, to flash vaporize the sample; the column, since the mixture components must remain gaseous and since vapor pressure depends on temperature; and the detector, in order to keep the mixture components from condensing.
11. The size of the liquid sample injected into the column must be small because after it is converted to a gas, its volume must still be small enough to fit onto the column all at once. If it were too large, it would not fit onto the column all at once and would bleed onto the column gradually over a period of time. This would mean that the eluting peaks, at best, would exhibit fronting and tailing, and at worst, would be broad and quite likely overlap with each other to the point of exhibiting poor or no resolution.
12. An open-tubular capillary column is a very long (30 to 300 ft), narrow-diameter tube in which the stationary phase is held in place by adsorption on the inside wall. Such a column is useful because it allows the use of a very long column (for better resolution) with minimal gas pressure required.

13. The packed column can be from 2 to 20 ft in length, typically has a diameter of $^1/_8$ or $^1/_4$ in., and has small particles, often coated with a thin layer of liquid stationary phase, packed in the tube. The open-tubular capillary column can be up to 300 ft in length, has an extraordinarily small diameter (capillary), and has the liquid stationary adsorbed on the inside surface of the tube. In terms of separation ability, the open-tubular capillary column is better because the mixture components contact more stationary phase (column is longer) while passing through the column. The amount injected for the open-tubular capillary column must be much less (0.1 mL maximum, as opposed to 20 mL for the $^1/_8$-in. packed column) because the column diameter is much less and a greater volume would overload it.

14. Analytical GC is the use of GC solely for analysis, qualitative or quantitative. Preparative GC is the use of GC for preparing pure samples for use in another experiment.

15. A changing of the column temperature in the middle of the run.

16. The temperature of the column affects both the vapor pressures of the mixture components and the solubilities of the mixture components in the stationary phase. Sometimes a low temperature separates some mixture components that are not separated at higher temperatures. However, the lower temperature may result in inordinately long retention times for other mixture components. Temperature programming allows the best of both worlds—complete separation of short-retention-time components at the low temperature, while shortening retention times of the others when the column temperature is raised later.

17. A temperature program should be used that begins at 80°C to facilitate separation of A and B, then increases to 150°C to decrease the retention times of C and D. This would ensure resolution of all components in a reasonable time.

18. Chromosorb is the trade name given to diatomaceous earth, the decayed silica skeletons of algae. It is the substrate material on which the liquid stationary phase is adsorbed in packed columns.

19. Low-molecular-weight alcohols are highly polar, thus FFAP or Casterwax would be useful in their separation.

20. Higher carrier gas flow rates would result in shorter retention times because when the mixture components are present in the mobile phase, they progress to the opposite end of the column at a faster rate.

21. A universal detector is one that generates an electronic signal for all mixture components regardless of their identity. More sensitive means that smaller quantities can be detected. More selective means that only some mixture components can be detected and that there are fewer interferences for these components.

22. (a) FID          (b) ECD          (c) TCD          (d) TCD
    (e) FID          (f) GC-MS        (g) GC-MS        (h) ECD
    (i) NPD          (j) GC-IR        (k) FPD          (l) PID
    (m) Hall

23. (a) FID                          (b) TCD
    (c) TCD, GC-MS                   (d) ECD, NPD, GC-MS, GC-IR
    (e) TCD                          (f) ECD
    (g) FID                          (h) TCD, ECD, GC-IR, PID
    (i) Hall                         (j) GC-MS, GC-IR

24. (a) Flame ionization detector     (b) Electron capture detector
    (c) Flame ionization detector     (d) Electron capture detector
    (e) Mass spectrometer detector    (f) Mass spectrometer detector
    (g) Thermal conductivity detector (h) Flame ionization detector
    (i) Electron capture detector     (j) Mass spectrometer detector
    (k) Thermal conductivity detector

25. (a) Flame ionization              (b) Thermal conductivity
    (c) Thermal conductivity          (d) Electron capture
    (e) Flame ionization

26. (a) More universal, does not destroy sample, safer to use

    (b) More sensitive

27. GC-MS stands for gas chromatography–mass spectrometry. The detector is a mass spectrometer. This detector determines the mass-to-charge ratio of fragments resulting from the destruction of component molecules by a high-energy electron beam and displays the mass spectrum, a graph of signal intensity vs. mass-to-charge ratio. This mass spectrum is a molecule fingerprint and can be fully recorded as the peak is being traced, adding a definitive dimension to qualitative analysis by GC.

28. If an unknown is totally unknown, retention time data are limited because there is uncertainty about whether untested compounds have the same retention times as those tested. However, if the unknown is not totally unknown and only a limited number of compounds is known to possibly be in the sample, retention time data can be quite valuable if all the known compounds have different retention times. In that case, it is a matter of matching up retention times of knowns with the peaks in the unknown's chromatogram.

29. (a) Yes. Benzene, ethylbenzene, and isopropylbenzene.

    (b) Toluene, n-propylbenzene

    (c) There may be other compounds, whose retention times have not been measured, that may exhibit the same retention times as toluene and n-propylbenzene. Also, there is no match for compounds B and D among the known liquids measured.

30. Retention time data by itself are never sufficient to identify mixture components unless it is known that the standards tested were the only possibilities.

31. (a) Carrier gas, mobile phase              (b) Retention time

    (c) Thermal conductivity detector          (d) Electron capture detector

    (e) Injection port, column, detector       (f) Peak area

    (g) Theoretical plates, resolution         (h) Temperature programming

    (i) Preparative GC                         (j) Flame ionization detector

    (k) Open-tubular column

32. Response factor = 0.939 cm$^2$/mg, % methylene chloride = 31%.

33. An internal standard is a compound that is added to the standard solutions in consistent amounts in order to eliminate the problem of variable injection volume. The inability to consistently inject a given volume accurately means that the analyte peak size is not reproducible. However, the ratio of the analyte peak size to the internal standard peak size *is* reproducible, and so this ratio is plotted on the y-axis of the standard curve vs. the analyte concentration.

34. The internal standard method uses an internal standard substance added in a constant amount to all standards and the sample. Area ratio of analyte peak to internal standard peak is plotted vs. concentration of analyte. The standard additions method uses the addition of the analyte in increasing amounts to the sample. Peak area is plotted vs. concentration added and the line is extrapolated to zero peak area to get the sample concentration.

35. (a) Internal standard—Series of standard solutions are prepared in which the analyte is present at increasing known concentrations and a second substance, the internal standard, is present at a constant concentration. This amount of internal standard is also added to the unknown.

    Standard additions—Standards are prepared by adding a constant known amount of the analyte to the unknown sample, with the chromatogram measured after each addition. Alternatively, a series of standards could be prepared with the unknown as the diluent.

    (b) Internal standard—The analyte peak and the internal standard peak are well resolved from each other and the solvent peak(s).

    Standard additions—The analyte peak grows proportional to the amount of analyte added.

    (c) Internal standard—Area ratio vs. concentration.

    Standard additions—Peak size vs. concentration added (extrapolation required).

36. (a) Pipet increasing amounts of ethyl alcohol into the flasks. Pipet a constant amount of isopropyl alcohol into each of the flasks. Dilute standards with an appropriate solvent. Unknown must have isopropyl added.
    (b) Ethyl alcohol and isopropyl alcohol peaks are well resolved. All other peaks need not be resolved (there would be many in the gasoline sample).
    (c) Area ratio vs. concentration.

37. Response factor method:
    (a) No sample or solution preparation required. One chromatogram of unknown sample required. However, considerable experimentation required to determine response factors (see text).
    (b) One chromatogram of unknown needed. Also, one chromatogram of each compound in the unknown in the pure state, in order to determine response factors.
    (c) No plotting required.
    Serial dilution method:
    (a) Prepare series of standards with analyte in increasing concentration. Unknown measured as is.
    (b) Only analyte peak need be resolved.
    (c) Peak size vs. concentration.
    Internal standard method and standard additions method: see question 35.

38. Peak tailing and fronting.

39. A leaky or plugged syringe, a worn septum, a leak in the pre- or postcolumn connections, or a contaminated detector.

40. An insufficiently conditioned column, a detector that has not achieved a stable temperature, or a faulty or contaminated detector.

41. Peaks from both injections will appear on the chromatogram and confuse the interpretation of the chromatogram.

42. Perform a rapid bakeout via temperature programming after the analyte peaks have eluted, use pure reagents, or replace or clean septa, carrier, or column.

43. 1.28%

44. The sample was injected directly without any prior preparation procedure that would require a calculation.

45. $2.10 \times 10^2$ ppm

# Chapter 13

1. High-performance (or pressure) liquid chromatography.

2. LC, liquid chromatography; LLC, liquid–liquid chromatography; LSC, liquid–solid chromatography; BPC, bonded phase chromatography; IEC, ion exchange chromatography; IC, ion chromatography; SEC, size exclusion chromatography; GPC, gel permeation chromatography; GFC, gel filtration chromatography.

3. High performance refers to the fact that complex mixtures can be resolved and quantitated accurately in a very short time.

4. HPLC (high-performance liquid chromatography) is an instrumental chromatography technique in which the mobile phase is a liquid. Refer to Figure 13.1 and the accompanying discussion for details of the HPLC system.

5. Refer to Figure 13.1. The mobile phase path begins in the solvent reservoir where a tube with a metal sinker/filter is immersed. A high-pressure pump draws the solvent out of the reservoir, through the pump, and past the sample injection device where the sample is introduced. It then proceeds to the guard column, where undesirable sample components are removed, and from there to the analytical column, where the components of interest are separated. The detector then detects mixture components as they elute. After passing through the detector, the mobile phase is channeled to waste.

6. Speed and overall performance.

7. (a) HPLC, high pressure pump; GC, pressure from compressed He cylinder.
   (b) HPLC, liquid (composition varies); GC, gas (usually helium).
   (c) HPLC, inside a short (about 1 ft) metal tube; GC, inside a glass or metal tube that can be about $1/4$ in. in diameter and up to 300 ft in length and can be a capillary tube, either $1/8$ or $1/4$ in. in diameter.
   (d) HPLC, all types; GC, partition and adsorption.
   (e) HPLC, not applicable; GC, separation of mixture components assisted by differences in vapor pressures.
   (f) HPLC, special injection valve; GC, mixture syringe injected into flowing carrier gas in heated injection port.
   (g) HPLC, solubility in liquid mobile phase vs. interaction with stationary phase; GC, vapor pressure and interaction with stationary phase.
   (h) HPLC, often standard analytical instruments but with flow cell; GC, many different designs that detect mixture components in the gas phase.
   (i) HPLC, computerized data system; GC, computerized data system.
   (j) HPLC, chromatogram with peaks; GC, chromatogram with peaks.
8. Mixture components move back and forth between the mobile phase and stationary phase as they pass through the column. In GC, the mobile phase is a gas, so the vapor pressure—the tendency to be in the gas phase—is important.
9. The pump operates by creating a vacuum to draw the mobile phase from the reservoir. If there are dissolved gases in the mobile phase, they can withdraw from solution when under this vacuum.
10. Mobile phases and samples contain small particulates that can damage the column and other hardware unless they are removed by filtration.
11. Degassing is the removal of dissolved gases from mobile phases and samples. It can be accomplished by: 1) reducing the pressure in the air space above the liquid in the container, 2) placing the container holding the liquid in an ultrasonic bath, or 3) both numbers 1 and 2 at the same time.
12. Helium sparging is a method for degassing a solvent consisting of a vigorous bubbling of helium gas through the solvent.
13. Mobile phases and samples contain small particulates that can damage the column unless they are removed by filtration.
14. Whether the samples or mobile phase contains an organic solvent dictates what filter material to use.
15. Paper may not be chemically inert to organic solvents. Nylon, on the other hand, is inert to organic solvents.
16. For those samples and mobile phases that contain an organic solvent.
17. A guard column is a short, less-expensive liquid chromatography column that is placed ahead of the analytical column in an HPLC system. The purpose of a guard column is to adsorb and retain mixture components that would contaminate the more expensive analytical column. In-line filters are relatively coarse filters (compared to prefilters) placed in the mobile phase line to filter out particulates that may be introduced on-line, such as from sample injection.
18. The in-line filter ahead of the column is a last-second filtering precaution, just in case particulates are introduced with the sample.
19. The HPLC pump must be capable of providing extremely high pressure and pulsation-free flow.
20. A reciprocating piston pump is a pump that utilizes a piston in a cylinder to pull and push the liquid mobile phase from the mobile phase reservoir through the HPLC system. Two check valves (backflow preventers) are in-line to help force the liquid in only one direction. See Figure 13.4.
21. A check valve is a device used in a flow stream to allow flow in only one direction. It is needed in the HPLC system to ensure that mobile phase flows in only one direction through the pump.
22. Isocratic elution—One mobile phase composition for entire run.
    Gradient elution—Mobile phase composition is altered during the run in some preprogrammed manner.

23. A gradient programmer is that part of an HPLC instrument that provides for the gradual changing of the mobile composition in the middle of the run.

24. The gradient elution method for HPLC is the method in which the mobile phase composition is changed in some preprogrammed way in the middle of the run. The device that accomplishes this is called the gradient programmer and is placed between the mobile phase reservoir and the pump. It is useful in experiments in which altering the mobile phase composition assists with the resolution of the mixture.

25. A parameter that alters retention time (and hence resolution) is altered in the middle of the run in each case.

26. Individual mixture components vary as to their solubilities in different mobile phases and thus will display characteristic retention times with each different mobile phase composition. Therefore, changing the mobile phase composition in the middle of the run will slow down some components and speed up others and change the resolution of all peaks.

27. Solvent strength refers to the ability of a particular mobile phase (solvent) to elute mixture components. For example, if use of a particular solvent results in short retention times for mixture components, the solvent is strong. If use of a solvent increases retention times, it is a weak solvent.

28. 1) The septum material may not be compatible with all mobile phases, thus creating the possibility for contamination, and 2) the system is under a pressure that is too high to make the septum-piercing method a viable possibility.

29. See Figure 13.7 and the accompanying discussion.

30. See Figure 13.7 and the accompanying discussion.

31. The analyst may either: 1) install a smaller loop on the injector, or 2) partially fill the current loop, using the syringe to measure the volume.

32. Normal phase—Stationary phase is polar, mobile phase is nonpolar. Reverse phase is just the opposite.

33. Reverse phase, since the mobile phase is polar and the stationary phase is nonpolar.

34. For reverse phase, common mobile phases are water, methanol, acetonitrile, and mixtures of these. Common stationary phases are phenyl, C8, and C18. For normal phase, common mobile phases are hexane, cyclohexane, and carbon tetrachloride. Common stationary phases are structures that include cyano, amino, and diol groups.

35. Bonded phase chromatography is a type of liquid–liquid chromatography in which the liquid stationary phase is chemically bonded to the support material (as opposed to being simply adsorbed). The stationary phase can be either polar or nonpolar, and thus both normal phase and reverse phase are possible.

36. Both nonpolar.

37. C18—reverse phase; adsorption chromatography—normal phase.

38. Both cation exchange resins.

39. Cation exchange resins have negatively charged bonding sites for exchanging cations, while anion exchange resins have positively charged sites for exchanging anions.

40. Ion chromatography is the modern name for ion exchange chromatography. The mobile is always a pH-buffered water solution, such as an acetic acid–sodium acetate water solution.

41. GPC utilizes nonpolar organic mobile phases, such as THF, trichlorobenzene, toluene, and chloroform, to analyze for organic polymers such as polystyrene. GFC utilizes mobile phases that are water-based solutions and is used to analyze for naturally occurring polymers, such as proteins and nucleic acids.

42. (a) Normal and reverse phase HPLC
    (b) Size exclusion HPLC
    (c) Normal and reverse phase HPLC

43. In reverse phase chromatography, the polar mixture components would elute first since they would be attracted by the polar mobile phase and repelled by the nonpolar stationary phase. In normal phase chromatography, nonpolar mixture components would elute first since they would be attracted by the nonpolar mobile phase and repelled by the polar stationary phase.

44. Mobile phase composition, stationary phase composition, flow rate, temperature of stationary phase.

45. (a) Isocratic elution refers to the same mobile phase composition used for the entire run, changed only by shutting down the pump and restarting it. Gradient elution refers to changing the mobile phase composition in the middle of the run.

(b) Normal phase chromatography refers to a procedure in which a nonpolar mobile phase is used in combination with a polar stationary phase. Reverse phase chromatography refers to a procedure in which a polar mobile phase is used in combination with a nonpolar stationary phase.

(c) The UV absorbance detector is sensitive, but not universal. The refractive index detector is universal, but not sensitive, cannot be used with gradient elution, and is subject to temperature effects.

46. The fixed-wavelength detector utilizes a glass filter as the monochromator. The variable-wavelength utilizes a slit/dispersing element/slit monochromator. The latter design is used in order to maximize sensitivity by setting the monochromator to the wavelength of maximum absorbance. If this is not important, then the fixed-wavelength design may be preferred because it is less expensive.

47. A diode array detector is a UV absorbance detector in which the light is not dispersed until after it has passed through the flow cell. The dispersed wavelengths spray across an array of photodiodes, which detect the absorbed wavelengths all at once, allowing for simultaneous readings that translate into the absorption spectrum. Thus, comprehensive qualitative analysis is possible, as well as rapid changeover of wavelengths between peaks to maximize peak size and delete interferences.

48. Detector, advantage, disadvantage
(a) UV, sensitive, not universal
(b) Refractive index, universal, not sensitive
(c) Fluorescence, very sensitive, not universal or highly applicable
(d) Conductivity, selective for ions, not universal—only for ions
(e) Amperometric, broad applicability, frequently needs servicing
(f) LC-MS, excellent for qualitative analysis, expensive
(g) LC-IR, sensitive and fast, not water compatible

49. Not all potential mixture components will absorb UV light.

50. See number 48 for some examples.

51. (a) Fluorescence    (b) refractive index    (c) conductivity    (d) UV absorbance
(e) refractive index    (f) fluorescence    (g) UV absorbance    (h) conductivity

52. Yes, because the absorptivity of mixture components varies.

53. A suppressor is a device that selectively removes ions from a flowing solution. In an ion exchange HPLC experiment in which the mobile phase contains ions, a suppressor must be used to remove these ions so that the ions of the mixture can be detected. A conductivity detector measures ions by the conductivity that they induce in the mobile phase, so the mobile phase ions constituent an interference.

54. FTIR, like UV absorbance, refractive index, etc., is a technique for liquid solutions but has the advantage in that it is fast, allowing a complete spectrum to be obtained as a given mixture component elutes, making it an extremely powerful tool for qualitative analysis.

55. See Section 13.7.

56. Plugged in-line filter is a typical cause of unusually high pressure. The solution is to backflush the filter or otherwise clean it.

57. See Section 13.8.4.

58. This is a symptom of channeling in the column. The problem is solved by replacing the column.

59. 1) Slow elution of chemicals adsorbed on the column, 2) temperature effects, such as with the refractive index detector, and 3) a contaminated detector

60. 4.51 mg

61. 0.0218%

# Chapter 14

1. The two classifications are potentiometry and amperometry. Potentiometry is a technique in which electrode potential is measured and related to concentration, while amperometry is a technique in which the current flowing at an electrode is what is measured and related to concentration.

2. Oxidation–reduction reactions involve electron transfer typically by direct collision between chemical species in solution. In electroanalytical chemistry, electron transfer occurs, but through electrical conductors rather than by direct collision.

3. A cell is a complete electroanalytical system consisting of an electrode at which reduction occurs, as well as an electrode at which oxidation occurs, and including the connections between the two. A half-cell is half of a cell in the sense that it is one of the two electrodes (and associated chemistry) in the system, termed either the reduction half-cell or the oxidation half-cell. The anode is the electrode at which oxidation takes place. The cathode is the electrode at which reduction takes place. An electrolytic cell is one in which the current that flows is not spontaneous, but rather due to the presence of an external power source. A galvanic cell is a cell in which the current that flows is spontaneous.

4. A galvanic cell operates of its own accord as a result of a spontaneous redox reaction. An electrolytic cell operates as a result of an external power source (e.g., a battery) in the circuit.

5. A salt bridge is a connection between half-cells that allows ions to diffuse between them, a requirement for a complete circuit so that a current can flow in the external circuit.

6. The standard reduction potential, symbolized by $E^\circ$, is a number reflecting the relative tendency of a reduction half-reaction to occur. In a table of standard reduction potentials, such as Table 14.1, those half-reactions at the top have positive numerical values and a strong tendency to occur. Those near the bottom have negative values and a tendency to go in the reverse direction.

7. (a) no                          (b) no                          (c) yes

8. −1.38 V

9. To the left because $E^\circ_{cell}$ is negative.

10. (a) +1.57 V                                    (b) Yes, because $E^\circ_{cell}$ is positive.

11. +1.1037 V. Reaction proceeds to the right.

12. +0.60 V

13. −1.57 V

14. The answers differ only in that the signs of the calculated $E^\circ$ values are different. The reason is that the reactions in the two problems are the same except reversed.

15. $E^\circ_{cell} = +0.51$ V; yes $E^\circ_{cell}$ is positive.

16. +2.11 V

17. −0.74 V

18. See Section 14.5.

19. +0.43 V

20. 3.14 V

21. 0.210 V (for the reduction reaction)

22. +1.78 V (for the reduction reaction)

23. −0.72 V

24. +2.205 V

25. −0.526 V (for $2 Cr^{3+} + 3 Ni \leftrightarrow 2 Cr + 3 Ni^{2+}$)

26. −1.333 V

27. −0.942 V

28. Potentiometry is the measurement of electrode potential in chemical analysis procedures for the purpose of obtaining qualitative and quantitative information about an analyte. The reference electrode is a half-cell that is designed such that its potential is a constant, making it useful as a reference point for potential measurements. Ground is the ultimate reference point in electronic measurements.

It is a common wire threading throughout a circuit, often connected to the frame of the instrument and to the ground prong on the electrical outlet. An indicator electrode is the electrode that gives the information sought—either the potential or current that leads to the answer to the analysis.

29. Equations (14.6) and (14.7) clearly show that the potential of the SCE is directly proportional to the [Cl⁻] inside the electrode. Since [Cl⁻] is a constant because the solution is saturated with KCl, E must be constant, a requirement of a reference electrode.

30. The potential of this electrode depends only on the chloride ion concentration inside. If the KCl is saturated, it is assured that the chloride ion concentration is constant, which in turn means that the potential of this electrode never changes, a requirement of a reference electrode.

31. The salt bridge for reference electrodes consists of porous fiber tips that provide for the diffusion of ions in and out of this half-cell.

32. The Nernst equation defining the potential of the silver–silver chloride electrode is Equation (14.9). Since the [Cl⁻] in such an electrode is a constant, the potential also must be a constant (the requirement of a reference electrode) because [Cl⁻] is the only variable on which the potential depends.

33. See Equation (14.10) and accompanying discussion.

34. The end result of the derivation is Equation (14.12). The explanation leading up to this equation clearly shows that the potential, E, is directly proportional to the pH of the solution into which the solution is dipped.

35. Either one buffer solution, with a pH near that of the solution being tested, or two buffer solutions, with pH values that bracket that of the solution being tested, are usually used. When the pH electrode is dipped into a buffer solution during the standardization step, the pH reading on the meter is adjusted to the given pH using the standardize knob, or other control, on the meter.

36. A combination pH electrode consists of a pH electrode and a reference electrode in a single probe.

37. Both provide an electrolyte solution for the electrical contact between two half-cells so that ions can freely diffuse between the two half-cells and a current can flow externally, allowing the desired measurement to be made.

38. An ion-selective electrode is a half-cell that is sensitive to a particular ion like the pH electrode is sensitive to the hydrogen ion.

39. (b) Nitrate concentration = $2.36 \times 10^{-4}$ M

40. In gel-filled electrodes, the interior reference solution is gelatinized. As such, there is no loss of solution through the salt bridge and no contaminating solution can enter. It cannot be refilled with reference solution. In double-junction electrodes, the usual electrode is inside another glass or plastic encasement and there are two junctions for contact to the external solution. The purpose of this design is to prevent contamination in either direction.

41. 1) So that ions can diffuse freely through the salt bridge, and 2) so that the electrode can be refilled with reference solution

42. Over the short term: tip, include the porous salt bridge, should be immersed in a soak solution (pH = 7 usually recommended). Over the long term: the protective plastic sleeve with which the electrode was shipped should be repositioned over the tip.

43. A platinum electrode is needed in some half-cells to provide a surface at which electrons can be exchanged. Such a surface is lacking where there is no solid metal as part of the half-reaction, such as with the ferrous–ferric half-reaction.

44. A potentiometric titration is one in which the end point is detected by measuring the potential of an electrode dipped into the titrated solution.

45. Potentiometry utilizes galvanic cells, while amperometry uses electrolytic cells.

46. Potentiometric methods measure a potential in a galvanic cell arrangement. Amperometric and voltammetric methods measure a current in an electrolytic cell.

47. See Section 14.6.

48. 1) Its surface is continuously renewed—thus avoiding problems with surface contamination, and 2) the solution is automatically stirred when the drop falls.

49. Polarography is the measurement of the current flowing at a dropping mercury electrode as the potential applied to this electrode is changed. Voltammetry is the measurement of the current flowing at a stationary electrode as the potential applied to this electrode is changed.

50. An amperometric titration is one in which the end point is determined by monitoring the current flowing at an electrode.

51. The moisture (water) in a sample.

52. See Section 14.8.

53. The prefix bi- is used because it uses two platinum prongs rather than one platinum and a reference electrode, as in a normal method.

54. To prevent water contamination from humid laboratory air.

55. Conditioning the solvent means to eliminate all water in the solvent by titrating it with the Karl Fischer reagent. This is done to keep from measuring this water rather than the water in the sample.

56. In the volumetric method, the titrant is added from an external reservoir. In the coulometric method, the titrant (iodine) is generated internally via an electrochemical reaction.

57. The titrant is standardized by titrating a known amount of water with the titrant and calculating the titer of the titrant, the weight of water consumed by 1 mL of titrant.

58. Titer is the weight of water consumed by 1 mL of the Karl Fischer titrant: 0.543 mg/mL.

59. 22%

60. The anode–cathode assembly generates the iodine. The dual-pin platinum electrode detects the end point.

61. The coulombs of current used to generate the iodine is proportional to the amount of iodine needed to consume the water. Titrant is not added from a buret.

# Chapter 15

1. Food products: soup, hot chocolate. Hygiene products: toothpaste, shaving gel.

2. See Section 15.2.2.

3. Capillary viscometry measures the time it takes for the test fluid to flow through a capillary tube. The viscosity is calculated from this time and the calibration constant of the tube.

4. Fluids of different viscosities flow through the capillary tube at different rates depending on their viscosities.

5. The Ostwald viscometer is the general name given to a capillary viscometer.

6. A Cannon–Fenske viscometer is a capillary type of viscometer. It utilizes the flow through a capillary tube as a means of measuring viscosity.

7. The Cannon–Fenske viscometer is pictured in Figure 15.3. The U-tube has a bend in the center. There are several bulbs in the U, which allows a greater volume of fluid to be tested; this also means a long and precise time measurement. The Ubbelohde viscometer is pictured in Figure 15.4. In this design, the U is completely vertical and, like the Cannon–Fenske viscometer, also has several bulbs for containing the fluid. The Cannon–Fenske viscometer is used for measuring the kinematic viscosity of transparent Newtonian liquids, especially petroleum products and lubricants. The Ubbelohde viscometer is also used for the measurement of kinematic viscosity of transparent Newtonian liquids, but by the suspended level principle.

8. $0.117 \text{ cS sec}^{-1}$

9. 13 cS

10. Rotational viscosity methods measure viscosity by measuring the torque required to rotate a spindle immersed in the fluid sample.

11. See Section 15.3.1.

12. DTA—Differential thermal analysis.
    DSC—Differential scanning calorimetry.
    TGA—Thermogravimetric analysis.

13. All the heat being added to the sample is being used to melt the sample rather than to raise its temperature.

14. If there is no physical or chemical change occurring in the sample as the temperature of the surrounding increases, there will be no difference in the two temperatures. If there is an endothermic physical or chemical change occurring, the temperature of the surroundings increases at a faster rate than that of the sample. If there is an exothermic physical or chemical change occurring, the temperature of the sample increases at a faster rate.

15. The negative peak means that the temperature of the surroundings is increasing at a faster rate than the temperature of a sample. This occurs, for example, when the sample melts.

16. The difference is $T_S - T_R$, in which $T_S$ is the temperature of the sample and $T_R$ is the temperature of a reference material. This difference is negative when an endothermic process occurs because $T_S$ lags behind $T_R$ due to the fact that the sample absorbs heat from the environment while its temperature remains the same. The opposite is true for an exothermic process.

17. In DTA, temperature is monitored. In DSC, heat flow is monitored. The data obtained are similar. DTA can be considered a more global term, covering all differential thermal techniques, while DSC is a DTA technique that gives calorimetric (heat transfer) information.

18. S refers to the sample. R refers to the reference material. There must be two locations because the heat flow to each must be monotored individually so as to produce the $\Delta T$ data.

19. Indium is a common reference material for DSC.

20. Endothermic and exothermic physical and chemical changes produce results in DSC. Melting is just one such change. Other examples would include chemical reactions that are either endothermic or exothermic. Physical changes would include deformations that do not involve melting.

21. See Section 15.3.4.

22. D = the sodium D line at 589 nm; 20 is the temperature of the substance.

23. To check for the purity (or concentration) of the distillate.

24. To select a specific wavelength because the refractive index varies with the wavelength. Also, the light source used in most commercial refractometers uses ordinary white light and not a sodium lamp.

25. Because the refractive index varies with the temperature.

26. 10.5% (1° Brix = 1% sucrose)

27. See Section 15.5.

28. Optical rotation refers to the rotation of the plane of polarized light as it passes through a sample.

29. $CH_3CHOHCH_2CH_3$, $CH_3CHClCH_2CH_3$, $CH_3CHBrCH_2CH_3$, $CH_3CH(OCH_3)CH_2CH_3$

30. Samples of 2-butanol are not optically active because they are actually equal mixtures of two enantiomers (racemic mixtures), so their optical rotations cancel out.

31. Each will rotate the plane of polarized light equally in opposite directions. Such a mixture is not optically active because the rotations cancel.

32. A polarimeter measures the degree of rotation of the plane-polarized light as this light passes through the sample.

33. 7.80 g/mL

34. 5.25 g/cm$^3$

35. 0.781 g/mL

36. 1.63 g/mL

37. This is due to voids caused by spaces between the particles or by the porosity of the material.

38. As the fill rate increases, the particles have less time to arrange or settle, resulting in larger volumes.

39. 1.32 g/mL

40. Mesh indicates that there are 60 openings (or apertures) per linear inch.

41. The particle size of the packing material is smaller than 80 mesh and larger than 100 mesh. In other words, it is the material that passed through the 80-mesh screen and was trapped on the 100-mesh screen.

42.

Sample Size: 1253 g

| Sieve No. | Mesh Size (mm) | Mass (kg) | Mass Fraction | Cumulative Mass Fraction Oversize | Relative Frequency |
|-----------|----------------|-----------|---------------|-----------------------------------|--------------------|
| 1 | 0.71 | 0.094 | 0.0750 | 0.0750 | |
| 2 | 0.5 | 0.063 | 0.0503 | 0.1253 | 0.2394 |
| 3 | 0.355 | 0.078 | 0.0623 | 0.1875 | 0.4293 |
| 4 | 0.25 | 0.167 | 0.1333 | 0.3208 | 1.2693 |
| 5 | 0.18 | 0.398 | 0.3176 | 0.6385 | 4.5377 |
| 6 | 0.125 | 0.288 | 0.2298 | 0.8683 | 4.1791 |
| 7 | 0.071 | 0.096 | 0.0766 | 0.9449 | 1.4188 |
| 8 | 0.045 | 0.069 | 0.0551 | 1.0000 | 2.1180 |
| Fines (residual) | 0 | | | | |
| Total | | 1.253 | 1.0000 | | |

43. Tension—Pulling forces.
    Compression—Pushing forces.
    Shear—Tearing forces.
44. Flexural tests, impact tests, and hardness tests
45. Ductile materials tend to stretch, while brittle materials do not.
46. Young's modulus provides an indication of how much a material will stretch under a given load.
47. Elastic range—A material that is deformed will return to its original dimensions if a stress is applied that is within the range.
    Plastic range—A material will remain deformed when the stress is removed.
48. Ductility
49. Hardness—Resistance to abrasion, deformation, or penetration.
50. Using the Moh hardness scale: quartz
51. Depth of penetration: Rockwell superficial and Rockwell standard hardness tests
52. Diameter of indentation: Brinell hardness. Length of diagonal: Vickers and Knoop microhardness.
53. To push the penetration past any surface imperfections.
54. Yes, because the Rockwell test is a penetration test, if the material is too thin, the supporting anvil will alter the readings. This is called the anvil effect.

# Chapter 16

1. (a) glucose, (b) amino acids, (c) nucleotides, (d) nucleotides, (e) amino acids, (f) glucose, and (g) amino acids
2. Starch, DNA, and some proteins are polymers with helical structures.
3. Biologically, carbohydrates are most important as energy sources. This is especially true for monosaccharides, disaccharides, and starches. Starches also function as energy *storage* molecules in plants. Cellulose is an essential component of plant cell walls.
4. Glucose linked to glucose forms maltose.
5. Triacylglycerols contain only glycerol and fatty acids.
6. See Figure 16.3.
7. Steroids lack fatty acids.
8. Triacylglycerols are very hydrophobic, while the liquid in blood (plasma) is mostly water and is therefore very hydrophilic. Hydrophobic molecules literally are water fearing or water hating.

9. See Figure 16.7.

10. (a) Primary structure is the linear amino acid sequence held together by peptide bonds. (b) Secondary structure is the regular, orderly folding of the primary structure in three-dimensional space. Repetitive structures such as the alpha helix or beta pleated sheets are examples of secondary structures. These structures are stabilized by backbone interactions, primarily hydrogen bonding, with no R groups involved. (c) Tertiary structures are unique, with each polypeptide chain possessing its own. These structures are complex three-dimensional conformations, held together entirely by R group interactions. The four types of R group interactions are hydrogen bonding, disulfide bonds, hydrophobic interactions, and ionic interactions. (d) Quaternary structures are found only in oligomeric proteins, those with multiple polypeptide chains, and consist of the complex three-dimensional folding of the various chains with each other. R group interactions, primarily hydrogen bonding, stabilize quaternary structures.

11. Hydrophobic R groups would force a protein to align itself away from water.

12. (a)Definitely true, since in order to have a quaternary structure, there *must* be at least two chains. (b)Also true, since R groups are responsible for stabilizing quaternary structures. (c) Since primary structural changes affect higher levels of structure, this is also true. (d) This is impossible to determine with the information given.

13. A nucleotide consists of a nitrogenous base linked to a monosaccharide linked to a phosphate functional group.

14. DNA bases carry the hereditary information and therefore must be protected.

15. The three types of RNA are transfer (tRNA), ribosomal (rRNA), and messenger (mRNA).

16. The two strands of a DNA double helix are held together by hydrogen bonds between the bases; two between A and T, and three between G and C.

17. DNA incorporates the monosaccharide deoxyribose, while RNA has ribose. One of the four bases is different: RNA has uracil in place of thymine in DNA. Finally, RNA has no double helix.

18. Agarose is much safer to work with, since acrylamide is a potential neurotoxin.

19. An electrophoresis buffer must have ions to carry current, and must not interfere with the stability of the sample.

20. Capillary electrophoresis, HPLC, UV-VIS spectroscopy, and gel electrophoresis are examples of instrumental analysis used to study biomolecules.

21. At pH 8.9, a peptide with an isoelectric point of 7.2 would have a negative charge, since the pH is above the isoelectric point.

22. Both anion exchange and reverse phase columns may be used to separate nucleotides.

23. Nucleotides and amino acids are chemically very similar, making any type of laboratory separation difficult.

24. An ion exchange column bears either positive (anion exchanger) or negative charges (cation exchanger). Sample components migrate through the column at different rates, based on magnitude of charge. For example, a molecule with a net charge of ++++ will migrate through a cation exchange column more slowly than a molecule with a net charge of +, since it will interact more strongly with the negative charges in the column. The different rates of movement allow for separation of the sample components.

25. A structural feature that many molecules that absorb UV light have is a conjugated pi bond skeleton.

# Index